BASIC
ELECTRONICS
FOR
ENGINEERING
TECHNOLOGY

BASIC ELECTRONICS FOR ENGINEERING TECHNOLOGY

NORVAL R. EKELAND
George Brown College
Toronto, Ontario, Canada

PRENTICE-HALL, INC., *Englewood Cliffs, New Jersey 07632*

Library of Congress Cataloging in Publication Data

Ekeland, Norval R
  Basic electronics for engineering technology.

  Includes index.
  1. Electronics. 2. Electric engineering.
I. Title.
TK7815.E33   621.3   80-23284
ISBN 0-13-060467-4

Editorial/production supervision by Ros Herion
Interior design by Gary Samartino
Cover design by RD Graphics
Manufacturing buyer: Joyce Levatino

Printed in the United States of America

10 9 8 7 6 5 4 3 2 1

PRENTICE-HALL INTERNATIONAL, INC., *London*
PRENTICE-HALL OF AUSTRALIA PTY. LIMITED, *Sydney*
PRENTICE-HALL OF CANADA, LTD., *Toronto*
PRENTICE-HALL OF INDIA PRIVATE LIMITED, *New Delhi*
PRENTICE-HALL OF JAPAN, INC., *Tokyo*
PRENTICE-HALL OF SOUTHEAST ASIA PTE. LTD., *Singapore*
WHITEHALL BOOKS LIMITED, *Wellington, New Zealand*

To my classes of 1977 through 1980, who were my best proofreaders and critics; in particular to my deaf students, whose tenacious will to learn inspired me to write the manuscript; and

to my wife, *Doreen*, whose uncomplaining dedication to an arduous typing task made the manuscript possible.

CONTENTS

PREFACE   *xiii*

1   ELECTRIC CURRENT   *1*

1-1   Energy and Electricity, *1*
1-2   Structure of Matter, *2*
1-3   Atom and Electron Size, *2*
1-4   The Bohr Atomic Theory, *3*
1-5   Conductors, *6*
1-6   The Electron, *7*
      Exercises, *8*

2   THREE BASIC ELECTRICAL UNITS   *9*

2-1   Current, *9*
2-2   Voltage, *14*
2-3   Resistance, *18*
2-4   Resistors and Temperature, *25*
2-5   Fuses, *26*
2-6   Schematic Drawings, *27*
2-7   Multiple and Submultiple Units, *27*
      Exercises, *30*

3   OHM'S LAW   *35*

3-1   Ohm's Law, *35*
3-2   Reference Point, *38*
      Exercises, *42*

4 ENERGY, POWER, AND SHOCK   *45*

4-1 Energy and Work, *45*
4-2 Power, *48*
4-3 Kilowatt-Hour, *51*
4-4 Shock, *51*
     Exercises, *52*

5 SERIES CIRCUITS   *55*

5-1 Series Circuits, *55*
5-2 Characteristics of Series Circuits, *55*
5-3 Voltage Drop, *58*
5-4 Fault-Finding, *58*
5-5 Voltage Ratios, *60*
5-6 Power Dissipation, *61*
5-7 Power Transfer; Constant Sources, *68*
     Exercises, *74*

6 PARALLEL CIRCUITS   *78*

6-1 Applications of Parallel Circuits, *78*
6-2 Principle of Operation, *78*
6-3 Parallel Circuit Arrangements, *78*
6-4 Characteristics of Parallel Circuits, *79*
6-5 Kirchhoff's Current Law, *81*
6-6 Total Resistance of Parallel Circuits, *81*
6-7 Power (Wattage) Dissipation, *88*
6-8 Fault-Finding, *91*
     Exercises, *93*

7 SERIES–PARALLEL CIRCUITS   *99*

7-1 Series–Parallel Circuits, *99*
7-2 Solving R_T for Complex Circuits, *101*
7-3 Calculation of I and V in Series–Parallel Circuits, *103*
7-4 Fault-Finding in Series–Parallel Circuits, *108*
7-5 Tracing and Redrawing Complex Circuits, *111*
7-6 Calculating Current and Voltage Values, *115*
     Exercises, *115*

8 VOLTAGE DIVIDERS AND WHEATSTONE BRIDGE   *122*

8-1 Introduction, *122*
8-2 Voltage Divider, *122*
8-3 Wheatstone Bridge, *128*
     Exercises, *132*

9   BATTERIES   *134*

9-1   Introduction, *134*
9-2   Principle of Battery Operation, *134*
9-3   Classification of Batteries, *134*
9-4   Connecting Cells Together, *137*
9-5   Testing, *139*
9-6   Primary Cells, *140*
9-7   Secondary Cells, *146*
9-8   Battery Applications, *153*
       Exercises, *155*

10   THEOREMS   *157*

10-1   Introduction, *157*
10-2   Kirchhoff's Loop Solution, *158*
10-3   Superposition Theorem, *165*
10-4   Thevenin's Theorem, *172*
10-5   Norton's Theorem, *184*
10-6   Relationship Between Norton's Equivalent Circuit and
        Thevenin's Equivalent Circuit, *192*
        Exercises, *193*

11   PRINCIPLES OF MAGNETISM   *199*

11-1   Introduction, *199*
11-2   Magnetism, *199*
11-3   Natural Magnetism, *199*
11-4   Artificial Magnetism, *200*
11-5   Atomic Theory of Magnetism, *201*
11-6   Three Categories of Artificial Magnets, *203*
11-7   The Magnetic Field, *205*
11-8   Characteristics of Flux Lines, *207*
11-9   Magnetism and Demagnetism, *208*
11-10 Air Gap, *208*
11-11 Care of Permanent Magnets, *209*
11-12 Magnetic Measurements, *210*
11-13 Magnetics Terms, *216*
        Exercises, *216*

12   ELECTROMAGNETISM   *219*

12-1   Electromagnetism, *219*
12-2   Magnetism and Current, *219*
12-3   Left-Hand Rule for Conductors, *221*
12-4   Flux and Conductors, *221*
12-5   Flux and Coils, *222*

12-6   Magnetic Measurements, *225*
12-7   Power Loss in Ferromagnetic Materials, *245*
12-8   Magnetic Circuits, *254*
       Exercises, *272*

13   **INDUCTANCE**   *275*

13-1   Inductance, *275*
13-2   Induce, *275*
13-3   Electromagnetic Induction, *276*
13-4   Counter Voltage (CEMF), *278*
13-5   Production of Induced Voltage, *278*
13-6   Lenz's Law, *285*
13-7   Flux Linkage, *285*
13-8   Self-Inductance, *285*
13-9   Magnitude of Counter Voltage, *285*
13-10   Faraday's Law, *286*
13-11   Unit of Inductance, *286*
13-12   Physical Determinants of Inductance, *288*
13-13   Practical Considerations, *290*
13-14   Mutual Inductance, *291*
13-15   Coefficient of Coupling, *292*
13-16   Inductors and L_M, *295*
13-17   Transformers, *301*
13-18   Voltage Ratio, *302*
13-19   Power in Primary and Secondary Windings, *304*
13-20   Current Ratio, *306*
13-21   Total Power in Power Transformers, *308*
13-22   Transformers and Power Transfer, *309*
13-23   Transformer Power Losses, *312*
13-24   Practical Considerations, *312*
13-25   Autotransformers, *312*
13-26 L/R Time Constants (Rise), *313*
13-27 L/R Decay Curves, *340*
13-28   Practical Considerations, *349*
13-29   Rise and Decay Curves, *349*
13-30   Time Constants and Frequency, *349*
       Exercises, *354*

14   **CAPACITANCE**   *360*

14-1   Introduction, *360*
14-2   Capacitor, *360*
14-3   Purpose of Capacitors, *361*
14-4   Capacitors and Conductance, *361*
14-5   Principle of Operation, *361*

14-6   Clarification, *363*
14-7   Electrostatic Field Measurements, *363*
14-8   Factors Affecting Capacitance, *365*
14-9   Calculating Capacity, *367*
14-10   Unit of Capacitance—Farad, *368*
14-11   Dielectric Strength, *369*
14-12   Effect on Capacity of Dielectric Constant (K) and
        Dielectric Strength, *370*
14-13   Working Voltage (W.V.D.C.), *371*
14-14   Capacity, Voltage, and Charge, *371*
14-15   Connecting Capacitors, *373*
14-16   Capacity and Developed Voltage, *375*
14-17   Capacitor Construction, *378*
14-18   Capacitor Types, *379*
14-19   Practical Considerations, *384*
14-20   Capacitive-Resistive (C-R) Time-Constant Circuits, *384*
14-21 C-R Discharge, *402*
14-22   Time Constant and Frequency, *411*
        Exercises, *418*

15   ALTERNATING CURRENT   *423*

15-1   Introduction, *423*
15-2   Rules of Magnetism, *423*
15-3   Generation of ac Voltage, *426*
15-4   Sine of the Angle, *429*
15-5   Instantaneous Values, *432*
15-6   Sine Wave Values, *437*
15-7   Frequency and Period, *441*
15-8   Time, Angle of Cut, and v, *443*
15-9   Frequency and Wavelength, *447*
15-10   Angular Velocity (ω); ($2\pi f$), *449*
        Exercises, *451*

16   AC CIRCUITS   *454*

16-1   Introduction, *454*
16-2   Reactance, *454*
16-3   Vector Diagrams, *455*
16-4   Phase Angles, *456*
16-5   Resistance in ac Circuits, *461*
16-6   Inductive Reactance, *462*
16-7   Capacitance Reactance, *465*
16-8   Power in ac Circuits, *469*
        Exercises, *473*

17   IMPEDANCE   476

17-1   Introduction, 476
17-2   Phasor Diagrams, 477
17-3   Series Circuits, 483
17-4   Parallel Impedance Circuits, 503
         Exercises, 516

18   RESONANT CIRCUITS   518

18-1   Introduction, 518
18-2   Series Resonant Circuits, 519
18-3   Parallel Resonance, 541
         Exercises, 561

19   FILTERS   565

19-1   Introduction, 565
19-2   Classifications, 565
19-3   Resonant Filters, 566
19-4   Nonresonant Filters, 571
19-5   Constant-K Filters, 586
19-6   m-Derived Filter, 593
19-7   Active Filters, 593
         Exercises, 595

20   COMPLEX NUMBERS   598

20-1   Complex Numbers and Impedance, 598
20-2   Complex Numbers and Series Circuits, 601
20-3   Complex Number Operations, 606
20-4   Mathematical Operations of Real and Imaginary Numbers, 608
20-5   Polar Form Mathematical Operations, 612
20-6   Complex Numbers and Parallel Circuits, 614
20-7   Complex Numbers and Series–Parallel Circuits, 618
20-8   Z of Multiple Parallel Circuits, 620
20-9   Thevenin's Theorem and Impedance, 622
20-10   Norton's Theorem and Impedance, 627
         Exercises, 631

21   TEST EQUIPMENT   635

21-1   The Multimeter, 635
21-2   Voltmeter Loading, 646
         Exercises, 649

APPENDIX   651
ANSWERS TO SELECTED PROBLEMS   669
INDEX   679

PREFACE

This is a teaching textbook. It is not merely an assemblage of statistics and facts. Many electronics textbooks are written either by an administrator who is divorced from the classroom and the student or by a member of a university faculty who does not have a feel for the needs of the technology or technician student. This textbook was developed in the classroom over a period of years. During this time different teaching techniques were tried; the one which best accelerated the learning response for each topic was incorporated in the text. The language and interpretation of ideas is straightforward. The contents have been proofread by students, who made suggestions for easier understanding. The style of the text is based on theories of learning and reinforcement, some of the techniques having been acquired from the responses of students to teaching methods.

The abundance of illustrations pays tribute to the notion that ideas are more quickly assimilated in picture form than in prose. Practical applications are cited along with theories to give meaning to otherwise dry facts. One example solution is offered for every possible type of problem which may be encountered.

The scope and penetration of the topics fulfill, and in numerous cases surpass, the requirements for certification of engineering technicians and encompass an array of topics which have been bypassed by one textbook or another. The units used in the mathematical calculations are those of the International System of Units (S.I. units).

The birth of this textbook took place during the first year that the author was given a class of forty students who, after a period of six weeks, were taken back to the beginning of the course because they were failing or near failing. The manuscript's power to communicate and effectively teach was proven during this period and in successive years with similar types of classes, when the text was credited with raising the marks of the majority of students to the A- and B-grade levels.

NORVAL R. EKELAND

BASIC ELECTRONICS FOR ENGINEERING TECHNOLOGY

chapter 1

ELECTRIC CURRENT

1-1 Energy and Electricity

The most fundamental law in physics states that energy can neither be created nor destroyed. The source of all energy in the world is the sun. Much of the energy reaching the earth from the sun has been stored in the form of oil, coal, radioactive materials and a long list of other storage facilities. If it were not for this stored energy, human beings could not exist on this planet. Heat, light, sound, wind pressure, and electricity are among the most common forms of energy. Electronics makes possible the conversion of each of these forms to another.

The device used to convert energy from one form to another is called a *transducer*.

A microphone is a transducer because it converts sound energy into electrical energy. A radio speaker is a transducer because it converts electrical energy into sound energy. The television picture tube, or oscilloscope cathode ray tube is a transducer because it converts electrical energy into light energy. The television studio camera tube is a transducer because it converts light energy into electrical energy. The list of transducers and their energy conversions could go on for pages and pages.

It is necessary to understand the nature of electric current and its source in order to understand how electricity is able to carry energy from one point to another and how it is able to convert energy from one form to another.

1

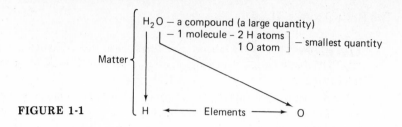

FIGURE 1-1

1-2 Structure of Matter

The *atom* is the base of all matter or substance. Matter is anything that takes up space and has weight. Therefore, everything in the world—gas, liquid, and solid—is matter or substance. Sometimes a substance will consist of only one of the 103 known elements in the world; at other times a substance or matter may consist of two or more of the known elements in the world. The latter substance is referred to as a compound. Two very common examples of compounds are table salt—written chemically as NaCl and made up of the elements sodium (Na) and chlorine (Cl)—and water, written chemically as H_2O. The latter consists of two parts of the element hydrogen (H) and one part of the element oxygen (O) (Fig. 1-1).

The *molecule* is the smallest particle into which any substance can be reduced and still retain its characteristics and have a separate stable existence. The atoms of many elements such as copper are quite stable when they exist alone and therefore are both atom and molecule at the same time. They are called monatomic molecules. On the other hand oxygen and hydrogen atoms by themselves are very unstable, but a molecule of the compound water (H_2O) which they form is very stable. Water is an example of a compound with triatomic molecules, meaning that each molecule consists of three atoms—two hydrogen and one oxygen. Other compounds may consist of atoms from several elements which have combined together to form polyatomic molecules of a new substance.

1-3 Atom and Electron Size

The atom is the smallest particle into which an element can be broken down and still retain all of its original characteristics (properties).

The atom is infinitesimal. If, for example, copper atoms were placed side by side in a straight line like billiard balls, there would be one hundred million (10^8) of them in a one-inch line. Electrons are even smaller. It takes over 40,000 electrons to span the diameter of one single copper atom. If it were possible to magnify the atom one hundred thousand billion times, the electrons would be the size of basketballs and would be spaced twelve to twenty miles apart.

1-4   The Bohr Atomic Theory

The following theory about the atom was put forward by a Danish physicist, Dr. Niels Bohr, in 1913 and was named the Bohr Atomic Theory in his honor.

The atom consists of a small, central, positively charged nucleus surrounded by a widely dispersed quantity of electrons arranged in imaginary concentric shells revolving around the nucleus (Fig. 1-3). The electrons may move anywhere in their respective orbits (energy rings) but not in between the orbits. The radius of motion changes slightly from time to time for any one electron or set of electrons so that instead of a thin line orbit or shell, a band (energy band) develops. The distance from the nucleus to the orbit depends on the energy level of the electron.

The greater the energy level contained by an electron, the further it will move away from the nucleus because it will have a greater average orbital velocity, and the more easily it can be removed from the atom. [Kinetic energy = mass $\times$ (velocity)^2.] It must be emphasized, however, that only the electrons in the outside (valence) ring can be removed from the atom and become free electrons. Electrons have a negative charge and are considered to be the unit of negative electricity.

The nucleus of the atom contains protons and neutrons closely packed in a small area (Fig. 1-2). The protons have mass, and each carries a unit of positive charge. The neutrons are particles with the same mass as protons but have no electrical charge (neutral). The total number of electrons in the shells of the atom is equal to the number of protons in the nucleus. Thus, the normal atom is electrically neutral.

Figure 1-3(a) represents the atomic structure of the simplest of all atoms, the hydrogen atom. It consists of one shell (K shell) containing one electron. The K shell is incomplete.

Figure 1-3(b) shows the helium atom. It consists of one shell (K shell) containing two electrons. The K shell is complete and a new shell must be formed to hold the electrons of atoms with atomic numbers greater than two. The atomic number of each atom in the periodic table is equal to the number of electrons contained in that atom (Table 1-1).

A generally accepted rule of thumb is that the maximum number of electrons which any shell can hold is twice the square of the number of that shell. There are exceptions to the rule, however (Table 1-1).

$$\text{Atom} \begin{cases} \text{Nucleus} - \text{Protons} \\ \qquad\qquad - \text{Neutrons} \\ \\ \text{Electrons in surrounding shells} \end{cases}$$

FIGURE 1-2

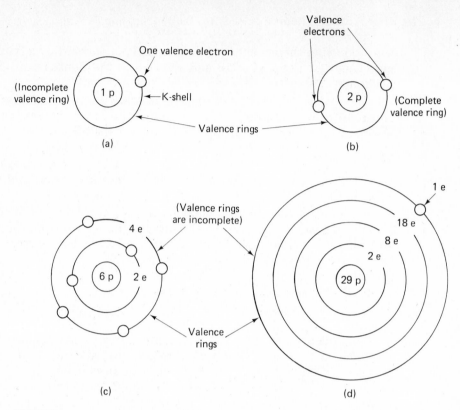

FIGURE 1-3 Atomic structure of various elements: (a) hydrogen atom (1 e); (b) helium atom (2 e); (c) carbon atom (6 e); and (d) copper atom (29 e).

$$\textbf{maximum electrons} = 2 \times n^2$$

where n = shell number

The above rule of thumb is applied in the following examples (see Fig. 1-4):

K shell is the first shell
Maximum number = 2×1^2 = 2 electrons

L shell is the second shell
Maximum number = 2×2^2 = 8 electrons

M shell is the third shell
Maximum number = 2×3^2 = 18 electrons

There is a maximum of seven shells of electrons surrounding a nucleus. The seventh shell is the Q shell. The last element in Table 1-1, Lawrencium, has a total of 103 electrons.

Atoms are normally electrically neutral because the number of protons is balanced by an equal number of electrons. If an atom should

4

Table 1-1 The Atomic Periodic Table

Atomic Number	Element	Electrons Per Shell					Atomic Number	Element	Electrons Per Shell						
		K	L	M	N	O			K	L	M	N	O	P	Q
1	Hydrogen	1					53	Iodine	2	8	18	18	7		
2	Helium	2					54	Xenon	2	8	18	18	8		
3	Lithium	2	1				55	Cesium	2	8	18	18	8	1	
4	Beryllium	2	2				56	Barium	2	8	18	18	8	2	
5	Boron	2	3				57	Lanthanum	2	8	18	18	9	2	
6	Carbon	2	4				58	Cerium	2	8	18	19	9	2	
7	Nitrogen	2	5				59	Praseodymium	2	8	19	20	9	2	
8	Oxygen	2	6				60	Neodymium	2	8	19	21	9	2	
9	Fluorine	2	7				61	Promethium	2	8	18	22	9	2	
10	Neon	2	8				62	Samarium	2	8	18	23	9	2	
11	Sodium	2	8	1			63	Europium	2	8	18	24	9	2	
12	Magnesium	2	8	2			64	Gadolinium	2	8	18	25	9	2	
13	Aluminum	2	8	3			65	Terbium	2	8	18	26	9	2	
14	Silicon	2	8	4			66	Dysprosium	2	8	18	27	9	2	
15	Phosphorus	2	8	5			67	Holmium	2	8	18	28	9	2	
16	Sulfur	2	8	6			68	Erbium	2	8	18	29	9	2	
17	Chlorine	2	8	7			69	Thulium	2	8	18	30	9	2	
18	Argon	2	8	8			70	Ytterbium	2	8	18	31	9	2	
19	Potassium	2	8	8	1		71	Lutetium	2	8	18	32	9	2	
20	Calcium	2	8	8	2		72	Hafnium	2	8	18	32	10	2	
21	Scandium	2	8	9	2		73	Tantalum	2	8	18	32	11	2	
22	Titanium	2	8	10	2		74	Tungsten	2	8	18	32	12	2	
23	Vanadium	2	8	11	2		75	Rhenium	2	8	18	32	13	2	
24	Chromium	2	8	13	1		76	Osmium	2	8	18	32	14	2	
25	Manganese	2	8	13	2		77	Iridium	2	8	18	32	15	2	
26	Iron	2	8	14	2		78	Platinum	2	8	18	32	16	2	
27	Cobalt	2	8	15	2		79	Gold	2	8	18	32	18	1	
28	Nickel	2	8	16	2		80	Mercury	2	8	18	32	18	2	
29	Copper	2	8	18	1		81	Thallium	2	8	18	32	18	3	
30	Zinc	2	8	18	2		82	Lead	2	8	18	32	18	4	
31	Gallium	2	8	18	3		83	Bismuth	2	8	18	32	18	5	
32	Germanium	2	8	18	4		84	Polonium	2	8	18	32	18	6	
33	Arsenic	2	8	18	5		85	Astatine	2	8	18	32	18	7	
34	Selenium	2	8	18	6		86	Radon	2	8	18	32	18	8	
35	Bromine	2	8	18	7		87	Francium	2	8	18	32	18	8	1
36	Krypton	2	8	18	8		88	Radium	2	8	18	32	18	8	2
37	Rubidium	2	8	18	8	1	89	Actinium	2	8	18	32	18	9	2
38	Strontium	2	8	18	8	2	90	Thorium	2	8	18	32	19	9	2
39	Yttrium	2	8	18	9	2	91	Protactinium	2	8	18	32	20	9	2
40	Zirconium	2	8	18	10	2	92	Uranium	2	8	18	32	21	9	2
41	Niobium	2	8	18	12	1	93	Neptunium	2	8	18	32	22	9	2
42	Molybdenum	2	8	18	13	1	94	Plutonium	2	8	18	32	23	9	2
43	Technetium	2	8	18	14	1	95	Americium	2	8	18	32	24	9	2
44	Ruthenium	2	8	18	15	1	96	Curium	2	8	18	32	25	9	2
45	Rhodium	2	8	18	16	1	97	Berkelium	2	8	18	32	26	9	2
46	Palladium	2	8	18	18	0	98	Californium	2	8	18	32	27	9	2
47	Silver	2	8	18	18	1	99	Einsteinium	2	8	18	32	28	9	2
48	Cadmium	2	8	18	18	2	100	Fermium	2	8	18	32	29	9	2
49	Indium	2	8	18	18	3	101	Mendelevium	2	8	18	32	30	9	2
50	Tin	2	8	18	18	4	102	Nobelium	2	8	18	32	31	9	2
51	Antimony	2	8	18	18	5	103	Lawrencium	2	8	18	32	32	9	2
52	Tellurium	2	8	18	18	6									

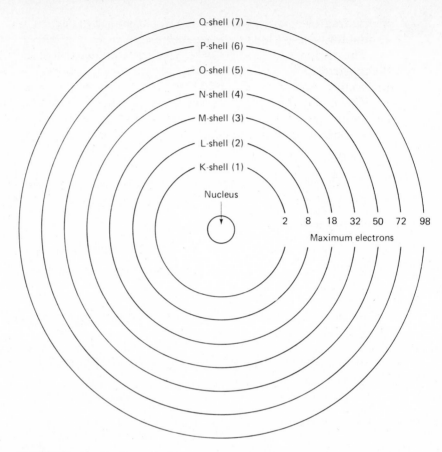

FIGURE 1-4 A theoretical atom showing the shells and maximum electrons.

acquire more electrons than its normal number, it ceases to be an atom because it then has a negative charge. It is then an ion and is called an *anion*. If the atom loses one of its electrons it becomes an ion with a positive charge and is called a *cation*.

1-5 Conductors

An electrical conductor is a substance in which valence electrons can be easily broken away from their atoms. Thus it is able to provide a large number of free electrons which become the circuit current. Some metals provide more free electrons per unit volume than others. The precious metals, gold, silver, and platinum are the best conductors, but they are very expensive and are seldom used except in integrated circuits, where gold has been used extensively. Copper is the most commonly used conductor because it is inexpensive compared to, for example, gold. Aluminum is also used because of its low cost but it has caused safety problems. Aluminum is popular because, although it has only 60% as many free electrons available as copper and therefore is not as good a conductor, it is lighter and less expensive.

The reason why one metal is a better conductor than another can be demonstrated by comparing the carbon atom, a poor conductor, with the copper atom, a good conductor.

The outer shell (ring) of any atom is called the *valence ring* and the electrons contained within it are called valence electrons. A comparison of the carbon atom to the copper atom reveals the following facts:

CARBON

1. The valence electrons are contained in the second shell (L).
2. The valence ring is complete (4 electrons).

COPPER

1. The valence electrons are contained in the fourth shell (N).
2. The valence ring has only one electron. It could hold a maximum of 32 electrons.

Analysis of the comparison shows that the better conductor has an atom with the following properties:

1. The valence electrons are located the furthest from the nucleus.
2. The valence ring is most incomplete.

1-6 The Electron

The destruction of a tree by a bolt of lightning and the operation of an electronic watch are both the result of the flow of electrons. The tree is split or set on fire because of the uncontrolled flow of billions of electrons, and the watch operates because of the precise control of relatively few electrons. It is the ability to control the flow of electrons which makes electrical and electronic technology possible.

Because electronics and electricity involve the creation and control of a flow of electrons, an understanding of the electron is essential.

Copper is a good conductor because the valence electrons are easily removed from copper atoms. The electrons cannot leave the valence ring, however, until they are supplied with additional energy of sufficient quantity that the electrons can rise from the valence band and enter the conduction band. The additional energy may come from different forms such as light (solar cell applications), heat (light-emitting diode applications), or battery voltage. These are only a few of many sources. Once the electrons are in the conduction band they move between the atoms and are pulled toward the positive side of the battery when the pressure of a battery voltage is applied to the circuit. All these free electrons moving in the same direction, toward the positive side of the battery, are called current flow.

In other words, in an ordinary electrical circuit, when voltage from one source or another breaks valence electrons away from their atoms, these free electrons are attracted to the positive side of the battery and constitute current flow.

SUMMARY: CHARACTERISTICS OF THE ELECTRON

1. The electron has a negative charge.
2. The charge on every electron is identical regardless of the atom from which it comes.
3. The electron orbits; therefore, it must possess kinetic energy.
4. There are as many orbital electrons in an atom as there are protons in the nucleus.
5. The electron travels in a curved orbit around the nucleus because of the electric force of attraction between the negatively charged electron and the positively charged protons in the nucleus (unlike charges attract).

EXERCISES

1. What is the purpose of a transducer?
2. Name three types of transducers.
3. Name four common forms of energy.
4. Define matter.
5. Define the difference between an element and a compound.
6. What is the smallest part to which any substance can be reduced and still retain its characteristics?
7. What is the difference between a molecule and an atom?
8. If a particular atom has 12 protons, how many electrons will it have?
9. Where are the protons found in an atom?
10. Where are the electrons which contribute to current flow located?
11. What are the electrons which contribute to current flow called?
12. a. What is an ion?
    b. What is the difference between a positive and a negative ion?
13. State the difference between the atom of a material which is a good conductor and of one which is a poor conductor.
14. Define conductor.
15. State three characteristics of electrons.

chapter **2**

THREE BASIC ELECTRICAL UNITS

The three basic electrical units are:

Current
Voltage
Resistance

2-1 Current

Current flow takes place whenever a source of voltage is connected to a closed circuit. The pressure of the voltage causes the electrons to move from the negative to the positive terminal. Because the positive side of the voltage source is deficient in electrons, the electrons will be attracted toward it. When an electron enters the positive side of the battery, one is released from the negative side of the battery. A continuous flow of electrons leaves from the negative side of the battery, flows through the external circuit, and returns back to the positive side of the battery. This flow of electrons is known as electron current flow, or conduction.

The electron is infinitesimal, as has been explained, but we must be able to measure its quantity. The coulomb is the name given to the basic unit of a quantity of electrons.

2-1.1 The Coulomb

There are 6.24×10^{18} electrons in one coulomb (Fig. 2-1). If there were 12.48×10^{18} electrons contained within an electric circuit there would be 2 coulombs of charge in that circuit.

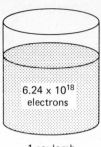

FIGURE 2-1 An imaginary container of 1 coulomb of electrons.

1 coulomb

$$\text{charge} = \frac{\text{total number of electrons}}{6.24 \times 10^{18}}$$

$$Q = \frac{n}{6.24 \times 10^{18}} \ \text{C (coulombs)}$$

where Q is the letter symol for electric charge (coulomb)

n is the total number of electrons involved

C is the number of coulombs of charge

PROBLEM 2-1

If there are 24.96×10^{18} free electrons in a piece of copper wire, how many coulombs of charge are there?

SOLUTION   *(cover solution and solve)*

$$Q = \frac{n}{6.24 \times 10^{18}}$$

$$Q = \frac{24.96 \times 10^{18}}{6.24 \times 10^{18}} = 4 \ \text{C}$$

Comparison: This problem can be compared to boxes of cookies. If a box of cookies contained 6.24×10^{18} cookies how many boxes of cookies would there be, if there were a total of 24.96×10^{18} cookies?

$$\text{Number of boxes} = \frac{n}{6.24 \times 10^{18}}$$

$$= \frac{24.96 \times 10^{18}}{6.24 \times 10^{18}} = 4 \ \text{boxes}$$

PROBLEM 2-2

An electric circuit has 65 coulombs of electrons pass through it. How many electrons are flowing in the circuit?

SOLUTION   *(cover solution and solve)*

$$Q = \frac{n}{6.24 \times 10^{18}}$$

$$n = Q \times (6.24 \times 10^{18})$$

$$= 65 \times (6.24 \times 10^{18})$$

$$= 405.6 \times 10^{18} \text{ electrons}$$

2-1.2 The Ampere

The coulomb is a quantity of electrons (charge) and nothing else. Since current is a flow of electrons and flow involves time, the unit of time must also be involved in the measurement of electron flow. The relationship between charge (number of coulombs) and time is called an ampere. The time involved is always stated in seconds.

DEFINITION OF AMPERE

One ampere is the rate of flow of electron current when one coulomb of electrons pass a given point in an electrical circuit in one second.

$$\text{current in amperes} = \frac{\text{number of coulombs}}{\text{time in seconds}}$$

$$I = \frac{Q}{t}$$

where I is current in amperes
t is time in seconds
Q is number of coulombs

2-1.3 Units of Current

The current in electronic equipment is seldom as large as one ampere. The most common unit of current which flows in electronic circuits is either the milliampere or the microampere.

$$1 \text{ milliampere (mA)} = \frac{1}{1000} \text{ A} = 10^{-3} \text{ A}$$

$$1 \text{ microampere } (\mu\text{A}) = \frac{1}{1{,}000{,}000} \text{ A} = 10^{-6} \text{ A}$$

Meters are readily available which can measure either unit of current.

PROBLEM 2-3

16×10^{18} electrons pass through a circuit in 6 s. What is the amount of current flow?

SOLUTION *(cover solution and solve)*

The number of electrons first have to be converted into coulombs.

$$Q = \frac{n}{6.24 \times 10^{18}}$$

$$= \frac{16 \times 10^{18}}{6.24 \times 10^{18}} = 2.564 \text{ C}$$

Next calculate the amount of current.

$$I = \frac{Q}{t}$$

$$= \frac{2.564}{6} \qquad = 0.427 \text{ A}$$

$$= (0.427) \times 10^3 = 427 \text{ mA}$$

PROBLEM 2-4

40 C of electrons pass through a battery in 3 hours. What is the value of current flow?

SOLUTION *(cover solution and solve)*

$$I = \frac{Q}{t}$$

$$= \frac{40}{3 \times 3600} \qquad = 0.0037 \text{ A}$$

$$= (0.0037) \times 10^6 = 3700 \text{ } \mu A$$

PROBLEM 2-5

How long will it take 2.4 millicoulombs of charge to pass through an electric circuit if the current is 80 amperes?

SOLUTION *(cover solution and solve)*

$$I = \frac{Q}{t}$$

$$t = \frac{Q}{I}$$

$$= \frac{2.4 \times 10^{-3}}{80} \qquad = 0.03 \times 10^{-3} \text{ s}$$

$$= 0.03 \text{ ms}$$

$$= (0.03 \times 10^{-3}) \times 10^6 = 30 \text{ } \mu s$$

2-1.4 Velocity of Current Flow

Current is the flow of electrons, and when it is measured against time, the quantity of flow may be stated in amperes or parts thereof. Note that nothing yet has been said about the speed of electron flow.

The actual velocity of the movement of electrons in an electric circuit is very slow. One particular electron will move only a fraction of an inch in a second. The effective velocity, however, is the speed of light, which is 300,000,000 meters per second, or 186,000 miles per second. This instantaneous velocity comes about because electrons moving at one end of a circuit will cause the electrons at the other end to move at the same instant. This reaction is comparable to the last marble moving the instant that the first marble is struck when marbles are placed tightly together in a long line.

2-1.5 Measuring Current

Meters used for measuring current are called ammeters, milliammeters or microammeters. If the value of current is large, such as in an automobile electrical circuit or in an electric stove circuit, the meter used to measure the current will be an ammeter. In electronics, milliammeters or microammeters are commonly used to measure current.

As a water pipe must be cut into two pieces in order to insert the water meter to measure the quantity of water flow, an electric circuit must also be opened and the current meter inserted in series with the circuit (see Fig. 2-2) to measure the current (electron) flow.

Two other rules must be followed when measuring current:

1. The range switch must be set on a range larger than the expected value of current to be measured.
2. The positive lead of the current meter must be connected to the positive side of the source of voltage; and the negative lead of the current meter connected to the negative side of the source of voltage.

FIGURE 2-2 How to connect a current meter in an electrical circuit. Note that the positive of meter connects to positive polarity of battery.

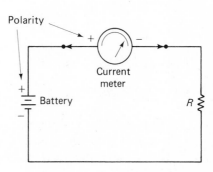

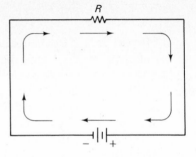

FIGURE 2-3 Current flows out of the negative side of the battery, through the external circuit, and returns to the positive side of the battery.

2-1.6 Direction of Current Flow

Current flow is electron flow, and since the negative polarity of a source of voltage is due to an excess of electrons, current is considered to flow from the negative side of the source of voltage through the circuit to the positive side of the battery (Fig. 2-3). This theory is referred to as the electron current flow theory.

Some textbooks refer to the direction of current flow as being opposite to the direction of electron flow. This direction of current flow is called conventional current flow and is commonly used in the electrical industry. This text uses the direction of electron flow as the direction of current flow.

2-2 Voltage

2-2.1 Static Charge

Valence electrons are considered to be in a static (fixed) position when they remain in their orbit in an atom. An electrical charge is produced in the atom of an element if some electrons are displaced from that atom through friction or through the effect of some outside source (Fig. 2-4). The resultant charge in that element is known as static electricity.

Everyone has experienced static electricity. For example, a person walking across a rug in a room with low humidity will displace electrons from the rug to the shoes. The excess electrons on the shoes will rush through the body to a metal object when it is touched, and the person will experience a nasty electric shock with an accompanying spark.

Another example of static charge is the attraction of a piece of paper to a comb after the comb has been pulled through one's hair. The friction between the comb and the hair builds up a negative charge on the comb. The paper is effectively positively charged, and the oppositely charged objects attract one another.

2-2.2 Law of Charges

If an element or compound contains an excess of electrons it is said to be negatively charged; if it contains a deficiency of electrons with respect to another object it is said to be positively charged.

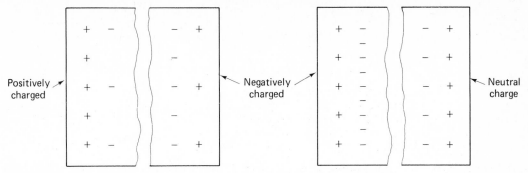

FIGURE 2-4 Charges, on objects, produced by friction. Note: The positive charge is due to a deficiency of electrons; the negative charge is due to an excess of electrons.

Two like-*charged* objects will repel one another; like charges repel.

One object charged negatively will attract another object charged positively, unlike charges attract.

2-2.3 Current Requirements

Although it is possible for a static charge to provide a great potential difference (as with lightning, which results from unequal charges between two clouds), the actual flow of electrons from a static charge is of such short duration that very little usable electrical energy is available. In order to maintain a steady flow of electrons through a conductor, or medium, a constant electric pressure must be applied. This pressure is known as voltage or electromotive force and is measured with meters called voltmeters.

2-2.4 Electric Pressure of Voltage

The letter symbol used for voltage is V.

One of the methods of producing voltage is a chemical system called a battery. There are other means of producing voltage, but the battery symbol will be used to indicate all sources of voltage for the time being.

Voltmeters are used to measure voltage in electric circuits. The voltmeter is always placed in parallel with (across) the component which it is to measure. Remember, *voltage* is pressure; pressure is always produced *across* a component.

Figure 2-5 shows a voltmeter placed across two components. The source of voltage is measured by V_{M1} and the load (lamp) is measured by V_{M2}.

Voltmeter V_{M1} will read the same voltage (6 volts) whether S_1 is open or closed but V_{M2} will read voltage only when S_1 is closed. There is no current flow through the load when switch S_1 is open; therefore, there will be no potential difference (voltage) developed across the lamp. As

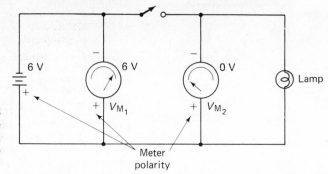

FIGURE 2-5 How to connect voltmeters. There is no voltage across the lamp because the switch is open. Note the polarity of the voltmeters to the battery polarity.

soon as S_1 is closed, current will flow through the lamp, because there will be a closed circuit. The lamp will light and V_{M2} will measure the *potential difference* developed *across* the lamp. Since there is no other load (resistance) in the circuit, V_{M1} and V_{M2} will read the same value, which is the value of the voltage source.

2-2.5 Definition of Voltage

Voltage may be defined as the pressure existing between any two points in a circuit or the difference in potential energy between any two points.

This pressure breaks electrons loose from atoms and causes the free electrons to flow in a specific direction, and current is said to flow in the circuit.

2-2.6 Polarity

Polarity of voltage is indicated by the plus sign for the positive polarity and by the minus sign for the negative polarity. Polarity is a point where there is an excess or deficiency of electrons with respect to another point.

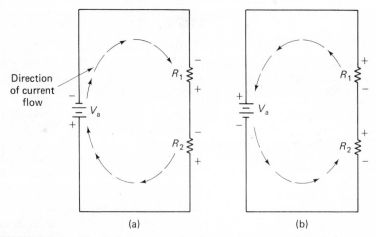

FIGURE 2-6 The voltage at the end of the resistor where the current enters is negative; where it leaves, it is positive.

The polarity of the voltage developed across components through which current flows has a definite direction, and this polarity will change with a change in the direction of the current flow [Fig. 2-6(a) and (b)].

A rule which can be used to establish polarity of voltage across a component through which current is flowing is the following: The end of the component into which the current enters is negative in polarity with respect to the end from which the current exists, which makes the exit point positive with respect to the entry point.

2-2.7 Voltage Reference Point

The polarity of a particular point in an electrical circuit may be negative or positive depending upon the point to which it is being compared. Therefore, polarity is dependent upon the reference point of measurement. For example, in Fig. 2-7, point B is negative with respect to point A or point P. But point B is positive with respect to point C or point N. An analogy can be made with cities and points of reference. For example, New York City is south of Montreal but it is north of New Orleans. Winnipeg is west of Toronto but east of Vancouver. Chicago is east of Los Angeles but is west of New York City.

2-2.8 Measuring Voltage

Three rules when measuring voltage are:

1. The voltmeter range must always be set at a value larger than what is expected to be measured.
2. The positive lead (larger or red lead) of the meter must always be placed on the positive polarity end of the component and the other lead of the meter must be placed on the negative end of the component, as shown in Fig. 2-8.
3. The voltmeter must be connected across the component.

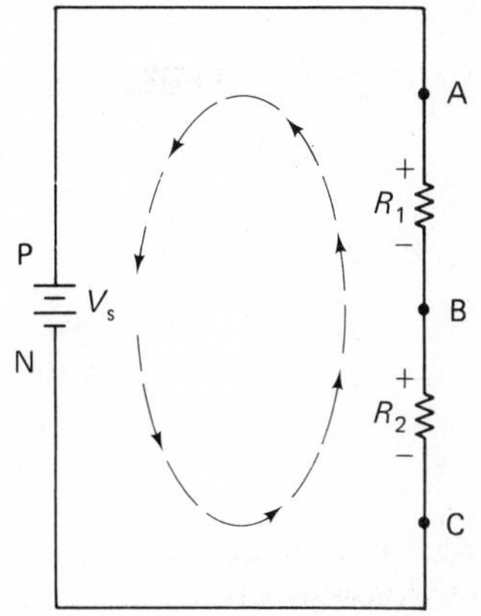

FIGURE 2-7   The voltage at point B can be either positive or negative depending on the point of reference "C" or "A", respectively.

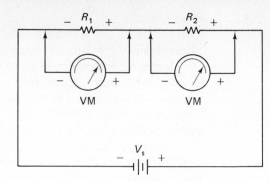

FIGURE 2-8 Voltmeters must be connected *across* the load (resistance). Note the polarity of the meter connections.

2-3 Resistance

Resistance is the third basic electrical unit and is also a property of electrical and electronic circuits. Resistance must be present in every electronic or electrical circuit or there would be no control over the quantity of electrons which flow in the circuit.

2-3.1 Definition of Resistance

Resistance may be defined as that property of a component which opposes current flow. A component, usually called a resistor, is said to have a resistance of one ohm if it permits one ampere of current to flow in the circuit when one volt of voltage is connected across the circuit.

$$\text{resistance} = \frac{\text{voltage}}{\text{current}}$$

$$R = \frac{V}{I}$$

where R is in ohms
V is in volts
I is in amperes

The basic unit of resistance is "OHM"

The symbol for resistance is the greek letter omega (Ω)

The drafting symbol for resistance is ———⌁⌁⌁———

PROBLEM 2-6

What is the resistance of a light bulb which allows 3 amperes of current to flow through it when 120 volts are connected across it?

SOLUTION   *(cover solution and solve)*

$$R = \frac{V}{I} = \frac{120}{3} = 40 \ \Omega$$

Ch. 2 Three Basic Electrical Units

2-3.2 Load

Some form of resistance must be connected to the output terminals of all electrical and electronic circuits. This resistance may be in any of many forms, such as a light bulb, an electrical appliance, a loudspeaker, an electric motor, or an actual resistance. For this reason, the general expression used for the resistance added to the output terminals of a circuit is *load*.

The size of the load primarily determines the amount of current which can flow in the circuit. Even though the word load refers to resistance, the expressions, small or large load actually relate directly to the value of current flow. For example, a large current flow is said to be a large load and a small current is a small load.

The expression small or large load is also relative to the type of circuit in which it is present. One ampere might be a very large load in electronic circuits where it is common to have milliamp and microamp values, but it would be a very small load in factory electrical circuits where there may be hundreds of amperes flowing.

PROBLEM 2-7

What value of current will be permitted to flow in a 120 volt circuit if a 6 ohm load is connected across it?

SOLUTION   *(cover solution and solve)*

$$I = \frac{V}{R} = \frac{120}{6} = 2 \text{ A}$$

2-3.3 Types of Resistors

Resistors themselves are made in several different ways. The simplest way to make a resistor is to wind resistive wire on a cylindrical core. This type of resistor is called a wire-wound resistor. Resistors of this type are used in applications of high heat and are referred to as high wattage resistors (greater than two watts).

A more common type of resistor, because it is used in low-heat applications, is the carbon resistor. The carbon resistor may be of composition, deposited or film-type construction.

Spraying a thin film of resistive material on a glass rod produces a *carbon film* type of resistor. Vaporizing a metal film on a ceramic rod produces a *metal film resistor*. The metal film resistor is usually more desirable than the resistive film resistor because it is more stable in changing ambient (surrounding) temperatures.

All of the carbon resistors and the metal film-type resistors are available in values ranging from a fraction of an ohm to several million ohms. [See Tables 2-3(a) and 2-3(b).]

The resistance value of a resistor has no relationship to its physical size; a 22 ohm resistor may be the same physical size as a 22 megohm

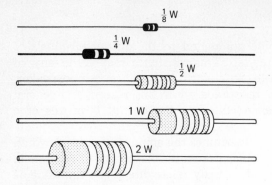

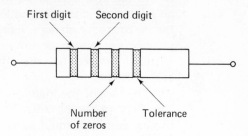

First digit   Second digit

Number
of zeros   Tolerance

FIGURE 2-9 The ability of a resistor to dissipate heat is dependent on its physical size. A 2 watt resistor is much larger than a 1/4 watt resistor.

FIGURE 2-10 How to hold a resistor in order to read its color-coded value, and what each band represents.

resistor. The physical size of a resistor is related only to its ability to dissipate heat. A 2 watt resistor will be much larger than a $\frac{1}{4}$ watt resistor and so on (Fig. 2-9).

2-3.4 The Color Code

The resistance values of resistors from $\frac{1}{8}$ watt to 2 watt sizes (Fig. 2-9) are marked on the resistors in bands of color (Fig. 2-10). The bands of color are closer to one end of the resistor than the other. The end of the resistor to which the bands of colors are closest should be held by the left hand, as shown in Fig. 2-10. Table 2-1 shows the specific value which each color represents.

The first band gives the first number (digit).

Table 2-1 Color Code (shows the number each color represents in the resistor color code)

Color	Value	% Tolerance
Black	0	
Brown	1	
Red	2	
Orange	3	
Yellow	4	
Green	5	
Blue	6	
Violet	7	
Gray	8	
White	9	
Gold		± 5%
Silver		± 10%
None		± 20%

Table 2-2   How to Read the Color Code of a Resistor

Examples	First Band	Second Band	Third Band	Value
1	Brown	Red	Orange	
	1	2	3	12000 ohms
2	Red	Green	Brown	
	2	5	1	250 ohms
3	Yellow	Violet	Black	
	4	7	0	47 ohms
4	Orange	Black	Brown	
	3	0	1	300 ohms

Sometimes the third band may be silver or gold. In such cases:
Gold indicates that the first two numbers are divided by 10.

5	Yellow	Green	Gold	
	4	5	÷ 10	4.5 ohms

Silver indicates that the first two numbers are divided by 100.

6	Orange	Gray	Silver	
	3	9	÷ 100	0.39 ohms

The second band gives the second number (digit).

The third band gives the number of zeros (multiplier) in the value.

2-3.5 Tolerance

Nearly all resistors today have a fourth band. This band designates the tolerance of the resistor, i.e., how much larger in value or how much smaller in value the resistor may measure from its color-coded value and still be considered acceptable. Table 2-2 shows the tolerance values.

EXAMPLE

Suppose the color-coded value of a resistor is

Brown	Black	Red	Silver
1	0	00	± 10%

The resistance value is 1000 ohms and the tolerance is ± 10%
Ten percent of 1000 is

$$1000 \times \frac{10}{100} = 100 \text{ ohms}$$

This resistor would be acceptable if its resistance measured any value between 900 ohms (-10%) and 1100 ohms (+10%). Five percent resistor values are calculated using the same principle.

Before the 1960s, 20% resistors were the most common resistors and the least expensive. The 10% and 5% resistors were less popular. Present day technology has made the 5% resistors almost as inexpensive as the larger tolerance types. Manufacturers such as Phillips Industries make only 5% resistors.

2-3.6 Standard Values

Resistors are manufactured in standard values. These values are known as EIA values and are shown in Tables 2-3(a) and 2-3(b).

2-3.7 Variable Resistors

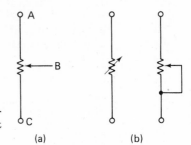

FIGURE 2-11 Schematic symbols: (a) potentiometer; (b) rheostat symbols.

POTENTIOMETERS

A variable resistor is one in which the value of resistance may be changed. The most common type of variable resistor is the potentiometer; its most familiar application is the volume control of a radio or television. Points A and C in Fig. 2-11 are connected across the audio signal source in radios, and by adjusting the position of B, different levels of volume may be obtained from the radio speaker. A study of Fig. 2-11 will reveal that as the resistance between points A and B increases, the resistance between points B and C decreases. In current control applications, points B and C may be connected together, in which case the potentiometer is called a rheostat.

The potentiometer comes in many different shapes, and can range in wattage value from a fraction of a watt to several watts.

Table 2-3(a) EIA, 10% Tolerance Resistors Available in Electronic Stores

Ohms	Ohms	Ohms	Ohms	Ohms	Ohms	Ohms	Meg.	Meg.
2.7	18	120	820	5.6K	39K	270K	1.5	10
3.3	22	150	1K	6.8K	47K	330K	1.8	12
3.9	27	180	1.2K	8.2K	56K	390K	2.2	15
4.7	33	220	1.5K	10K	68K	470K	2.7	18
5.6	39	270	1.8K	12K	82K	560K	3.3	22
6.8	47	330	2.2K	15K	100K	680K	3.9	—
8.2	56	390	2.7K	18K	120K	820K	4.7	—
10	68	470	3.3K	22K	150K	Meg.	5.6	—
12	82	560	3.9K	27K	190K	1	6.8	—
15	100	680	4.7K	33K	220K	1.2	8.2	—

Table 2-3(b)   EIA, 5% Tolerance Resistors

Ohms	Ohms	Ohms	Ohms	Ohms	Ohms	Ohms	Ohms	Meg.
2.7	16	100	620	3.9K	24K	150K	910K	5.1
3	18	110	680	4.3K	27K	160K	Meg.	5.6
3.3	20	120	750	4.7K	30K	180K	1	6.2
3.6	22	130	820	5.1K	33K	200K	1.1	6.8
3.9	24	150	910	5.6K	36K	220K	1.2	7.5
4.3	27	160	1K	6.2K	39K	240K	1.3	8.2
4.7	30	180	1.1K	6.8K	43K	270K	1.5	9.1
5.1	33	200	1.2K	7.5K	47K	300K	1.6	10
5.6	36	220	1.3K	8.2K	51K	330K	1.8	11
6.2	39	240	1.5K	9.1K	56K	360K	2	12
6.8	43	270	1.6K	10K	62K	390K	2.2	13
7.5	47	300	1.8K	11K	68K	430K	2.4	15
8.2	51	330	2K	12K	75K	470K	2.7	16
9.1	56	360	2.2K	13K	82K	510K	3	18
10	62	390	2.4K	15K	91K	560K	3.3	20
11	68	430	2.7K	16K	100K	620K	3.6	22
12	75	470	3K	18K	110K	680K	3.9	—
13	82	510	3.3K	20K	120K	750K	4.3	—
15	91	560	3.6K	22K	130K	820K	4.7	—

Taper is another consideration when purchasing potentiometers. Figure 2-12 shows seven different tapers available in potentiometers and some suggested applications for each type of taper.

ADJUSTABLE RESISTORS

Adjustable resistor is another term used for a variable resistor, but unlike the potentiometer, once its value has been adjusted it is generally left in that position for a considerable length of time. These resistors are generally used in high wattage applications. See Fig. 2-13.

2-3.8 Conductance (G)

Whenever a voltage source is connected to a load (resistance value), a current will flow and it is said that the circuit is conducting. In other words, the term conductance means the ease with which current can flow in a circuit when a voltage is applied. The term conductance is the antonym of, or opposite in meaning to resistance. A circuit with a high conductance value will have a low resistance value.

Conductance is therefore the reciprocal of resistance.

$$G = \frac{1}{R} \text{ siemens}$$

The circuit in Fig. 2-14(b) has the higher conductance because it permits the greater amount of current flow.

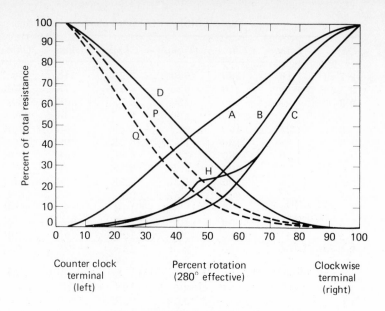

Counter clock
terminal
(left)

Percent rotation
(280° effective)

Clockwise
terminal
(right)

A — For potentiometer or rheostat where uniform
resistance change is required.

B — Semi-logarithmic curve for tone control or
audio circuit control.

C — Logarithmic curve for audio circuit control.

D — Tapered at both ends.

H — Tapped logarithmic curve for audio level control
for automatic bass compensation.

P — Semi-logarithmic curve—reverse taper for use
in contrast and picture control circuits in
television.

Q — Semi-logarithmic curve—reverse taper for use
in contrast and picture control circuits in
television.

FIGURE 2-12   Standard tapers.

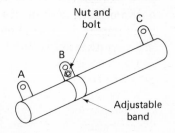

FIGURE 2-13   Adjustable resistor.

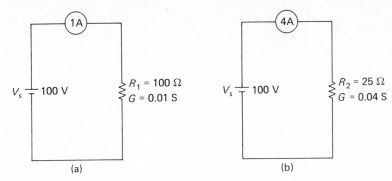

FIGURE 2-14 Two separate circuits showing resistance and their respective conductances.

EXAMPLE

An electric blanket has a resistance of 20 ohms. What is the conductance of the heating element in the blanket?

$$G = \frac{1}{R}$$

$$= \frac{1}{20} = 0.5 \text{ S (siemens)}$$

The letter symbol for conductance is **G**

The SI unit for conductance is *siemens*

The symbol for siemens is **S**

2-4 Resistors and Temperature

The metals from which conductors are made have a specified amount of resistance per unit length and area, which can be measured if the wire is several hundred meters in length. There are tables in engineering handbooks which show the amount of resistance per unit area and length for copper, gold, silver and other metals used in electric circuits. A reference temperature is always stated for the resistance of each metal, because most conductive materials and most resistive materials from which resistors are made change in value with changes in temperature. The reference temperature is generally 20°C. Engineering handbooks also have tables showing the amount of resistance change for each metal, in ohms per degree centigrade change. This latter statistic is referred to as the *temperature coefficient* of the material.

The temperature coefficient of conductors and components can usually be disregarded in non-critical applications such as economy-type radios. Temperature coefficient must be considered, however, in sensitive or precise equipment, in applications where conductors may be extremely long, or in environments in which there are extreme changes in temperature.

Resistors or wires which exhibit an increase in resistance with an increase in ambient (surrounding) temperature are said to have a *positive temperature coefficient*. Those components whose resistance decreases with an increase in temperature are said to have a *negative temperature coefficient*.

Some devices, such as thermistors and glo-bar resistors are intentionally made so that their resistance values change considerably as temperature changes. Some of these devices have a *negative temperature coefficient* and others have a *positive temperature coefficient*.

2-5 Fuses

2-5.1 Fuses

Symbol for fuse ⸺⚬〰⚬⸺

A fuse is a protective device which protects the power supply and the other circuitry against excessive current demand. The fuse is generally installed directly in series with the line cord and the power supply. The power supply is that circuit which may reduce or increase the line voltage and which converts the ac voltage to dc voltage for the operation of the remaining circuitry.

2-5.2 Slow-Blow Fuse

The slow-blow or time-delay fuse is used in applications in which there is an occasional surge (a momentary large amount) of current such as may be caused by the turning off or on of a motor or an air conditioner. An 18,000 B.T.U., 220 volt air conditioner will draw nearly 50 A the first instant that the compressor is switched on, but will draw 18 A or less once it has been started. A 20 A slow-blow (time-delay) fuse will not burn out during the short 50 A surge in current, but will protect against an extended overload of more than 20 A.

Slow-blow fuses have current and voltage ratings similar in range to fast-blow fuses, but come in limited lengths. They are available in the glass tube and screw-in type of casing and mounting.

2-5.3 Practical Considerations

First and foremost, a fuse should never be replaced by one of a higher current rating than the one which was supplied with the equipment or recommended for a particular application. The fuse current rating in newly built equipment should never exceed the working current by more than 10% to 20%.

FUSE FAILURE

There are at least five reasons for fuse failure:

1. A short in the circuit.

2. Fuses tend to lose their current-carrying capacity over a period of time and may fail because of age.

3. An unexplainable momentary large surge of voltage. The current surge of a motor-operated appliance or power tool may blow fuses frequently. The fast-blow fuse should be replaced with a slow-blow fuse.

4. Frequent failure of a fuse in a new installation may indicate that the fuse supplied is underrated.

5. The fuse was a factory defect.

A fuse has a zero resistance value if measured with a standard ohm-meter; therefore, its insertion into a circuit does not upset the operation of that circuit. An open fuse, of course, measures infinite resistance, and full circuit voltage will exist across it, whereas zero volts will exist across a fuse which is intact.

2-5.4 Fuses and Buss Numbers

The most common fuses used in electronic and other small current devices are designated by a number called a Buss number, which is stamped into one metal end of the fuse. Some of the more popular Buss numbers are: AGS; AGC; AGX; AGA; MDV; MDL; SFE.

The above Buss numbers are only a few of the many numbers by which fuses may be purchased. There is no systematic relationship between the designated Buss number and the physical size, voltage rating or current capacity of a fuse. The numbering system has grown haphazardly over the past 50 years, starting with the automobile industry which began with the AG system to code fuse sizes. Additional physical sizes, voltage ratings and current capacities were required as new industries emerged and grew over the years. The result was that the Buss number identifying system was completely lost.

2-6 Schematic Drawings

The expression *schematic diagram* means a drawing, and is comparable to the term *blueprint* in the construction industry. Symbols are used in electronic schematic diagrams in the way that symbols are used in blueprints for doors and windows. A list of symbols is given in the appendix, but they need to be learned only as they are required.

There are two ways of showing connection and no connection between components. Figure 2-15a and b shows both methods.

2-7 Multiple and Submultiple Units

The basic units—ampere, volt, and ohm—are practical values in most electrical power circuits, but in many applications these units are either

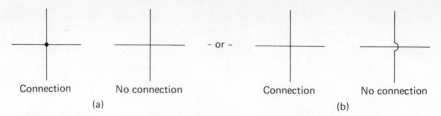

Connection	No connection		Connection	No connection
(a)		- or -	(b)	

FIGURE 2-15 Two methods of indicating a connection and no connection on schematic diagrams.

too small or too big. As an example, resistances can be a few million ohms, the output of a high-voltage supply in a television receiver is about 24,000 volts, and current through tubes and transistors is generally thousandths or millionths of an ampere. In such cases, it is helpful to use multiples and submultiples of the basic units. As shown in Table 2-4, these units are based on the decimal system of tens, hundreds, thousands, etc. The common conversions for V, I, and R are given.

2-7.1 Explanation of Milli and Micro

The concepts of milli and micro are difficult for many people to grasp until they are explained in terms of something with which they are familiar, such as an apple pie.

Suppose an apple pie were cut into 1000 pieces. Each piece of pie could be called a millipie (Fig. 2-16). It is for this same reason that one thousandth, 1/1000, of an ampere is called a milliampere. Milli means one one-thousandth of the whole thing. The same piece of pie could be written mathematically as 1/1000 of a pie or 10^{-3} of a pie. If you had 12 pieces of millipie, you would have 12 pieces of a pie, each piece being 1/1000 of the whole pie. Two hundred and fifty millipies would mean that you had 250 pieces of millipie, each piece being 1/1000 of the whole pie. Mathematically, 250 millipies could be written (250×10^{-3}) pies. Notice that when writing mathematically, the number (10^{-3}) merely replaces the word milli, because 10^{-3} equals milli.

A quarter of a pie is written in decimal values as 0.25 of a pie. It is

Table 2-4 Conversion Factors*

Prefix	Symbol	Relation to Basic Unit	Examples
mega	M	$1,000,000 = 1 \times 10^{6}$	$8 \text{ M}\Omega = 8,000,000 \text{ }\Omega$ $= 8 \times 10^{6} \text{ }\Omega$
kilo	k	$1,000 = 1 \times 10^{3}$	$18 \text{ kV} = 18,000 \text{ V}$ $= 18 \times 10^{3} \text{ V}$
milli	m	$0.001 = 1 \times 10^{-3}$	$48 \text{ mA} = 48 \times 10^{-3} \text{ A}$ $= 0.048 \text{ A}$
micro	μ	$0.000001 = 1 \times 10^{-6}$	$28 \text{ }\mu\text{V} = 28 \times 10^{-6} \text{ V}$ $= 0.000028 \text{ V}$

*See list of prefixes and the powers of ten in the appendix.

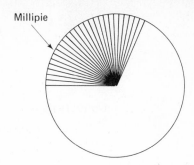

FIGURE 2-16 An imaginary pie cut into 1000 pieces (only a few slices are shown). Each piece is a millipie.

often difficult to work with decimals, so whole numbers are used as much as possible in electronics.

CONVERTING DECIMALS INTO WHOLE NUMBERS

A quarter of a pie (0.25 pie) can be expressed with whole numbers by converting the quantity into terms of millipies. In other words, how many millipies would one have if one had 0.25 of a whole pie? The conversion is made by multiplying by 10^3 (1000).

$$= 0.25 \text{ pie} \times 10^3$$

$$= 0.25 \times 1000 = 250 \text{ millipies}$$

Any decimal number can be converted into a whole number by multiplying the decimal by 10^3 (1000) which converts the value into whole "milli" numbers.

$$0.12 \text{ A} \times 10^3 = 120 \text{ mA}$$

$$0.036 \text{ A} \times 10^3 = 36 \text{ mA}$$

Often a number is so very small that multiplying it by 10^3 will not make it a whole number. For example:

$$0.000059 \text{ A} = 0.000059 \times 10^3 = 0.059 \text{ mA}$$

The next smallest piece of "pie" is micro. Micro means 10^{-6}. The number 0.000059 can be multiplied by 10^6 to convert it to a whole number. Thus:

$$0.000059 \text{ A} = 0.000059 \times 10^6 \text{ A}$$

$$= 59 \ \mu\text{A}$$

CONVERTING WORDS INTO MATH EXPRESSIONS

Thirty-six milliamperes could never be entered into an equation as 36 mA. The word must be converted into its mathematical equivalent; 36 mA would be placed into an equation as 36×10^{-3} A.

For the same reason that a number such as 36 mA cannot be written into an equation, neither can 59 μA nor 36 μA. The word must be converted into its math equivalent.

$$59 \ \mu A = 59 \times 10^{-6} \ A$$
$$36 \ \mu A = 36 \times 10^{-6} \ A$$

SUMMARY:

1. Converting decimals into whole numbers.
$$0.019 \ A = 0.019 \times 10^3 = 19 \ mA$$
$$0.39 \ A = 0.39 \times 10^3 = 390 \ mA$$
$$0.8956 \ A = 0.8956 \times 10^3 = 895.6 \ mA$$

2. Converting words into numbers.
$$28 \ mA = 28 \times 10^{-3}$$
$$56 \ mA = 56 \times 10^{-3}$$
$$987 \ mA = 987 \times 10^{-3}$$

EXERCISES

1. What two conditions must exist before there can be current flow?

2. From what battery polarity terminal will current flow in an electrical circuit and to what battery polarity will it return?

3. Define the term "coulomb."

4. There are 24.96×10^{18} electrons flowing in a circuit. How many coulombs does this represent?

5. There are 18.72×10^{20} electrons flowing past a given point in a circuit every 2 s. How many amperes are flowing in the circuit?

6. How much current is flowing in a circuit if 18 C of electrons pass a given point in 4 s?

7. Sixty amperes of current are flowing in a circuit. How many seconds will it take 4.8 mC of charge to pass a given point?

8. How many coulombs of charge pass through a resistor every second if the current flow is 6 mA?

9. State the effective velocity of current flow.

10. Should a current meter be inserted in series with a resistor or across it in order to measure the current flow through it?

11. Name two other rules other than that in question 10 which must be observed when measuring current.

12. Fill in the blanks in Table P2-1.

Table P2-1

	I	Q	t
a.	10 A	2 C	
b.	10 mA		5 s
c.		5 C	10 ms
d.	2 μA	8 μC	
e.		5 mC	13 μs
f.		10 mC	1 ms
g.	220 A	336 C	
h.	1100 A	248 C	

13. Which two objects will attract one another, like-charged objects or oppositely charged objects?

14. Define the term voltage.

15. Write the letter symbol for voltage.

16. A 90 V source is connected in series with a 30 kΩ resistance. How much current will flow?

17. Complete Table P2-2 for a series circuit.

Table P2-2

	V	I	R
a.		10 μA	8 kΩ
b.	1 mV		100 k
c.		0.1 mA	1 k
d.		50 mA	5.6 kΩ
e.	0.1 V		1 k
f.		10 A	100 kΩ
g.	8.66 V	2.2 A	
h.	3 mV	1 mA	

18. What two electrical units must be present in an electrical component to produce voltage?

19. Label the negative and positive polarities across resistors R_1 and R_2 in Fig. 2-17.

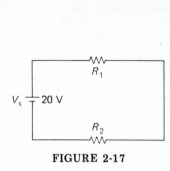

FIGURE 2-17

FIGURE 2-18

20. a. State the polarity at the following points with respect to C in Fig. 2-18.

   + (1) Point B

   − (2) Point D

   + b. With respect to N, what is the polarity at Point C?

   − c. With respect to P, what is the polarity at Point C?

21. State three rules which must be observed when measuring voltage.

22. Does a large load indicate

   a. a small resistance,

   b. a large amount of resistance in the circuit,

   c. a small current flow,

   d. a large current flow.

23. Name two common EIA sizes of low wattage resistors.

24. Write the resistance value and tolerance value for each of the resistors in Table P2-3.

Table P2-3

Resistor	1st	2nd	3rd	4th
		Bands		
a.	Red	Red	Yellow	Silver
b.	Yellow	Violet	Brown	None
c.	Orange	Grey	Black	Gold
d.	Brown	Black	Black	Gold
e.	Green	Blue	Blue	Silver
f.	White	Black	Green	Silver
g.	Violet	Orange	Yellow	Silver

25. Write the color code for the resistors in Table P2-4:

Table P2-4

Resistor	Bands		
	1st	2nd	3rd
a. 470			
b. 390			
c. 2700			
d. 33,000			
e. 5.6 MΩ			

26. Calculate the minimum and maximum values which the resistors in Table P2-5 may measure with a 10% tolerance.

Table P2-5

Resistor	Value	
	Minimum	Maximum
2700		
3 MΩ		
100		
1200		
39,000		
470,000		

27. A charged insulator has an excess of 25×10^{18} electrons. Give its charge in coulombs, with polarity.

28. Another insulator has a deficiency of 50×10^{18} electrons. Give its charge in coulombs, with polarity.

29. A charge of 8 C flows past a given point every 2 s. How much is the current in amperes?

30. A current of 4 A charges an insulator for 2 s. How much charge is accumulated?

31. Convert the following to siemens of conductance:

a.   0.04 Ω

b.   0.5 Ω

c.   1000 Ω

32. Convert the following to ohms of resistance:
    a. 0.001 S
    b. 2 mS
    c. 100 S

33. If 1.25×10^{18} electrons per second flow through a bulb's filament, how much is the current?

chapter **3**

OHM'S LAW

3-1 Ohm's Law

The relationship between the three electrical properties; current, voltage, and resistance, is known as Ohm's law. The law was named after the physicist, George Simon Ohm, who discovered the relationship between V, I, and R in 1827.

The definition for resistance is really one form of Ohm's law, and the example which was used before—that a light bulb would have 40 Ω resistance if 3 A of current passed through it when 120 V was applied—is an application of Ohm's Law.

The basic form of Ohm's Law is:

$$\textbf{voltage} = \textbf{current} \times \textbf{resistance}$$

$$V = IR$$

The interpretation of the above law is that the current which will flow in any circuit is directly proportional to the applied voltage and inversely proportional to the resistance in that circuit.

Ohm's Law is one of the most important laws in electricity. Everyone who is designing or doing any practical work in electronics and electricity is continually using Ohm's law.

3-1.1 Three Ways to Write Ohm's Law

There are three mathematical forms of Ohm's law. (Memorize only one form and from it derive the other two.)

$$\text{current} = \frac{\text{voltage}}{\text{resistance}} \qquad\qquad \text{resistance} = \frac{\text{voltage}}{\text{current}}$$

$$I = \frac{V}{R} \qquad\qquad V = I \times R \qquad\qquad R = \frac{V}{I}$$

The first form is interpreted as: one ampere is the amount of current which will flow through a one-ohm resistance which has one volt applied across it.

3-1.2 Application of Ohm's Law

PROBLEM 3-1

A light bulb is connected across a 12-volt battery (Fig. 3-1). What is the amount of current flow in the bulb if its resistance is 6 ohms?

SOLUTION   *(cover solution and solve)*

1. Draw the circuit (Fig. 3-1).

2. Use the formula $I = \dfrac{V_s}{R}$

$$= \frac{12}{6} = 2 \text{ A}$$

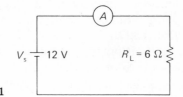

FIGURE 3-1

PROBLEM 3-2

2 A of current flows through a 50-Ω resistor, R_2, connected in series with a battery. What is the value of the applied voltage V_a?

SOLUTION   *(cover solution and solve)*

The applied voltage V_a is:

$$V_a = I \times R$$

$$= 2 \times 50 = 100 \text{ V}$$

PROBLEM 3-3

What is the resistance of the load if 3 A of current flows in a circuit when a 120-V source is connected to it? (See Fig. 3-2.)

SOLUTION   *(cover solution and solve)*

1. Draw the circuit.

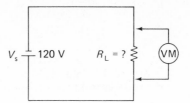

FIGURE 3-2

2. $R = \dfrac{V_s}{I}$

$\quad = \dfrac{120}{3} = 40 \ \Omega$

3-1.3 Graphical Representation of Ohm's Law

The graph in Fig. 3-4 shows a graphical representation of Ohm's Law with a fixed resistance of 4 Ω. The graph may be plotted with the use of a test circuit is shown in Fig. 3-3. As the formula $I = V/R$ shows, the current will double if the voltage doubles; the current will triple if the voltage is tripled (providing the resistance is kept at a constant value).

The graph in Fig. 3-4 showing current plotted against voltage may be plotted in the following manner by using data obtained with the circuit in Fig. 3-3:

1. There can be only two variables, therefore, by using data obtained with the circuit in Fig. 3-3 the resistance must be held constant. Suppose the resistance is held constant at 4 Ω.

2. Draw a table (Table 3-1) showing the value of current as voltage is increased from 1 V to 16 V:

Table 3-1

V	I
1 V	1/4 A
2 V	1/2 A
3 V	3/4 A
4 V	1 A
8 V	2 A
12 V	3 A
16 V	4 A

3. Plot the voltage and current values on a piece of linear graph paper with the dependent variable (current) on the Y-axis and the independent variable (voltage) on the X-axis (Fig. 3-4).

A graph may be plotted showing how current changes with a change in voltage. The reverse cannot be done, however, because current does not

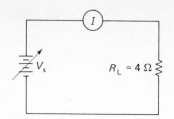

FIGURE 3-3   A test circuit used to obtain the data for plotting the curve in Fig. 3-4.

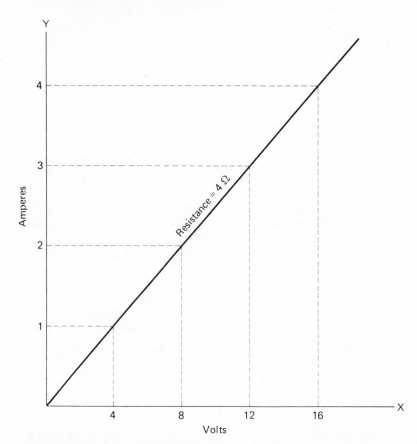

FIGURE 3-4   How current varies with voltage when a 4-Ω resistor is connected in the circuit. Note the linear curve.

change the pressure, i.e., voltage, in a one-resistor circuit, as current is dependent upon voltage, but voltage is not dependent upon current.

3-2   Reference Point

3-2.1   Voltage Polarity—At a Given Point

As discussed in section 2-2.7, the polarity at a given point in a circuit can be either positive or negative.   Figure 3-5 is an example of how voltage polarity is relative to a reference point.

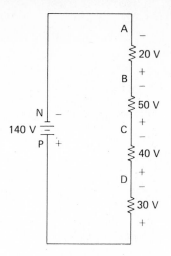

FIGURE 3-5

1. With respect to N:
   B is (+)
   C is (+).
2. With respect to P:
   B is (−)
   C is (−).
3. With respect to D:
   B is (−)
   A is (−).

Compare **1.** with **2.** In **1.** B is (+) with respect to point N, but in **2.** and **3.** B is negative with respect to point P and point D respectively.
In summary: there is no absolute voltage polarity.

3-2.2 Voltage Magnitude and Reference Points

As voltage polarity must have a reference point (point of reference), voltage magnitude also must have a point of reference.
For example look at Fig. 3-6.

1. With respect to point N:
   B is +20 V
   C is +70 V.
2. With respect to point P:
   B is −120 V
   C is − 70 V.
3. With respect to point D:
   B is − 90 V
   A is −110 V.

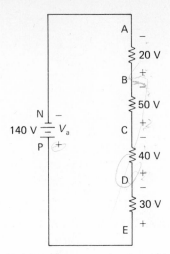

FIGURE 3-6

Note in **1.** that point B is +20 V with respect to point N, but point B is –120 V with respect to point P in **2.** In **3.**, point B is –90 V with respect to point D. The difference in voltage is due to the change in reference points.

3-2.3 Grounds

The schematic symbol for ground is (⏚). The word ground usually means reference point. If, for example, point P were grounded (Fig. 3-7), it would mean that all points showing the ground symbol are connected together, and instead of saying with respect to point P one would say with respect to ground. The same expression would apply to point N or point D if they were grounded.

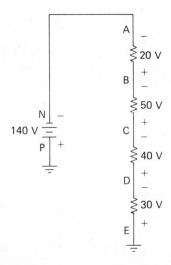

FIGURE 3-7

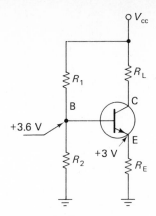

FIGURE 3-8

3-2.4 Less Negative, More Negative, or Positive Expressions

There are situations, when measuring voltage in electronic equipment, when the expressions more positive or more negative are used. Both expressions must be understood by the electronics technician. One such situation which may be found is in transistor circuits such as in Fig. 3-8. Figure 3-8 shows an NPN transistor which in order to operate must have a positive voltage on its base with respect to the emitter. The base voltage measures +3.6 V and the emitter measures +3 V; this makes the base more positive by 0.6 V with respect to the emitter, and the transistor will operate. If the polarity of the two voltages were reversed, however, the base would then be negative with respect to the emitter and the transistor would not operate. The technician must be able to make such an analysis.

Figure 3-9 shows voltages across each of the resistors, but the voltages at the different points will vary with respect to ground and will be different from that across each other resistor.

With the expressions less negative or less positive or more negative or more positive, there is always a point of reference.

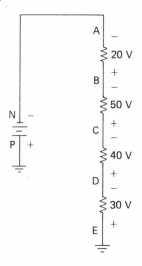

FIGURE 3-9

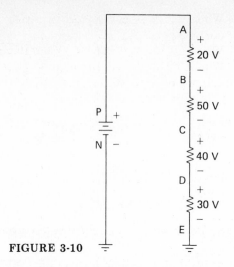

FIGURE 3-10

For example, the voltage at point B in Fig. 3-9 is -120 V with respect to ground, and point C is -70 V with respect to ground. A common expression which says the same thing is, point C is 50 V less negative than point B.

The battery polarity in Fig. 3-10 is the opposite of that in Fig. 3-9, so each point is now positive with respect to ground. For example, point B is +120 V with respect to ground and point C is +70 V with respect to ground. The expression would now be that point C is 50 V less positive than point B.

EXERCISES

1. State the three forms of Ohm's Law.

2. State which unit in each of the following pairs is larger:
   a. volt or kilovolt
   b. ampere or milliampere
   c. ohm or megohm
   d. volt or microvolt
   e. siemens or microsiemens
   f. watt or killowatt
   g. volt or millivolt
   h. megohm or kilohm

3. A 90-V source is connected in series with a 30 kΩ resistance.
   a. Draw the schematic diagram.
   b. How much current flows through the resistance?
   c. How much current flows through the voltage source?
   d. If the resistance is tripled, how much is the current in the circuit?

4. Fill in the blank spaces in the following statements pertaining to series circuits.

a. When 10 V is connected in series with a 2.5 Ω resistance, the current is _____ A.

b. When 10 V produces 5 A, the resistance is _____ Ω.

c. When 12 A flows through a 4-Ω resistance, the voltage drop across the resistor is _____ V.

d. The resistance of 270,000 is _____ kΩ.

5. A circuit has 8 A of current flowing in it. What will be the current flow if the resistance is kept unchanged, but the voltage is doubled? _____ A.

6. A circuit has 8 A of current flowing in it. What will be the current flow if the voltage is kept the same but the resistance is doubled? _____ A.

7. Convert the following units using powers of 10 where necessary:

a. 12 mA to A

b. 5000 V to kV

c. $\frac{1}{2}$ MΩ to Ω

d. 100,000 Ω to MΩ

e. $\frac{1}{2}$ A to mA

f. 9000 μS to S

g. 1000 μA to mA

h. 5 kΩ to Ω

8. A source of applied voltage produces 2 mA through a 10 MΩ resistance. How much is the applied voltage?

9. a. How much resistance allows 30 A of current to flow when 60 volts is applied?

b. How much resistance allows 1 μA current with 10 kV applied?

c. Why is it possible to have less current in b. with the higher applied voltage?

10. Calculate the current, I, in ampere units, for the following examples:

a. 45 V applied to a 68 kΩ resistance.

b. 240 V across a 1.2 MΩ resistor.

11. Calculate the voltage for the following examples:

a. 68 μA through 47 kΩ resistance.

b. 2.3 mA through 22 MΩ resistance.

c. 237 A through 0.012 Ω resistance.

12. Refer to Fig. 3-10 and state the magnitude and polarity of the voltages at the following points with respect to the stated reference points.

with respect to ground:

	D is	+30 V.
a.	C is	_____
b.	A is	_____
c.	P is	_____

with respect to point D:

d. P is   _____

e. B is   _____

f. N is   _____

g. A is   _____

with respect to N:

h. A is   _____

i. B is   _____

j. C is   _____

k. ground is   _____

with respect to B:

l. D is   _____

m. E is   _____

n. ground is   _____

o. P is   _____

p. N is   _____

chapter 4

ENERGY, POWER, AND SHOCK

4-1 Energy and Work

As has been stated earlier, all energy comes from the sun. The world is receiving new energy from the sun every day, some of which heats the earth and produces light, but much of which is stored in one facility or another, such as vegetation growth. Fossil fuel energy is an accumulation of energy which has been stored in the earth over millions of years. The point is that on earth, energy can neither be created nor destroyed; it can only be transformed from one form to another. In other words, the total energy after a transformation must exactly equal the total energy before the transformation.

The process of transforming energy from one form to another is called *work*. The amount of energy transformed is the amount of work done. Consequently, energy and work are numerically the same. They have the same symbol, W, the same formula, and the same SI unit, the joule (J), named in honor of James P. Joule, an English scientist of the 1800s.

Consider a person full of energy, standing at the bottom of a set of stairs. After carrying a parcel up several flights of stairs against the force of gravity, he or she will be very tired and will have very little energy left, because work has been done. The work was the conversion of the chemical energy stored in the body tissues to the mechanical energy of carrying X number of pounds of weight up X number of stairs.

In summary, energy is the potential, capacity, or ability to do work. The chemical energy of a battery has the ability or potential to do work. When the battery in a flashlight is used to produce light, the chemical energy is converted into heat energy in the bulb, which in turn produces

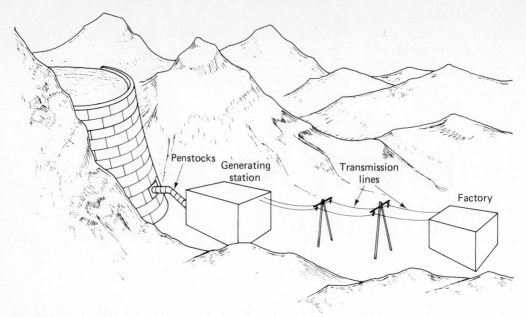

FIGURE 4-1 How potential energy (water) is converted into electrical energy and then possibly into mechanical energy in the factory.

light (the bulb is a transducer), and work is done because energy conversion has taken place.

Energy exists in several forms: among them are heat, light, chemical, mechanical, and electrical energy. When electricity is used, some form of energy, chemical, heat, or mechanical, is being converted into some other form of energy through electrical energy. Work is being done at each stage of conversion.

An example of energy conversion is a hydroelectric system. A high dam is built across a river, and a large lake is formed behind the dam. The water stored in the lake possesses *potential energy* (the energy of position). The water is directed down to a generating station through pipes called penstocks. This falling water has *kinetic energy* (the energy of motion), and causes generator turbines to turn. The rotating generator turbines produce electrical energy which is carried to a factory through transmission lines where it is converted into light, heat, or mechanical energy (see Fig. 4-1). Work is done each time one form of energy is converted into another form. A heater converts electrical energy to heat energy. A light bulb converts electrical energy to light energy. An electric motor converts electrical energy to mechanical energy. The heater, the light bulb, and the electric motor all are transducers.

4-1.1 Measurement of Electric Energy (W)

The symbol for electric energy is W

The SI unit of energy whether electrical or mechanical is the joule (J)

Ch. 4 Energy, Power, and Shock

One joule of electric energy is that amount of energy which has the ability to move one coulomb of electric charge between two points when a potential difference of one volt is applied.

1 joule of energy = 1 coulomb × 1 volt

$$W = QV$$

where W is energy in joules
               Q is coulombs of charge
               V is potential difference in volts

PROBLEM 4-1

A standard 12-V car battery can store 2.4×10^5 coulombs of electrons. How much energy does this particular size of battery possess?

SOLUTION   *(cover solution and solve)*

$$W = QV$$

$$= (2.4 \times 10^5) \times 12$$

$$= 28.8 \times 10^5 \text{ joules}$$

Therefore, this battery can store about 2.8 million joules of energy to operate the car's electrical system.

PROBLEM 4-2

How many coulombs of electrons (charge) would a 6-V battery have to store in order to have the same electrical energy as the above 12-V battery (28.8×10^5 joules)?

SOLUTION   *(cover solution and solve)*

$$\text{from} \quad W = QV$$

$$\text{then} \quad Q = \frac{W}{V}$$

$$= \frac{28.8 \times 10^5}{6}$$

$$= 4.8 \times 10^5 \text{ coulombs}$$

Thus a 6-V battery has to store twice as many electrons as does a 12-V battery to have the same amount of energy.

4-1.2 Measurement of Work

Work is the word used to describe the conversion of one form of energy to another. The amount of energy transformed is the amount of work done;

therefore, as mentioned previously, energy and work are numerically the same.

The symbol for work is W

The SI unit for work is J

In each of the above problems the amount of energy stored by each battery is the amount of work each battery could do. Both the 12-V and 6-V battery could do 2.88 million (2.88×10^6) joules of work.

PROBLEM 4-3

How much work must be done (or how much electrical energy must be supplied to a kettle) to raise 2 kilograms (kg) of water from 20°C to 100°C? (To raise 1 kg of water 1°C, 4185 joules of heat energy is required.)

SOLUTION   (cover solution and solve)

$$W = Km \, \Delta t$$

where   K is the constant 4185J/ 1kg/1°C
m is the mass in kg
Δt is the change in temperature in °C

$$W = 4185 \times 2 \times (100 - 20)$$

$$= 4185 \times 2 \times 80 = 669{,}600 \text{ joules}$$

669,600 joules of work would have to be done.

4-2 Power

Time is involved in any activity, including work. Power is the rate of performing work; therefore, power is expressed in X number of joules per second.

DERIVING THE POWER EQUATION

$$\text{power } (P) = \frac{\text{work } (W)}{\text{time } (t)} \tag{1}$$

where   work (W) is in joules
time (t) is in seconds

$$\text{Joules (J)} = QV \tag{2}$$

Substitute (2) for (1) and

$$\text{power} = \frac{QV}{t} \tag{3}$$

Ch. 4 Energy, Power, and Shock

But $I = \dfrac{Q}{t}$ or $Q = It$ (see section 2-1.1) $\hspace{2cm}$ (4)

Substitute (4) in (3) and

$$\text{power} = \frac{(It) \times V}{t} = I \times V \ \ \text{W (watts)} \hspace{2cm} (5)$$

Power is measured in watts, in honor of the Scottish scientist James Watt.

power = voltage $\times$ current (expressed in watts)

$P = V \times I$ (watts)

4-2.1 Power Wattage Dissipation

Whenever current flows through a component heat is produced. The classic example of this statement is a light bulb, the purpose of which is to produce light, not heat. The light bulb filament has resistance, the value of which can be measured with an ohmmeter. It is the current flow through this resistance which produces heat. The amount of heat produced or dissipated from a component is energy used, in joules, and since time is involved, the value is expressed in watts.

Often the value of current through a component is not known, but the wattage can still be computed by the following formulas:

The basic equation is:

$$P = I \times V \hspace{2cm} (6)$$

$$\text{but} \ \ I = \frac{V}{R} \hspace{2cm} (7)$$

Substitute (7) in (6) and

$$P = \frac{V}{R} \times V = \frac{V^2}{R} \hspace{2cm} (8)$$

In a similar manner, if the value of current and resistance is known, then the wattage formula will be:

$$P = I \times V \hspace{2cm} (6)$$

$$\text{but} \ \ V = I \times R \hspace{2cm} (9)$$

Substitute (9) in (6) and

$$P = I \times (I \times R) = I^2 R \hspace{2cm} (10)$$

SUMMARY:

The three formulas for wattage or heat to be dissipated, sometimes referred to as Ohm's Law for wattage formulas are:

power = voltage $\times$ current or $P = V \times I$

$$\text{power} = \frac{\text{voltage}^2}{\text{resistance}} \quad \text{or} \quad P = \frac{V^2}{R}$$

$$\text{power} = \text{current}^2 \times \text{resistance} \quad \text{or} \quad P = I^2 R$$

PROBLEM 4-4

How much current will a 240-V electric stove draw when

1. the 1000-W element is on?
2. the 750-W element is on?
3. both elements are on?

SOLUTION *(cover solution and solve)*

1. From $P = I \times V$

    then $I = \dfrac{P}{V}$

    $= \dfrac{1000}{240} = 4.17 \text{ A}$

2. $\qquad I = \dfrac{P}{V}$

    $= \dfrac{750}{240} = 3.13 \text{ A}$

3. When both elements are on, the current required is equal to the sum of the current drawn by each element.

    $$I_{\text{total}} = I_{1000\text{w}} + I_{750\text{w}}$$

    $$= 4.17 + 3.13$$

    $$= 7.3 \text{ A}$$

4-2.2 Horsepower

The watt is a small unit of power.

$$1 \text{ W} = \frac{1}{746} \text{ HP (horsepower)} \quad \text{or} \quad 1 \text{ HP} = 746 \text{ W}$$

EXAMPLE

If an electric motor is rated at 5 HP it will draw $746 \times 5 = 3730$ W of power.

4-3 Kilowatt-Hour

The consumer of electric energy pays for power in units of kilowatt-hours (kWh). This power is measured by an instrument known as a watt-hour or kilowatt-hour meter, and the power is sold per kilowatt-hour.

One watt-hour of energy is consumed when one watt of power is drawn for one hour. The prefix kilo means 1000; therefore, 1 kWh of power is drawn if 1000 W of power is drawn for 1 h, or if 100 W of power is drawn for 10 h. Thus the amount of energy, or kilowatt-hours, consumed is the product of power and time, the time being measured in hours.

PROBLEM 4-5

What is the cost of operating a refrigerator for 12 h at a rate of 5 cents per kWh? Assume the compressor motor is a 1/4-HP motor.

SOLUTION   *(cover solution and solve)*

a.
$$1/4\text{-HP motor} = 1/4 \times 746$$
$$= 186.5 \text{ W}$$

b.
$$\text{Number of kWh} = 186.5 \times 12 = 2.238 \text{ kWh}$$

Cost is $2.238 \times 5 = 11.19$ cents

PROBLEM 4-6

How much will it cost to leave a 60-W light burning for 10 days if electricity costs 6 cents per kWh?

SOLUTION   *(cover solution and solve)*

a.
$$\text{Amount of power used in kWh} = \frac{\text{watts} \times \text{time}}{1000}$$

$$= \frac{60 \times (24 \times 10)}{1000}$$

$$= \frac{14,400}{1000} = 14.4 \text{ kWh}$$

b.
$$\text{Cost} = 14.4 \text{ kWh} \times 6 = 86.4 \text{ cents}$$

4-4 Shock

A danger for all who work in electronics—don't take it lightly

4-4.1 Electricity—The Paradox

The one most common aid in our life is electricity. It is the one thing which has done the most to lift the quality of life in the world yet is also one of the most dangerous threats to life.

The average person associates shock with voltage, but actually it is current that does the damage. People have survived contact with several thousand volts, but others have been killed by less than fifty volts. It is the amount and duration of electric current travelling through the body which determines the severity of the shock. Current which enters or passes through the heart and respiratory centers is particularly dangerous. Thus, the path which the current takes through the body is also a key factor in death by electrical shock. Moisture is also an important factor. The skin resistance of the average person is approximately 20,000 Ω, but this resistance will be less than 1000 Ω when one is in a bath.

Table 4-1 Current and Shock

Current in mA	Severity of Shock
1 mA or less	Little or no sensation
5 mA	Painful shock
10 to 15 mA	Freezing to source
30 mA	Breathing becomes difficult—possibly unconsciousness
70 to 100 mA	Ventricular fibrillation (heart loses its rhythm and becomes incapable of doing its job)
100 to 200 mA	Almost certain death (heart will be paralyzed)
over 200 mA	Victim is thrown away from source—not necessarily fatal

RESCUE PROCEDURE

1. Turn off the power or push the victim away from the power source with a dry stick or board.

2. A person thrown to the floor by an electric shock should be persuaded to lie there. Do not restrain the victim as a struggle could add strain to the heart and cause cardiac arrest.

3. It is imperative that artificial respiration be applied to expired victims as soon as they have been removed from the power source. Over 70% of victims who receive artificial respiration within three minutes after shock recover. One more minute's delay reduces the recovery rate to 58%. If artificial respiration is not given within five minutes, death will result.

EXERCISES

1. Write the symbol for energy.

2. State the SI unit for work.

3. Energy is the potential, capacity, or ability to do work. State one example which will demonstrate this definition of energy.

4. A 9-V transistor radio battery can store 3600 C of electrons. How much energy (J) does this particular size of battery possess?

Ch. 4 Energy, Power, and Shock

5. How many electrons (charge) would a 12-V car battery have to store in order to contain 2.8 million J of energy?

6. How much work must be done (or how much electrical energy must be supplied to a kettle) to raise 4 kg of water from $10°C$ to $90°C$? (4185 J of heat energy is required to raise 1 kg of water $1°C$.)

7. How much current will be drawn by each of the following elements of a stove when connected to a 240-V power line?
   a. 1500-W element _____ A
   b. 800-W element   _____ A
   c. 1000-W element   _____ A

8. The element of a 120-V electric heater is rated at 1500 W. What is the resistance of the element?

9. How many watts does 1-HP represent?

10. Define the term one watt-hour of energy.

11. What is the cost of operating a refrigerator for 24 h if the cost of electricity is 8 cents per kWh? (Assume the compressor motor is a 1/4-HP motor.)

12. How much would it cost to leave a 100-W light bulb burning for 10 days, if electricity costs 9 cents per kWh?

13. The same large amount of current which passes between the finger and thumb of person A also passes between the hand and foot of person B. All other things being equal, which person, A or B, is most likely to be electrocuted?

14. State the approximate skin resistance values which may be expected for:
   a. wet skin
   b. dry skin

15. What is the usual result of a five-minute delay in giving artificial respiration to expired electrical shock victims?

16. Which would most often be the fatal current to an individual?
   a. 300 mA
   b. 30 mA

17. A 100-V dc motor is rated at 2.5 HP. How much current will this motor draw?

18. A dc motor draws 10 A of current. How many coulombs of current will pass through the motor in 10 min (minutes)?

19. How much voltage will be required to force 1.5 A of current through a light bulb having 6 Ω resistance?

20. What is the resistance of a 60-W light bulb operating on 120 V?

21. What is the amount of heat produced in a 1000-Ω transistor collector load resistor through which 15 mA is passing?

22. A current of 3.5 A flows through a 2400-Ω resistor.
    a. How much heat is produced in the resistor?
    b. What is the potential difference across the resistor?

23. a. What is the hot resistance of a 100-W, 120-V light bulb?
    b. How much current will this bulb draw?
    c. How much will it cost to operate this bulb for 8 h, if power costs 8 cents per kWh?

24. a. What current does a 5-HP, 220-V single phase motor draw?
    b. How much does it cost to run this motor for 8 h if power costs 6 cents per kWh?

chapter 5

SERIES CIRCUITS

5-1 Series Circuits

5-1.1 Series Resistive Circuits and Ohm's Law

The circuits discussed in the chapter on Ohm's Law were series circuits, i.e., the positive side of the battery was connected to the load and the negative side of the battery was connected to the opposite end of the same load, as shown in Fig. 5-1.

There are many applications of series circuits. For example, series circuits are used in virtually all tube, transistor and integrated circuits (I.C.s). In these applications, one or more load resistors are connected in series with the plate of the tubes, the collector of the transistors, or the I.C. circuits. In these applications, the change in voltage occurring across the load resistors is removed in the form of amplified signals and is passed on to successive stages of amplification. Other applications of series circuits are light bulbs in series with house power. Electric stoves and appliances are all applications of series circuits. The resistances in these applications are the light bulbs, the stove elements, and the motors in the appliances.

5-2 Characteristics of Series Circuits

The circuit in Fig. 5-2 has a source voltage of 100 V and a resistance of 10 Ω, resulting in a current flow of 10 A (from Ohm's Law).

If the 100-V battery were connected to a 40-Ω resistor load (Fig. 5-3), the current flow would be 2.5 A, but the voltage developed across the load would still be 100 V ($V_R = I \times R$).

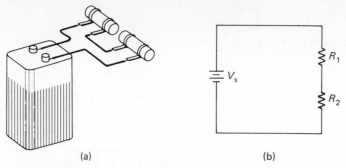

(a)   (b)

FIGURE 5-1   (a) Pictorial view of a simple series circuit. (b) Schematic diagram of (a).

The 10-Ω resistor is said to be a larger load than the 40-Ω resistor because it allows more current flow.   From the above facts we can establish:

The *first significant characteristic* of a series circuit is that current in a series circuit can travel by only one path, as indicated by each figure in this chapter.   The path is from the negative side of the battery through the load and back to the positive side of the battery.   This fact brings forth a *second characteristic* which is that the current in a series circuit is the same at each and every point in the circuit.

If the same 100-V battery is connected in series with the same R_1 and R_2 from Figs. 5-2 and 5-3, the circuit shown in Fig. 5-4 will be formed.

The total current will now be limited by not just R_1 or just R_2, but by both resistors.

$$R_T = R_1 + R_2$$

$$R_T = 10 + 40 = 50 \ \Omega$$

We now discover a *third characteristic*: the total resistance (R_T) of a series circuit is equal to the sum of all the series resistors ($R_T = R_1 + R_2 + R_3 + \cdots + R_n$).

The opposition to the current in Fig. 5-4 increases in direct proportion to the increase in resistance.   The total resistance is now 50 Ω, so the current is reduced to 2 A.

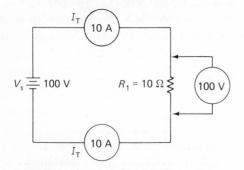

FIGURE 5-2   A series circuit.

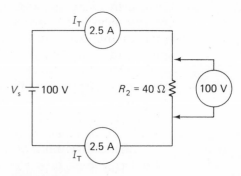

FIGURE 5-3   A series circuit.

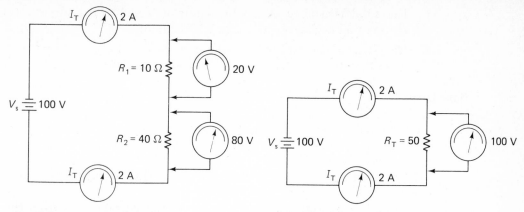

FIGURE 5-4 A series circuit.

FIGURE 5-5 The equivalent circuit of Fig. 5-4.

$$I = \frac{V}{R} = \frac{100}{50} = 2 \text{ A}$$

The 2 A will be the value of current which will flow through the two resistors and into and out of the battery. The equivalent circuit is shown in Fig. 5-5.

Note the magnitude of voltage developed across each resistor in Fig. 5-4.

The voltage developed across R_1 is

$$V_{R_1} = I_R \times R_1$$
$$= 2 \times 10 = 20 \text{ V}$$

The voltage developed across R_2 is

$$V_{R_2} = I_R \times R_2$$
$$= 2 \times 40 = 80 \text{ V}$$

Also note that the sum of the voltages developed across R_1 (20 V) and R_2 (80 V) is equal to the source voltage of 100 V. This fact leads to the *fourth characteristic* of a series circuit, which is that the sum of the voltages developed across all the resistors in a series circuit is exactly equal to the value of the source of voltage. This characteristic was observed and set down as a law by Gustav Kirchhoff in 1847.

Kirchhoff's Voltage Law states that the sum of all of the voltage drops in a series circuit is equal to the sum of all of the sources of voltage in that circuit.

SUMMARY:

1. The current has only one path by which it can travel.

2. The current is the same magnitude at any and every point in the circuit.

3. The total resistance of a series circuit is equal to the sum of the values of all the resistors in that series circuit.

4. The sum of the voltages developed across all the resistors in a series circuit is exactly equal to the source of voltage.

5-3 Voltage Drop

A voltmeter connected from point A to point C (Fig. 5-6) will measure 100 V, but the voltage from point B to C will measure only 80 V. The voltage has been dropped by 20 V by resistor R_1. The expression for this is that the resistor has a voltage drop across it. The voltage drop across R_1 is 20 V.

If the switch in Fig. 5-6 (S_1) is opened, the current flow from the voltage source will be zero, and there will be no voltage drop across R_1 and no voltage drop across R_2.

5-4 Fault-Finding

5-4.1 Open Components

If switch S_1 in Fig. 5-6 is opened, there will be no current flow in the circuit. This is analogous to automobile traffic: there will be no traffic flow on a road if a lift bridge on that road is open (lifted). There can be no current flow if a component in a series circuit is open. No current flow means that zero voltage drops across the components in that series circuit.

$$V_R = I \times R \qquad \text{If } I = 0$$

$$\text{then } V_R = 0 \times R = 0 \text{ V}$$

Because the pressure of the voltage source is still present, the total value of the voltage source will be produced across the open switch or component.

The best practical example of this type of fault and its consequences

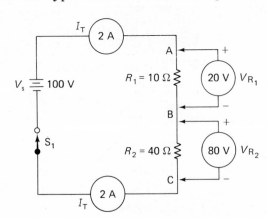

FIGURE 5-6 Voltages across each component when the switch is closed.

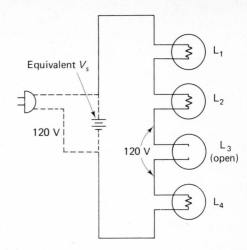

FIGURE 5-7 An open light bulb
(burned out) in a series light-string.

is a string of Christmas tree lights (Fig. 5-7). All the tree lights go out
when one bulb socket is left open, because no current can flow. However,
120 V would be measured with a voltmeter across the terminals of the
open socket.

The nasty shock which one would receive if one foolishly poked one's
finger in the socket of the burned-out light bulb would be full proof that
120 V were being produced across the socket terminals.

The fact that 120 V will be produced across the open component is
also proof that there can be voltage without current. The fact that the
other bulbs do not light means that there is no current flow through them,
yet the nasty shock from the socket of the open bulb proves that the full
120-V supply of voltage is present.

5-4.2 Short Circuits

A component is said to be shorted if its resistance has decreased to 0 Ω,
or if its value is much less than its stated value. A closed switch is equivalent to a total short, because the resistance of it is 0 Ω.

The string of Christmas tree lights can also be used as an example to
demonstrate the effect of a short (Fig. 5-8). When the circuit is normal,
all lights will light properly. If, however, a wire is connected across the
terminals of lamp 2 (L_2), the light in L_2 will go out but the other bulbs
will light more brightly. The wire across lamp 2 shorts out its resistance,
and the current will not go through the resistance of the lamp but will
flow through the zero resistance of the wire. The voltage developed across
the wire, and of course across the lamp, will be zero.

$$V = I \times R$$

If $R = 0$

then $V = 0$

The other lamps will light more brightly because there will be less

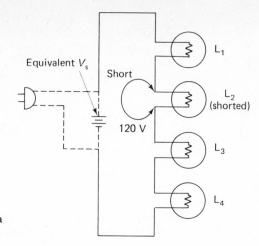

FIGURE 5-8   A shorted lamp in a
series light-string.

total resistance in the series circuit. More current will flow and more
light will be produced.

5-5 Voltage Ratios

It has been noted that the larger the resistance, the greater the voltage
developed across that resistance. In fact, voltage is directly proportional
to the resistance in a series circuit.

If the voltage drops for each of the resistors in Fig. 5-9 were calculated, they would be

$$V_{R_1} = 40 \text{ V}$$

$$V_{R_2} = 60 \text{ V}$$

$$V_{R_3} = 100 \text{ V}$$

These voltage values could be calculated by Ohm's Law or by the use
of a *voltage ratio formula*. There are situations in which the voltage ratio
formula is much more convenient to use; in other instances, it is the only
method which can be used. To use the voltage ratio formula:

1.   Find the total resistance

$$R_T = R_1 + R_2 + R_3$$

$$= 20 + 30 + 50 = 100 \ \Omega$$

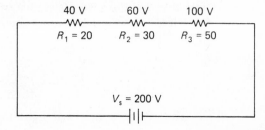

FIGURE 5-9   A series circuit.

2. The voltage developed across each component is equal to the ratio of its resistance value to total resistance, multiplied by the source voltage.

EXAMPLE

$$V_{R_1} = \frac{R_1}{R_T} \times V_s$$

$$= \frac{20}{100} \times 200 = 40V$$

$$V_{R_2} = \frac{R_2}{R_T} \times V_s$$

$$= \frac{30}{100} \times 200 = 60\ V$$

$$V_{R_3} = \frac{R_3}{R_T} \times V_s$$

$$= \frac{50}{100} \times 200 = 100\ V$$

5-6 Power Dissipation

5-6.1 Wattage Calculation

The value of resistance required in an electronic or electrical circuit is only one of two values which must be known before a resistor can be used or ordered. The other value which must be known is the wattage of the resistor. The wattage represents the amount of heat which a resistor must be able to dissipate. A resistor will burn out, for example, if its calculated value is 1 W (i.e., if it must be able to dissipate 1 W of heat), but a 1/2-W resistor is used.

Resistors are made in different physical sizes (Fig. 2-9) to keep temperature low. The larger the physical size of the resistor, even though the resistance value is constant, the greater the amount of heat it will radiate or dissipate. The principle is the same as that which operates in hot water heating or electrical heating radiators in a room. The larger the room to be heated, the larger the heating radiators must be. This is because a larger radiator unit has a larger surface area over which air can flow, and a greater amount of heat can be removed from it.

The wattage rating of a resistor may be calculated by one of three power equations:

$$\text{Wattage } (P) = V \times I \tag{1}$$

$$= I^2 R \tag{2}$$

$$= \frac{V^2}{R} \tag{3}$$

The equation which is used is always that which contains the values of the two known variables.

5-6.2 Resistor Selection

Equations (1), (2), and (3) above show that the required minimum wattage rating of a resistor will depend upon the value of current flowing through a component, or the value of voltage developed across the component, or the value of resistance of that component. The minimum wattage rating of a resistor may therefore change if the value of one of the above electrical units changes.

A 50-Ω 1/2-W resistor will possibly burn out if it is connected in a circuit requiring a 50-Ω 1-W resistor. However, a 50-Ω 2-W or 5-W resistor can be used safely in such a circuit, and the resistor will last a long time because it will operate much cooler than would a 1-W resistor. The only problem is that the larger the wattage, the greater the cost of the component and the greater the space which it occupies.

5-6.3 Practical Wattage Values

The practical wattage value, i.e., the wattage value actually connected in a circuit, should be twice that of the calculated value because of the confined areas of some installations where there is little or no air movement for cooling.

Resistors are manufactured in standard EIA values of 1/4-, 1/2-, 1-, 2-, 5-, 10-W, and greater. They must be ordered in these sizes.

PROBLEM 5-1

A 9-V car radio, requiring 1.5 A to operate properly, is to be installed in a car with a 12-V electrical system. Calculate the resistance and wattage values for the voltage dropping resistor (R_1) which must be connected in series with the radio and battery.

SOLUTION   (cover solution and solve)

1. Draw the circuit (Fig. 5-10).

2. Resistor R_1 must drop the voltage to the car radio by 3 V. The current through R_1 will be 1.5 A.

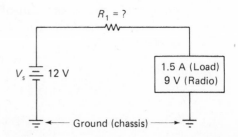

FIGURE 5-10   A voltage-dropping resistor in series with a 9-V radio.

3. $R_{R_1} = \dfrac{V_{R_1}}{I_{R_1}}$

$= \dfrac{3}{1.5} \quad = 2\,\Omega$

4. $P_{R_1} = I \times V$

$= 1.5 \times 3 = 4.5\ \text{W}$

5. This radio might be installed under the dash where there is very little air circulation, so the practical wattage value of R_1 should be $2 \times 4.5 = 9$ W. A 9-W resistor should therefore be ordered.

PROBLEM 5-2

Make the following calculations for the circuit in Fig. 5-11:

a. Show with arrows on Fig. 5-11 the direction of the current flow.

b. Mark the voltage polarity at the ends of each resistor.

c. Calculate the value of voltage drop across each resistor.

d. Calculate the value of the source voltage.

e. Calculate the minimum wattage rating of each resistor.

f. How much power is the battery supplying to the circuit?

g. With respect to point P, what is the value and polarity of the voltage at points A, B, C, D, E, and F?

h. With respect to point N, what is the value and polarity of the voltage at points A, B, C, D, E, and F?

SOLUTION   (cover solution and solve)

a. See Fig. 5-11. (Note the direction of the current flow.)

b. See Fig. 5-11.

c. Voltage drops:

$$V_{R_1} = I_R \times R_1$$

$$= (3 \times 10^{-3}) \times (2 \times 10^{3}) \quad = 6\ \text{V}$$

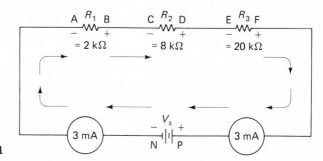

FIGURE 5-11

$$V_{R_2} = I_R \times R_2$$

$$= (3 \times 10^{-3}) \times (8 \times 10^3) \ = 24 \text{ V}$$

$$V_{R_3} = I_R \times R_3$$

$$= (3 \times 10^{-3}) \times (20 \times 10^3) = \underline{60 \text{ V}}$$

$$\text{Total} \quad 90 \text{ V}$$

d. Value of source voltage equals 90 V.

e. All three independent variables, V, I, and R are now known; therefore, any of the three power formulas could be used to determine the wattage dissipation of the resistors. In this solution a different power formula will be used for each resistor.

$$P_{R_1} = I_R^2 \times R$$

$$= (3 \times 10^{-3})^2 \times (2 \times 10^3)$$

$$= (3 \times 10^{-3}) \times (3 \times 10^{-3}) \times (2 \times 10^3)$$

$$= 18 \times 10^{-3} \text{ W}$$

$$= 18 \text{ mW}$$

$$P_{R_2} = \frac{V^2}{R}$$

$$= \frac{(24 \times 24)}{8 \times 10^3}$$

$$= 72 \times 10^{-3} \text{ W}$$

$$= 72 \text{ mW}$$

$$P_{R_3} = I \times V_{R_3}$$

$$= (3 \times 10^{-3}) \times 60$$

$$= 180 \times 10^{-3} \text{ W}$$

$$= 180 \text{ mW}$$

f. One of two methods may be used to determine the power supplied by the battery:

Method 1:

battery power = battery voltage $\times$ battery current

$$= 90 \times (3 \times 10^{-3})$$

$$= 270 \times 10^{-3} \text{ W}$$

$$= 270 \text{ mW}$$

Method 2:

The power supplied by the battery is only that power which is used by the circuit.

$$P_{\text{batt}} = P_{R_1} + P_{R_2} + P_{R_3}$$
$$= 18 \text{ mW} + 72 \text{ mW} + 180 \text{ mW}$$
$$= 270 \text{ mW}$$

g. With respect to point P:

$$A = -90 \text{ V}$$
$$B = -84 \text{ V}$$
$$C = -84 \text{ V}$$
$$D = -60 \text{ V}$$
$$E = -60 \text{ V}$$
$$F = 0 \text{ V}$$

h. With respect to point N:

$$A = 0 \text{ V}$$
$$B = + 6 \text{ V}$$
$$C = + 6 \text{ V}$$
$$D = +30 \text{ V}$$
$$E = +30 \text{ V}$$
$$F = +90 \text{ V}$$

SUMMARY:

Observations about Problem 5-2 can be summarized as follows:

1. The sum of the voltage drops across all of the resistors connected in series is equal to the value of the source voltage (Kirchhoff's Law for Voltage).
2. The largest resistor has the largest voltage developed across it.
3. The voltage developed across each resistor is directly proportional to the value of the resistance. (Voltage is directly proportional to resistance.)
4. The *voltage drop across* a resistor is synonymous with the *voltage developed across* the resistor.
5. The voltage can be developed across a resistor only if current flows through it.
6. Voltage is always measured with respect to a reference point.
7. The supply voltage is always measured across the open component, but when a component is open, there is no current flow in the circuit.

8. A voltmeter must always be connected across the component whose voltage is to be measured.

9. A current meter (ammeter or milliammeter) is always connected in series with the circuit.

10. Load expresses a large or small amount of current flow in a circuit. A relatively large amount of current flow is a large load.

11. The greatest amount of heat is produced by the largest resistor.

PROBLEM 5-3

What is the resistance of a 1500-W, 120-V toaster element?

SOLUTION *(cover solution and solve)*

The known variables are:

$$P = 1500 \text{ W}$$

$$V = 120 \text{ V}$$

Use equation $P = \dfrac{V^2}{R}$

$$R = \dfrac{V^2}{P}$$

$$= \dfrac{120 \times 120}{1500} = 9.6 \ \Omega$$

PROBLEM 5-4

Three resistors, $R_1 = 60 \ \Omega$, $R_2 = 320 \ \Omega$, and $R_3 = 80 \ \Omega$ are connected in series with a battery. A voltmeter connected across R_2 reads 160 V. What is the value of the battery voltage?

SOLUTION *(cover solution and solve)*

1. Calculate the current flow through R_2.

$$I_{R_2} = \dfrac{V}{R_2} = \dfrac{160}{320} = 0.5 \text{ A}$$

2. The current is the same value in every component in a series circuit; therefore, the voltage across resistors R_1 and R_3 can now be calculated.

$$V_{R_1} = I \times R_1$$

$$= 0.5 \times 60 = 30 \text{ V}$$

$$V_{R_3} = I \times R_3$$

$$= 0.5 \times 80 = 40 \text{ V}$$

3. The source voltage is equal to the sum of the voltages across R_1, R_2, and R_3 (Kirchhoff's Law).

$$V_s = V_{R_2} + V_{R_1} + V_{R_3}$$
$$= 160 + 30 + 40 = 230 \text{ V}$$

PROBLEM 5-5

Two resistors, $R_1 = 10 \text{ K}\Omega$ and $R_2 = 7.5 \text{ K}\Omega$ are connected in series to a 70-V source of voltage.
a. What is the voltage drop across the R_2 (7.5-KΩ resistor)?
b. What heat is produced in R_2?

SOLUTION   (cover solution and solve)

1. Draw the circuit and label.
   a. Method 1:
      (1) Find $R_T = R_1 + R_2$

      $$= 10 \text{ K} + 7.5 \text{ K} = 17.5 \text{ K}$$

      (2) Total current flow through the circuit is

      $$I_T = \frac{V}{R_T} = \frac{70}{17500} = 0.004 \text{ A}$$

      Current values are always stated in whole numbers; therefore, convert to mA:

      $$I_T = \frac{70}{17500} \times 10^3 = 4 \text{ mA}$$

      (3) Voltage across R_2 is

      $$V_{R_2} = I_R \times R_2$$
      $$= (4 \times 10^{-3}) \times (7.5 \times 10^3) = 30 \text{ V}$$

      Method 2 (voltage ratio method):

      $$V_{R_2} = \frac{R_2}{R_1 + R_2} \times V_s$$

      $$= \frac{7.5 \text{ K}}{10 \text{ K} + 7.5 \text{ K}} \times 70$$

      $$= \frac{7.5 \text{ K}}{17.5 \text{ K}} \times 70 \qquad\qquad = 30 \text{ V}$$

   b. Wattage of R_2:
      Method 1:

      $$P = I^2 R$$
      $$= (4 \times 10^{-3}) \times (4 \times 10^{-3}) \times (7.5 \times 10^3)$$
      $$= 120 \text{ mW}$$

Method 2:

$$P = V_{R_2} \times I_{R_2}$$
$$= 30 \times (4 \times 10^{-3})$$
$$= 120 \text{ mW}$$

5-7 Power Transfer; Constant Sources

5-7.1 Maximum Power Transfer

The design of a circuit—for example, the output of a stereo amplifier to a speaker—often incorporates the problem of maximum transfer of power from one stage to another. All active devices such as transistors or tubes, and other devices such as batteries, power supplies, and generators have internal D.C. resistances. Power is always lost when current flows through resistance; therefore, there is always some loss of power when current is transferred from one stage to another, e.g., when current passes from a power source to the electronic equipment to which it is supplying power.

The problem in transferring power from point A to point B lies in getting maximum power transfer. The maximum amount of power is transferred only when the load resistance (terminating resistance) is equal to the internal resistance of the source (Thevenin's Resistance). This fact is easily proven by the following example case studies:

case 1: load resistance is equal to the resistance of the power source

PROBLEM 5-6

Suppose a 100-V source of power—for example, a battery with an internal resistance of 50 Ω (an exaggerated value)—is connected to a 50-Ω load as shown in Fig. 5-12. Calculate the amount of power which is transferred to the load from the source.

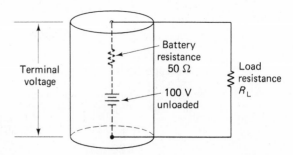

FIGURE 5-12 A battery, its unloaded voltage, and internal resistance connected across a load.

SOLUTION   (cover solution and solve)

Total circuit resistance (R_T) is

$$R_T = R_B + R_L$$
$$= 50 + 50$$
$$= 100 \; \Omega$$

The circuit current will be

$$I = \frac{V}{R} = \frac{100 \text{ V}}{100} = 1 \text{ A}$$

Power developed by the battery will be

$$P = V \times I$$
$$= 100 \times 1$$
$$= 100 \text{ W}$$

Power developed across the load will be

$$P = I^2 R$$
$$= (1)^2 \times 50$$
$$= 50 \text{ W}$$

Fifty watts of power is transferred to the load.

---

case 2: load resistance is double the resistance of the source of power

PROBLEM 5-7

Try to prove that if the value of the load were doubled, the power across the load would double.

SOLUTION   (cover solution and solve)

If $R_L = 2 \times 50 = 100 \; \Omega$ (twice the battery resistance), then

$$R_T = R_B + R_L$$
$$= 50 + 100$$
$$= 150 \; \Omega$$

The circuit current will be

$$I = \frac{V}{R} = \frac{100 \text{ V}}{150} = 0.667 \text{ A}$$

Power developed by the battery will be

$$P = V \times I$$
$$= 100 \times 0.667$$
$$= 66.7 \text{ W}$$

Power developed across the load will be

$$P = I^2 R$$
$$= (0.667)^2 \times 100$$
$$= 44.44 \text{ W}$$

Therefore, 44 W of power is transferred, compared to 50 W; less power is transferred to the load when the load resistance is greater than the source resistance.

---

case 3: load resistance is half the resistance of the resistance of the power source

PROBLEM 5-8

Reducing the value of the load resistance will increase the amount of current which will flow through the load. One would think that this change in load resistance would increase the amount of power transferred to the load, because the increased current is squared. Try to prove this deduction.

SOLUTION   *(cover solution and solve)*

$$\text{If} \quad R_\text{L} = \frac{1}{2} \times 50 = 25 \ \Omega$$

$$\text{then} \quad R_\text{T} = R_\text{B} + R_\text{L}$$
$$= 50 + 25$$
$$= 75 \ \Omega$$

The circuit current will be

$$I = \frac{V}{R} = \frac{100 \text{ V}}{75} = 1.33 \text{ A}$$

Power developed by the battery will be

$$P = V \times I$$
$$= 100 \times 1.33$$
$$= 133 \text{ W}$$

Power developed across the load will be

$$P = I^2 R$$

$$= (1.33)^2 \times 25$$

$$= 44W$$

Therefore, 44 W is transferred, compared to 50 W. Less power is transferred to the load when the load resistance is less than the source resistance.

Conclusion: The maximum amount of power is transferred when the load resistance is equal to the source resistance.

5-7.2 Power Efficiency

Take note that, as shown in the following case-study summary, the percentage of power transferred, and therefore the efficiency of the circuit, is increased only when the load resistance is greater than the source resistance. The reason for this phenomenon is mainly that the amount of power developed by the battery is less.

case 1: $R_L = R_{source}$

$$\text{efficiency} = \frac{P_{R_L}}{P_{batt}} = \frac{50}{100} \times 100 = 50\%$$

case 2: $R_L = 2 \times R_{source}$

$$\text{efficiency} = \frac{P_{R_L}}{P_{batt}} = \frac{44}{66} \times 100 = 67\%$$

(This is the value of maximum efficiency. Note that the power developed by the battery is the least.)

case 3: $R_L = \frac{1}{2} \times R_{source}$

$$\text{efficiency} = \frac{P_{R_L}}{P_{batt}} = \frac{44}{133} \times 100 = 33\%$$

Conclusion: The *percentage* of the amount of developed power transferred to the load and the efficiency of the circuit can be increased by making the load resistance greater than the source resistance.

5-7.3 Impedance Matching

The problem in most design is not efficiency, but getting the maximum amount of power transferred from point A to point B. Previous case studies show that this takes place when the load resistance is equal to the power source resistance. The condition in which the load resistance is equal to the source resistance is called *impedance match*. Making the load resistance appear equal to the source resistance is called *impedance matching*. The problem of impedance matching is of such importance in electronics design that special circuits are designed, or used, to make the load resistance appear equal to the source resistance. The transformer which connects the speaker to the power output stage of a radio or stereo, for example, is designed to match the low speaker resistance to the larger-power amplifier internal resistance. Common collector bipolar transistor circuits and common drain *mosfet* circuits are used exclusively to match the impedances of circuits between which they are inserted. The methods of impedance match will be discussed in advanced studies. Maximum power transfer to the load is not always of prime importance. There are situations in which it is more important to maintain a constant voltage or constant current from a source.

5-7.4 Constant Voltage Source

There are situations in the electronics lab in which the voltage output of the attached signal generator must remain constant while different values of loads are connected to it. A signal generator exhibiting this characteristic would be called a *constant voltage source.*

The voltage output of a circuit will remain constant and therefore will be called a *constant voltage source* when the load resistance is so much greater than the source resistance that the source resistance value will become negligible in the overall total circuit resistance.

There is no particular ratio of load to voltage-source internal resistance at which a voltage source may be considered a constant-voltage source. The larger the ratio of R_L to source resistance the more accurate the results. Some people accept a ratio of R_L to source resistance of 5/1 as a minimum ratio. Some situations may require a ratio of 10/1 or more. A voltage source may be considered as a constant voltage source if its output voltage varies less than 1% under different loads. A constant voltage source is sometimes produced artificially in the lab. This can be done by connecting a resistor, smaller in value than the expected loads, across the source terminals (in parallel with the terminals). The source resistance is then effectively less than the resistance of the smallest load resistance, and a reasonable constant voltage output can then be obtained across a variety of loads. A constant voltage source is often referred to as a Thevenin's circuit. (See Chapter 10, Sec. 10-4.)

EXAMPLE

When the load resistance was 50 Ω in case 1 (Sec. 5-7.1), the voltage output of the battery and the voltage across the load were each 50 V.

$$V_{batt} = I \times R$$
$$= 1 \times 50$$
$$= 50 \text{ V}$$

Reducing the load resistance to 25 Ω as in case 3 (sec. 5-7.1) would cause the battery voltage to drop to 33 V.

$$V_{batt} = I \times R$$

$$= 1.33 \times 25$$

$$= 33 \text{ V}$$

33 volts is a large change from 50 volts.

Suppose the load resistance were made 5000 Ω, which is many times greater than the battery internal resistance of 50 Ω. The voltage across the load would now be 100 V, because 50 Ω is a negligible value compared to 5000 Ω. The battery voltage will now remain at 100 V even if the load resistance were changed as much as 1000 Ω.

5-7.5 Constant Current Source

A source of power (current) to a circuit may be considered as a constant-current source if the load resistor connected across its terminals is much smaller than the internal resistance of the power source. The closer the load values to a short circuit relative to the source resistance the more accurate will be the results. Another way of stating the same fact is that a source of current is considered to be a constant-current source if its internal resistance is very much larger than the value of the load resistance connected across it. The most important fact to both statements is that the source to load resistance ratio must be very large. A minimum ratio of source resistance to load resistance of 10/1 to 100/1 will produce increasingly accurate results. A constant-current source supply can be produced artificially by connecting a large resistance in series with the power source and the load. The current delivered to a variety of loads of a small resistance range will then be fairly constant. A constant current source is often referred to as a Norton's circuit. (See Chapter 10, Sec. 10-5.)

EXAMPLE

The current output of the battery, and thus the current through the load in case 1 (Sec. 5-7.1) when $R_L = R_{source}$ was 1 A.

If the load resistance is increased to 150 Ω as in case 2, the current being delivered to the load will drop to 0.667 A, which is a 33% decrease—a very large change.

Suppose the load resistance were reduced to 3 Ω. The current would now be 2 A because the 3-Ω load is negligible compared to the much larger 50-Ω resistance of the source.

$$I = \frac{V}{R} = \frac{100}{(50+3)} = 2 \text{ A}$$

The current delivered to the load will now remain virtually constant if the load resistance is doubled or halved, and the source will be considered to be a constant current source.

EXERCISES

1. A resistance connected in series with a source of voltage is called a load. Which would be the largest load if connected in a series circuit?
   a. 400 Ω
   b. 40 Ω
   c. 4 Ω

2. How many paths may a current follow when travelling in a series circuit?

3. Refer to Fig. 5-13 and calculate:
   a. total resistance.
   b. the current flowing through meters A, B, and C
   c. the voltage across R_1 and R_2.
   d. the voltage drop across resistor R_1.
   e. the voltage drop across R_2 if the switch S_1 is opened.

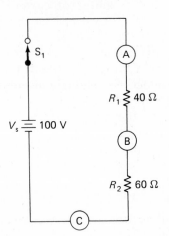

FIGURE 5-13

4. Use a formula to state the relationship between current, voltage and resistance.

5. State Kirchhoff's Law for Voltage.

6. Refer to Fig. 5-13:
   a. Switch S_1 is open. How much voltage will be measured across:
      (1) S_1.
      (2) R_1.
      (3) R_2.

b. Switch S_1 is closed. How much voltage will be measured across the switch?

7. Refer to Fig. 5-13. Assume that R_2 is shorted and is reduced to 10 Ω.

   a. What will be the new circuit current value?

   b. Does the shorted resistor cause the circuit current to increase or decrease?

   c. Has the voltage increased or decreased across:

      (1) Resistor R_1.

      (2) Resistor R_2.

8. What would be the results of connecting a 5-W resistor in a circuit requiring a 1-W resistor?

9. Refer to Fig. 5-14 and determine the following:

   a. the value of the current flow through meters M_1, M_2, and M_3.

   b. the voltage drop across each resistor.

   c. the amount of heat, in watts, developed across each resistor.

   d. the value of the supply of voltage.

   e. use arrows to show the direction of current flow in Fig. 5-14.

   f. how much power is the 80 V supplying to the circuit.

   g. with respect to point N, what is the value and polarity of the voltage at points A, B, C, D, E, and F.

   h. whether the largest or smallest resistor has the largest voltage developed across it.

   i. how much R_3 drops the supply voltage to the rest of the circuit.

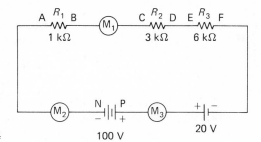

FIGURE 5-14

10. Fill in the blanks in questions 10a. to 10d. for the values given for Fig. 5-15.

   a. I_T 3 A

       R_T 25 Ω

       R_1 15 Ω

       V_T ———

       V_{R_1} ———

       I_{R_1} ———

       V_{R_2} ———

       I_{R_2} ———

       R_2 ———

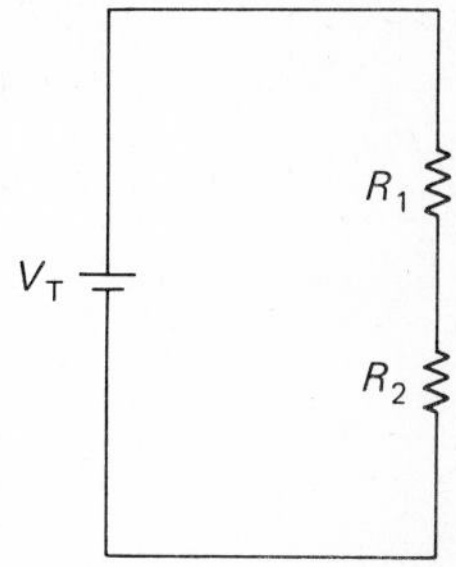

FIGURE 5-15

b. V_{R_1} __60 V__
   I_{R_1} __3 A__
   R_2 __25 Ω__
   R_1 _____
   V_{R_2} _____
   I_{R_2} _____
   V_T _____
   I_T _____
   R_T _____

c. I_T __4A__
   R_1 __6 Ω__
   R_2 __10 Ω__
   V_T _____
   R_T _____
   V_{R_1} _____
   I_{R_1} _____
   V_{R_2} _____
   I_{R_2} _____

d. V_{R_1} __30 V__
   R_2 __20 Ω__
   V_T __50 V__
   I_T _____
   R_T _____
   I_{R_1} _____
   R_1 _____
   V_{R_2} _____
   I_{R_2} _____

11. A speaker has an impedance of 16 Ω. What must be the output impedance of an audio amplifier in order to deliver maximum power to the speaker?

12. An unloaded A.C. generator with an internal resistance of 6 Ω develops 120 V_{rms}.

    a. What must be the resistance of the load for the load to receive the maximum amount of power from the generator?
    b. What will be the voltage delivered to the load in a., when the load is receiving maximum power?
    c. What will be the efficiency of the generator if the load resistance is 2 Ω?
    d. What will be the efficiency of the generator if the load resistance is 18 Ω?
    e. Will the generator be most efficient when the load resistance is:

       (1) much larger than the generator resistance?
       (2) equal to the generator resistance?
       (3) much less than the generator resistance?

    f. What two major effects will a load resistance much smaller than maximum power transfer value have on the generator?

13. The terminal voltage of a car battery drops from 12.6 V when unloaded, to 9 V when the starter, drawing 200 A, is switched on.

    a. What is the resistance of the starter?
    b. What is the resistance of the battery?
    c. If the starter in the car requires maximum power in order to turn over the engine, what should be the resistance of the starter to get maximum power transferred, if the battery characteristics remain unchanged?
    d. How much power will be delivered to the starter when the battery voltage is 9 V loaded?

       (1) _____ W
       (2) _____ HP

    e. What power will be delivered to the starter in part c. in

       (1) _____ W
       (2) _____ HP

f. The internal resistance of every battery increases as the battery becomes discharged. How much power would be delivered to the starter if the starter resistance remained the same, but the battery resistance increased to the value of the starter resistance?

(1) _____ W

(2) _____ HP

14. What would be the loaded battery terminal voltage in **13.** f?

15. What is meant by the expression constant-voltage source?

16. What is meant by the expression constant-current source?

17. What would change a constant-current source to a constant-voltage source?

18. What should be the value of the internal resistance of the source of voltage in order to get the maximum amount of power to a 1-HP motor, operating from a 120-V loaded power-supply?

chapter **6**

PARALLEL CIRCUITS

symbol for parallel is ∥

6-1 Applications of Parallel Circuits (∥)

Regardless of how sophisticated, simple, or complicated circuits are, and regardless of the number of circuits in a piece of electronic equipment, virtually all are connected in parallel with one another (see Fig. 6-1). For example, all lights, plugs, and appliances in every home and apartment are connected in parallel with one another (Fig. 6-2). The headlights, tail lights, ignition system, and all the accessories in a car are all connected in parallel with one another.

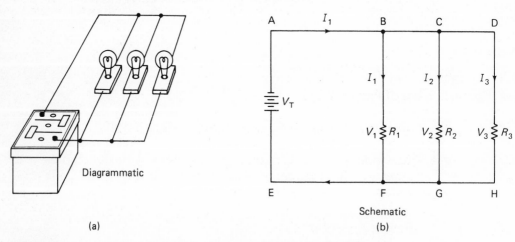

Diagrammatic

(a)

Schematic

(b)

FIGURE 6-1   (a) Pictorial view of a simple parallel circuit. (b) Schematic diagram of (a).

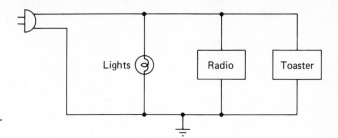

FIGURE 6-2 Sample of an electrical circuit in a house.

The advantage of parallel arrangements is that the change in current flow in one branch does not affect the current flow in the other branches.

6-2 Principle of Operation

A parallel circuit is identified by the fact that it is a circuit in which current has more than one path by which it can flow. Figure 6-3(d) is a parallel circuit and can be used to demonstrate the principle of a parallel circuit.

The four circuits in Fig. 6-3 show four different conditions in a parallel circuit and show how each condition affects the total circuit current.

6-3 Parallel Circuit Arrangements

Figure 6-4 also shows parallel circuits. In fact, these circuits are the same circuit as in Fig. 6-3(d), but are drawn differently.

The features (parameters) that make Fig. 6-3 and Fig. 6-4 identical are:

1. Each diagram has two branches. (The current divides at point A and flows in two directions, then recombines at point B.)

$$I_T = I_1 + I_2$$

2. The component values are identical.

6-4 Characteristics of Parallel Circuits (∥)

Several observations may be made from Fig. 6-3 and Fig. 6-4:

1. The source voltage is common (the same value) to all branches. In other words, each component is connected directly to the source voltage.
2. A change in current flow in one branch does not affect the current flow in any other branch.
3. The number of paths by which current can flow is equal to the number of branches in the circuit.

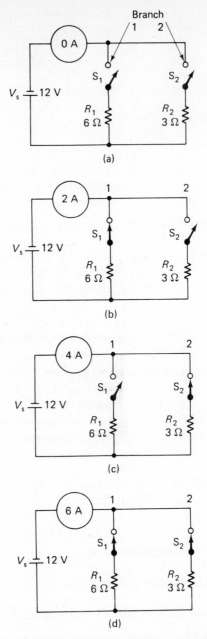

FIGURE 6-3 (a) Switches S_1 and S_2 are both open. There is zero current flow from the battery. A parallel circuit with both branches open. (b) S_1 is closed, but S_2 is still open. Two amps of current will flow from the battery and through R_1 in branch 1. Example: $I_1 = V/R = 12/6 = 2$ A. (c) S_1 is open, and S_2 is closed. Four amps of current will flow from the battery and through R_2 in branch 2. Example: $I_2 = V/R = 12/3 = 4$ A. (d) S_1 is closed, and S_2 is closed. Two amps of current flow in branch 1. Four amps of current flow in branch 2. A total of 6 A of current flows from and to the battery.

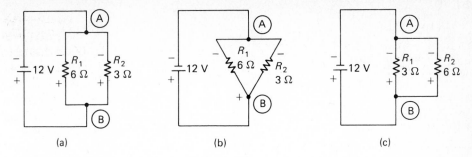

FIGURE 6-4   Three identical parallel circuits.

4. The current flow from the source is equal to the sum of the branch currents.

$$I_T = I_1 + I_2 + I_3 + \cdots + I_n$$

6-5 Kirchhoff's Current Law

Characteristic 4, above, of a parallel circuit was also observed and proven by Gustav Kirchhoff in 1847; it was set down as a law and named Kirchhoff's Law in his honor.

Kirchhoff's Current Law: At any junction point, the algebraic sum of all currents entering that point is equal to the algebraic sum of all currents leaving that point

6-6 Total Resistance of Parallel Circuits

6-6.1 Current and Ohm's Law Method

As calculated earlier for Fig. 6-3, the current through Branch 1 is 2 A, and the current through Branch 2 is 4 A. The total current flowing out of and into the battery is

$$I_T = I_1 + I_2$$
$$= 2 + 4 = 6 \text{ A}$$

As the 6 A of current is being supplied by the 12-V battery, the use of Ohm's Law will show that the total resistance of the two branches must be 2 Ω. The 2 Ω is the equivalent resistance of the 2 parallel resistors.

$$R_T = \frac{V_s}{I_T} = \frac{12}{6 \text{ A}} = 2 \ \Omega$$

The equivalent resistance for the circuit in Fig. 6-3(d) and Fig. 6-4 is shown in Fig. 6-5.

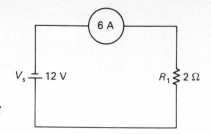

FIGURE 6-5 Equivalent circuit of
Figs. 6-3 and 6-4.

6-6.2 Characteristic 5

Notice that the 2-Ω total resistance is much smaller than the resistance in either branch in Figs. 6-3(d) or 6-4. This fact leads to the observation of a fifth characteristic of parallel circuits: the total resistance of a parallel circuit is always less than the value of the smallest resistance in any branch.

6-6.3 Calculating the Total Resistance of Resistors in Parallel (||)

The concept of conductance can be used to show that the total resistance of resistances connected in parallel can be calculated by yet another method other than the Ohm's-Law method.

$$\text{total conductance } (G_T) = G_1 + G_2 + G_3 + \cdots + G_n \tag{1}$$

$$\text{or} \quad \frac{1}{R_T} = \frac{1}{R_1} + \frac{1}{R_2} + \frac{1}{R_3} + \cdots + \frac{1}{R_n}$$

The solution to this reciprocal formula is very simple with the use of a calculator, as shown in the following problem.

PROBLEM 6-1

Four resistors—a 40-Ω resistor, R_1, a 60-Ω resistor, R_2, a 20-Ω resistor, R_3, and a 12-Ω resistor, R_4—are connected in parallel. What is the total resistance of the four resistors? Solve with the reciprocal formula and a calculator.

SOLUTION   (*cover solution and solve*)

$$\frac{1}{R_T} = \frac{1}{R_1} + \frac{1}{R_2} + \frac{1}{R_3} + \frac{1}{R_4}$$

$$= \frac{1}{40} + \frac{1}{60} + \frac{1}{24} + \frac{1}{12}$$

Solve with the calculator:
Enter 40

$$\text{reciprocal} = 0.0250$$

Press (+)
Enter 60

$$\text{reciprocal} = 0.0166$$

Press (+)
Enter 24

$$\text{reciprocal} = 0.0417$$

Press (+)
Enter 12

$$\text{reciprocal} = 0.0833$$

Press (=)

$$\frac{1}{R_T} = 0.1667$$

$$\text{reciprocal} = R_T = 6 \ \Omega$$

6-6.4 Classical Formula for Two Resistors in Parallel

An alternative method, when only two resistors are connected in parallel, may be derived from the reciprocal formula

$$\frac{1}{R_T} = \frac{1}{R_1} + \frac{1}{R_2} \tag{2}$$

Find the common denominator in equation (2) and then invert the fraction. Equation (2) now becomes equation (3).

$$R_T = \frac{R_1 R_2}{R_1 + R_2} \tag{3}$$

Equation (3) means that the total resistance of two resistors in parallel is equal to the product of the resistances divided by the sum of the resistances. This equation is sometimes referred to as the classical formula.

PROBLEM 6-2

Use the classical formula to calculate the R_T of Fig. 6-3(d).

SOLUTION   *(cover solution and solve)*

$$R_T = \frac{R_1 R_2}{R_1 + R_2}$$

$$= \frac{6 \times 3}{6 + 3} = 2 \ \Omega$$

The value of 2 Ω was the same value obtained when Ohm's Law and the current were used to find the value of R_T in Sec. 6-6.1.

6-6.5 Equal-Valued Parallel Resistors

If the values of resistors connected in parallel are all the same, then the R_T may also be calculated by:

$$\text{total resistance } (R_T) = \frac{\text{the value of one resistor}}{\text{number of resistors in parallel}}$$

$$R_T = \frac{R_1}{\text{number in parallel}}$$

EXAMPLE

The total resistance of 2 equal-value resistors as shown in Fig. 6-6 will be

$$R_T = \frac{250}{2} = 125 \ \Omega$$

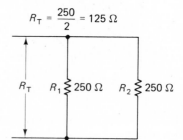

FIGURE 6-6   Two equal resistors connected in parallel.

PROBLEM 6-3

Determine the total resistance of a 60-Ω resistor, R_1, connected in parallel with a 40-Ω resistor, R_2. Use the classical formula.

SOLUTION   *(cover solution and solve)*

$$R_T = \frac{R_1 \times R_2}{R_1 + R_2}$$

$$= \frac{60 \times 40}{60 + 40} = 24 \ \Omega$$

PROBLEM 6-4

Find the total resistance of Fig. 6-7.

SOLUTION   *(cover solution and solve)*

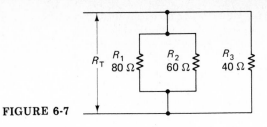

FIGURE 6-7

Use the reciprocal formula and a calculator

$$\frac{1}{R_T} = \frac{1}{R_1} + \frac{1}{R_2} + \frac{1}{R_3}$$

$$= \frac{1}{80} + \frac{1}{60} + \frac{1}{40}$$

$$= 0.0125 + 0.0167 + 0.0250$$

$$\frac{1}{R_T} = 0.0542$$

$$R_T = 18.45 \ \Omega$$

PROBLEM 6-5

Find the value of R_1 for Fig. 6-8.

SOLUTION *(cover solution and solve)*

Use the reciprocal formula and a calculator

$$\text{from} \quad \frac{1}{R_T} = \frac{1}{R_1} + \frac{1}{R_2}$$

$$\frac{1}{R_1} = \frac{1}{R_T} - \frac{1}{R_2}$$

$$= \frac{1}{2000} - \frac{1}{6000} \quad \left(\begin{array}{l}\text{Recall the method used}\\\text{in Problem 6-1.}\end{array}\right)$$

$$= 0.0005 - 0.000167$$

$$\frac{1}{R_1} = 0.000333$$

$$R_1 = 3000 \ \Omega$$

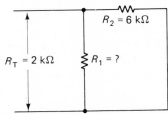

FIGURE 6-8

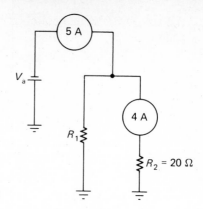

FIGURE 6-9

PROBLEM 6-6

1. Find the voltage across R_1 (Fig. 6-9).
2. Find I_{R_1}.
3. Find V_a.

SOLUTION   (cover solution and solve)

1. $V_{R_2} = I_{R_2} \times R_2$

   $= 4 \times 20 = 80$ V

   R_1 is in parallel with R_2, so $V_{R_1} = V_{R_2} = 80$ V

2. $I_{R_1} = I_T - I_{R_2}$

   $= 5$ A $- 4$ A $= 1$ A

3. This is a parallel circuit, so

   $$V_{R_2} = V_{R_1} = V_a = 80 \text{ V}$$

PROBLEM 6-7

If all the resistors in Fig. 6-10 are equal, what is the value of R_3?

SOLUTION   (cover solution and solve)

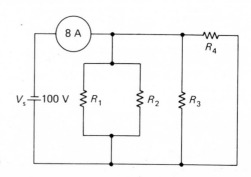

FIGURE 6-10

1. All the resistors are equal in value, so the value of current through each resistor will be the same.

$$I_T = 8 \text{ A}$$

The total number of resistors is 4. The current through each resistor is

$$\frac{I_T}{\text{number of resistors}} = \frac{8}{4} = 2 \text{ A}$$

2. The 100 V is common to all resistors, so the resistance of R_3 is

$$\frac{V_{R_3}}{I_{R_3}} = \frac{100}{2} = 50 \text{ } \Omega$$

PROBLEM 6-8

What is the value of the applied voltage (V_a) in Fig. 6-11?

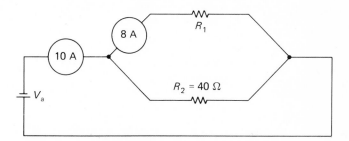

FIGURE 6-11

SOLUTION   *(cover solution and solve)*

We must know 2 values (I and R) in order to find V.

$$R_2 = 40 \text{ } \Omega$$
$$I_{R_1} = 8 \text{ A}$$
$$I_{R_2} = I_T - I_{R_1}$$
$$= 10 \text{ A} - 8 \text{ A} = 2 \text{ A}$$
$$V_{R_2} = I_{R_2} \times R_2$$
$$= 2 \times 40 = 80 \text{ V}$$

It is a parallel circuit, so the voltage is common to all components. Therefore

$$V_{R_2} = V_{R_1} = V_a = 80 \text{ V}$$

6-7 Power (Wattage) Dissipation

6-7.1 Ohm's Law for Wattage

Ohm's Law for wattage is as applicable to parallel circuits as it is to series circuits. The same formulas apply.

$$P = V \times I \qquad P = \frac{V^2}{R} \qquad P = I^2 R$$

The known electrical units determine which of the three formulas is used at the time. For example, in Fig. 6-12, a minimum of two electrical units is known in each branch; therefore, any one of the three wattage (power) formulas may be used.

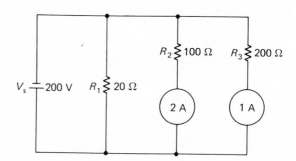

FIGURE 6-12

EXAMPLE

$$P_{R_1} = \frac{V^2}{R} = \frac{200 \times 200}{20} = 2000 \text{ W}$$

$$P_{R_2} = I^2 \times R = 2^2 \times 100 = 400 \text{ W}$$

$$P_{R_3} = I \times V = 1 \times 200 = 200 \text{ W}$$

6-7.2 Total Power Dissipation

The total power dissipated by the above three resistors is the sum of the three wattages.

$$P_T = 2600 \text{ W}$$

The total power delivered by the source voltage may be calculated by still another method:

1. First calculate the value of current through each resistor.
2. Add all the currents together, and finally;
3. Use the equation, $P = V \times I$.

EXAMPLE

1. $I_{R_1} = \dfrac{V}{R_1} = \dfrac{200}{20} = 10 \text{ A}$

2. $I_T = I_{R_1} + I_{R_2} + I_{R_3}$

3. $\quad = 10 + 2 + 1 = 13 \text{ A}$

4. $P_T = I_T \times V_T$

   $\quad = 13 \times 200 = 2600 \text{ W}$

6-7.3 Resistance Value Versus Heat Dissipation

Consideration of the above example will reveal that in parallel circuits the greatest heat is produced by the smallest resistor. This characteristic is the opposite of that in series circuits. However, the total power supplied by the source of voltage is still the sum of the heat dissipation of all the components.

SUMMARY:

1. Components are in parallel when the electrons approaching a point can divide and travel by different paths, i.e., current can travel by two or more paths.

2. The voltage value is the same across all the components in parallel circuits.

3. The value of current flowing in each branch may be different from each of the other branches, and is determined by Ohm's Law.

4. The current supplied by the source voltage (I_s) is the sum of all the individual branch currents.

5. The total resistance of all of the branches is less than the value of the smallest resistor.

6. The total resistance of two *unequal* resistors may be calculated by

$$R_T = \frac{R_1 R_2}{R_1 + R_2}$$

7. If resistors are equal in value, the total resistance may be calculated by

$$R_T = \frac{\text{value of one resistor}}{\text{number of resistors in parallel}}$$

8. The greatest heat is developed in the smallest resistor.

9. The power which the source of voltage must supply is equal to the sum of the wattages developed in each resistor (load).

If each of the resistors in Fig. 6-13 is rated at 300 W which one would overheat?

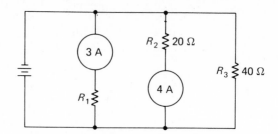

FIGURE 6-13

SOLUTION   (cover solution and solve)

$$V_{R_1} = V_{R_2} = I \times R$$

$$= 4 \times 20 = 80 \text{ V}$$

$$P_{R_1} = I \times V$$

$$= 3 \times 80 \qquad = 240 \text{ W}$$

$$P_{R_2} = I^2 R$$

$$= 4^2 \times 20 \qquad = 320 \text{ W}$$

$$P_{R_3} = \frac{V^2}{R}$$

$$= \frac{80 \times 80}{40} \qquad = 160 \text{ W}$$

Resistor R_2 would overheat because 320 W of heat is being produced in a 300-W resistor.

1. What is the total resistance of Fig. 6-14?
2. What is the total conductance?

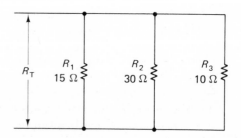

FIGURE 6-14

SOLUTION   *(cover solution and solve)*

1. Use the reciprocal formula and a calculator

$$\frac{1}{R_T} = \frac{1}{R_1} + \frac{1}{R_2} + \frac{1}{R_3}$$

$$\frac{1}{R_T} = \frac{1}{15} + \frac{1}{30} + \frac{1}{10}$$

$$= 0.0667 + 0.0333 + 0.1000$$

$$\frac{1}{R_T} = 0.2$$

$$R_T = 5$$

2. Conductance is the inverse of resistance.

$$\therefore G = \frac{1}{R_T} = \frac{1}{5} = 0.2 \text{ S}$$

6-8 Fault-Finding

6-8.1

CASE 1:

The following current would flow in the circuit in Fig. 6-15 when S_1 is closed:

$$I_{R_1} = \frac{V_s}{R_1} = \frac{10}{2} = 5 \text{ A}$$

$$I_{R_2} = \frac{V_s}{R_2} = \frac{10}{5} = 2 \text{ A}$$

$$I_{R_3} = \frac{V_s}{R_3} = \frac{10}{10} = \underline{1 \text{ A}}$$

The total current would be   8 A

6-8.2

CASE 2:

If S_1 were opened and all the other branches were the same as in case 1, there would be no current flow through R_1, and the total current from V_s would be reduced to 3 A (i.e., $I_{R_2} + I_{R_3}$). The voltage across R_1 (V_{R_1}) would be zero, and the full V_s voltage would be across S_1. In other words, there is no current flow in an open branch circuit. The full supply voltage will be across the open switch because no voltage is developed across the resistive load.

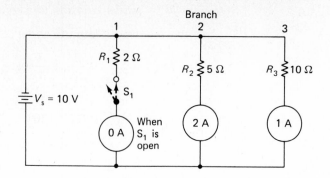

FIGURE 6-15 Switch S_1 is open.

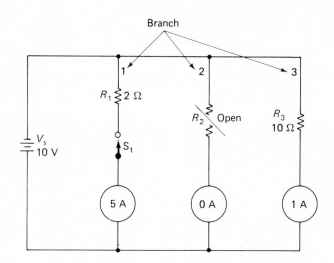

FIGURE 6-16 Resistor R_2 is open.

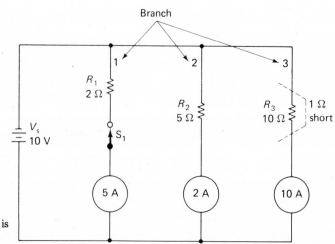

FIGURE 6-17 Resistor R_3 is shorted.

CASE 3:

If R_2 in Fig. 6-16 is opened, and branch 1 and branch 3 are the same as in case 1, there will be no current flow in branch 2, and the total current from V_s will be reduced to $(I_{R_1} + I_{R_3})$ or 6 A. The full supply voltage will still be measured across R_2 even though there is no current flow through it.

CASE 4:

If the resistance of R_3 in Fig. 6-17 has decreased to 1 Ω (the expression is that it is shorted), the resulting current through R_3 will be 10 A. A short in a circuit, therefore, causes the current to increase, which can create serious problems in the circuit.

$$I_{R_3} = \frac{V_s}{R_3} = \frac{10}{1} = 10 \text{ A}$$

$$I_{R_1} = 5 \text{ A}$$

$$I_{R_2} = 2 \text{ A}$$

$$I_T = 17 \text{ A}$$

A short in R_3 has caused the total current to increase, but the voltage across R_3 will remain at the supply-voltage value (it will not change).

SUMMARY:

1. There is no current flow in a branch in which there is either an open switch or an open component. The full supply voltage will be measured across the open component.

2. The current flow in a branch in which there is a shorted component will increase. This will cause the source current to also increase, but the voltage across the branch will remain at the supply-voltage value, because the branch is connected directly to the supply voltage.

EXERCISES

1. State the five characteristics of parallel circuits.

2. State Kirchhoff's Law for current.

3. Does the resistance of a parallel circuit increase or decrease if another resistor is placed parallel with the existing parallel resistors?

4. a. Use the classic formula for calculating the total resistance of parallel circuits to determine the resistance of Fig. 6-18.

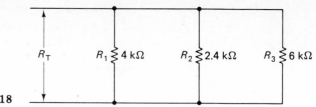

FIGURE 6-18

b. What would be the total resistance of Fig. 6-18 if R_2 were to become open?

c. What would be the total resistance of the circuit if R_3 were to become shorted?

5. The total resistance of two resistors connected in parallel is 4000 Ω. The value of one of the resistors, R_1 is 12,000 Ω. Determine the value of the second parallel resistor.

6. a. Calculate the voltage developed across R_2 in Fig. 6-19.

b. What is the value of V_a in Fig. 6-19?

c. Calculate the heat (watts) which R_1 and R_2 must dissipate.

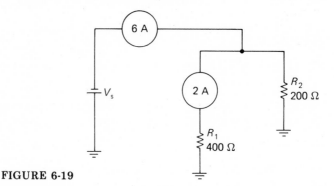

FIGURE 6-19

7. a. All the resistors in Fig. 6-20 are equal. What is the value of resistor R_5?

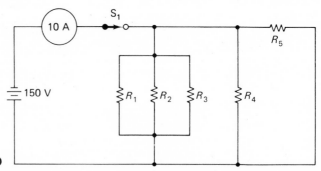

FIGURE 6-20

b. Calculate the power being supplied to the circuit by the battery.

c. Calculate the minimum wattage size requirement of each resistor in Fig. 6-20.

d. What will happen to the value of the current supplied by the battery in Fig. 6-20 if R_4 becomes shorted?

e. How will the value of the total current flow be affected if R_1 becomes open?

f. What voltage will be measured across R_2 if the switch (S_1) is opened?

g. What voltage would be measured across R_2 if R_2 were open?

h. What voltage will be measured across the switch (S_1) if the switch is open?

8. How may one determine when 2 components are in parallel?

9. Two resistors, a 200-Ω and a 10-Ω, are connected in parallel. Which resistor will have the greatest heat developed in it?

10. How much current will flow through the milliammeter in Fig. 6-21?

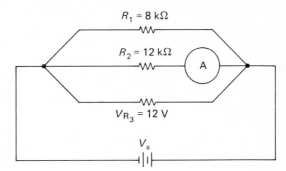

FIGURE 6-21

11. Find the value of R_2 in Fig. 6-22.

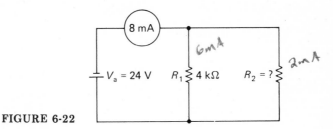

FIGURE 6-22

12. What is the value of source voltage in Fig. 6-23?

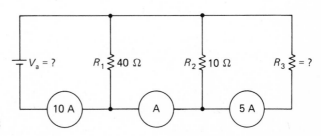

FIGURE 6-23

13. Find the value of I_1 and I_2 if I_T in Fig. 6-24 is 6 mA.

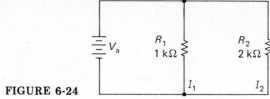

FIGURE 6-24

14. If R_T in Fig. 6-25 is 2 kΩ, find R_1.

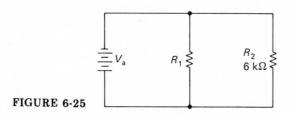

FIGURE 6-25

15. Find the value of R_1 in Fig. 6-26.

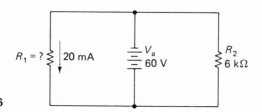

FIGURE 6-26

16. How will closing the switch in Fig. 6-27 affect the voltage across R_1?

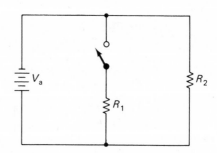

FIGURE 6-27

17. a. Find R_T in Fig. 6-28.
  b. Calculate the voltage across the 6-kΩ resistor in Fig. 6-28.

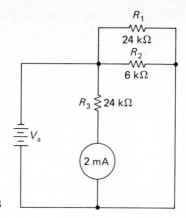

FIGURE 6-28

18. Find the voltage of R_1 in Fig. 6-29.

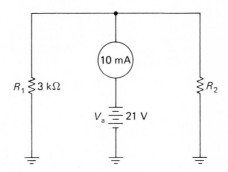

FIGURE 6-29

19. Refer to Fig. 6-30.
   a. Calculate the resistance of R_1.
   b. Calculate the value of V_s.

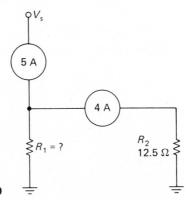

FIGURE 6-30

20. From the schematic diagram in Fig. 6-31 find:

R_T = _____ I_5 = _____
I_T = _____ P_1 = _____
P_T = _____ P_2 = _____
I_1 = _____ P_3 = _____
I_2 = _____ P_4 = _____
I_3 = _____ P_5 = _____
I_4 = _____

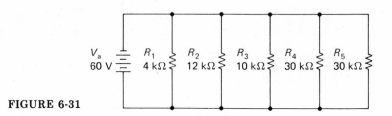

FIGURE 6-31

chapter **7**

SERIES-PARALLEL CIRCUITS

7-1 Series–Parallel Circuits

Electrical and electronic circuits are not limited to simple series or simple parallel circuits. These two types of circuits are most often combined into what is known as series–parallel circuits (Fig. 7-1). Most electronic equipment consists of numerous series–parallel circuits.

Figure 7-2 shows what appears to be two different series–parallel circuits, but in fact it shows the same circuit drawn differently. The resistors R_3 and R_4 between points A and B are in parallel because the current (I_T) divides at point A, flows through each branch, and recombines at point B to return to the source voltage.

Figure 7-2c is a simplified circuit of Fig. 7-2. The parallel circuit R_3, R_4 in Fig. 7-2a has been reduced to one series resistance R_{TP} of 4 Ω, and

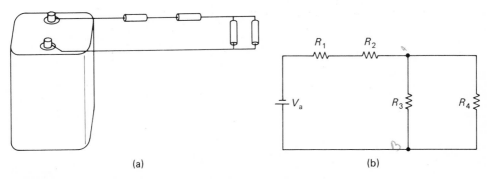

(a)   (b)

FIGURE 7-1   (a) Pictorial view of a simple series circuit. (b) Schematic diagram of (a).

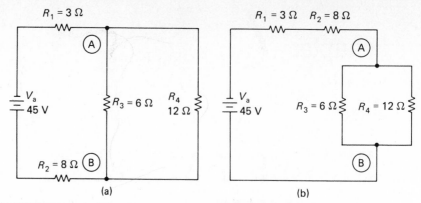

FIGURE 7-2a and b   Identical series-parallel circuits.

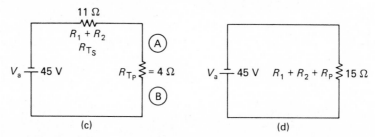

FIGURE 7-2c and d   (c) Simplified circuit of Fig. 7-2a. (d) Equivalent circuit of Fig. 7-2a.

the other series resistors R_1 and R_2 have been combined to $R_{T_S} = (R_1 + R_2) = 11\ \Omega$.

Figure 7-2d is the final and equivalent circuit for Fig. 7-2a. The series resistors in Fig. 7-2c have been combined into one resistor in Fig. 7-2d.

Figure 7-3a is another series–parallel circuit. The current (I_T) from the negative side of the battery arrives at point A where it divides; part of it flows down through branch 1 consisting of R_1 and R_2, and the remainder flows through branch 2 consisting of R_3 and R_4. The two branch currents recombine at point B to form I_T which returns back to the source voltage.

7-1.1 Sample Solution

METHOD 1 SOLUTION FOR THE CIRCUIT IN FIG. 7-3a:

The equivalent circuit for Fig. 7-3a can be found by adding the resistances in branch 1 and branch 2 (Fig. 7-3b) and then using the classic equation for parallel resistance to find R_T (Fig. 7-3c).

$$R_T = \frac{R_1 \times R_2}{R_1 + R_2}$$

$$= \frac{30 \times 15}{30 + 15} = \frac{450}{45} = 10\ \Omega$$

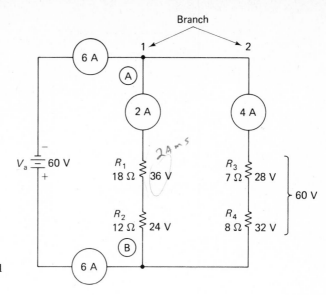

FIGURE 7-3a   Series-parallel circuit.

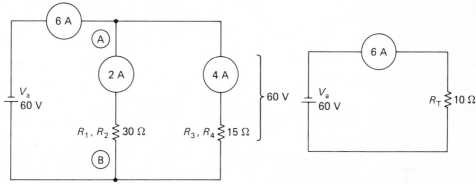

FIGURE 7-3b   Simplified   circuit for Fig. 7-3a.

FIGURE 7-3c   Equivalent   circuit for Fig. 7-3a.

METHOD 2 SOLUTION:

The R_T for Fig. 7-3a could also be found by first finding each branch current value and then using these current values and Ohm's Law to find R_T.

$$R_T = \frac{V_a}{R_T} = \frac{60}{6} = 10\ \Omega$$

7-2 Solving R_T for Complex Circuits

Regardless of how complex or elaborate series–parallel circuits may be, there is a standard method one can use to simplify them. The circuit in Fig. 7-4a is one sample of a complex series–parallel circuit.

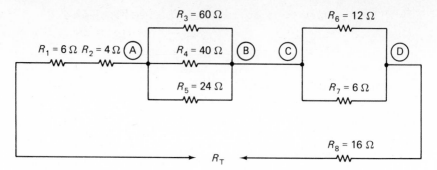

FIGURE 7-4a   A complex series-parallel circuit.

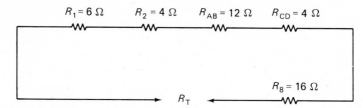

FIGURE 7-4b   Simplified circuit of Fig. 7-4a.

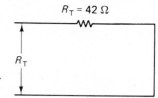

FIGURE 7-4c   The equivalent circuit for Fig. 7-4a.

METHOD FOR SOLUTION:

1. Reduce each parallel circuit to a single resistance. See Fig. 7-46.

   a. The resistors between points A and B are in parallel because the circuit current will divide three ways at point A and recombine at point B. The total resistance of R_3, R_4, and R_5 is

   $$\frac{1}{R_{AB}} = \frac{1}{60} + \frac{1}{40} + \frac{1}{24}$$

   $$= 0.0166 + 0.0250 + 0.0417$$

   $$R_{AB} = 12\ \Omega$$

   b. Resistors R_6 and R_7 are in parallel. Their total resistance is (use the reciprocal formula and a calculator)

   $$R_{CD} = \frac{R_6 \times R_7}{R_6 + R_7}$$

   $$= \frac{6 \times 12}{6 + 12} = 4\ \Omega$$

2. Draw the equivalent circuit. (See Fig. 7-4b.)

3. Add all the series resistors in Fig. 7-4b together for the total resistance (R_T).

$$R_T = R_1 + R_2 + R_{AB} + R_{CD} + R_8$$

$$= 6 + 4 + 12 + 4 + 16$$

$$= 42\ \Omega$$

4. Draw the equivalent series circuit for the total resistance. (See Fig. 7-4c).

7-3 Calculation of I and V in Series–Parallel Circuits

There is also a standard procedure to find the current and voltage in complex series–parallel circuits.

Figure 7-4 will again be used as the sample circuit, but with a voltage source and current meters inserted in series with the resistors; see Fig. 7-5a.

METHOD FOR SOLUTION OF CURRENT AND VOLTAGE:

1. The total resistance of the circuit must first be determined. Therefore the method outlined in section 7-2 must be followed, with a simplified circuit (Fig. 7-4a to 7-4c) drawn for each step.

2. Each of the current values and voltage values may now be determined by using the simplified circuits in the order of Fig. 7-4c, 7-4b, and 7-4a as shown in Fig. 7-5b to 7-5d.

3. Calculate the total current (use Fig. 7-5b).

$$I_T = \frac{V}{R} = \frac{168}{42} = 4\ \text{A}$$

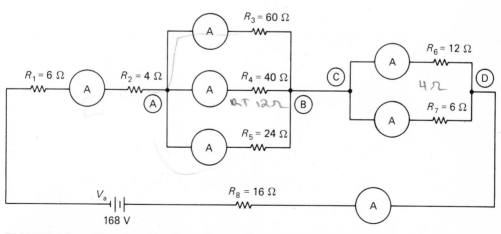

FIGURE 7-5a The same circuit as Fig. 7-4a but with a voltage source and with current meters added.

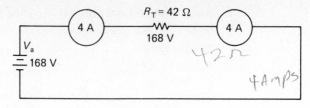

FIGURE 7-5b The equivalent circuit for Fig. 7-5a showing supply voltage and circuit current values.

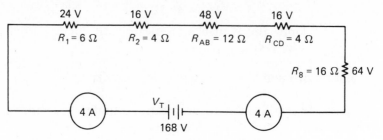

FIGURE 7-5c The simplified circuit of Fig. 7-5a.

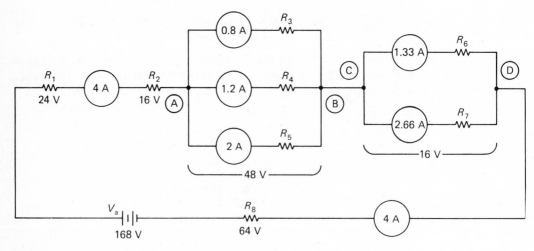

FIGURE 7-5d This is the same circuit as Fig. 7-5a but with the value of the voltage and currents labeled on the components.

4. Determine the voltage across each series–equivalent resistor in Fig. 7-5c (the same circuit as Fig. 7-4b):

$$V_{R_1} = I_T \times R_1 = 4 \times 6 = 24 \text{ V}$$

$$V_{R_2} = I_T \times R_2 = 4 \times 4 = 16 \text{ V}$$

$$V_{R_{AB}} = I_T \times R_{AB} = 4 \times 12 = 48 \text{ V}$$

$$V_{R_{CD}} = I_T \times R_{CD} = 4 \times 4 = 16 \text{ V}$$

$$V_{R_8} = I_T \times R_8 = 4 \times 16 = \underline{64 \text{ V}}$$

total voltage (V_T) = 168 V

Note that the sum of the voltages across all of the resistors is equal to the source voltage.

5. Determine the current through resistors R_3, R_4, and R_5 which forms R_{AB} (see Fig. 7-5a). Resistors R_3, R_4, and R_5 are in parallel between points A and B. The voltage, $V_{R_{AB}}$, (see step 4) is therefore common to each of the three resistors R_3, R_4, and R_5. The current through them may be calculated with Ohm's Law.

$$I_{R_3} = \frac{V_{AB}}{R_3} = \frac{48}{60} = 0.8 \text{ A}$$

$$I_{R_4} = \frac{V_{AB}}{R_4} = \frac{48}{40} = 1.2 \text{ A}$$

$$I_{R_5} = \frac{V_{AB}}{R_5} = \frac{48}{24} = \underline{2.0} \text{ A}$$

$$\text{total current} = 4.0 \text{ A}$$

Note that the sum of the currents through the parallel resistors is equal to the source current (Kirchhoff's Law for current).

6. Find the current through R_6 and R_7 which forms R_{CD} (see Fig. 7-5a). The procedure is exactly the same as in 3 above. Resistors R_6 and R_7 are in parallel between points C and D; therefore the voltage across R_6 and R_7 will be 16 V (see step 4) and the current through them will be

$$I_{R_6} = \frac{V}{R_6} = \frac{16}{12} = 1.33 \text{ A}$$

$$I_{R_7} = \frac{V}{R_7} = \frac{16}{6} = \underline{2.66} \text{ A}$$

$$\text{total current} = 4.00 \text{ A}$$

SUMMARY:

The voltage across each resistor and the value of current flow through each resistor is now known.

$$\text{Voltage across } R_1 = 24 \text{ V}$$
$$R_2 = 16 \text{ V}$$
$$R_3 = 48 \text{ V}$$
$$R_4 = 48 \text{ V}$$
$$R_5 = 48 \text{ V}$$
$$R_6 = 16 \text{ V}$$
$$R_7 = 16 \text{ V}$$
$$R_8 = 64 \text{ V}$$

the current through R_1 = 4.00 A

$$R_2 = 4.00 \text{ A}$$
$$R_3 = 0.80 \text{ A}$$
$$R_4 = 1.20 \text{ A}$$
$$R_5 = 2.00 \text{ A}$$
$$R_6 = 1.33 \text{ A}$$
$$R_7 = 2.66 \text{ A}$$
$$R_8 = 4.00 \text{ A}$$

These values may now be labelled on the components in the original circuit (Fig. 7-5a) as shown in Fig. 7-5d.

PROBLEM 7-1

Refer to Fig. 7-6a. Find the

a. current through each resistor.
b. voltage across each resistor.

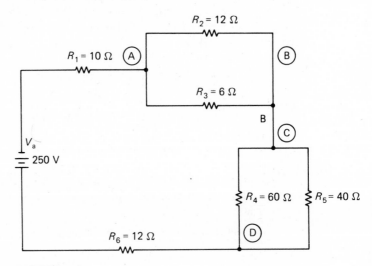

FIGURE 7-6a

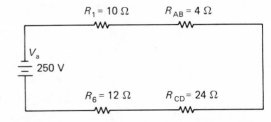

FIGURE 7-6b The simplified circuit of Fig. 7-6a.

SOLUTION   (cover the solution and solve)

1. Find the equivalent resistance of each parallel circuit.

$$R_{AB} = \frac{R_2 \times R_3}{R_2 + R_3} = \frac{12 \times 6}{12 + 6} = 4 \ \Omega$$

$$R_{CD} = \frac{R_4 \times R_5}{R_4 + R_5} = \frac{60 \times 40}{60 + 40} = 24 \ \Omega$$

2. Draw the simplified circuit (see Fig. 7-6b).
3. Find the equivalent circuit by adding the resistors.

$$R_T = R_1 + R_{AB} + R_{CD} + R_6$$

$$= 10 + 4 + 24 + 12 = 50 \ \Omega$$

4. Find the source current (I_T).

$$I_T = \frac{V}{R_T} = \frac{250}{50} = 5 \ A$$

5. Find the voltage across each resistance in the simplified circuit (Fig. 7-6b).

$$\text{voltage across } R_1 = I_T \times R_1 = 5 \times 10 = 50 \ V$$
$$R_{AB} = I_T \times R_{AB} = 5 \times 4 = 20 \ V$$
$$R_{CD} = I_T \times R_{CD} = 5 \times 24 = 120 \ V$$
$$R_6 = I_T \times R_6 = 5 \times 12 = \underline{60 \ V}$$
$$\text{total voltage} = 250 \ V$$

Kirchhoff's Voltage Law has been satisfied.

6. Find the current flow through each resistor in Fig. 7-6a. Resistors R_2 and R_3 are in parallel between points A and B; therefore, the voltage across R_2 and R_3 is 20 V.

$$\text{current through } R_2 = \frac{V_{AB}}{R_2} = \frac{20}{12} = 1.66 \ A$$

$$R_3 = \frac{V_{AB}}{R_3} = \frac{20}{6} = \underline{3.33 \ A}$$

$$\text{total current} = 5.0 \ A$$

7. Find the current through R_4 and R_5. The same procedure is used as in 6 above.

$$\text{current through } R_4 = \frac{V_{CD}}{R_4} = \frac{120}{60} = 2A$$

$$R_5 = \frac{V_{CD}}{R_5} = \frac{120}{40} = \underline{3A}$$

total current = 5 A

7-4 Fault-Finding in Series–Parallel Circuits

The knowledge of the procedure for calculating total resistance, voltage, and current values in complex series–parallel circuits may now be used to determine faults in these circuits. The fault-finding procedure is much the same as for parallel or series circuits.

7-4.1 Fault-Finding in the Circuit in Fig. 7-7

The circuit in Fig. 7-7 will be used as a sample circuit in which faults may be assumed. The following case studies will give some insight into the effect that certain faults can produce in a circuit.

CASE 1:

Switch S_1 is closed and the circuit is working normally. See Fig. 7-7a. If S_1 is closed the following currents will flow:

$$\text{Branch 1.} \quad I_1 = \frac{V_s}{R} = \frac{10}{2} = 5 \text{ A}$$

$$\text{Branch 2.} \quad I_2 = \frac{V_s}{R} = \frac{10}{(1+4)} = 2 \text{ A}$$

$$\text{Branch 3.} \quad I_3 = \frac{V_s}{R} = \frac{10}{(1+9)} = \underline{1 \text{ A}}$$

total current = 8 A

CASE 2:

Switch S_1 is open. (See Fig. 7-7b.)
There will be no current flow through R_1 when the switch is open. The voltage across R_1 (V_{R_1}) will be zero and the full supply voltage will be across the switch. In other words, there is no current flow in an open branch circuit, and the full supply voltage is across the open component because no voltage will be developed across the resistive load in series with the open component.

CASE 3:

Resistor R_2 is open. (See Fig. 7-7c.)
If R_2 is open, there will be no current flow in branch 2, and the total current from the voltage source will be reduced to ($I_{R_1} + I_{R_3}$), or 6 A. The full supply voltage will be across R_2 because there is no current in branch 2, and there will be zero volts across R_3.

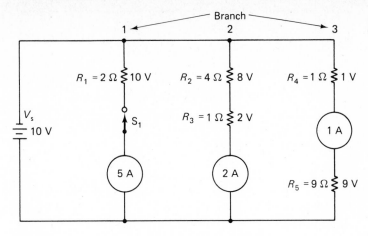

FIGURE 7-7a   Case 1: The circuit is normal.

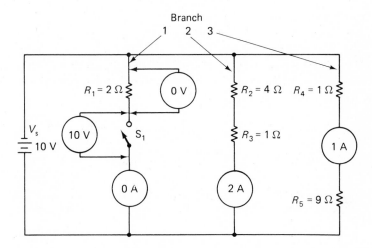

FIGURE 7-7b   Case 2: The switch, S_1, in branch 1 is open.

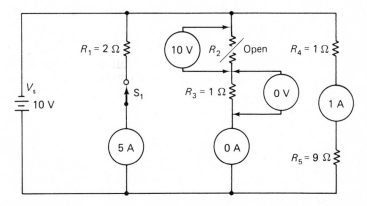

FIGURE 7-7c   Case 3: A resistor in branch 2 is open.

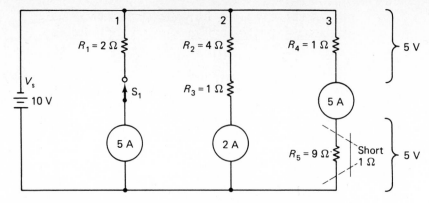

FIGURE 7-7d The current flow through branch 3 when R_5 has a 1-Ω short. Also note the new voltage value across R_4 and R_5.

CASE 4:

Resistor R_5 is shorted. (See Fig. 7-7d.)
The current flow in branch 3 will increase if the 9-Ω resistor R_5 develops, for example, a 1-Ω short. The total circuit current will also increase in that event from 8 to 12 A. The voltage developed across R_5 will decrease to 5 V but the voltage across resistor R_4 in series with R_5 will increase to 5 V.

7-4.2 Fault-Finding in the Circuit in Fig. 7-8

The circuit shown in Fig. 7-8 is a series–parallel circuit of a different type than the circuit in Fig. 7-7a. The knowledge gained from the discussion in section 7-4.1 will now be applied to the circuit in Fig. 7-8 to determine the effects created in the circuit under various faulty conditions.

CASE 1:

Resistor R_1 becomes shorted.
If R_1 becomes shorted in Fig. 7-8 the total circuit resistance decreases

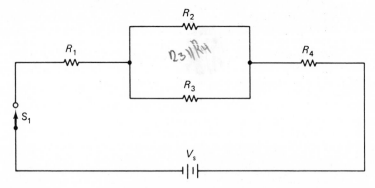

FIGURE 7-8 Circuit for Case Study 1 through Case Study 5.

and the circuit current increases. The voltage across R_4, R_2, and R_3 will increase, but the voltage across R_1 will decrease.

CASE 2:

Resistor R_4 becomes shorted.
This fault produces symptons similar to those produced by R_1. If R_4 becomes shorted, the voltage across R_2, R_3, and R_1 will increase but the voltage across R_4 will decrease.

CASE 3:

Resistors R_1 or R_4 become open.
The supply voltage will be measured across the open component, but zero voltage will be measured across all the other resistors if R_1 or R_4 open.

CASE 4:

Resistors R_2 or R_3 decrease in value.
The total circuit resistance will decrease in value if R_2 or R_3 is reduced in value or becomes shorted, but the total current will increase. The result is that the voltage across R_1 and R_4 will increase in value, but the voltage across the parallel network will decrease in value.

CASE 5:

Resistors R_2 or R_3 increase in value.
The parallel network resistance will increase in value if either R_2 or R_3 become open or increase in value. The circuit current will decrease and the voltage across R_1 and R_4 will decrease, but the voltage across the parallel network will increase in value.

SUMMARY:

1. The voltage measured across an open switch or component in a series circuit is always the supply voltage. Zero volts will be measured across any component connected in series with an open switch or component.
2. The voltage decreases across a component which becomes shorted in a series circuit, but the voltage across the other components will increase.

7-5 Tracing and Redrawing Complex Circuits

Electronic circuits are generally very complex. Regardless of the complexity, no circuit can be more than a number of series–parallel arrangements. Many schematic diagrams are impossible to understand until the circuits are traced and redrawn to the familiar series–parallel format.

It is essential that everyone, from the engineer to the technician, is able to trace circuits and to reduce them to the simplified form.

As one unconsciously follows a procedure to undress at night, or follows certain directions to get home, one must also follow a set of rules to trace complex circuits.

The procedure is first to trace the circuit and then to redraw it in a simplified series–parallel format. This is the procedure with even the most sophisticated electronic equipment. It is not uncommon for persons to guess the type of circuit and its operation instead of tracing and redrawing it, but technical disaster is generally the reward.

7-5.1   Tracing and Redrawing Procedure (Fig. 7-9a)

1. Label each end of each component with a letter (see Fig. 7-9a).

2. Begin tracing from one end of the source of supply (use P for Fig. 7-9a) and make a rough sketch as the tracing progresses (Fig. 7-9b). Redraw the circuit to a series–parallel format. Series resistors are usually drawn in a horizontal plane; parallel resistors are drawn in a vertical plane.

3. Remember:

    a. If current has only one path to take when it reaches a component, then that component is a series component.

    b. If current reaches a point where it can divide, and take more than one path, then that point marks the beginning of a parallel arrangement.

7-5.2   Simplifying the Circuit

The tracing procedure as outlined in Sec. 7-5.1 will now be used to rearrange the circuit shown in Fig. 7-9a to its actual series–parallel configuration as shown in Fig. 7-9d.

1. Label each end of each component with a letter. (See Fig. 7-9a.)

2. Begin at one end of the battery (use P for Fig. 7-9a) and trace from one letter to the next letter as shown in the example below.

    a. A connects to P (series–horizontal).

    b. The other end of A is B.

    c. Letters C, G, E, and S all connect to B (parallel–vertical).

    Redraw the circuit as the tracing progresses as shown in Fig. 7-9b.

3. Label the free end of each resistor with its appropriate letter (Fig. 7-9b).

4. Now join the free ends of these resistors to the other resistors as the circuit is traced (Fig. 7-9c).

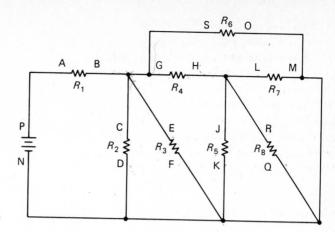

FIGURE 7-9a   A complex circuit.

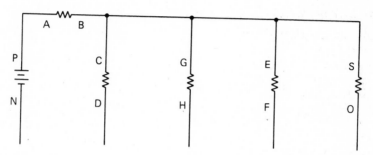

FIGURE 7-9b   A redrawing in progress of the complex circuit in Fig. 7-9a.

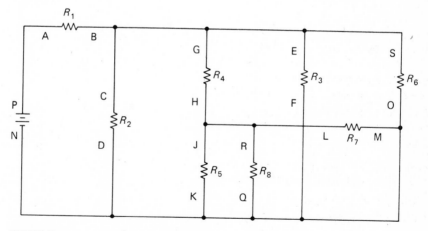

FIGURE 7-9c   A nearly completed redrawn circuit for Fig. 7-9a.

a.   H connects to J, L, and R (Fig. 7-9c).

b.   D, F, and O all connect to N.

c.   M, Q, and K also connect to N. (N is the circuit reference at the negative end of the battery.)

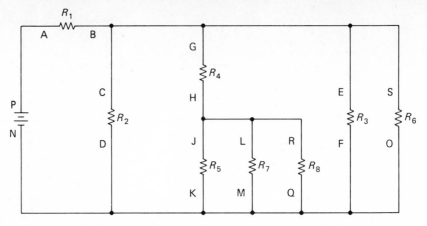

FIGURE 7-9d   The simplified circuit of Fig. 7-9a.

5. Redrawn circuits seldom are finished circuits and must be redrawn again.   Figure 7-9c for example, shows the resistor LM drawn in a horizontal and confusing position. The drawing therefore must be redrawn.

6. Figure 7-9c shows that letters J, R, and L are all connected together and that letters K, Q, O, and M are connected together; therefore resistors JK, RQ and LM are all in parallel.

7. Redraw Fig. 7-9c to Fig. 7-9d.

7-5.3 The Equivalent Circuit

Once the simplified circuit has been drawn, all the parallel resistors must be reduced to their series-equivalent value (Fig. 7-9e), and the circuit must be redrawn with the equivalent series value labelled with original letters (Fig. 7-9f). This is done so that the location of individual resistors may always be identified.

The process must be repeated until all the components are redrawn as a series circuit (Fig. 7-9g or Fig. 7-9h).

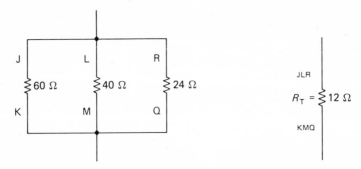

FIGURE 7-9e   Resistors R_5, R_7, and R_8 extracted from Fig. 7-9d.

FIGURE 7-9f   The equivalent circuit of Fig. 7-9e.

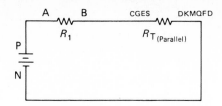

A   B   CGES   DKMQFD

R_1 $R_{T\,(Parallel)}$

P

N

FIGURE 7-9g   A series equivalent circuit of Fig. 7-9a.

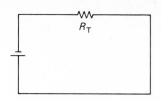

R_T

FIGURE 7-9h   The final equivalent circuit of Fig. 7-9a.

7-6   Calculating Current and Voltage Values

It is now a simple matter, with the use of Ohm's Law, to calculate current and voltage values through and across each component. One must simply work backward from the last drawing to the first drawing. See section 7-3.1.

EXERCISES

Note: The following problems pertain to series–parallel circuits. Most of them will require simplification of the circuit with the circuit-tracing procedure before they can be solved.

1. Find the total resistance for Fig. 7-10.

2. Refer to Fig. 7-11 and calculate

R_T = __10K__ V_{R_2} = __12v__
I_T = __3mA__ V_{R_3} = __12v__
P_T = __90w__ P_1 = __54 mW__
I_2 = __2mA__ P_2 = __24 mW__
I_3 = __1mA__ P_3 = __12 mW__
V_{R_1} = __18v__

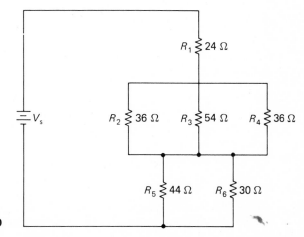

R_1 ⩗ 24 Ω

V_s R_2 ⩗ 36 Ω R_3 ⩗ 54 Ω R_4 ⩗ 36 Ω

R_5 ⩗ 44 Ω R_6 ⩗ 30 Ω

FIGURE 7-10

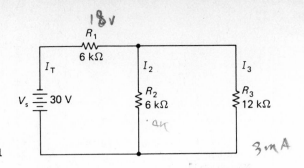

FIGURE 7-11

3. Find the resistance between points A and B in Fig. 7-12.

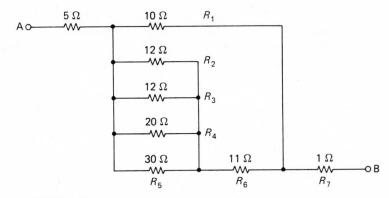

FIGURE 7-12

4. Refer to Fig. 7-13 and find
   a. the value wattage of R_4.
   b. the wattage of

$R_1 =$ _____
$R_2 =$ _____
$R_3 =$ _____
$R_4 =$ _____

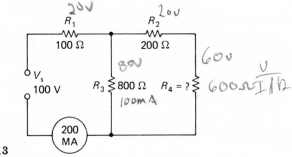

FIGURE 7-13

5. The total current, I_T in Fig. 7-14 is 5 A.
   a. Find the value of resistor R_1.
   b. Calculate I_{R_4}.

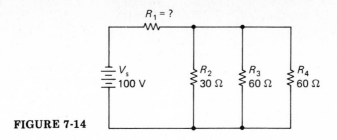

FIGURE 7-14

6. If the total resistance in Fig. 7-15 is 18 kΩ, what is the resistance of R_1?

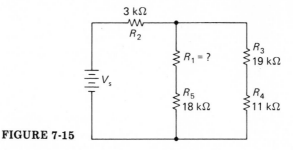

FIGURE 7-15

7. Refer to Fig. 7-16 and find the following:
   a. I_1 = _____
   b. I_T = _____
   c. V_a = _____
   d. R_T = _____
   e. Which resistor dissipates the greatest amount of heat?

17.5✓

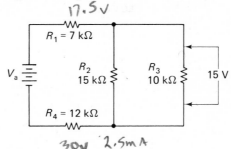

FIGURE 7-16

30v   2.5m A

8. Find R_1 and I_1 in Fig. 7-17.

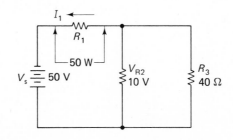

FIGURE 7-17

9. Refer to Fig. 7-18 and calculate

$R_T = $ _____

$I_T = $ _____

$P_T = $ _____

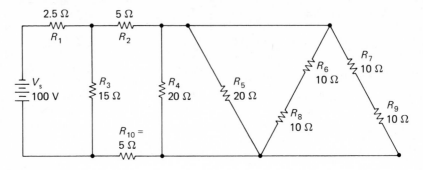

FIGURE 7-18

10. What is the applied voltage in Fig. 7-19?

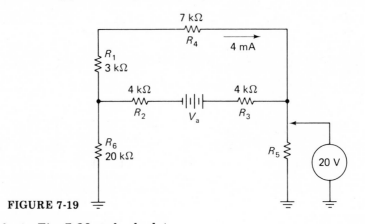

FIGURE 7-19

11. Refer to Fig. 7-20 and calculate

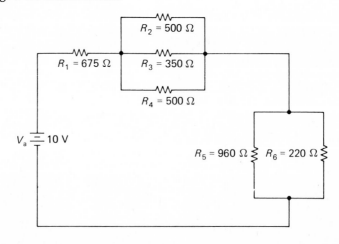

FIGURE 7-20

a. total resistance (R_T).
b. total current (I_T).
c. V_{R_1} = _____ P_{R_1} = _____
    V_{R_4} = _____ P_{R_5} = _____
    V_{R_6} = _____ P_{R_2} = _____

12. Calculate the voltage developed across R_7 in Fig. 7-21.

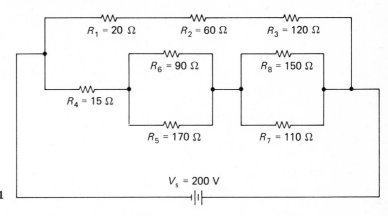

$R_1 = 20\ \Omega$ $R_2 = 60\ \Omega$ $R_3 = 120\ \Omega$

$R_6 = 90\ \Omega$ $R_8 = 150\ \Omega$

$R_4 = 15\ \Omega$

$R_5 = 170\ \Omega$ $R_7 = 110\ \Omega$

$V_s = 200$ V

FIGURE 7-21

13. Refer to Fig. 7-22 and calculate

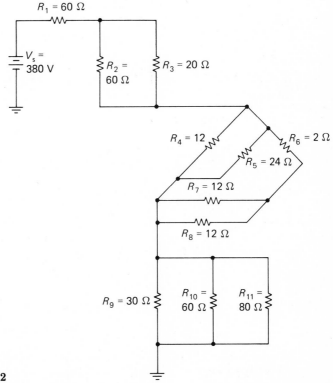

$R_1 = 60\ \Omega$

$V_s = 380$ V

$R_2 = 60\ \Omega$ $R_3 = 20\ \Omega$

$R_4 = 12$ $R_6 = 2\ \Omega$

$R_5 = 24\ \Omega$

$R_7 = 12\ \Omega$

$R_8 = 12\ \Omega$

$R_9 = 30\ \Omega$ $R_{10} = 60\ \Omega$ $R_{11} = 80\ \Omega$

FIGURE 7-22

a. R_T = _____
b. I_T = _____
c. I_{R_2} = _____

14. Find the value of I_T in Fig. 7-23.

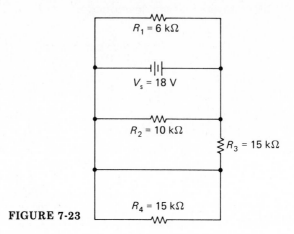

FIGURE 7-23

15. Which resistor in Fig. 7-24 dissipates the least amount of power?

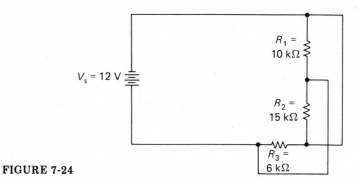

FIGURE 7-24

16. Simplify and calculate the total resistance for Fig. 7-25.

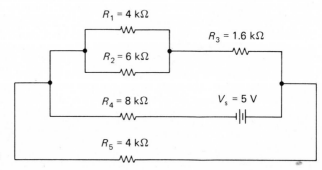

FIGURE 7-25

17. Simplify and calculate the total resistance for Fig. 7-26.

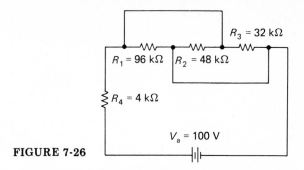

FIGURE 7-26

18. Simplify and calculate the total resistance for Fig. 7-27.

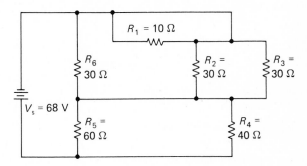

FIGURE 7-27

19. Simplify and calculate the total resistance for Fig. 7-28.

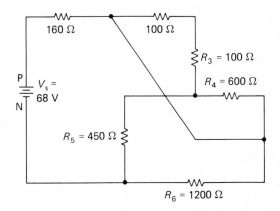

FIGURE 7-28

chapter **8**

VOLTAGE DIVIDERS AND WHEATSTONE BRIDGE

8-1 Introduction

Many pieces of electronic equipment, both older (tube equipment) and more modern (transistor I.C. circuits), use voltages of different values obtained from what is referred to as a voltage divider (V.D.) or bleeder circuit. This method is used to avoid the high cost of multiple sources of voltage. The voltage divider is simply several resistors connected in series with one another, creating a series–parallel circuit application.

8-2 Voltage Divider

It is important that the various voltages obtained remain as constant in value as possible, regardless of the possible shutdown of some circuits or of a change in the load connected to the voltage divider. Voltage regulation, therefore, is necessary. The simplest way to obtain voltage regulation and at the same time cause voltage division is to create a bleeder resistor network V.D. (Fig. 8-1). The word bleeder refers to the current which is common to all the dividing resistors. The voltage divider thus has three purposes:

1. To divide the voltage supply into different voltages.
2. To keep the power supply voltage from rising excessively if all the loads are disconnected.
3. To provide voltage regulation (but it is not a good voltage regulator).

Figure 8-1 shows three resistors connected in series to a 30-V battery or power supply. The value of each of the divider resistors in Fig. 8-1 is

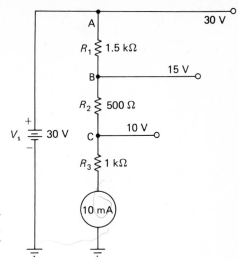

FIGURE 8-1   An unloaded voltage divider (no load at terminals). (Voltage dividers are merely applications of series resistive circuits.)

different, so one can see by visual inspection that each resistor drops the voltage by an amount which is in proportion to its value.

Point A to ground is 30 V (no drop).

Point B to ground is 15 V (R_1 drops V_s by 15 V).

Point C to ground is 10 V (R_2 drops V_s by 5 V).

The resistors R_1, R_2, and R_3 in Fig. 8-1 are called the bleeder resistors and they divide the voltage as shown. If a load were connected to point B or to point C (Fig. 8-2), the voltage at these points would change from the voltages shown in Fig. 8-1 because of the additional current coming from

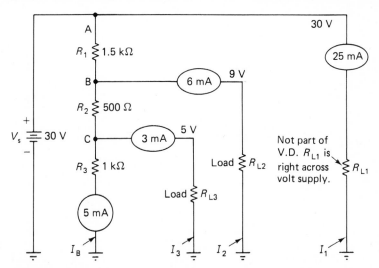

FIGURE 8-2   A loaded voltage divider. Note that the bleeder current and the voltages at points B and C have changed from Fig. 8-1.

the loads which would flow through R_1 and R_2. The summing-the-current method is used to select bleeder resistors of the correct value to supply the desired voltage.

8-2.1 Voltage Divider Design

Three values must be known before a voltage divider can be designed:

1. The voltage value of each load (voltage requirement of each load).
2. The maximum current demand of each load. This is usually determined by actually measuring the current drawn by each load before the design is begun.
3. The appropriate value of bleeder current when all loads are disconnected.

The value of the bleeder current (I_B) chosen for the bleeder circuit is very important. If the I_B value is too small, then the voltage regulation, which it is supposed to provide, will be insufficient because load current variations in R_{L_2} and R_{L_3} (Fig. 8-2) could override the bleeder current. If the bleeder current is higher than necessary, there will be needless waste of power because the heat produced in the bleeder resistors will be dissipated and lost in space. A compromise, therefore, must be made. The compromise used by designers is an arbitrary bleeder current value of 10% to 25% of the total current drawn by the loads.

PROBLEM 8-1

Design a voltage divider circuit which will supply

a.   25 V and 60 mA to R_{L_1},
b.   15 V and 30 mA to R_{L_2}, and
c.   10 V and 10 mA to R_{L_3}.

SOLUTION   (*cover solution and solve*)

1. Draw the circuit (Fig. 8-3).
2. Determine the value of bleeder current (I_B).

total of load currents =

$$I_{RL1} + I_{RL2} + I_{RL3} =$$

60 mA + 30 mA + 10 mA = 100 mA

bleeder current value (use the arbitrary percentage of 10%) =

10% of 100 mA = 10 mA

3. Calculate the value of the current through R_1, R_2, and R_3. Once the value of I_B is determined, this value must be added to

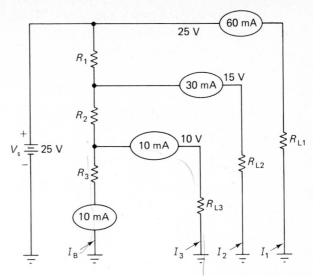

FIGURE 8-3 A voltage divider circuit.

the value of the load currents which flow through each of the bleeder resistors.

$$I_{R_3} = I_B \qquad\qquad\qquad\qquad\qquad\qquad = 10\,\text{mA}$$

$$I_{R_2} = I_B + I_{R_{L3}} = 10\,\text{mA} + 10\,\text{mA} \qquad\qquad = 20\,\text{mA}$$

$$I_{R_1} = I_B + I_{R_{L3}} + I_{R_{L2}} = 10\,\text{mA} + 10\,\text{mA} + 30\,\text{mA} = 50\,\text{mA}$$

4. Once the current values through each of the bleeder resistors is determined, the voltage drop across each resistor must be derived in order to calculate the value of the resistors.

$$V_{R_1} = V_{R_{L1}} - V_{R_2} = 25 - 15 = 10\text{ V}$$

$$V_{R_2} = V_{R_{L2}} - V_{R_{L3}} = 15 - 10 = 5\text{ V}$$

$$V_{R_3} = V_{R_{L3}} \qquad\qquad\qquad = 10\text{ V}$$

5. Resistances of R_1, R_2, and R_3.

$$R_1 = \frac{V_{R_1}}{I_{R_1}} = \frac{10\text{ V}}{50\text{ mA}} = 200\ \Omega$$

$$R_2 = \frac{V_{R_2}}{I_{R_2}} = \frac{5\text{ V}}{20\text{ mA}} = 250\ \Omega$$

$$R_3 = \frac{V_{R_3}}{I_{R_3}} = \frac{10\text{ V}}{10\text{ mA}} = 1000\ \Omega$$

The closest practical resistance value is chosen.

$$R_1 = 200\ \Omega$$

$$R_2 = 240\ \Omega$$

$$R_3 = 1000\ \Omega$$

6. The wattage value of R_1, R_2, and R_3 must also be known before the resistors can be ordered. Any one of the three wattage formulas may be used. Here, $P = I \times V$ is used.

$$P_{R_1} = I_{R_1} \times V_{R_1}$$
$$= (50 \times 10^{-3}) \times 10 \quad = 500 \text{ mW}$$
$$P_{R_2} = I_{R_2} \times V_{R_2}$$
$$= (20 \times 10^{-3}) \times 5 \quad = 100 \text{ mW}$$
$$P_{R_3} = I_{R_3} \times V_{R_3}$$
$$= (10 \times 10^{-3}) \times 10 \text{ V} = 100 \text{ mW}$$

Practical wattage values should be roughly twice the calculated value, using the closest standard value.

$$P_{R_1} = 1 \text{ W}$$
$$P_{R_2} = \frac{1}{2} \text{ W}$$
$$P_{R_3} = \frac{1}{2} \text{ W}$$

8-2.2 Voltage Regulation

Voltage regulation has been defined as maintaining a constant voltage output despite changes in load current. This is an ideal objective which is difficult to achieve. Voltage regulation in practice is maintaining the load voltage at a value which differs from the no-load voltage by less than a given percentage.

$$\% \text{ voltage regulation} = \frac{V_{NL} - V_L}{V_L} \times 100$$

where V_{NL} is the no-load voltage (load is disconnected)
V_L is the voltage when the circuit is under load

The percentage of voltage regulation can best be calculated under actual operating conditions and not by theoretical computations.

8-2.3 Potentiometer (Variable Voltage Divider)

The simplest type of voltage divider is the potentiometer, commonly known in radio and television as the volume control. The potentiometer circuitry shown in Fig. 8-4 has a potentiometer connected across a 6-V battery.

If the shaft of the potentiometer in Fig. 8-4 is set so that the wiper

Ch. 8 Voltage Dividers and Wheatstone Bridge

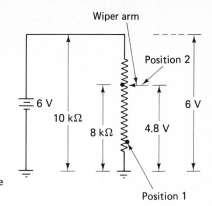

FIGURE 8-4 A potentiometer (voltage divider) set in Position 2.

arm is in Position 1, the resistance from A to ground will be 1000 Ω and the voltage from A to ground will be 1000/10,000 × 6 = 0.6 V.

If the shaft is set so that the wiper arm is in Position 2, the resistance to ground will be 8000 Ω and the voltage from B to ground will be 8000/10,000 × 6 = 4.8 V.

Any other value of voltage may be obtained by merely adjusting the potentiometer shaft.

8-2.4 Voltage Divider and Voltage Polarity

It is desirable sometimes to obtain a negative voltage from the same power supply which is supplying positive voltage. This can be done by connecting some point along the voltage divider circuit to ground (Fig. 8-5). The ground is now used as the reference point for all voltages obtained from the voltage divider.

PROBLEM 8-2

Suppose the circuit in Fig. 8-5 was to be used as a power supply for a field effect and bipolar transistor amplifier. The -5-V terminal may be used as the gate (reverse bias) supply voltage and would draw no current. Allow 10% for bleeder current (I_{R_3}). Design the voltage divider circuit.

SOLUTION   *(cover solution and solve)*

$$\text{total load current } I_\text{L} = 140 \text{ mA}$$

$$\text{bleeder current 10\% of } I_\text{L} = 14 \text{ mA}$$

Calculated values for each resistor

	Resistance	Wattage
bleeder resistor $R_3 = \dfrac{15 \text{ V}}{14 \text{ mA}}$ = 1071.4 Ω	210 mW
V.D. resistor $R_2 = \dfrac{10 \text{ V}}{34 \text{ mA}}$ = 294 Ω	340 mW

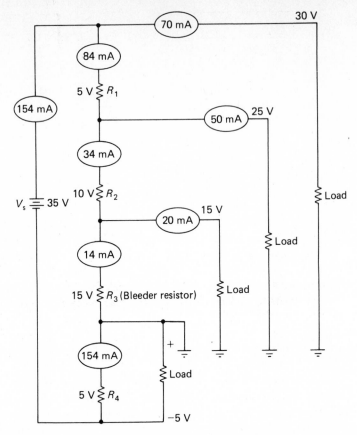

FIGURE 8-5 A voltage divider to supply positive and negative voltages.

$$\text{V.D. resistor } R_1 = \frac{5 \text{ V}}{84 \text{ mA}} = 59.5 \ \Omega \quad 420 \text{ mW}$$

$$\text{V.D. resistor } R_4 = \frac{5 \text{ V}}{154 \text{ mA}} = 32.47 \ \Omega \quad 775 \text{ mW}$$

The next step before constructing the circuit would be to select the resistors closest in value to the calculated resistance values and double the calculated wattage ratings.

8-3 Wheatstone Bridge

The bridge circuit, the simplest of which is called a Wheatstone bridge (see Fig. 8.6a), was named after Charles Wheatstone, who used this system to measure resistance in the 1850s.

The Wheatstone bridge principle is very important because it forms the basic circuit for a number of very accurate test instruments such as vacuum tube voltmeters and resistance, inductance, and capacity bridges.

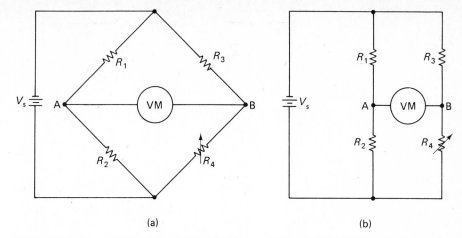

(a)   (b)

FIGURE 8-6   A basic Wheatstone bridge. (a) and (b) are identical
circuits but are drawn differently.

The accuracy depends upon the accuracy of the resistors in the arms and
the sensitivity of the meter. The principle is also used in differential transistor amplifier circuits and many other applications such as electronic
thermometers.

The principle of operation of the balanced bridge may be explained
with the following example circuits.

Figure 8-7 shows four resistors in a series parallel arrangement, with
R_1 equal to R_3, and R_2 equal to R_4. The result is that the voltage across
R_1 equals the voltage across R_2, producing zero voltage between points
A and B.

The circuit in Fig. 8-8 is similar to Fig. 8-7 except that R_3 is now 5 Ω
instead of 10 Ω and R_4 is now 20 Ω instead of 10 Ω. The current in
Branch 2, however, is of such a value that the voltage across R_3 is still
10 V and the voltage across R_4 is still 40 V. A voltmeter connected between points A and B would read 0 V, and the circuit is said to be balanced.   In both Fig. 8-6 and Fig. 8-7, the current through Branch 1 is
common to both R_1 and R_2, and the current in Branch 2 is common to
R_3 and R_4. The above statements may be written into the following ratio

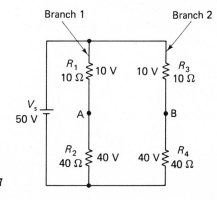

FIGURE 8-7

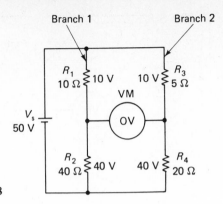

Branch 1 Branch 2

R_1
10 Ω 10 V

R_3
10 V 5 Ω

VM

V_s
50 V 0V

R_2
40 Ω 40 V

R_4
40 V 20 Ω

FIGURE 8-8

formulas:

$$\frac{V_{R_1}}{V_{R_2}} = \frac{V_{R_3}}{V_{R_4}} \quad \text{or} \tag{1}$$

$$\frac{I_1 R_1}{I_1 R_2} = \frac{I_2 R_3}{I_2 R_4} \tag{2}$$

The currents may be cancelled on each side of the equality sign because they are common in each leg when the circuit is balanced.

$$\frac{\cancel{I_1} R_1}{\cancel{I_1} R_2} = \frac{\cancel{I_2} R_3}{\cancel{I_2} R_4} \tag{3}$$

The resistance ratio formula for a balanced bridge may now be written

$$\frac{R_1}{R_2} = \frac{R_3}{R_4} \tag{4}$$

cross multiply and

$$R_1 R_4 = R_2 R_3 \tag{5}$$

Any one of four equations may now be derived from Eq. (5).

$$R_1 = \frac{R_2 R_3}{R_4} \qquad R_4 = \frac{R_2 R_3}{R_1}$$

$$R_2 = \frac{R_1 R_4}{R_3} \qquad R_3 = \frac{R_1 R_4}{R_2}$$

The student should be able to derive his own resistance ratio formula from which any one of the four unknown values may be obtained. The resistance ratio formula may be set up in any balanced situation by making the ratio of the two resistors in Branch 1 equal to the ratio of the two resistors in Branch 2.

To what value of resistance must R_2 be set to balance the bridge in Fig. 8-9?

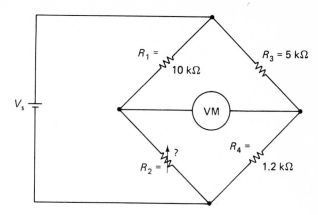

FIGURE 8-9

SOLUTION   (*cover solution and solve*)

The basic formula is

$$\frac{R_1}{R_2} = \frac{R_3}{R_4}$$

from which $R_1 R_4 = R_2 R_3$

and $R_2 = \dfrac{R_1 R_4}{R_3}$

$$= \frac{10{,}000 \times 1200}{5000}$$

$$= 2400 \ \Omega$$

PROBLEM 8-4

What value of resistance must R_4 be if R_1 is set at 8 kΩ when the bridge is balanced? (See Fig. 8-10.)

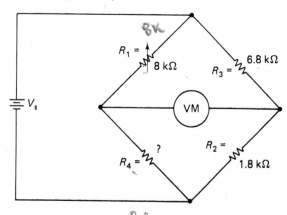

FIGURE 8-10

$\dfrac{R_1}{R_4} = \dfrac{R_3}{R_2} =$

The basic formula is

$$\frac{R_1}{R_4} = \frac{R_3}{R_2} \qquad R_4 R_3 = R_1 R_2$$

$$R_4 = \frac{R_1 R_2}{R_3} \qquad R_4 = \frac{R_1 R_2}{R_3}$$

$$= \frac{8000 \times 1800}{6800}$$

$$R_4 = 2118 \ \Omega$$

EXERCISES

1. Calculate the values of R_1, R_2, and R_B for the voltage divider circuit in Fig. 8-11 with loads as shown. Allow 10% bleeder current. Do not calculate the value of the loads R_{L_1}, R_{L_2} and R_{L_3}.

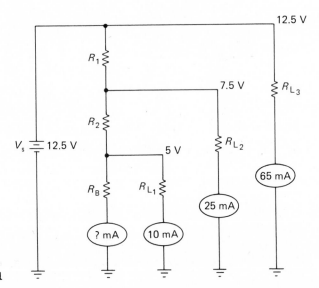

FIGURE 8-11

2. Design a voltage divider to supply 10 V at 10 mA, 25 V at 25 mA, and 40 V at 15 mA from a 40-V power supply. Allow a bleeder current of 10%. Do not calculate the value of the loads R_{L_1}, R_{L_2} and R_{L_3}.

3. Design a voltage divider to divide a 30-V power supply voltage so as to provide power for the following loads: 10 V at 12 mA, and 18 V at 18 mA. The 30 V is connected directly to a 70 mA load. Allow a bleeder current of 10%. Round off. (Do not calculate the value of the loads.)

4. Design a voltage divider circuit to supply power to the following volt-age loads. Draw the circuit and allow 10% bleeder current. Do not calculate the value of the loads.

   +15 V at 140 mA
   +12.5 V at 100 mA
   +7.5 V at 40 mA
   −2.5 V at 0 current

5. Refer to the balanced Wheatstone bridge in Fig. 8-12 and calculate R_X when

$$R_1 = 100\ \Omega$$
$$R_2 = 200\ \Omega$$
$$R_3 = 450\ \Omega$$

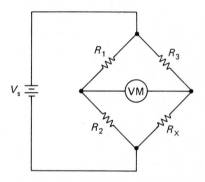

FIGURE 8-12

6. Refer to the balanced Wheatstone bridge in Fig. 8-12 and determine R_2 when

$$R_1 = 1000\ \Omega$$
$$R_3 = 900\ \Omega$$
$$R_X = 340\ \Omega$$

7. Refer to the balanced Wheatstone bridge in Fig. 8-12. Determine the value of R_3 if

$$R_X = 600\ \Omega$$
$$R_2 = 450\ \Omega$$
$$R_1 = 1200\ \Omega$$

chapter **9**

BATTERIES

9-1 Introduction

The battery was the first source of power for radio equipment. Since the radio is the granddaddy to present-day electronic technology such as computers and microprocessors, batteries can be considered as the power source which has made present-day electronics possible.

9-2 Principle of Battery Operation

The base of all batteries is the cell. Two or more cells connected together, whether in a series or parallel arrangement, constitute a battery. The common flashlight battery is really a flashlight cell. It will supply approximately 1.5 V of voltage. The amount of current which can be drawn from it per hour depends upon the square area of each of the two dissimilar plates and the volume of the electrolyte (acid). In other words, the current a cell is capable of delivering is directly proportional to the square area of the active plates.

A cell, in operation, can be distinguished from a battery in that the output voltage of a battery is more than a multiple of 2 V, but the output of a cell is 2 V or less.

9-2.1 Cell Operation

The basis of operation is basically the same for all types of cells: that is, the conversion of chemical energy into electrical energy. A cell is said to store electricity, but in truth it stores chemical energy which is converted into electrical energy when it is connected to a conductive circuit.

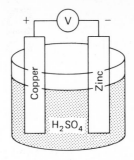

FIGURE 9-1   A basic battery cell.

The principle of battery operation was discovered around the year 1800 by an Italian physicist, Alessandro Volta. The battery was originally referred to as the voltaic cell, as is the primary cell by some texts today. Volta suspended a strip of zinc opposite a strip of copper in a container partly filled with vinegar (acetic acid). The resulting chemical action produced a D.C. voltage of 1.1 V between the two dissimilar metals.

Another simple form of battery cell is a strip of copper placed opposite a strip of zinc in a glass vessel containing a sulphuric acid solution diluted with distilled water in a ratio of 1:20 (Fig. 9-1). The acid solution is known as the electrolyte. The copper and zinc are known as dissimilar metals because they are at different positions on the atomic periodic table (copper is element number 29; zinc is number 30). The resulting chemical action when they are immersed in the acid causes electrons to be removed from the copper atoms, leaving them positively charged (as positive ions). Electrons are deposited on the zinc atoms, making them negatively charged (negative ions). The difference in potential between the positive charge on the copper electrode and the negative charge on the zinc electrode produces a potential difference of approximately 1.1 V between the two electrodes. The copper electrode is the positive terminal and the zinc electrode is the negative terminal.

9-3 Classification of Batteries

Batteries have two classifications:

  1. Primary (dry) cells—not rechargeable
  2. Secondary cells—rechargeable

9-3.1 Primary Cells

The simple battery as described above is of the primary cell classification, because one of the metals is "eaten away" as chemical action takes place, necessitating its replacement when it becomes discharged. (A flashlight battery falls within this classification.) Nearly all primary cells have the electrolyte in the form of a paste, hence the name *dry cell* (Fig. 9-2). Other chemical arrangements other than the simple battery as described above are used in dry cells. During discharge, metals can be "eaten away"

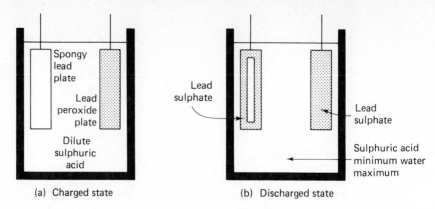

(a) Charged state   (b) Discharged state

FIGURE 9-2   The chemical change of a lead acid battery from charge to discharge conditions.

by the acid, and gas bubbles (often hydrogen) form around the positive terminal inside the battery, increasing the internal resistance of the battery. Current drawn through the large internal resistance by the load produces a large internal voltage drop, causing the external voltage between the positive and negative terminals to be reduced to a useless value; the battery is said to be discharged. Dry cells are usually low current types which are thrown away when the formation of gas or the corrosion of the metal plates makes them non-rechargeable and therefore unrecoverable. All dry cells do have a chemical within them called a "depolarizing agent" which reduces the formation of the gas bubbles, but with time the effectiveness of the depolarizing agent is gradually reduced and the battery is said to be discharged or weak.

9-3.2 Secondary Cells

Secondary batteries are used principally where large current demands are present or where rechargeable batteries are desired. Secondary cells operate on a principle similar to that applicable to primary cells, except that in secondary cells the dissimilar (two different types of) metals are not "eaten away," but merely have their chemical makeups changed as the chemical reaction takes place under discharge conditions. This chemical makeup is then reversed, and the original condition attained again by a charge. The common automobile battery (lead-acid battery) is a classic example of this battery classification. (See Fig. 9-2.)

CHARGED WET LEAD-ACID BATTERY

The plates are lead (Pb) and lead peroxide (PbO_2) and the electrolyte is $2(H_2SO_4)$. The chemical constituents are:

$$Pb + PbO_2 + 2(H_2SO_4)$$

WHEN DISCHARGED

The plates are lead sulphate $[2(PbSO_4)]$ and the electrolyte is $2(H_2O)$. When the discharged battery with the lead sulphate $2(PbSO_4)$ plates is connected to a source of voltage of a greater value than the battery itself, the $2PbSO_4$ plates revert to their original plates of Pb and PbO_2, and the electrolyte reverts to $2(H_2SO_4)$.

9-4 Connecting Cells Together

Cells often must be connected together to obtain higher voltages and current capabilities, since a chemical cell will generally deliver less than 2 V, and only small amounts of current.   There are two basic ways of connecting cells:

1. Series
2. Parallel

9-4.1 Series-Aiding

A series-aiding connection has the negative terminal of cell A connected to the positive terminal of cell B, and the negative terminal of cell B is connected to the positive terminal cell C, and so on (Fig. 9-3a).

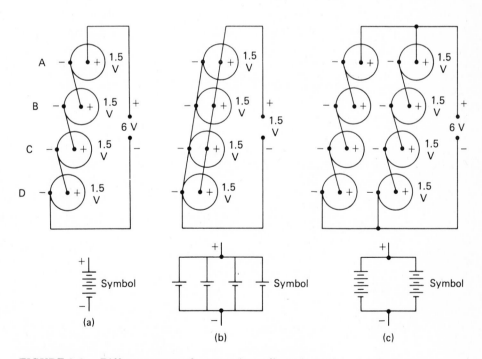

FIGURE 9-3   Different ways of connecting cells or batteries. (a) Series connection. (b) Parallel connection. (c) Series–parallel.

Four 1.5-V cells connected series-aiding as shown in Fig. 9-3a will produce 6 V across terminals A and D. The maximum current available, however, will be the same from the four cells together as from one cell.

SUMMARY:

When cells or batteries are connected series-aiding:

1. The total voltage obtained is the sum of the voltages of the cells or batteries connected together.
2. The current supply remains the same as from one cell, and it cannot exceed the rating of the current capacity of the least cell.

9-4.2 Connecting Cells in Parallel

Batteries are connected in parallel when all the positive terminals are connected together and all the negative terminals are connected together [Fig. 9-3(b)].

Four cells connected in parallel as shown in Fig. 9-3(b) will produce 1.5 V across terminals A and B, but the maximum current which the combination can deliver will be four times the current available from one cell.

SUMMARY:

When batteries are connected in parallel:

1. The total voltage obtainable is the same as for one cell.
2. The total current supply is the sum of the currents available from the individual cells. (Note: NiCd batteries must not be connected in parallel without a protective circuit.)

9-4.3 Series–Parallel Connections

Cells may also be connected in series–parallel, as shown in Fig. 9-3(c), where two sets or banks of series-connected cells are connected in parallel. The terminal voltage is the same as the voltage appearing across each bank, but the available current is twice that from one bank. Multiple voltage, high current batteries contain cells connected in a series–parallel arrangement.

9-5 Testing

9-5.1 Voltage Test of Batteries

As has been explained earlier, a discharge of batteries causes one of three things to happen:

1. Gas bubbles form around one of the posts (primary cells).

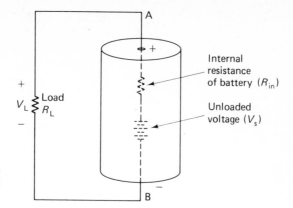

FIGURE 9-4 Battery voltage must be measured with the cell under load $[V_L = V_s - (I \times R_{in.})]$.

2. The chemical makeup of the two dissimilar plates changes so that they become similar types of metals (secondary cells).

3. One electrode is eaten away (primary cells).

Regardless of what occurs, the actual effect is an increase in the internal resistance (R_{in}) of the battery. If the voltage of the battery is measured without a load (open terminals), there is no current flow through the internal resistance (R_{in}), and consequently there is no voltage drop internally in the battery. The no-load voltage measured across the battery terminals (A and B in Fig. 9-4) will be the nominal value, for example, 1.5 V. If, on the other hand, a load, R_L is connected, there is current flow through the internal resistance of the battery as well as the load, and a voltage drop occurs across R_{in}. The resulting load voltage across terminals A and B will drop by the amount of the voltage drop internally across R_{in}. Charged batteries have a very small internal resistance and the internal voltage drop is negligible; thus, the voltage across open terminals A and B will be near the given voltage value for the battery. As the battery becomes discharged, the internal resistance increases and the voltage drop across R_{in} increases. The resulting voltage across the loaded terminals A and B becomes progressively lower until a certain value is reached, whereupon the battery is declared weak or dead and must be discarded, or recharged if it is a secondary battery.

The most important point to remember when testing batteries is that batteries must always be tested under load. The size of the load is dependent upon the type of battery and its voltage value.

1. To test a 12 V car battery one should measure the voltage across its terminals with the starter in operation (approximately a 200-A drain). The battery is charged if the 12-V battery measures 9 V or more under this 200-A load.

2. A 1.5-V flashlight battery (cell), by contrast, should be tested with a 10-Ω resistor connected across its terminals (150-mA drain). (A 7.5-Ω resistor is ideal but is not always available.) The voltage

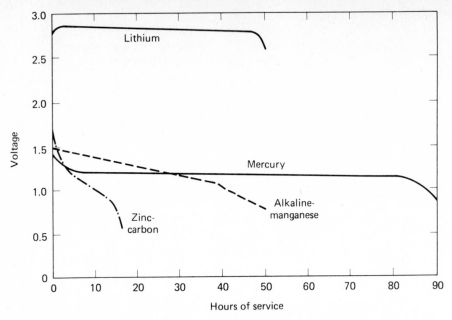

FIGURE 9-5 Discharge curve of a "D" size cell at 200-mA average drain.

across the terminals should measure greater than 1.2 V. A lesser voltage indicates a weak battery which should be discarded.

3. The load for a multiple-cell flashlight-type battery should be 10 Ω times the number of cells connected in series. If the 1.5-V cells are connected in parallel, the output voltage will still be 1.5 V, and 10 Ω should still be used as a load.

9-5.2 Hydrometer Test

The charged condition of a wet lead-acid battery may be checked with a hydrometer. A hydrometer measures the specific gravity of the electrolyte. The higher the specific gravity, the greater the amount of sulphuric acid in the electrolyte solution and, therefore, the higher the battery charge.

A hydrometer reading from 1200 to 1240 indicates low charge

A hydrometer reading from 1240 to 1275 indicates medium charge

A hydrometer reading from 1275 to 1300 indicates high charge

9-6 Primary Cells

The primary cell is nonrechargeable, but because of its low cost, its convenience, and the demand for batteries for lightweight portable transistor equipment, the primary cell is more popular than the secondary battery.

9-6.1 Primary Cell Comparisons

Table 9-1 shows the different types of primary cells and their characteristics. Comparison of weight to energy storage (wattage hours) shows that the zinc carbon battery is the heaviest of all the primary cells, and has the largest physical size.

DISCHARGE CURVES

Comparison of discharge curves discloses that the alkaline-manganese and zinc-carbon cells have sloping discharge curves, i.e., their terminal voltages gradually fall off. Mercury and lithium cells, on the other hand, maintain a steady output voltage until they are expended and then they suddenly become dead. (See Fig. 9-6).

STORAGE

All types of primary batteries can be stored equally well (disregarding shelf life) unless temperatures are extreme.

Comparison of the cost and service life of size AA cells shows that zinc-carbon cells are the most expensive in the long run. For example:

A 75¢ alkaline cell lasts four times as long as a 30¢ zinc-carbon cell.

A 40¢ zinc-chloride cell will last twice as long as a 30¢ zinc-carbon cell.

9-6.2 Carbon-Zinc Batteries (Leclanche Cell)

The carbon-zinc battery, commonly referred to as a flashlight battery, is the most popular cell in the dry cell classification because of its low cost.

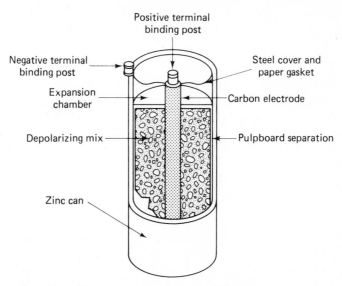

FIGURE 9-6   A typical carbon–zinc battery cell.

Table 9-1 Electrical Characteristics of Primary Cells

	Carbon-Zinc	Zinc-Chloride	Alkaline-Manganese	Mercury	Silver-Oxide	Lithium	Magnesium
Nominal Cell Voltage	1.5 V	1.5 V	1.4 V	1.35 and 1.4 V	1.5 V	2.95 V	2.0 V
Shape of Discharge Curve	Slope	Slope	Slope	Flat	Flat	Flat	Fairly Flat
Shelf Life @ 68°F to 80% of Initial Capacity (In Years)	2–3	2–3	3–5	2–3	2–3	3–5	2–3
Energy Output in Watt hours Per Lb	35	44	20–35	56	50	100–150	40
Per In.^{2}	2.3	3	2–3.5	6.4	8	8–15	4
Practical Drain Currents Greater Than 50 mA	100 mA per in.^{2}	150 mA per in.^{2}	200 mA per in.^{2}	no	no	yes	200–300 mA per in.^{2}
Internal Resistance (Z)	Low	Low	Very Low	Low	Low	Less Than 1 Ω	Low
Capacity @ 50 h, Current Drain	6 AH	—	10 AH	—	—	10 Ampere-Hour	—
Capacity vs. Temperature	Poor @ Low Temp.	Good @ Low Temp.	Fair to Good @ Low Temp.	Good @ High Temp. Poor @ Low Temp.	Poor @ Low Temp.	Excellent	Fair @ Low Temp.

The carbon-zinc cell is sometimes called the Leclanche cell because the principle of its operation was discovered by George Leclanche in 1868.

The carbon-zinc battery consists of

1. A zinc case (negative polarity)
2. A carbon center rod (positive polarity)
3. An ammonium chloride (NH_4Cl) electrolyte
4. A manganese dioxide depolarizing agent

Chemical activity between these components remains relatively dormant until the battery is connected to a circuit and delivers current. Discharge of the cell (operation) causes the positive ammonium (NH_4) ions to be repelled by the negative zinc ions when the zinc combines with the NH_4Cl to form $ZnCl_2$. The ammonium ions move through the electrolyte to the carbon rod where the ammonium (NH_4) is converted into ammonium gas (NH_3) and hydrogen gas (H). The hydrogen gas which accumulates around the carbon rod creates a condition called polarization. The effect of polarization is to increase the internal resistance of the battery. This produces a large internal voltage drop when the battery is delivering current, and the battery is said to become weak. Polarization is overcome by adding the chemical, manganese dioxide (MnO_2),—called a depolarizing agent—to the electrolyte paste. The MnO_2 depolarizing agent neutralizes the hydrogen around the carbon rod, keeping the battery internal resistance to a low value. The NH_3 gas combines with the water in the electrolyte paste to form ammonium hydroxide.

9-6.3 Zinc-Chloride Batteries

The zinc-chloride cell is really an improved zinc-carbon (Leclanche) cell. The advantage of the zinc-chloride cell is its greater efficiency, which means that the current output of a zinc-chloride cell is usually higher than that of a zinc-carbon cell. It will also operate at higher current drains for longer periods than will the zinc-carbon battery of the same physical size. The zinc-chloride battery also has the advantage of having good assurance against electrolyte leaks and much better operation at low temperatures.

9-6.4 Alkaline-Manganese Batteries

The alkaline-manganese battery is better known to the public as the "Duracell." It is sometimes referred to as an alkaline battery or manganese battery (do not confuse this battery with the magnesium battery) and is the next most popular dry cell to the zinc-carbon cell.

The alkaline battery supplies 1.4 V when new. It has two to three times the shelf life and about three times the capacity of a zinc-carbon cell, and has a higher reliability. Refer to Table 9-1 for other comparisons.

The high energy, stable voltage and long shelf life of the mercury cell has made possible such small equipment as watches, exposure meters, cameras, hearing aids, and pacemakers. Such equipment usually uses the tiny cells the size and shape of buttons, but the cells also come in cylindrical and flat-pellet configurations.

Mercury cells are of two basic types:

1. Voltage (nominal) of 1.35 V
2. Voltage (nominal) of 1.4 V

The 1.35-V types are generally used as voltage references for medical and scientific applications. The 1.4-V cells are used mainly in domestic applications where a flat voltage characteristic is required, and the current drain requirement is low to moderate.

The constituents of mercury cells are:

Anode (−) terminal—high purity amalgamated zinc. (Center terminal.)

Cathode (+) terminal—mercuric oxide and graphite. (Case terminal.)

Electrolyte—solution of alkaline hydroxide or potassium hydroxide.

Cell case—anti-corrosive nickel-plated steel.

The mercury cell has a longer shelf life than zinc-carbon and alkaline cells (up to five years' shelf life), and has a flat discharge characteristic curve, which means that the output voltage remains constant until the cell suddenly stops working.

Mercury cells have the disadvantage of having a lower terminal voltage and current capacity than the zinc-carbon and alkaline cells, and they also are more expensive.

9-6.6 *Silver-Oxide Cells*

The silver-oxide cell, because of its costly materials, is more expensive than the mercury cell. It is of the alkaline type, with an output voltage of 1.5 V at low current drains. Even though its long-term current drain capacity is low, it can deliver high currents for a short time, but this practice is not recommended. Because the current capacity of this cell is low, its use is restricted to low current applications such as watches, (used extensively), calculators, and hearing aids. Because of the silver-oxide cell's higher output voltage (due to its low internal resistance), some retailers replace mercury cells with silver-oxide types. Unless the application is a low current drain application, however, the practice can lead to degraded performance of the equipment.

9-6.7 Lithium Cells

The lithium cell is a relative newcomer to the dry-cell family. Its advantages are that it:

1. Provides high energy density (high watt/hours in proportion to weight and physical volume), i.e., 150 W-h per kg
2. Has a long shelf life and operating life (possibly ten years)
3. Has a flat voltage characteristic curve
4. Has an excellent performance in environmental extremes ($-18°C$ to $+74°C$)
5. Has the highest cell voltage (2.95) of any batteries of the dry-cell family
6. Operates efficiently at high discharge rates

The lithium cell is the lightest and the smallest in size and stores a great amount of energy. The lithium cell has nearly four times the energy density of a mercury cell. At $20°C$ and 1 A current drain, one lithium D-size cell has the energy density of 30 zinc-carbon cells. Lithium cells also deliver almost twice as much voltage per cell than any other type of primary cell. The capacity of lithium batteries is excellent at either low or high temperatures, whereas the capacity of most other batteries is either fair or even poor at low temperatures. Lithium batteries as yet are very expensive and therefore are used in only specialized military and industrial applications. A still greater disadvantage of the lithium cell is its tendency to explode if moisture penetrates the seal.

9-6.8 Charging Primary Cells

The best method of recharging primary cells is *don't*. Many charger units are available to the consumer which purport to recharge any kind of primary cell. These chargers are generally cheap units of very simple design and cannot handle the charge characteristics of numerous types of dry cells. The discharge of a dry cell usually involves the eroding away of one of the metallic electrodes. There is no way that this metal can be replaced by a charging process. At best a dry cell can only be rejuvenated, but not charged. For this reason the most suitable candidates for rejuvenation are those cells which have been only slightly discharged. Recharging any sealed cell, however, can be dangerous because of the gases which are produced within the cell. Excessive gassing which may result from an excessive charge rate may result in cell explosion and damage to equipment, or even personal injury. This is especially true of tightly sealed alkaline, mercury or silver-oxide cells.

9-6.9 Solar Battery

This is a specialized battery which produces voltage when light (principally sunlight) falls between the strip of selenium material and the strip of iron of which it is constructed; another type of solar cell uses silicon, doped with an impurity adding either a 3 valence or a 5 valence element to silicon with an impurity. The amount of voltage is small—approximately 0.5 V—and the current is very small—approximately 600 μA. Both voltage and current disappear as soon as the light is removed. Solar cells are being used on coastal buoys, in some space objects where sunlight is always present, and in the Arctic where they are used to charge batteries.

9-7 Secondary Cells

There are four types of secondary cells, each type with two or more subtypes. The four secondary cell types are:

1. Lead-acid
2. Nickel-cadmium
3. Silver-zinc
4. Silver-cadmium

The lead-acid battery and nickel-cadmium battery are the two major types of secondary cells in use.

9-7.1 Lead-Acid Battery

The name "lead-acid battery" is synonomous with "car battery." Unless it is mounted upright, it will spill acid and eat away everything it touches.

There are two major types of lead-acid batteries, but both have low internal resistance and can deliver hundreds of amps of current with a small internal voltage drop. They also have the highest volt-per-cell value (2.1 V) of all rechargeable batteries, and cost the least per watt. A 12-volt lead-acid battery consists of six cells and its open circuit voltage will measure 12.6 V ± 0.2 V.

Although the lead acid battery has the disadvantages of being very heavy and containing corrosive sulphuric acid, it can withstand very high rates of charge and discharge. For example, the starter of a large V-8 engine may draw as much as 300 A of current on the initial crank, and this current can be supplied by this type of battery with only an 80 AH (ampere-hour) capacity. It can also be recharged in 30 min on a 50-A charger, from a completely discharged state. A continuous practice of allowing severe discharge and recharging, however, could seriously reduce the charge-discharge cycle capability of the battery.

The two major types (each has its subtypes) of lead-acid batteries are:

1. Wet cell
2. Gell cell

As the name implies, this is a liquid type lead-acid battery. It was developed in 1859 by Gaston Plante, and until the past five to eight years it was used in I suppose every car built. The wet lead-acid battery is most popular in automobiles, but it is also used as standby power in power systems in the range of 2 kW to 2 MW, for large record-storage computers. The wet lead-acid battery is popular as standby power because of its relatively low cost, and under conditions of good maintenance in a battery room it will provide reliable service for 10 to 20 years.

There are five types of wet lead-acid batteries:

1. Pure-lead
2. Lead-antimony flat plate
3. Lead-calcium flat plate
4. High-performance
5. Exide ironclad tubular positive plate

1. Pure-Lead Batteries. Pure-lead batteries are made with pure lead plates one-inch thick. They have a very low shelf discharge rate because there is no other element with which to react to provide internal battery action. Consequently, these batteries lose only 12% of their charge in one year.

2. Lead-Antimony Flat Plate Batteries. The lead antimony flat plate battery:

1. is the least expensive of the wet lead-acid batteries.
2. has the shortest life, but will last many years if properly maintained in a battery room.

This type of battery should not be left discharged too long nor its acid level allowed to drop below the top of the plates, or a process called sulphation will set in and the battery will not recharge to its designed voltage value.

Sulphation. Sulphation is one of the most common causes of lead-acid battery failure. Sulphation is the condition in which both plates permanently become lead sulphate ($PbSO_4$) either totally or in part.

There are two causes of sulphation:

1. The battery is left discharged for a period of time.
2. The battery is operated with the electrolyte below the top of the plates (lack of water).

In actual practice a lead-acid cell should never be allowed to become completely discharged. Complete discharge can cause permanent damage to the battery, because the lead sulphate crystals which form on the plates of the discharged battery cause the plates to bend or buckle due to the change in their chemical composition. Another reason for recharging a lead-acid battery as soon as possible after its discharge is to prevent the soft lead-sulphate crystals from hardening with age. The soft crystals change back into sulphuric acid, lead, and lead peroxide under a slow charge. The crystals will not respond to charge once they have hardened; the battery is then said to be permanently sulphated.

3. **Lead-Calcium Flat Plate Batteries.** The lead-calcium flat plate battery is identical to the lead-antimony battery except that the lead is alloyed with calcium instead of antimony. An alloy is used instead of pure lead in both types of batteries in order to give the grids of the plates the necessary strength to withstand vibration and jar. The calcium battery has a longer shelf life (does not discharge as rapidly) and loses less electrolyte through evaporation than the antimony lead-acid battery. This latter characteristic reduces the need for maintenance. The calcium cell has the disadvantage that its life cycle is reduced when it is operated at temperatures above $27°C$ or when it is subjected to frequent charge and discharge cycles.

4. **High-Performance Batteries.** The high-performance battery is the newest type of lead-acid battery. Its plate construction is similar to the lead-antimony flat plate battery except that the plates are made very thick and large. Its internal resistance is reduced by copper bars which are cast into the internal lead connecting straps. This battery has a current density of over 3000 A for 15 min, and takes up to 53% less space for the same capacity than either the lead-antimony or lead-calcium type batteries. It has an expected life of 14 to 15 years under battery room maintenance conditions.

5. **Exide Ironclad Tubular Positive Plate Batteries.** The exide ironclad tubular positive plate battery is a handmade battery. Its main advantage is its power density. This type of battery concentrates more discharge capacity per cubic meter of cell size than any other lead-acid battery. Its other advantages are:

1. The electrolyte evaporating rate is between that of a lead-calcium battery and a lead-antimony battery.
2. It has a discharge capacity ten times that of a lead-calcium battery.
3. It has a high tolerance against high temperatures, and is the most highly recommended for high-temperature countries such as Africa.

GELLED ELECTROLYTE LEAD-ACID BATTERY

This cell has a gelled acid electrolyte. It was first placed on the market in 1965, but was little known until recent improvements have made it popular. These batteries may be used in any position and provide a safe, almost maintenance-free source of power when properly sealed and vented. They offer reliable and economical high performance. The maintenance-free characteristic of these batteries results from the sealed construction made possible by the use of gelled electrolyte. The gelled electrolyte eliminates the need to add electrolyte or water. These batteries are ideally suited to installations in all types of portable electronic equipment because they may be charged or discharged in any position. They are also ideally suited to the requirements of standby power systems. There are two types of gelled electrolyte batteries:

1. Lead-antimony
2. Lead-calcium

Lead-Antimony Gelled Battery. The initial price of the lead-antimony gel cell battery is lower than that of the gelled lead-calcium battery, but it has a much poorer shelf life. The shelf discharge rate is very high and may be as high as 25% of its charge per month at room temperature. It also has a higher rate of electrolyte evaporation, and a higher gassing rate while on charge than the lead calcium gel cell.

Lead-Calcium Gelled Battery. The lead-calcium gell cell has the lowest shelf discharge rate of the lead acid types but it should be stored in a cool environment in order to reduce its rate of shelf discharge. For example, it loses about 2% per month of its charge at room temperature. It loses only 0.5% of its charge at $-18°C$, but 12% per month at $35°C$. The gel cell has a low electrolyte evaporation loss and therefore has a longer service life than the lead-antimony battery.

The higher initial cost of the lead-calcium gel cell battery is offset by its longer service life and shelf life.

9-7.2 Nickel-Cadmium Batteries

There are numerous types of nickel-cadmium (Ni-Cd) batteries. One manufacturer in France alone produces 13 types of nickel-cadmium batteries. Unfortunately, there are no manufacturers of Ni-Cd batteries in North America—only distributors under different trade names. As is the case with other types of batteries, manufacturers have enough knowledge about NiCd batteries to fill a textbook, but very little is written on the subject.

No absolute rule nor one specification necessarily applies to a NiCd battery, because there are so many types with different characteristics. The following discussion is merely a precis of the subject.

The Ni-Cd battery will give two to three times the discharge current that a lead-acid battery of the same amp-hour rating will deliver. Thin-plate types are excellent for applications where very high peak discharge currents are required for short periods of time, such as in starting a jet engine. Ni-Cd batteries are not used in automobiles because of their higher cost, but are used mainly for standby service where a low discharge rate is important when the battery is not in use. They also are used where sealed units are desirable, such as in lawn equipment and portable drills..

There are two types of nickel-cadmium batteries:

1. Sintered-Plate Sealed Nickel-Cadmium Battery
2. Pocket-Plate or Sintered-Plate Vented Nickel-Cadmium Battery

Both types have a low internal resistance and therefore can deliver high currents with only a small internal voltage loss. Neither type produces an appreciable amount of gas while in storage; therefore, they can be sealed and operated in any position. Both types can be cycled (charge-discharge) at least several thousand times.

SINTERED-PLATE SEALED NICKEL-CADMIUM BATTERY

The sealed nickel-cadmium battery is used in applications ranging from consumer goods, such as cordless shavers and portable test equipment, to industrial, aviation, and aerospace equipment. A powder holds the active material used for both anode and cathode.

Characteristics and Advantages of Sealed Sintered Types.

1. The sealed Ni-Cd battery is a very rugged battery and has an indefinite shelf life in any state of charge.
2. It performs well over a wide temperature range of $-40°C$ to $+50°C$.
3. The voltage characteristic of the sealed Ni-Cd battery is quite flat, even at high discharge rates.
4. The sealed Ni-Cd battery may be charged at a fast rate, and is unaffected by moderately high rates of overcharge. These cells may be recharged as many as a thousand times without loss of capacity. The cell voltage also remains nearly constant throughout the discharge cycle, maintaining about 1.2 V per cell until discharge is complete. Sealed Ni-Cd cells are also well suited to standby power applications of less than 2 kW where they may be left in an overcharged condition for many months without damage to the cells.
5. Sealed cells may be connected in series to form higher voltage batteries, but they should not be connected in parallel without a protective device, such as a blocking diode, connected between

them. The reason for this precaution is that the internal resistance of Ni-Cd batteries varies over a wide range. The absence of the protective device can cause a battery with a very low internal resistance to charge at an exceedingly high rate and blow up. At the same time, a battery with high internal resistance, connected in parallel with the low internal resistance battery, will charge only a very small amount or not charge at all.

Some Ni-Cd battery suppliers, such as General Electric and SAFT, color-code the bottoms of their batteries. The color coding indicates a range of capacity. Only batteries of the same color code should be connected together for either charge or discharge purposes.

6. Sealed Ni-Cd batteries are made in a large variety of shapes and sizes including AA, C, and D sizes.

7. These batteries can supply current at a much higher rate than lead-acid cells.

8. They are corrosion-free.

9. They have high energy density (large current capacity per kilogram of weight).

Charging the Ni-Cd Battery. The C/10 (0.1 C) rate (ampere-hour capacity rating) is the recommended rate at which sealed Ni-Cd cells can be safely charged. The number 10 is the number of hours required at maximum charge efficiency to fully charge a completely discharged cell.

Disadvantage of Sintered-Plate Ni-Cd Batteries. Sintered-plate Ni-Cd batteries have a disadvantage other than cost and low cell voltage of a phenomenon called *memory*.

Memory occurs only after rigid repetitive fixed percentage discharge and charge cycles or after a battery is left on overcharge for a few months. Memory is the condition in which the battery does not respond to its design capacity when power demand is increased, but operates at the reduced power output close to the previous habitual percentage discharge and charge cycle. The battery voltage is depressed when it is in memory condition, and an allowance must be made in voltage-sensitive circuits for this drop.

Memory effect can be corrected, and the battery restored to its normal capacity by one complete discharge followed by a full charge cycle.

POCKET- OR SINTERED-PLATE VENTED NICKEL-CADMIUM BATTERY

Two types of construction are used in vented Ni-Cd batteries; sintered plate construction, as has already been discussed, or pocket-plate construction. The latter type is the most common.

Vented Ni-Cd cells are generally used where high capacities are required. Their capacity can range from 10 amp-hours to several hundred amp-hours. The thin plate types especially can deliver discharge rates many times their amp-hour rating.

Maintenance of these cells is minimal, as they seldom require water.

9-7.3 Other Types of Secondary Cells

Two other types of rechargeable batteries are specialized types. They are used in applications requiring a lighter, more powerful power source than the nickel-cadmium battery, such as in portable televisions, cameras, video-tape recorders, airborne telemetering equipment, and missiles. The two types are

1. Silver-zinc
2. Silver-cadmium

Both types have no memory effect and can supply currents of 100 A with an internal voltage drop of only 0.4 V. Their cost, however, is high because of the silver content of one of the electrodes.

SILVER-ZINC BATTERY

This battery has the highest energy density of all the secondary batteries. The silver-zinc battery has four times the energy density of a Ni-Cd battery and five times the energy density of the gelled lead-acid battery. Its maximum life, however, is only one-third that of a silver-cadmium cell. (See Fig. 9-7).

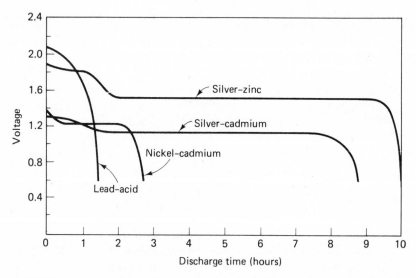

FIGURE 9-7   Discharge curves for secondary batteries of equal weight and current drain.

9-7.4 Recharging Secondary Batteries

Secondary batteries are recharged by connecting them across a source of D.C. voltage, called a battery charger or a converter (which is actually a charger). In the case of calculators and many transistor radios, a charger is called an adaptor.

The output voltage of any type of charging device must always be greater than the open circuit voltage rating of the battery. A rule of thumb is that the charger output voltage should be 15% greater than the voltage rating of the battery. However, the battery manufacturers' voltage and charge recommendations should be consulted and followed whenever possible.

A battery will be damaged if the battery charger output voltage is too large. Excess battery-charge current will flow into the battery and overheat it. The battery may also be damaged after it is charged if the trickle charge current is too large. Completely discharging or overcharging any lead-acid battery will at best seriously affect its life span, and at worst will permanently damage it.

CAPACITY

The capacity (C) of all rechargeable batteries is rated at a discharge rate of so many amp-hours. For example, a lead-acid battery of a certain size may have a discharge capacity of 20 amp-hours (C/20) at $80°C$, which means that it will deliver 1 A per h for 20 h, or 20 A for 1 h, or 40 A for $\frac{1}{2}$ h, and so on.

9-7.5 Comparison of the Two Major Secondary Cells

The lead-acid battery is $\frac{1}{3}$ to $\frac{1}{4}$ the cost of the Ni-Cd battery. However, it also has only $\frac{1}{3}$ to $\frac{1}{4}$ the discharge current rate.

The lead-acid battery has an energy-density close to the Ni-Cd, but the lead-acid battery has the highest voltage per cell (2.1 V versus 1.2 for Ni-Cd). (See Table 9-2.)

9-8 Battery Applications

9-8.1 Replacement of Dry Cells

One type of battery should not be replaced with a different type without a full understanding of why a particular type was originally chosen. The performance of the equipment could be depreciated with an ill-chosen type. For example, a mercury battery should never be replaced by a silver-oxide battery even though they are physically the same, because the current drain capacity of the silver-oxide is much smaller.

Much the same type of caution applies to other types of batteries. Table 9-3 lists a variety of primary and secondary batteries and their applications.

Table 9-2 Electrical Characteristics of Secondary Cells

	Lead-Acid Gelled or Liquid	Nickel-Cadmium	Silver-Cadmium	Silver-Zinc
Nominal Cell Voltage	2.1 V	1.25 V	1.1 V	1.5 V
Energy Output in Watt Hours				
Per Lb	9	12–16	22–34	40–50
Per In.^{2}	1.1	1.2–1.5	1.5–2.7	2.4–3.2
Shelf Life @ 68°F. to 80% of Capacity	8 months with lead-calcium grids	2 weeks–1 month	3 months	3 months
Internal Resistance (Z)	Low	Low	Very Low	Very Low
Discharge Curve	Slope	Flat	Flat	Flat
Discharge-Charge Cycle Life	200 to 500	500 to 2000	150 to 300	80 to 100

Table 9-3 Battery Types and Their Applications

Battery Type	Features	Applications
PRIMARY CELLS		
Carbon-Zinc	Low initial cost, easy availability	Radios, toys, novelties, flashguns, slide viewers
Alkaline	Wide temperature range, gradual voltage drop, good shelf life, high capacity, low resistance	Radios, electronic flash, flash-lights, tape recorders, radio-controlled models
Mercury	Excellent high-temperature performance, flat discharge, low noise	Hearing aids, preamplifiers, watches, cameras, test instruments, some calculators
Silver-Oxide	Flat characteristic, good low temperature performance	Watches, hearing aids, cameras, light meters
SECONDARY CELLS		
Lead-Acid	High capacity, rechargeable, high cycle life	Television, lights, power tools, transmitters, garden tools, recorders, fire and burglar alarms
Nickel-Cadmium	Rechargeable, good temperature range, high capacity, high cost	Shavers, tooth-brushes, electronic flash, recorders, radios, calculators

EXERCISES

1. What class of battery; primary or secondary, is rechargeable?

2. What form of energy is stored in batteries?

3. Describe the term electrolyte.

4. What material is used in each of the electrodes of the dry cell commonly used in flashlights?

5. a. What is meant by polarization as applied to a common dry cell?
   b. How may the effect of polarization be counteracted?

6. Describe how to charge a primary cell.

7. What is the chemical composition of the following components of a lead-acid battery?

	When Charged	When Discharged
a. the positive plate	_____	_____
b. the negative plate	_____	_____
c. the electrolyte	_____	_____

8. a. How many lead-acid cells are connected together to form a 12-V automobile battery?
   b. Are the above cells connected in series or parallel?

9. Mercury cells have two different voltage ratings: 1.35 V and 1.4 V. State the application of each.

10. What is the major advantage of the nickel-cadmium cell over the lead-acid cell?

11. What is the major advantage of the gel cell over the liquid lead-acid cell?

12. A battery is capable of delivering 5 A of current for 12 h. What is its amp-hour rating?

13. A cell whose internal resistance is 0.14 Ω delivers 400 mA to a load of 3 Ω. What is the terminal voltage delivered by the cell?

14. A cell when delivering 2 A of current has an output voltage of 1.4 V, but has an output of 1.6 V when there is no load connected. What is the internal resistance of the cell?

15. A 24-V battery measures 24 V on open circuit.
    a. What will be its terminal voltage when connected to a 390-Ω load if the battery internal resistance is 10 Ω?
    b. How much power is lost in the cell?
    c. How much current would flow if the battery should accidentally become shorted?

16. Can the internal resistance of a battery be measured with an ohmmeter? Explain your answer.

17. Six cells, each rated at 2 V and each capable of delivering 20 A for 2 h are connected together to form a battery.
   a. If the cells are connected in parallel
      (1) What is the battery terminal voltage?
      (2) What is the amp-hour capacity rating of the battery?
   b. If the cells are connected in series
      (1) What is the battery terminal voltage?
      (2) What is the amp-hour rating of the battery?

18. a. Describe how to properly test a 1.5-V flashlight battery.
    b. What is the minimum acceptable voltage from a 1.5-V flashlight cell?

19. Is a zinc-carbon cell suitable for a cassette tape recorder? Explain your answer.

20. What are the two most common causes of sulphation of wet lead-acid batteries?

21. What are the effects of sulphation?

22. What is the approximate voltage of one cell of each of the following batteries when it is charged?
   a. Lead-acid
   b. Carbon-zinc
   c. Alkaline manganese
   d. Mercury
   e. Nickel-cadmium
   f. Silver-oxide
   g. Gelled lead-acid

23. Why is low internal resistance desirable in a battery cell?

24. What is the hydrometer reading of a highly charged lead-acid battery?

25. What is the advantage of a secondary cell over a primary cell?

26. a. Describe what is meant by memory effect as applied to batteries.
    b. What type of cell is prone to the phenomenon of memory effect?

chapter **10**

THEOREMS

10-1 Introduction

The word *theorem* to a great number of students seems to connote a
completely different method from Ohm's Law of solving resistive circuits.
This misconception has been reinforced by the approach which many
textbooks take toward the subject. Be assured that there is nothing
mysterious and very little new in deriving resistive circuit solutions from
theorems.

Complex series-parallel circuits are called networks. Networks with
two sources of voltage, as simple as they may appear to be, cannot always
be solved with the application of Ohm's Law alone. Ohm's Law must
often be aided with an additional method for solution, or law of procedure, known as a theorem. In other words, the rule of a particular theorem
is used in conjunction with Ohm's Law to simplify the solution of difficult network problems.

The dictionary defines the word theorem as a statement or rule which
can be proved true, and so is established as a law or principle.

There are many different theorems, but the following four theorems
suffice to solve most resistive circuit problems.

1. Kirchhoff's two Laws (of Voltage and of Current)
2. Superposition Theorem
3. Thevenin's Theorem
4. Norton's Theorem

Although Kirchhoff's Laws alone may be used to solve network circuits, one or the other of Kirchhoff's Laws is also incorporated in the

solution of the other three theorems. For this reason Kirchhoff's Loop Method for solution will be discussed first. An identical circuit will be used for each of the theorems as the base circuit, so that it can be shown that the same answer will always be obtained regardless of the method used. Comparison of methods is also more easily done when the same circuit is used for each method.

The base circuit used for the four theorems (Fig. 10-5) is similar to that which may be found in the electrical system of a car (but the values are different). The alternator (V_1) in a car is always connected in parallel with the battery (V_2). The R_1 could be the alternator internal resistance and R_2 the battery internal resistance. R_L could be the headlights of the car when they are turned on. The current through R_L and the voltage across R_L would be that through and across the headlights respectively.

There are many different ways of drawing a circuit, and even simple circuits can be made to appear complex. The key to solving circuits is to redraw the circuit to be solved in the format best suited for the understanding and solution of a particular theorem. Each theorem has its own particular format.

10-2 Kirchhoff's Loop Solution

Gustav Kirchhoff established two laws between 1800 and 1850. The Voltage Law introduced in Chapter 5 (Series Circuits) is a characteristic of a series circuit.

10-2.1 Kirchhoff's Voltage Law

Kirchhoff's Voltage Law states that:

1. The sum of all voltage drops in a closed series circuit is equal to the sum of all the voltage sources in that series circuit.

or

2. The algebraic sum of the voltage drops and the voltage sources in a closed circuit is equal to zero.

EXAMPLE ───────────────────────────────────

See Fig. 10-1.
From point A

$$V_{R_1} + V_{R_2} + V_{R_3} = V_1 + V_2$$
$$5 \;+\; 15 \;+\; 20 \;= 10 + 30$$
$$40 = 40$$

The Kirchhoff Voltage Law expression which is most commonly applied to mathematical solutions is: The algebraic sum of the voltage

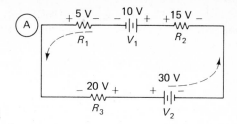

FIGURE 10-1

drops and the voltage sources in a closed circuit is equal to zero. (Note the slight difference in wording.)

$$V_{R_1} + V_{R_2} + V_{R_3} - V_1 - V_2 = 0$$
$$5\ +\ 15\ +\ 20\ -\ 10\ -\ 30 = 0$$

To solve networks with Kirchhoff's Law, closed loops are first formed:

Kirchhoff's Voltage Law is used to establish the closed loops.

Kirchhoff's Current Law is used to form the voltage formulas in the closed loops.

The current law was introduced in Chapter 6 (Parallel Circuits) as a characteristic of parallel circuits.

10-2.2 Kirchhoff's Current Law

The sum of all currents entering a point is equal to the sum of all currents leaving that point.

EXAMPLE

See Fig. 10-2.

$$\text{Current entering } \textcircled{A} = 9\ A$$
$$\text{Current leaving } \textcircled{A} = I_1 + I_2 + I_3$$
$$= 2\ +3\ +4\ = 9\ A$$
$$\text{Current entering } \textcircled{B} = I_1 + I_2 + I_3$$
$$= 2\ +3\ +4\ = 9\ A$$
$$\text{Current leaving } \textcircled{B} = 9\ A$$

See Fig. 10-3.

10-2.3 Advantage of Kirchhoff's Method

Kirchhoff's Laws will now be used to solve networks by what is known as Kirchhoff's Loop Method.

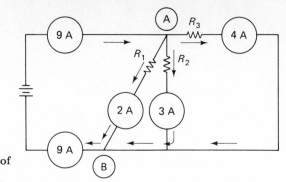

FIGURE 10-2   An example of Kirchhoff's current law (KCL).

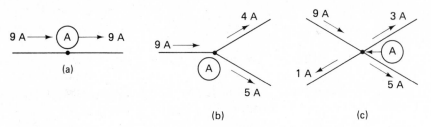

(a)

(b)

(c)

FIGURE 10-3   More examples of Kirchhoff's current law (KCL).

As will be seen upon the completion of the solution of any one of the example problems, the advantage of Kirchhoff's Method is that the current through each resistive component is automatically determined while arriving at the solution. The voltage drops are then easily obtained.

PROBLEM 10-1

Use Kirchhoff's Loop Method to find the value of current flow and the voltage drop across R_L in Fig. 10-4.

SOLUTION   (cover solution and solve)

1. The drawing in Fig. 10-4 is in a form (format) which makes it difficult to complete closed loops. It must be redrawn in a format which leads to the easiest solution with Kirchhoff's Laws.

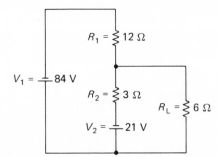

FIGURE 10-4

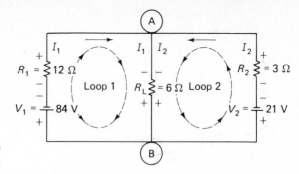

FIGURE 10-5 Note that the source voltages are no-load voltages.

Figure 10-5 is the standard format drawing (note R_L in the center between the two voltage sources) for Kirchhoff's Law.

2. Label the current on the schematic from V_1 as I_1 and from V_2 as I_2. The current through the load (R_L) will be different and could be labeled I_3, but the expression ($I_1 + I_2$) will be used in place of I_3 to reduce the number of unknown currents to two, so that the simultaneous equation method may be used to solve for the currents (Fig. 10-5).

3. Form the closed loops. Each loop includes one of the no-load voltage sources, the source resistance, and the load.

4. Label the assumed direction of the current flow through each loop (the final answer may show the direction of one of the currents to be opposite).

5. Label the assumed voltage polarity across each component.

6. Establish a common reference point from which to begin and finish each loop (see point B in Fig. 10-5). (The R_L must be included in each loop.)

7. Form the loops. Start from the reference point B and work to the left to form loop 1; at the same time write an equation including the polarity and magnitude of each voltage in each loop.

Loop 1:

$$84 - 12I_1 - 6(I_1 + I_2) = 0$$

Simplify

$$84 - 12I_1 - 6I_1 - 6I_2 = 0$$
$$84 - 18I_1 \qquad - 6I_2 = 0$$
$$84 - 18I_1 - 6I_2 \qquad = 0 \qquad (1)$$

8. Start from the reference point B and work to the right to form loop 2; at the same time write the equation for loop 2.

Loop 2:

$$21 - 3I_2 - 6(I_1 + I_2) = 0$$

Simplify

$$21 - 3I_2 - 6I_1 - 6I_2 = 0$$

$$21 - 9I_2 - 6I_1 \qquad = 0$$

$$21 - 6I_1 - 9I_2 \qquad = 0 \qquad (2)$$

9. Solve for I_2 and I_1. Equations (1) and (2) are linear equations. The unknown current values may be extracted by determinants or simultaneous equations. The multiplication and elimination method will be used here.

   a. Pair equations (1) and (2).

   $$84 - 18I_1 - 6I_2 = 0 \qquad (1)$$

   $$21 - 6I_1 - 9I_2 = 0 \qquad (2)$$

   b. Multiply Eq. (2) by -3 to eliminate I_1.

   $$84 - \cancel{18I_1} - 6I_2 = \quad 0$$

   $$\underline{-63 + \cancel{18I_1} + 27I_2 = \quad 0} \qquad (3)$$

   Add

   $$21 \qquad\qquad + 21I_2 = \quad 0$$

   $$21I_2 = -21$$

   c. Extract I_2

   $$I_2 = \frac{-21}{21} = -1 \text{ A}$$

   The negative sign of I_2 $(= -1)$ indicates that the direction of the current I_2 is opposite that assumed at the beginning.

   d. Now substitute -1 for I_2 in Eq. (1).

   $$84 - 18I_1 - 6(-1) = \quad 0$$

   $$84 - 18I_1 + 6 \qquad = \quad 0$$

   $$-18I_1 \quad = -90$$

   $$I_1 \quad = +5 \text{ A} \qquad (4)$$

10. Table the unknown current values. (Note that the I_2 current flows opposite the assumed direction. Therefore, the $(-)$ sign must be used with all calculations.)

    $$I_2 = -1 \text{ A}$$

    $$I_1 = 5 \text{ A}$$

    $$I_2 + I_1 = (-1) + (+5) = 4 \text{ A} \quad (\text{KCL})$$

    $$V_{R_2} = (I_1 + I_2) \times R_L = 4 \times 6 = 24 \text{ V} \qquad (5)$$

11. Verify with Kirchhoff's Voltage Law.

Loop 1:

$$84 - 18I_1 - 6I_2 = 0 \qquad (1)$$

Substitute

$$84 - 18(+5) - 6(-1) = 0$$
$$84 - 90 + 6 \qquad = 0$$
$$0 = 0$$

Loop 2:

$$21 - 6I_1 - 9I_2 = 0 \qquad (2)$$

Substitute

$$21 - 6(+5) - 9(-1) = 0$$
$$21 - 30 + 9 \qquad = 0$$
$$0 = 0$$

PROBLEM 10-2

Use Kirchhoff's Loop Method to solve for the current through and voltage across R_L in Fig. 10-6.

SOLUTION *(cover solution and solve)*

The solution to the circuit in Fig. 10-6 would be difficult unless it were drawn in the standard format for Kirchhoff's solution.

1. Redraw Fig. 10-6 as shown in Fig. 10-7.

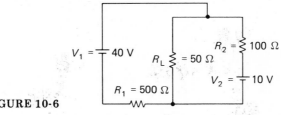

FIGURE 10-6

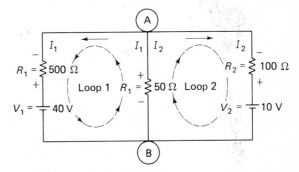

FIGURE 10-7 Figure 10-6 redrawn to the standard format for a Kirchhoff's solution.

2. Repeat steps 2–11 in Problem 10-1.

Answers:

$$I_1 = 68.75 \text{ mA}$$

$$I_2 = 43.8 \text{ mA}$$

$$I_1 + I_2 = 68.75 + 43.8 = 112.55 \text{ mA}$$

$$V_{R_L} = 5.62 \text{ V}$$

10-2.4 Practical Application of Kirchhoff's Laws

One or both of Kirchhoff's Laws is used in the solution for current and voltage values in nearly every electrical circuit, simple or complex. The reason for this fact is that Kirchhoff's Voltage Law is a characteristic of series circuits and Kirchhoff's Current Law is a characteristic of parallel circuits.

PROBLEM 10-3

Use Kirchhoff's Law on the automobile electrical charging system shown in Fig. 10-8 to determine:

1. The current being delivered by the alternator
2. The current flow through the load
3. The current flow in the battery
4. If the battery is charging or discharging

The alternator voltage when unloaded is 18 V.
The unloaded battery voltage is 12.6 V.
R_B is the internal battery resistance.
R_A is the internal alternator resistance.
R_L is the total resistance of 4 lamps in parallel.

SOLUTION   (cover solution and solve)

Repeat steps 2 to 10 in Problem 10-1, but also extend the calculations to answer the questions to this problem.

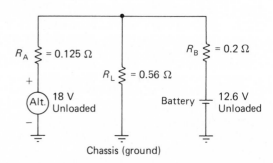

FIGURE 10-8 An   automobile
battery-charge circuit.

Answers:

1. Alternator current = 32 A
2. Load current = 25 A

3. and 4.   The battery is being charged with 7 A flowing into it.

The output voltage of the alternator is in parallel with the load (headlights) and the battery. The voltage across all three should be approximately 14 V.

10-3 Superposition Theorem

10-3.1 Limitations

The Superposition Theorem is another method of solving networks (complex series–parallel circuits). The Superposition Theorem, however, can be used only with resistive networks which are constant in value (linear). The Superposition Theorem will not generally hold true with impedance values, with nonlinear resistive values which change with heat or voltage, nor with vacuum tubes or transistors. Its use is also limited to those circuits which have more than one source of voltage. It can be used to determine the current through or voltage across only one branch of the circuit, for example, the load (R_L) branch.

10-3.2 Advantage of the Superposition Method

The dubious advantage of the Superposition Method over Kirchhoff's Loop Method is that it is possible to avoid using simultaneous equations or determinants to obtain the solution.

10-3.3 Principle

The Superposition Method is based on the principle that the current in any part of a circuit is the sum of the currents produced by each voltage acting separately while the other voltages are absent from the circuit. The theorem derives its name from the fact that these separate currents are superimposed upon one another in the final analysis. The Superposition theorem states that: In any network containing more than one voltage source, the current flow through any branch of the network is the algebraic sum of the component currents which would be produced by each voltage source if each source were connected alone in the circuit.

10-3.4 Solution Steps

There are five steps in applying the Superposition Theorem to analyze complex circuits.

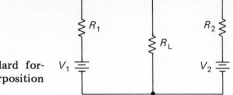

FIGURE 10-9 The standard format drawing for superposition
solutions.

1. Redraw the circuit to the standard format for the Superposition
   Theorem of the solution (Fig. 10-9).

2. Redraw the circuit for each voltage source, but with all other
   voltage sources removed and replaced with a short.

3. Determine the current delivered by each source and the current
   flowing in each resistor for each of the separate drawings made in
   step 2.

4. Superimpose the currents obtained in step 3 onto the standard
   format diagram produced in step 1.

5. Combine the superimposed currents to determine the final currents.

PROBLEM 10-4

Use the Superposition Theorem to determine the current through and
voltage across the load R_L in Fig. 10-10.

SOLUTION *(cover solution and solve)*

1. Redraw Fig. 10-10 to the superposition format drawing (Fig.
   10-11).

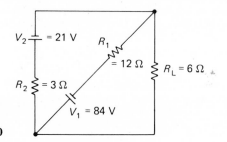

FIGURE 10-10

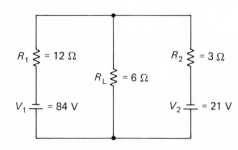

FIGURE 10-11 Standard format
drawing for the superposition
solution.

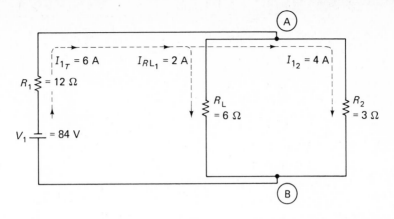

FIGURE 10-12   The circuit of Fig. 10-11 with only the V_1 voltage source.

2. There are two sources of voltage in this circuit. Therefore, two
    new circuits must be drawn, each circuit having only one source of
    voltage, with all other voltage sources removed and replaced by a
    short (Figs. 10-12 and 10-13).

3. Calculate the current delivered by each voltage source in each
    diagram.

    Current delivered by V_1 (Fig. 10-12)

$$I_T = \frac{V_1}{R_1 + R_{AB}}$$

$$I_T = \frac{84}{12 + \left(\dfrac{6 \times 3}{6 + 3}\right)} \qquad = 6 \text{ A}$$

Current through each resistor (Fig. 10-12)

$$I_{R_L} = \frac{V_{AB}}{R_L} = \frac{I_T \times R_{AB}}{R_L}$$

$$= \frac{6 \times 2}{6} \qquad = 2 \text{ A}$$

$$I_{R_2} = \frac{V_{AB}}{R_2} = \frac{I_T \times R_{AB}}{R_2}$$

$$= \frac{6 \times 2}{3} \qquad = 4 \text{ A}$$

Current delivered by V_2 (Fig. 10-13)

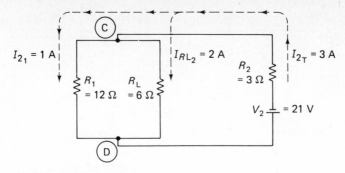

FIGURE 10-13 The circuit of Fig. 10-11 with only the V_2 voltage source.

$$I_T = \frac{V_2}{R_2 + R_{CD}}$$

$$= \frac{21}{3 + \left(\dfrac{6 \times 12}{6 + 12}\right)} = \frac{21}{3 + 4} = 3 \text{ A}$$

Current through each resistor (Fig. 10-13)

$$I_{R_L} = \frac{V_{CD}}{R_L} = \frac{I_T \times R_{CD}}{R_L}$$

$$= \frac{3 \times 4}{6} \qquad = 2 \text{ A}$$

$$I_{R_1} = \frac{V_{CD}}{R_1}$$

$$= \frac{3 \times 4}{12} \qquad = 1 \text{ A}$$

4. Superimpose the currents from Figs. 10-12 and 10-13 onto Fig. 10.14.

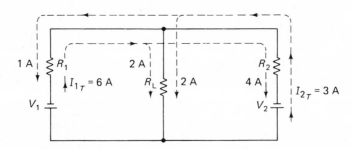

FIGURE 10-14 The currents of Figs. 10-12 and 10-13 superimposed on one diagram.

5. Combine the superimposed currents for the final answer.

$$I_{R_1} = 5 \text{ A} \qquad V_{R_L} = 24 \text{ V}$$

$$I_{R_L} = 4 \text{ A}$$

$$I_{R_2} = 1 \text{ A}$$

10-3.5 Practical Application of the Superposition Theorem

The Superposition Theorem could be applied to most of the circuits to which Kirchhoff's Law is applied.

The automobile charging circuit which was solved with Kirchhoff's Laws in Problem 10-3 may also be solved with the Superposition Theorem. Problem 10-5 is an example solution for a charging circuit.

PROBLEM 10-5

Use the Superposition Method on the automobile electrical circuit shown in Fig. 10-8 to solve for:

a. The current delivered by the alternator (18 V unloaded)
b. The load current
c. The current flow in the battery circuit
d. Whether the battery is charging or discharging
e. The voltage output of the alternator
f. The voltage across the headlights
g. The voltage across the battery terminals

SOLUTION   *(cover solution and solve)*

1. Redraw Fig. 10-8 as two circuits, each showing only one voltage source (Figs. 10-15 and 10-16).
2. Determine the current flow from each source of voltage in each diagram.

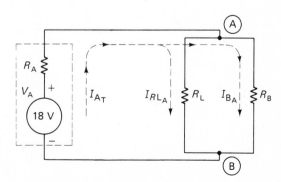

FIGURE 10-15 Alternator current through the load and battery resistance.

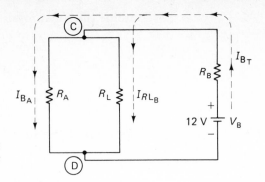

FIGURE 10-16 Battery current through the load and the alternator resistance.

Alternator circuit (Fig. 10-15)

a. Find R_T (Fig. 10-15). $(R_L \| R_B)$

$$R_{AB} = \frac{R_L \times R_B}{R_L + R_B}$$

$$= \frac{0.56 \times 0.2}{0.56 + 0.2} = 0.147 \ \Omega$$

$$R_T = R_A + R_{AB}$$

$$= 0.125 + 0.147 = 0.272 \ \Omega$$

b. Find current flow from the alternator (Fig. 10-15).

$$I_T = \frac{V_a}{R_T} = \frac{18}{0.272} = 66.176 \ A$$

c. Find the current flow through each resistor.

$$V_{AB} = I_T \times R_{AB}$$

$$= 66.176 \times 0.147$$

$$= 9.728 \ V$$

The voltage across AB is also across R_L and R_B.

$$I_{R\,L} = \frac{V_{AB}}{R_L} = \frac{9.728}{0.56} = 17.37 \ A$$

$$\text{and} \quad I_{R\,B} = \frac{V_{AB}}{R_B} = \frac{9.728}{0.2} = 48.64 \ A$$

Battery Circuit (Fig. 10-16)

a. Find R_T (Fig. 10-16). $(R_L \| R_A)$

$$R_{CD} = \frac{R_L \times R_A}{R_L + R_A}$$

$$= \frac{0.56 \times 0.125}{0.56 + 0.125} = \frac{0.07}{0.685} = 0.102 \ \Omega$$

$$R_T = R_B + R_{CD}$$
$$= 0.2 + 0.102 \qquad = 0.302 \ \Omega$$

b. Find the battery current (Fig. 10-16).

$$I_B = \frac{V_B}{R_T} = \frac{12.6}{0.302} = 41.72 \ \text{A}$$

c. Find the current through each resistor in Fig. 10-16.

Voltage across $R_{CD} = V_{R_L} = V_{R_A}$

$$V_{CD} = I_T \times R_{CD}$$
$$= 41.72 \times 0.102 = 4.25 \ \text{V}$$

$$I_L = \frac{V_{R_L}}{R_L} = \frac{4.25}{0.56} = 7.99 \ \text{A}$$

$$I_{R_A} = \frac{V_{RA}}{R_A} = \frac{4.25}{0.125} = 34 \ \text{A}$$

3. Superimpose the currents from Figs. 10-15 and 10-16 on Fig. 10-17.

4. Combine the superimposed currents for the final answer.

a. I_A has opposite currents (subtract).

$$I_A = 66.176 - 34 = 32.176 \ \text{A} \quad \text{(see Fig. 10-17)}$$

b. I_L currents are in same direction (add).

$$I_L = 17.37 + 7.99 = 25.36 \ \text{A}$$

c. I_B has opposite currents. The larger current is flowing into the battery. The battery is charging.

$$I_B = 48.64 - 41.72 = 6.92 \ \text{A}$$

e. The output voltage of the generator is

$$18 \ \text{V} - (I_A \times R_A) = 18 - (32.176 \times 0.125) = 13.978 \ \text{V}$$

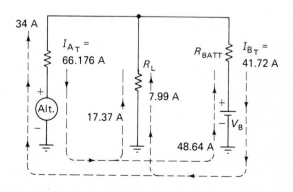

FIGURE 10-17   Battery and alternator currents superimposed on one diagram.

f. The voltage across the headlights is

$$I_L \times R_L = 25.36 \times 0.56 = 14.2 \text{ V}$$

g. The voltage measured across the battery terminals is

$$V_B = 12.6 - (I_B \times R_B)$$
$$= 12.6 - (6.92 \times 0.2) = 13.984 \text{ V}$$

In operation, the three voltages will have the same value. The difference in the calculations is due to two decimal places used.

10-4 Thevenin's Theorem

10-4.1 Practical Applications of Thevenin's Theorem

Thevenin's Theorem is possibly the most useful of the four theorems discussed because it has the most practical applications. This theorem is used whenever the unloaded output voltage across A and B and the impedance of power supplies is required, or when the input impedance of active devices such as transistors is required. The testing of batteries with and without a load is actually a practical application of Thevenin's Theorem. In such measurements two values are determined:

1. The unloaded output voltage (Thevenin's Voltage) which may be applied to a load
2. The internal resistance or impedance of the device (Thevenin's Resistance)

10-4.2 Advantage One of Thevenin's Theorem

Thevenin's Theorem can be considered as the ultimate in network (complex circuitry) reduction, because the solution reduces all circuits, less the load, to two components:

1. The equivalent resistance across the open terminals A and B, called Thevenin's Resistance (R_{Th}).
2. The voltage across the open terminals A and B, called Thevenin's Voltage (V_{Th}). This is the maximum voltage which can be applied across any load.

Thevenin's method of solution is often employed to determine these two component values before a circuit design is constructed and tested.

10-4.3 Thevenin's Theorem

Thevenin's Theorem states that upon the removal of the load, any network containing any number of voltage sources and complex resistive

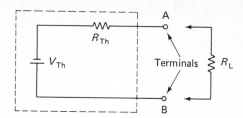

FIGURE 10-18 Thevenin's Equivalent Circuit (T.E.C.).

circuits can be reduced to a single voltage source in series with a single resistance and the two terminals across which the load may be connected. See Fig. 10-18.

There is no such thing as a Thevenin's circuit or a Norton's circuit. The conditions determine the type of circuit. Thevenin's voltage is the voltage measured across the open terminals of the voltage source or a load of infinite resistance. A circuit is therefore called a Thevenin's (or equivalent) circuit if the load resistance connected across its terminals is very large compared to the internal resistance (Thevenin's resistance) of the source voltage. The output voltage will also be relatively constant (a constant-voltage source) under these conditions.

It is quite common to read the instruction "Thevenize the circuit" or "Thevenize around the load." Each of these expressions is merely an instruction to calculate the internal resistance (impedance) of a piece of electronic equipment. The student may be misled by these instructions to believe that some special procedure must be used. This is not true; the internal resistance (impedance) of any one piece of electronic equipment will always be the same value regardless of the math procedure used to determine its value or the name given to it. The internal resistance will generally be called Thevenin's Resistance when using Thevenin's Theorem, or Norton's Resistance (Section 10-5) when using Norton's Theorem; otherwise it is just called the internal resistance. The procedure used and the value obtained will always be the same in all cases.

10-4.4 Advantage Two of Thevenin's Method

Another advantage of Thevenin's Theorem is that the current through and voltage across the different loads which may be connected to terminals A and B can be calculated without recalculating the complex circuit each time.

10-4.5 Solution Steps

There are eight steps in determining Thevenin's Equivalent Circuit.

1. Remove the load and label the open terminals A and B.
2. Redraw the given circuit to the standard format drawing for Thevenin's solution.

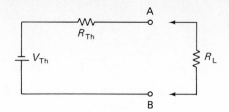

FIGURE 10-19   Thevenin's Equivalent Circuit (T.E.C.).

3. Find the voltage developed across each of the components in the closed loops with the voltage sources.

4. Find Thevenin's Voltage (V_{Th}).

5. Find Thevenin's Resistance (R_{Th}).

6. Draw Thevenin's Equivalent Circuit (Fig. 10-19).

7. Find I_{R_L}.

8. Find V_{R_L}.

PROBLEM   10-6

1. Calculate Thevenin's Equivalent Circuit for Fig. 10-20.

2. Determine
    a. The current flow through the load
    b. The voltage across the load

SOLUTION   *(cover solution and solve)*

1. Calculate Thevenin's Equivalent Circuit.
    a. Remove R_L and label the open terminals as A and B.
    b. Draw Fig. 10-20a in the standard format for Thevenin's solution (Fig. 10-20b). (R_L is removed.)
    c. Find Thevenin's Voltage.
        (1) Find the current flowing in the closed loop. The batteries are series-opposing.

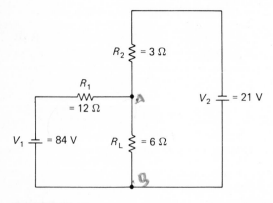

FIGURE 10-20a

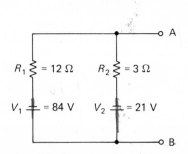

FIGURE 10-20b   Standard format
for Thevenin's solution.

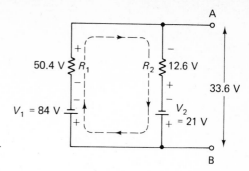

FIGURE 10-21 Deriving Thevenin's Voltage (V_{Th}).

(a) Effective voltage

$$V_{eff} = V_1 - V_2$$

$$V_{eff} = 84 - 21 \quad = 63 \text{ V}$$

(The current will flow from V_1.)

(b) Total resistance (R_T)

$$R_T = R_1 + R_2$$

$$= 12 + 3 = 15 \text{ } \Omega$$

(c) Current flow in the closed loop

$$I_T = \frac{V_{eff}}{R} = \frac{63}{15} = 4.2 \text{ A}$$

(2) Calculate the voltage drop across each of the resistors. Establish the voltage polarity and label (Fig. 10-21).

$$V_{R_1} = I_{R_1} \times R_1$$

$$= 4.2 \times 12 = 50.4 \text{ V}$$

$$V_{R_2} = I_{R_2} \times R_2$$

$$= 4.2 \times 3 \quad = 12.6 \text{ V}$$

(3) Determine Thevenin's Voltage (the voltage across terminals A and B).

$$V_{AB} = V_1 + V_{R_1} \text{ (series-opposing)}$$

$$= 84 - 50.4 = 33.6 \text{ V}$$

or $V_{AB} = V_2 + V_{R_2}$ (series-aiding)

$$= 21 + 12.6 = 33.6 \text{ V}$$

d. Calculate Thevenin's Resistance (R_{Th})

(1) Remove the voltage sources and replace with a short.

(2) Calculate R_{Th}.

R_1 is in parallel with R_2 ($R_1 \| R_2$) (see Fig. 10-20b).

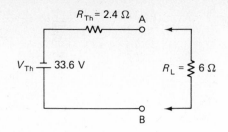

FIGURE 10-22 Thevenin's Equivalent Circuit (T.E.C.).

$$R_{AB} = \frac{R_1 \times R_2}{R_1 + R_2} = R_{Th}$$

$$= \frac{12 \times 3}{15} = 2.4\ \Omega$$

e. Draw Thevenin's Equivalent Circuit and label (Fig. 10-22).

2.

a. Current through R_L (I_{R_L})

$$I_{R_L} = \frac{V_{Th}}{R_{Th} + R_L}$$

$$= \frac{33.6}{2.4 + 6} = 4\ \text{A}$$

b. Voltage across R_L

$$V_{R_L} = I_{R_L} \times R_L$$

$$= 4 \times 6 = 24\ \text{V}$$

PROBLEM 10-7

1. Calculate Thevenin's Equivalent Circuit for the bridge circuit in Fig. 10-23.

2. Determine the current through and voltage across the R_L.

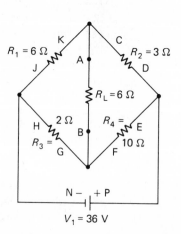

FIGURE 10-23 Bridge circuit.

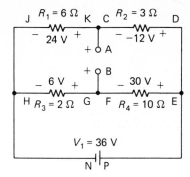

FIGURE 10-24   Simplified circuit (standard format for Thevenin's Voltage).

SOLUTION   (cover solution and solve)

1. Thevenin's Equivalent Circuit.
   a.   Remove R_L from terminals A and B.
   b.   Draw Fig. 10-23 in the standard format (Fig. 10-24)
   c.   Find Thevenin's Voltage.
      (1)   Current values
         (a)   Current flow through R_1 and R_2

$$I = \frac{V}{R_T} = \frac{36}{6 + 3} = 4 \text{ A}$$

         (b)   Current flow through R_3 and R_4

$$I = \frac{V}{R_T} = \frac{36}{12} = 3 \text{ A}$$

      (2)   Calculate voltage across each component, establish voltage polarity and label (Fig. 10-25).

$$V_{R_1} = I \times R_1 = 4 \times 6 = 24 \text{ V}$$
$$V_{R_2} = I \times R_2 = 4 \times 3 = 12 \text{ V}$$
$$V_{R_3} = I \times R_3 = 3 \times 2 = 6 \text{ V}$$
$$V_{R_4} = I \times R_4 = 3 \times 10 = 30 \text{ V}$$

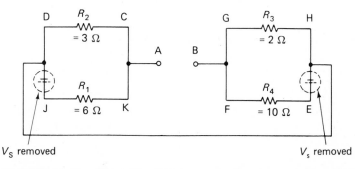

FIGURE 10-25   This is Fig. 10-24 rearranged so as to compute Thevenin's Resistance.

(3)   Determine the voltage across terminals A and B (V_{AB}).
On the left side of terminals A and B

> Terminal A is +24 V
>
> Terminal B is +6 V

The difference in voltage between terminals A and B is Terminal A is +18 V with respect to terminal B.

$$V_{AB} = 18 \text{ V}$$

and

Thevenin's Voltage = V_{AB} = 18 V

d. Calculate Thevenin's Resistance.
(1)   Figure 10-24 is difficult to analyze for resistance values. It must be redrawn as in Fig. 10-25 after replacing V_s with a short (removing V_s from the circuit). Trace Fig. 10-24 and draw and label the components in Fig. 10-25 as the tracing progresses.
(2)   Tracing from Fig. 10-24:

> A connects to C and K
>
> B connects to G and F
>
> D and J connect together through the shorted battery
>
> H and E connect together through the shorted battery
>
> DJ and HE connect together

(3)   Calculate R_{Th}.
R_2 and R_1 are in parallel

$$R_{T_1} = \frac{3 \times 6}{3 + 6} = 2 \ \Omega$$

R_3 and R_4 are in parallel   ($R_3 \| R_4$)

$$R_{T_2} = \frac{2 \times 10}{12} = 1.67 \ \Omega$$

Thevenin's Resistance (R_{Th}) =

$$R_{AB} = R_{T_1} + R_{T_2}$$
$$= 2 + 1.67 = 3.67 \ \Omega$$

e. Draw Thevenin's Equivalent Circuit and label (Fig. 10-26).

2.
a. Current through R_L

$$I_{R_L} = \frac{V_{Th}}{R_{Th} + R_L} = \frac{18}{3.67 + 6} = 1.86 \text{ A}$$

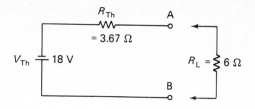

FIGURE 10-26   T.E.C.

b. Voltage across R_L

$$V_{R_L} = I_{R_L} \times R_L$$
$$= 1.86 \times 6 \qquad\qquad = 11.17 \text{ V}$$

PROBLEM 10-8

1. Calculate Thevenin's Equivalent Circuit for the attenuator circuit in Fig. 10-27.
2. Determine the current through and the voltage across the load (R_L).

SOLUTION   *(cover solution and solve)*

1. Calculate Thevenin's Equivalent Circuit.
    a. Remove R_L from terminals A and B.
    b. Draw the circuit in standard format (Fig. 10-28).

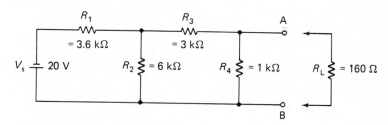

FIGURE 10-27   Attenuator circuit.

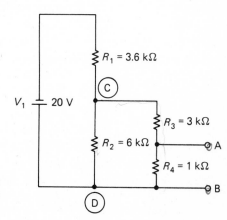

FIGURE 10-28   Standard format circuit for an attenuator circuit for Thevenin's Voltage.

c. Find Thevenin's Voltage.

(1) The easiest solution is to first find the resistance across R_{CD}. $[R_2 \| (R_3 + R_4).]$ Then use the voltage ratio formula.

$$R_{CD} = \frac{R_2 \times (R_3 + R_4)}{R_2 + (R_3 + R_4)}$$

$$= \frac{6000 \times 4000}{6000 + 4000}$$

$$= \frac{6000 \times 4000}{10000} = 2400 \ \Omega$$

(2) Voltage across R_{CD}

$$V_{CD} = \frac{R_{CD}}{R_{CD} + R_1} \times V_1$$

$$= \frac{2400}{2400 + 3600} \times 20 = 8 \ V$$

(3) Determine the voltage across terminals A and B. The voltage V_{AB} is equal to the voltage across $R_4(V_{R_4})$.

(4) The voltage V_{CD} is across $R_3 + R_4$; therefore, the voltage across $R_4(V_{R_4})$ can be determined by the voltage ratio formula.

$$V_{R_4} = \frac{R_4}{R_4 + R_3} \times V_{CD}$$

$$= \frac{1000}{1000 + 3000} \times 8 = 2 \ V$$

(5) Thevenin's Voltage = $V_{AB} = V_{R_4} = 2 \ V$.

d. Calculate Thevenin's Resistance.

(1) Replace V_1 with a short (remove V_1) and redraw Fig. 10-28 in a form which will facilitate resistance analysis. (R_1 is in parallel with R_2.) (See Fig. 10-29.)

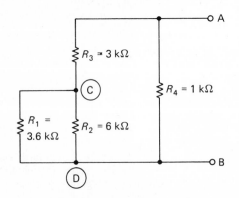

FIGURE 10-29 Standard format circuit to obtain Thevenin's Resistance for an attenuator circuit.

180

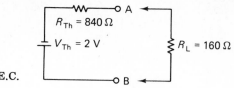

FIGURE 10-30   T.E.C.

(2)   Calculate R_{Th}.

$$R_{\text{CD}} = \frac{3600 \times 6000}{3600 + 6000} = 2250 \ \Omega$$

$$R_{T_S} = R_{\text{CD}} + R_3 \qquad \text{(where } R_{T_S} = \text{subtotal)}$$
$$= 2250 + 3000 = 5250 \ \Omega$$

$$R_{\text{AB}} = \frac{5250 \times 1000}{5250 + 1000} \equiv 840 \ \Omega$$

Thevenin's Resistance = R_{AB} = 840 Ω

e.   Draw Thevenin's Equivalent Circuit and label (Fig. 10-30).

2. a.   Current through R_L

$$I_{R_L} = \frac{V_{\text{Th}}}{R_{\text{Th}} + R_L} = \frac{2}{1000} \ \text{A}$$

$$= \frac{2}{1000} \times 10^3 = 2 \ \text{mA}$$

b.

$$V_{R_L} = I_{R_L} \times R_L$$
$$= (2 \times 10^{-3}) \times 160 \qquad = 320 \ \text{mV}$$

10-4.6 *Practical Applications of Thevenin's Method*

The solutions to the above example problems are referred to as the calculation method of Thevenin's Theorem.   Calculations cannot be made, however, if equipment is already constructed and the components are sealed in a cabinet.   Thevenin's two component values and equivalent circuit can be determined very easily with meters in the electronics lab. There are three other methods besides the load and no-load method. They are:

1. Matched load method
2. Two different loads method
3. Open and short method

Thevenin's Voltage is always the open terminal voltage of any equipment.   Thevenin's Resistance (internal resistance, or internal impedance of

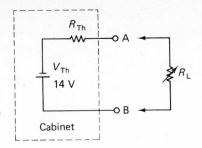

FIGURE 10-31 Matched load method to obtain Thevenin's Resistance.

equipment) is the most commonly required value. The following procedure describes the steps to be followed in determining the internal impedance of any piece of equipment. The equipment used in the following examples could be an audio generator or power supply.

MATCHED LOAD METHOD (FIG. 10-31)

1. Measure the voltage across open terminals AB (e.g., 14 V).
2. Connect a load across AB and adjust it until the voltage across it reads one-half the open terminal voltage (e.g., 7 V).
3. Because V_{RL} now equals $1/2$ V_{Th}, the other $1/2$ of V_{Th} is across R_{Th}. Because R_{Th} is in series with R_L, R_{Th} must equal R_L.
4. Now remove R_L and measure its resistance. This value will be the internal resistance (R_{Th}) of the equipment under test.

This method cannot be used in low resistance and high voltage circuits because of the large current which will flow when a resistance, R_L, equal to R_{Th} is connected across the equipment.

TWO DIFFERENT LOADS METHOD (FIG. 10-32)

This method can be used in any type of equipment, including low resistance and high voltage circuits such as power supplies. The procedure is as follows:

1. Connect a load R_{L_1} across terminals AB (e.g., 1 kΩ).

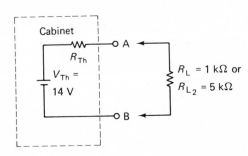

FIGURE 10-32 Two different-load method to obtain Thevenin's Resistance.

2. Measure the terminal voltage across the 1 kΩ resistor and the terminal current through the 1 kΩ resistor (e.g., 4.8 V and 4.8 mA).

3. Remove R_{L_1} and connect R_{L_2} of, for example, 5 kΩ.

4. Measure the terminal voltage and current of R_{L_2}, e.g., 10.285 V and 2.056 mA.

5. The internal resistance of the equipment (R_{Th}) is

$$R_{Th} = \frac{\Delta V}{\Delta I}$$

$$= \frac{10.285 - 4.8}{4.8 - 2.057}$$

$$= \frac{5.485}{2.743} = 2000 \ \Omega$$

OPEN AND SHORT METHOD (FIG. 10-33)

This method cannot be used in low resistance, high voltage circuits, such as power supplies, because of the possible damage to the equipment caused by the high current which would flow. However, this method (as well as all others) may be used in equipment such as generators. The procedure is as follows:

1. Measure the voltage across the open terminals AB (e.g., 14 V).

2. Measure the current flowing through the milliammeter when terminals AB are shorted by the current meter (e.g., 7 mA).

3. The internal resistance of the equipment (R_{Th}) will be

$$R_{Th} = \frac{V_{Th}}{I_{mA}} \quad \text{where} \quad \begin{array}{l} V_{Th} \text{ is Thevenin's voltage} \\ I_{mA} \text{ is current in milliamperes} \end{array}$$

$$= \frac{14}{7 \times 10^{-3}}$$

$$= 2000 \ \Omega$$

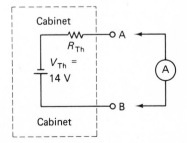

FIGURE 10-33 Open and short method to obtain Thevenin's Resistance.

10-5 Norton's Theorem

10-5.1 Practical Applications

Norton's Theorem is not as useful in solving networks as is Thevenin's Theorem, because its applications are limited to constant-current sources such as the collector circuit of transistors, in which the collector current is relatively unaffected by the collector voltage. Knowledge of the principle of Norton's Theorem is necessary, however, because of its applications and because of its relationship to Thevenin's Theorem.

As there is no such thing as a Thevenin's circuit because the conditions determine the type of circuit, there is no such thing as a Norton's circuit. Norton's current is the current that would flow across the terminals of a power source if the terminals were shorted. A circuit is therefore called a Norton's circuit when the internal resistance of the power source (Norton's resistance) is many times larger than the load resistance. This statement may be stated in reverse by saying a source of power is a Norton's circuit (a constant-current source) when the resistance of the load is very small compared to the internal resistance of the source. The closer that a load becomes to a short circuit relative to the internal resistance of the current source the more accurate the results. Regardless of which way the statements are made, the important fact is that the ratio of the source resistance to the load resistance must be very large. The output current will be relatively constant with a variety of loads under these conditions. A ratio of 1/10 may be considered a minimum ratio for wide tolerance circuits but may be unsatisfactory for critical circuits.

10-5.2 Norton's Theorem

Norton's Theorem states that when the load is replaced by a short in any two-terminal network or device containing any number of current or voltage sources and resistances, the network or device is equivalent to a single source of constant current shunted by a single resistance. The size of the current source is equal to the current which will flow between the terminals when they are shorted, and the shunting resistor is the internal resistance of the network or device (Norton's resistance). See Fig. 10-34.

10-5.3 Norton's Solution

NORTON'S RESISTANCE (R_{Th})

Norton's Theorem states that any circuit is equivalent to a constant current source in parallel with the equivalent resistance (internal) of that circuit. In practice or in theory, the equivalent resistance of any circuit has only one value regardless of how it is calculated. Norton's Resistance (R_{Th}) is calculated in the same manner in which Thevenin's Resistance was calculated. In either case, Ohm's Law is employed. When the equiva-

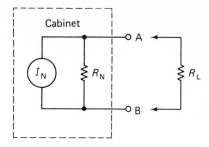

FIGURE 10-34 Norton's Equivalent Circuit (N.E.C.).

lent resistance is used in Thevenin's Circuit it is called Thevenin's Resistance. When the same value is calculated with the same procedure but used in Norton's Circuit it is called Norton's Resistance.

NORTON'S CURRENT

Norton's current is the current which will flow between the output terminals A and B of a device or network when a short is placed across the terminals. A current meter of any range has such a low resistance that it is considered to be a short, and when it is connected across the output terminals of a device it effectively shorts those terminals; therefore, the expression "shorting the output terminals" actually means placing a current meter between the terminals.

10-5.4 Steps in Norton's Solution

1. Remove the load and replace it with a short (current meter).
2. Redraw the given circuit in the standard format for Norton's solution (Fig. 10-35).
3. Find Norton's Current.
4. Find Norton's Resistance.

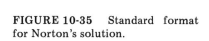
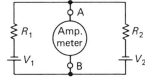

FIGURE 10-35 Standard format for Norton's solution.

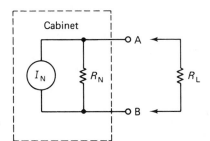

FIGURE 10-36 Norton's Equivalent Circuit (N.E.C.).

5. Draw Norton's Equivalent Circuit (Fig. 10-36).

6. Find the current through and the voltage across the load.

PROBLEM 10-9

See Fig. 10-37.

1. Calculate Norton's Equivalent Circuit
2. Calculate
   a. The current through the load
   b. The voltage across the load

SOLUTION   *(cover solution and solve)*

1. Calculate Norton's Equivalent Circuit.
   a. Remove the load and label the output terminals A and B.
   b. Draw Fig. 10-37 in the standard format for Norton's solution with the load removed (Fig. 10-38).
   c. Find Norton's Current (I_N).
      (1) Place a short (current meter) across terminals A and B (Fig. 10-39).
      (2) Calculate I_1 and I_2.

$$I_1 = \frac{V_1}{R_1} = \frac{84}{12} = 7 \text{ A}$$

$$I_2 = \frac{V_2}{R_2} = \frac{21}{3} = 7 \text{ A}$$

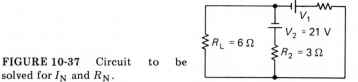

FIGURE 10-37   Circuit   to   be
solved for I_N and R_N.

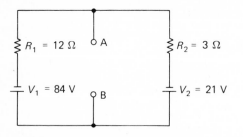

FIGURE 10-38   Circuit   in   Fig.
10-37 with load removed.

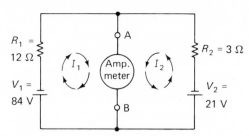

FIGURE 10-39   Standard format
circuit   for   Norton's   Current
solution.

(3) Norton's Current is the sum of I_1 and I_2. Currents I_1 and I_2 flow through the current meter in the same direction. Therefore, they add.

$$I_N = I_1 + I_2$$
$$= 7 + 7 = 14 \text{ A}$$

d. Norton's Resistance.
   (1) Remove the short from terminals A and B. Replace the source voltages with a short and redraw Fig. 10-39 as in Fig. 10-40 in order to calculate the total resistance.
   (2) R_1 is in parallel with R_2. $(R_1 \| R_2)$

$$R_T = \frac{R_1 \times R_2}{R_1 + R_2}$$

$$= \frac{12 \times 3}{12 + 3} = 2.4 \ \Omega$$

Norton's Resistance = $R_T = 2.4 \ \Omega$

e. Draw Norton's Equivalent Circuit. (See Fig. 10-41.)

2.
a. Find the current through the load.
   (1) Connect the load across Norton's Equivalent Circuit (Fig. 10-41).
   (2) Find I_{R_L} by using the *inverse current ratio formula* [Eq. (6)].

$$I_{R_L} = \frac{R_N}{R_N + R_L} \times I_N \tag{6}$$

(3) $$I_{R_L} = \frac{2.4}{2.4 + 6} \times 14 \doteq 4 \text{ A}$$

b. Find the voltage across the load R_L.

$$V_{R_L} = I_{R_L} \times R_L$$
$$= 4 \times 6 = 24 \text{ V}$$

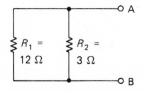

FIGURE 10-40 Standard format circuit for Norton's Resistance solution.

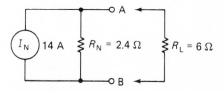

FIGURE 10-41 Norton's Equivalent Circuit (N.E.C.) of the circuit in Fig. 10-37.

PROBLEM 10-10

See Fig. 10-42.

1. Calculate and draw Norton's Equivalent Circuit.
2. Find
   a. I_{R_L}
   b. V_{R_L}

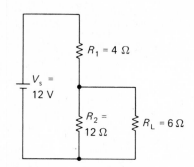

FIGURE 10-42 Circuit to be solved for I_N and R_N.

SOLUTION   *(cover solution and solve)*

1. Calculate Norton's Equivalent Circuit.
   a. Remove the load and label the output terminals A and B.
   b. Redraw Fig. 10-42, with R_L removed, in the standard format drawing (Fig. 10-43).
   c. Find Norton's Current.
      (1) Place a short (current meter) across terminals A and B (Fig. 10-44).
      (2) Calculate Norton's Current (I_N). The current meter shorts out R_2. Therefore, R_1 is the only resistance opposing V_1. (I_N is the current flowing through the current meter.)

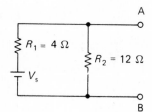

FIGURE 10-43 Standard format for Norton's Current solution.

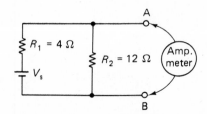

FIGURE 10-44 Determining Norton's Current.

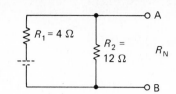

FIGURE 10-45 Standard format for Norton's Resistance solution.

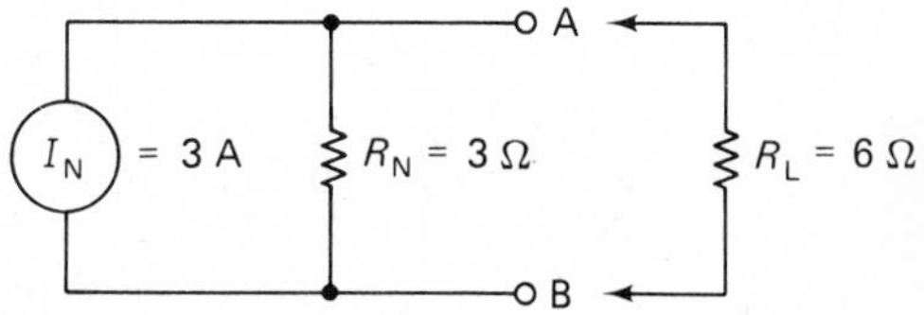

FIGURE 10-46 Norton's equivalent circuit for the circuit in Fig. 10-42.

$$I_N = \frac{V_1}{R_1} = \frac{12}{4} = 3 \text{ A}$$

   d. Find Norton's Resistance.
     (1) Remove the short from terminals A and B and replace V_s with a short. The circuit now becomes Fig. 10-45.
     (2) R_1 is in parallel with R_2. $(R_1 \| R_2)$

$$R_N = \frac{R_1 \times R_2}{R_1 + R_2}$$

$$= \frac{4 \times 12}{4 + 12} = 3 \ \Omega$$

   e. Draw Norton's Equivalent Circuit (Fig. 10-46).

2.

   a. Find current through the load. [Use the inverse current ratio formula—Eq. (6).]

$$I_{R_L} = \frac{R_N}{R_N + R_L} \times I_N$$

$$= \frac{3}{3 + 6} \times 3 = 1 \text{ A}$$

   b. Find the voltage across the load.

$$V_{R_L} = I_{R_L} \times R_L$$

$$= 1 \times 6 \quad = 6 \text{ V}$$

PROBLEM 10-11

See the attenuation circuit in Fig. 10-47.

1. Calculate and draw Norton's Equivalent Circuit.

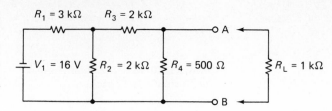

FIGURE 10-47 An attenuator circuit to be solved for I_N and R_N.

2. Find
   a. I_{R_L}
   b. V_{R_L}

SOLUTION (cover solution and solve)

1. Calculate Norton's Equivalent Circuit.
   a. Remove R_L from terminals A and B.
   b. Redraw Fig. 10-48 (with R_L removed) in the standard format for Norton's solution (Fig. 10-48).
   c. Find Norton's Current.
      (1) Place a short (current meter) across terminals A and B (Fig. 10-48). Note R_4 is now also shorted out.
      (2) The voltage across CD is that which will force current through R_3. The current through R_3 is that current which will flow through the current meter, and is Norton's Current.

   (a) $R_T = R_1 + (R_2 \times R_3)$

   $$= 3000 + \left(\frac{2000 \times 2000}{2000 + 2000} \right)$$

   $$= 3000 + 1000 = 4000 \ \Omega$$

   (b) $I_T = \dfrac{V_1}{R_T} = \dfrac{16}{4 \times 10^3} = 4 \text{ mA}$

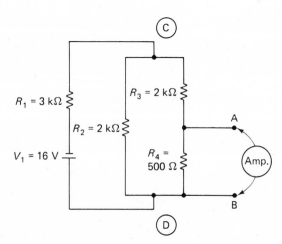

FIGURE 10-48 Standard format circuit for Norton's Current solution for an attenuator circuit.

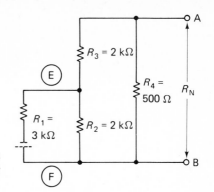

FIGURE 10-49 Standard format for Norton's Resistance solution for an attenuator circuit.

$$V_{CD} = I_T \times R_{CD}$$

$$= (4 \times 10^{-3}) \times (1 \times 10^3) = 4 \text{ V}$$

(3) Norton's Current (I_N)

$$I_N = \frac{V_{CD}}{R_3} = \frac{4}{2 \times 10^3} = 2 \text{ mA}$$

d. Find Norton's Resistance.
  (1) Redraw Fig. 10-47 as a circuit for resistance solution (Fig. 10-49) but with the short across terminals A and B removed, and the source voltage replaced by a short.
  (2) Find the resistance across

$$R_{EF} = R_1 \| R_2$$

$$R_{EF} = \frac{3000 \times 2000}{3000 + 2000} = 1200 \ \Omega$$

  (3) Find the resistance across A and B (which is Norton's Resistance). $[(R_{EF} + R_3) \| R_4]$

$$R_{AB} = (R_{EF} + R_3) \times R_4$$

$$= (1200 + 2000) \| 500$$

$$= \frac{3200 \times 500}{3700} = 432 \ \Omega$$

Norton's Resistance R_N is 432 Ω.
e. Draw Norton's Equivalent Circuit (Fig. 10-50).

2.
  a. Find the current through the load (I_{R_L}). [Use the inverse current ratio formula—Eq. (6).]

$$I_{R_L} = \frac{R_N}{R_N + R_L} \times I_N$$

$$= \frac{432}{432 + 1000} \times 2 \text{ mA} \qquad = 0.6 \text{ mA}$$

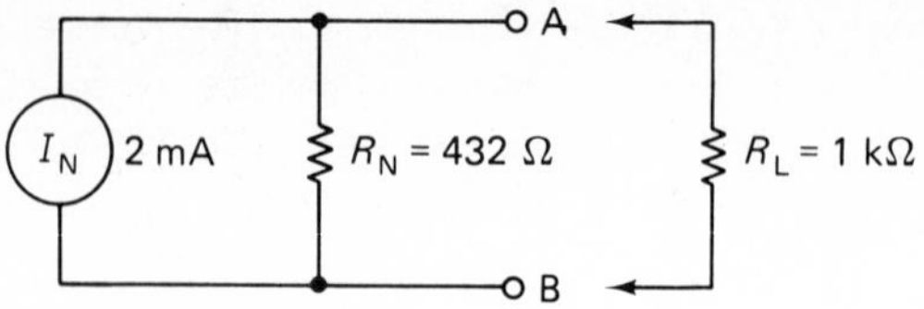

FIGURE 10-50   Norton's Equivalent Circuit (N.E.C.) for the attenuator circuit in Fig. 10-47.

   b. Find the voltage across the load (V_{R_L}).

$$V_{R_L} = I_{R_L} \times R_L$$
$$= (0.6 \times 10^{-3}) \times (1 \times 10^3) = 0.6 \text{ V}$$

10-6   Relationship Between Norton's Equivalent Circuit and Thevenin's Equivalent Circuit

Problems 10-9 and 10-6 use the same base circuit and values.

The solution to Norton's Equivalent Circuit in Problem 10-9 was

$$I_N = 14 \text{ A}$$
$$R_N = 2.4 \ \Omega$$

The solution to Thevenin's Equivalent Circuit in Problem 10-6 was

$$V_{Th} = 33.6 \text{ V}$$
$$R_{Th} = 2.4 \ \Omega$$

Note that R_N and R_{Th} are the same value.

Also note the value of Thevenin's Voltage divided by Thevenin's Resistance.

$$\frac{V_{Th}}{R_{Th}} = \frac{33.6}{2.4} = 14 \text{ A}$$

The current value obtained is equal to I_N.

The values obtained in Problem 10-9, multiplied together, produce Thevenin's Voltage (Problem 10-6).

$$V_{Th} = I_N \times R_N$$
$$= 14 \times 2.4 = 33.6 \text{ V}$$

SUMMARY:

1. Kirchhoff's two laws:
    a. Kirchhoff's Voltage Law
    b. Kirchhoff's Current Law
    are used in whole or part in the solution of all electrical and electronic circuits with more than two components.

2. Norton's Theorem and Thevenin's Theorm are used with two different types of circuits.

3. Thevenin's Equivalent Circuit is actually the circuit of a constant-voltage source, as was explained in Sec. 5-7.4. In actual practice, Thevenin's Equivalent Circuit is valid only when the load, R_L, is at least three times the value of the R_{Th} resistance. The most common constant-voltage source circuits are voltage-regulated power supplies.

4. Norton's Equivalent Circuit is actually the constant-current source circuit, as was discussed in Sec. 5-7.5. In practice, the load connected across Norton's Equivalent Circuit must be less than one-third the R_N resistance for the circuit to be useful as a common current source. The most common constant-current source circuit is the output of a transistor amplifier.

5. Norton's Resistance (R_N) is identical in value and is determined with the same procedure as is Thevenin's Resistance (R_{Th}).

6. Conversion from Thevenin's Voltage (V_{Th}) to Norton's Current (I_N) and vice versa can be made as follows:

$$\text{Thevenin's Voltage} \quad V_{Th} = R_N \times I_N$$

$$\text{Norton's Current} \quad I_N = \frac{V_{Th}}{R_{Th}}$$

EXERCISES

1. Refer to Fig. 10-51 and calculate (using first the values given in I and then in II below):
   a. Thevenin's Voltage
   b. Thevenin's Resistance
   c. Draw Thevenin's Equivalent Circuit and label values
   d. Load current
   e. Load voltage

I	II
$R_1 = 6\ \Omega$	$R_1 = 8\ \Omega$
$R_2 = 3\ \Omega$	$R_2 = 4\ \Omega$

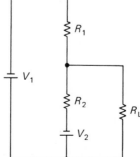

FIGURE 10-51

I	II
$R_L = 8\ \Omega$	$R_L = 12\ \Omega$
$V_1 = 60$ V	$V_1 = 90$ V
$V_2 = 24$ V	$V_2 = 30$ V

2. Refer to Fig. 10-52 and calculate (using first the values given in part I and then in part II):

a. Thevenin's Voltage
b. Thevenin's Resistance
c. Draw Thevenin's Equivalent Circuit and label values
d. Load current
e. Load voltage

I	II
$R_L = 16\ \Omega$	$R_L = 6\ \Omega$
$R_2 = 6\ \Omega$	$R_2 = 2\ \Omega$
$R_1 = 12\ \Omega$	$R_1 = 4\ \Omega$
$V_2 = 48$ V	$V_2 = 15$ V
$V_1 = 120$ V	$V_1 = 45$ V

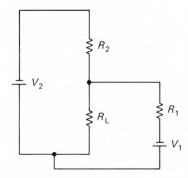

FIGURE 10-52

3. Refer to Fig. 10-53 and calculate (using first the values given in part I and then in part II):

a. Thevenin's Voltage
b. Thevenin's Resistance
c. Draw Thevenin's Equivalent Circuit and label values
d. Load current
e. Load voltage

I	II
$R_1 = 7\ \Omega$	$R_1 = 2\ \Omega$
$R_2 = 8\ \Omega$	$R_2 = 4\ \Omega$
$R_3 = 4\ \Omega$	$R_3 = 12\ \Omega$
$R_4 = 6\ \Omega$	$R_4 = 12\ \Omega$
$R_L = 36\ \Omega$	$R_L = 24\ \Omega$
$V_1 = 30$ V	$V_2 = 24$ V

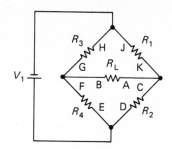

FIGURE 10-53

4. Refer to Fig. 10-54 and calculate (using first the values given in part I and then in part II):
   a. Thevenin's Voltage
   b. Thevenin's Resistance
   c. Draw Thevenin's Equivalent Circuit and label values
   d. Load current
   e. Load voltage

I	II
$R_1 = 14 \ \Omega$	$R_1 = 2 \ \Omega$
$R_2 = 16 \ \Omega$	$R_2 = 4 \ \Omega$
$R_3 = 8 \ \Omega$	$R_3 = 12 \ \Omega$
$R_4 = 12 \ \Omega$	$R_4 = 12 \ \Omega$
$R_L = 72 \ \Omega$	$R_L = 24 \ \Omega$
$V_1 = 60 \ V$	$V_1 = 24 \ V$

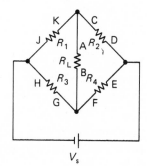

FIGURE 10-54

5. Refer to Fig. 10-55 and calculate (using first the values given in part I and then in part II):
   a. Thevenin's Voltage
   b. Thevenin's Resistance
   c. Draw Thevenin's Equivalent Circuit and label values
   d. Load current
   e. Load voltage

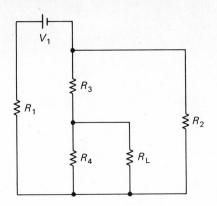

FIGURE 10-55

	I		II
R_1	= 360 Ω	R_1	= 3 kΩ
R_2	= 600 Ω	R_2	= 6 kΩ
R_3	= 300 Ω	R_3	= 4 kΩ
R_4	= 100 Ω	R_4	= 1.2 kΩ
R_L	= 1200 Ω	R_L	= 12 kΩ
V_1	= 20 V	V_1	= 18 V

6. Refer to Fig. 10-56 and calculate (using first the values given in part I and then in part II):

   a. Thevenin's Voltage
   b. Thevenin's Resistance
   c. Draw Thevenin's Equivalent Circuit and label values
   d. Load current
   e. Load voltage

	I		II
R_4	= 200 Ω	R_4	= 600 Ω
R_3	= 600 Ω	R_3	= 2 kΩ
R_2	= 1200 Ω	R_2	= 3 kΩ

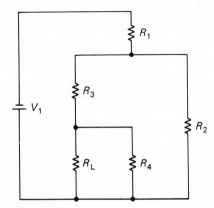

FIGURE 10-56

I	II
$R_1 = 720\ \Omega$	$R_1 = 1.5\ \text{k}\Omega$
$R_L = 2400\ \Omega$	$R_L = 6\ \text{k}\Omega$
$V_1 = 40\ \text{V}$	$V_1 = 9\ \text{V}$

7. Refer to Fig. 10-57 and calculate (using first the values given in part I and then in part II):

    a. Norton's Current
    b. Norton's Resistance
    c. Draw Norton's Equivalent Circuit
    d. Load current
    e. Load voltage

I	II
$R_1 = 8\ \Omega$	$R_1 = 8\ \Omega$
$R_2 = 2\ \Omega$	$R_2 = 4\ \Omega$
$R_L = 2\ \Omega$	$R_L = 1\ \Omega$
$V_1 = 64\ \text{V}$	$V_1 = 48\ \text{V}$
$V_2 = 20\ \text{V}$	$V_2 = 12\ \text{V}$

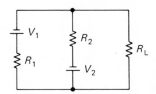

FIGURE 10-57

8. Refer to Fig. 10-58 and calculate (using first the values given in part I and then in part II):

    a. Norton's Current
    b. Norton's Resistance
    c. Draw Norton's Equivalent Circuit
    d. Load current

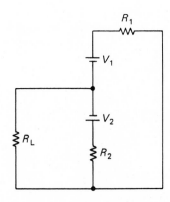

FIGURE 10-58

e. Load voltage

I	II
$R_1 = 4\ \Omega$	$R_1 = 8\ \Omega$
$R_2 = 1\ \Omega$	$R_2 = 4\ \Omega$
$R_L = 1\ \Omega$	$R_L = 1\ \Omega$
$V_1 = 30$ V	$V_1 = 48$ V
$V_2 = 10$ V	$V_2 = 12$ V

chapter **11**

PRINCIPLES OF MAGNETISM

11-1 Introduction

The subject of magnetism has been made very confusing over the years by the many different ways which have been devised to measure it and the number of different names used to label it. We will consider the newly accepted International System (SI) of units and refer to the now outdated CGS system only for the purpose of comparison.

11-2 Magnetism

Magnetism is an electrical property, not a component, which makes electronics and electricity possible. Nearly all electronic equipment will have a component of one form or another which operates by way of the property of magnetism. Each of the following components operates because of its magnetic property: transformer, motor, generator, relay (solenoid), loudspeaker, and ac to dc adaptor.

11-3 Natural Magnetism

Magnetism was found in the mineral magnetite and was used by the Chinese over 2,600 years ago for navigational purposes (Fig. 11-1). Magnetite, an iron-oxide mineral, was discovered in Asia Minor and was called Lodestone, a word meaning leading stone (Fig. 11-2). It is now mined in the United States, Norway, and Sweden as a byproduct of other mining activities.

The earth itself is a natural magnet, having north and south magnetic poles which are displaced from the north and south geographic poles by

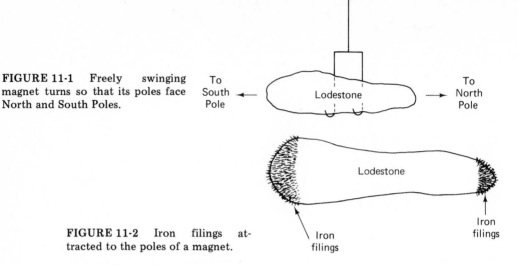

FIGURE 11-1 Freely swinging magnet turns so that its poles face North and South Poles.

To South Pole ← Lodestone → To North Pole

Lodestone

Iron filings

FIGURE 11-2 Iron filings attracted to the poles of a magnet.

Iron filings

some 10° to 15° depending upon where on the earth the measurement is made. The N-magnetic pole is located on the Prince of Wales Island a few miles inside of the Arctic Circle, and 1,200 miles from the geographic north pole. The angle of displacement is called the "magnetic declination" and on navigational maps is recorded in degrees and is called "magnetic variation." The angle of magnetic variation changes from year to year and place to place in the world. All navigational compasses must be corrected for it because of the difference in location between the N-magnetic and N-geographic poles.* The presence of the earth's magnetic field can be dramatically demonstrated by pointing an unmagnetized screwdriver in the direction of the earth's north pole (in the northern hemisphere) and tapping it with some heavy object for a short time (Fig. 11-3). One will then be able to pick up light metal objects, such as small steel screws with the magnetic field of the screwdriver, or if a compass is brought up to it, the N-pole of the compass will swing toward that same end of the screwdriver.

11-4 Artificial Magnetism

The reason the screwdriver blade was first unmagnetized was that all the domains of the atoms (referred to as molecules at one time) in the steel blade were haphazardly arranged (Fig. 11-4a). After being subjected to the pull of the earth's magnetic north pole, the domains aligned themselves in an orderly and reinforcing north-to-south arrangement as shown in Fig. 11-4b.

*References. (1) Arthur N. Strahler, *Physical Geography*, 2nd ed. (New York: John Wiley & Sons, Inc., 1962); (2) *Isogenic Map of the Magnetic North Polar Area—1948* (Ottawa, Ontario, Canada: Canada Department of Mines and Resources, 1948); (3) *Royal Canadian Air Force Observer Training Manual*, vol. II (Ottawa, Ontario, Canada: Air-Officer Commanding Training Command, 1953).

Ch. 11 Principles of Magnetism

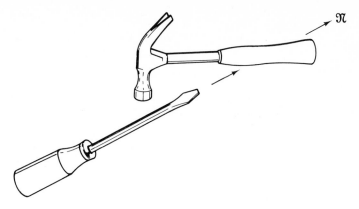

FIGURE 11-3 The magnetization of a screw driver.

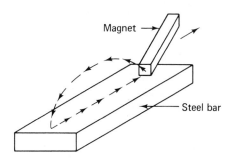

Stroking method (magnetic induction)

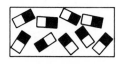

Unmagnetized steel
(haphazard domains)

(a)

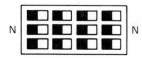

Magnetized steel
(aligned domains)

(b)

FIGURE 11-4 Magnetization of a
steel bar (magnetization by induction).

If a piece of ferrous material (iron) is brought close to a strong magnetic field, that material will become magnetized, as the screwdriver was
magnetized by the earth's magnetic field. Figure 11-5 shows the magnetic
field around a conductor in electromagnetism.

11-5 Atomic Theory of Magnetism

It is often asked why some materials magnetize very easily while others
are difficult to magnetize, and still others do not magnetize at all. The accepted theory today involves the atom and the electrons which spin
around the nucleus of the atom. It is thought that the electrons in an
atom spin in much the same way as the earth spins on its axis as it orbits

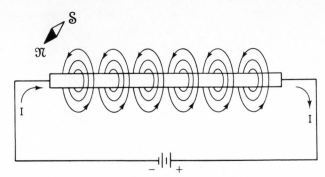

FIGURE 11-5 Lines of force around a conductor.

the sun, and that it is the spin of the electrons that produces the magnetic field around them. Thus, each electron acts as a tiny magnet which by itself is too weak to have any significant magnetic effect. However, most atoms have many electrons revolving about their nuclei.

Reference to the atomic periodic table shows that:

nickel has a total of 28 electrons

iron has 26 electrons

tungsten has 74 electrons

copper has 29 electrons

Why is it that the *first two* elements, iron and nickel, are *magnetic materials*, but elements such as tungsten and copper which have more electrons are *not magnetic* materials? The answer is the reinforcing effect of the direction of spin of the electrons. If, for example, *tungsten*, which has a total of 74 electrons, had 37 of the electrons spinning clockwise and 37 spinning counterclockwise, the magnetic field of one spin direction would cancel the magnetic field of the opposite spin direction and the material could never be magnetized. If, on the other hand, the *iron* atom, which has a total of 26 electrons, had 10 electrons spinning in a clockwise direction and 16 spinning in a counterclockwise direction, a net magnetic field of 6 electrons would be spinning in the counterclockwise direction. The result would be an atom with a tiny magnetic field of a definite magnetic polarity. The greater the net number of electrons spinning in a particular direction, the stronger will be this tiny magnetic field.

11-5.1 Domain

There is an interaction between the net magnetic lines of force of atoms. The interaction can be an attracting effect or a repelling effect between the atoms depending upon the direction of their magnetic fields. Whenever the magnetic fields of a relatively small group of atoms (10^{14} to 10^{15} atoms) are parallel and their poles are oriented in the same direction, they have a reinforcing effect and attract one another. Such a group of atoms is called a *domain*.

A *domain* of atoms has a definite north and south polarity and an intense magnetic field strength, compared to the magnetic field of one atom. One domain of atoms, however, will have a polarity and a field strength different from its neighboring domains. In other words, the domains lie in haphazard directions with respect to one another in un-magnetized materials (Fig. 11-4a), and the materials appear not to possess any magnetism. The domains will align themselves in an orderly north-to-south magnetic pole direction and the material will become magnetized when the material is brought into the vicinity of an external magnetic field (*magnetism by induction*). A very strong magnetizing force will eventually align all the domains in a north-to-south pole direction; the material is then said to be magnetically saturated.

Materials in which the clockwise direction of spin of one-half of the electrons is cancelled out by the counterclockwise spin of the other half of the electrons will have no magnetic domains and are nonmagnetic materials.

11-6 Three Categories of Artificial Magnets

Magnetism is an invisible force which cannot be detected until a magnet material or compass is brought into its vicinity (Fig. 11-6). Any material which has magnetic domains (ferromagnetic) but is not magnetized until a magnetizing force is brought up to it is called an artificial magnet.

There are three main categories of artificial magnets:

1. Permanent magnets
2. Temporary magnets
3. Electromagnets

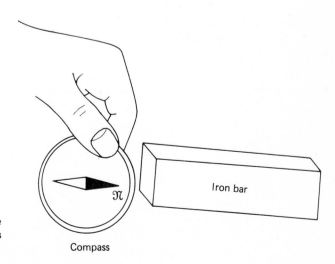

FIGURE 11-6 The bar must be magnetized, because the compass needle is attracted to it.

Compass

11-6.1 Permanent Magnets

Materials (ferromagnetic) which become magnetized in the presence of a magnetic field (force) and retain the magnetism even after the magnetizing force has been removed are called permanent magnets (Fig. 11-7). Hard iron (hard in terms of its tempering) can become a permanent magnet if it is heat-treated in the presence of a strong magnetic force. Permanent magnets are made of hard iron. The best permanent magnet material known today is alnico (an aluminum, nickel, and cobalt alloy). Alnico alloy has no magnetic field until it is brought into the presence of a magnetic force, but will lose very little of its magnetic strength if it is heat-treated at the same time. Permanent magnetism is a property highly desirable in loudspeaker magnets.

11-6.2 Temporary Magnets

Any material which acquires the strength of the magnetizing force while in its presence but which becomes demagnetized as soon as the magnetizing force is removed is called a temporary magnet. Soft iron is one of the best metals for a temporary magnet. Soft iron is used in transformer cores where the magnetic field must increase and decrease and change in magnetic polarity in unison with the ac current applied to the transformer.

11-6.3 Electromagnets

Electromagnetism is magnetism which is present only when an electric current is passing through the turns of a wire which is wound on a piece of soft iron.

A relay (sometimes called a solenoid) which closes and opens when a switch in series with it and a source of voltage is closed and opened is an example of an electromagnet. Magnetism is present in the soft iron core when the switch is closed but disappears the instant that the switch is opened. A car horn and a door bell are other applications of electromagnets.

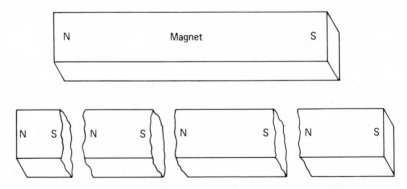

FIGURE 11-7   Each piece has a weaker field than the whole.

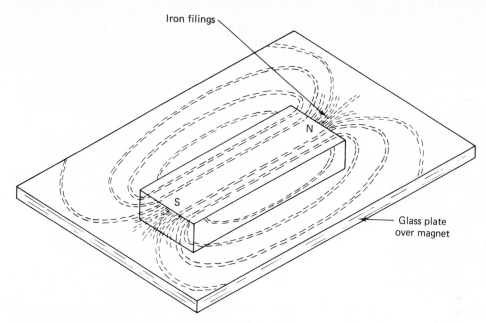

Iron filings

N

S

Glass plate
over magnet

FIGURE 11-8   Pattern formed by iron filings.

11-7   The Magnetic Field

Regardless of the category of magnet which one may consider, all have certain common properties and characteristics.

As the magnetic domains in magnetic materials have two poles (north and south), magnetic materials which contain magnetic domains will also have two poles. If iron filings are sprinkled on a piece of glass placed over a magnet (Fig. 11-8), the magnetic force will exhibit itself as the iron filings form complete loops. The lines followed by the iron filings are called lines of force, lines of flux, or flux lines, and the region in which the material (iron filings) is acted upon by the magnetic force is called the *magnetic field*.

Note that in Fig. 11-8, the fact that the iron filings align themselves end to end indicates north-to-south polarization. The concentrations of filings at the ends of the magnet indicate that

1. The greatest field strength is at the ends (Fig. 11-9).
2. The field weakens as it leaves the ends of the magnet.
3. The divergence of the lines of force indicates that there is a repelling force between force lines, and in a less concentrated position they lose the strength of their reinforcing effect.

11-7.1   Magnetic Attraction and Repulsion

The best example of magnetic attraction is the earth's north magnetic pole attracting the north-seeking pole of the geographic compass. The

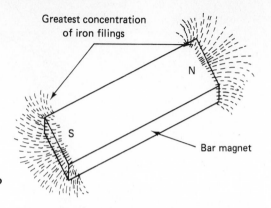

Greatest concentration
of iron filings

N

S

Bar magnet

FIGURE 11-9   Iron filings cling to
the poles of a magnet.

north pole of a navigator's compass is technically referred to as the north-
seeking pole, because only unlike poles attract.

Figure 11-10a is a photograph of the orientation of iron filings on a
piece of paper placed over two bar magnets which are separated so that
the S-pole of one magnet is facing the N-pole of the other (unlike poles).
The resulting pattern is shown in the drawing (Fig. 11-10b). The magnetic
field of the left magnet leaves in an arbitrary direction from the N-pole
and enters the S-pole of the right bar magnet, and the fields at the outside
ends pass through the air space and combine. The long continuous magnetic lines between the outside ends try to contract and thereby tend to
pull the two magnets together. The resulting attraction and pulling together of the two bar magnets establishes the rule that

unlike magnetic poles attract

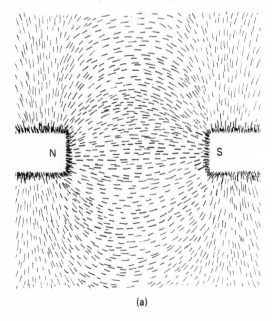

FIGURE 11-10   (a) Pattern of iron filings
around unlike poles (N and S). (b) Unlike poles;
the outside continuous lines tend to pull magnets together.

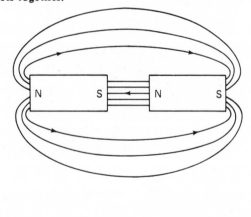

(a)

(b)

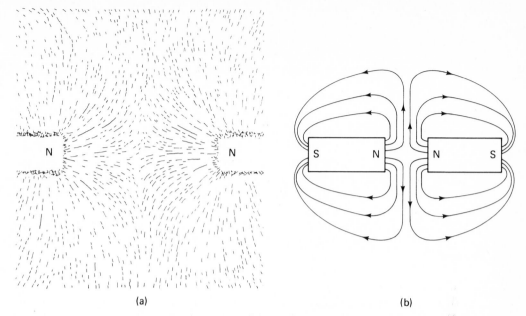

(a)   (b)

FIGURE 11-11   (a) Pattern of iron filings around like magnetic poles (N and N). (b) Fields from like poles repel, tending to push the magnets apart.

Figure 11-11a is a photograph of the field formed by the iron filings placed over two bar magnets with their N-poles facing one another (like poles). The drawing in Fig. 11-11b more clearly illustrates the resulting magnetic fields. As can be seen, the magnetic fields of the two magnets are not merged as they were before. There is a repelling effect, causing the two magnets to push away from one another. This establishes the rule that

like magnetic poles repel

11-8 Characteristics of Flux Lines

A study of the effect of iron filings in the presence of a magnet (magnetic field) and the attraction and repulsion of magnets can be summarized into the characteristics of the lines of force.

1. Magnetic lines of force have polarity and direction. They exit from the north pole of a magnet and enter the south pole.
2. The strongest magnetic fields are at the ends of the magnet.
3. Magnetic lines of force form complete loops.
4. The magnetic loops have tension, which keeps them as short as possible.
5. Magnetic lines of force repel one another, which cause them to diverge once they leave the confines of the magnet.

6. Magnetic lines of force weaken as they extend out from the magnet.

7. Like magnetic poles repel one another and unlike poles attract one another.

11-9 Magnetism and Demagnetism

The alloy "Alnico 5" is the material used today for making general purpose magnets. The metal is first sized and shaped for the application in which it is to be used, and then it is placed between the jaws of a very large horseshoe-shaped electromagnet. A switch carrying tens of amperes of dc current to the electromagnet is closed for a second or two, and the alnico becomes magnetized and ready for its installation.

Demagnetizing (degaussing) is accomplished by placing the magnetized object inside a coil which is connected to a 110-V ac voltage. The changing magnetic field in the coil demagnetizes the object.

11-10 Air Gap

The air space between the poles of a magnet (Fig. 11-12) is called the air gap. A bar magnet will have a very long air gap because of the distance between its poles. The longer the bar, the weaker will be the magnetic field between the poles external to the magnet (Fig. 11-12a).

11-10.1 Ring Magnets (Toroids)

A magnet shaped in the form of a doughnut is called a ring magnet (Fig. 11-13a). A piece of soft iron shaped in the form of a doughnut and wound with wire is called a toroid electromagnet (Fig. 11-13b).

In both a ring magnet and a toroid, closed magnetic loops are formed which produce a maximum concentration of magnetic lines of force in-

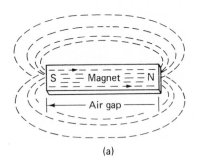

(a)

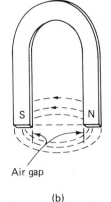

Air gap

(b)

FIGURE 11-12   (a) Long air gap between poles. (b) short air gap between poles.

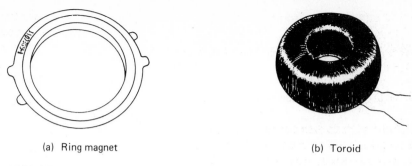

(a) Ring magnet   (b) Toroid

FIGURE 11-13   Two types of ring magnets.

side the doughnut and a minimum magnetic flux on the outside. The toroid not only has maximum magnetic strength but it is less prone to the induction of unwanted currents from undesirable external magnetic fields. Memory cores in older computers used toroid coils to store data.

11-11   Care of Permanent Magnets

A magnet loses a great deal of its magnetism if it is dropped onto a hard surface, struck by a heavy object, heated to a very high temperature, or if its magnetic field is left to project into space from the magnet itself. The latter problem is overcome by placing a keeper (a piece of soft iron) over the ends of the magnet (Figs. 11-14, 11-15, and 11-16).

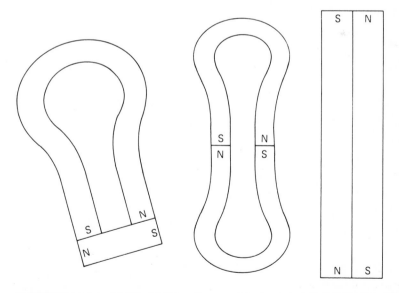

FIGURES 11-14, 11-15, 11-16   Types of keepers to minimize magnetic deterioration.

11-12.1 Magnetic Flux (φ)

Magnetic flux, sometimes called lines of induction, is the total number of magnetic lines of force leaving or entering the pole of a magnet.

the symbol for magnetic flux is the greek letter phi φ

the SI unit of magnetic flux is the weber (wb) 1 weber = 10^8 lines of force (flux)

11-12.2 Flux Density (B)

The unit of magnetic flux has little meaning other than to describe a condition of magnetic force, because the area containing these lines of force may be the size of a pin head or the area of one city block. A given number of lines in the pin head would be a much stronger magnetic force than the same number of lines in the city block area. The number of lines of force indicates magnetic field strength only if the unit area is specified. See Fig. 11-17.

flux density (B) is the number of lines of force per square meter

$$\text{flux density} = \frac{\text{flux}}{\text{square area}} \tag{2}$$

$$B = \frac{\phi}{A}$$

where B is the flux density in units called teslas (T)

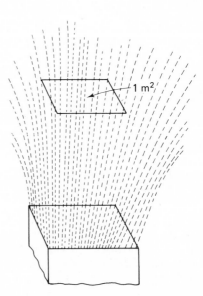

FIGURE 11-17   Flux density is the number of lines of magnetic flux in an area of 1 m^2.

ϕ is the number of webers of magnetic flux

(1 Wb = 10^8 lines of force)

A is the area in square meters (m^2)

From Eq. (2) it can be seen that 1 tesla (T) is the amount of 1 weber (10^8 lines of flux) contained within an area of 1 square meter (m^2).

PROBLEM 11-1

Suppose a magnet has a pole which has an area of 3.4×10^{-4} m^2 and a total flux of 900 μWb. What is the flux density at the end of that pole?

SOLUTION   (cover solution and solve)

$$A = (3.4 \times 10^{-4})\, m^2$$

$$\phi = 900\ \mu W = (900 \times 10^{-6})$$

$$B = \frac{\phi}{A}$$

$$= \frac{900 \times 10^{-6}}{3.4 \times 10^{-4}}$$

$$B = 2.647\ teslas\ (T)$$

PROBLEM 11-2

How much flux is required to provide a flux density of 3×10^{-2} T in a cross section of a rectangular bar 1×10^{-3} m wide and 5×10^{-3} m deep?

SOLUTION   (cover solution and solve)

area of cross section = width $\times$ depth

$$= (1 \times 10^{-3}) \times (5 \times 10^{-3})$$

$$= 5 \times 10^{-6}\ m^2$$

$$\phi = B \times A$$

$$= (3 \times 10^{-2}) \times (5 \times 10^{-6})$$

$$= 15 \times 10^{-8}\ Wb$$

$$= 0.15\ \mu Wb$$

11-12.3 Relative Permeability (μ_r)

Permeability is derived from the word permeate which means to penetrate and pass through.

There are three expressions of permeability.

1. Relative permeability (μ_r)
2. Absolute permeability (μ)
3. Permeability of air (μ_0)

The last two expressions are used when discussing electromagnetic circuits, but relative permeability is an expression used to define the characteristic of materials whether they are used in permanent, temporary, or electromagnetic applications.

Relative permeability has no unit. It is the *measure* of how much better a given material is, than air, as a path for magnetic lines of force. The word *relative* implies a comparison; thus, relative permeability indicates how much better a material conducts magnetic lines of force than does free space (air), or a vacuum which has a relative permeability value of 1.

$$\mu_r = \frac{\mu}{\mu_0} \tag{3}$$

where μ_r is relative permeability
μ is absolute permeability of a material
μ_0 is permeability of air

The above equation will be discussed later. It is used now only to show the relationship and to show why air has a relative permeability

Table 11-1 Relative Permeability (μ_r) of Various Materials

	μ_r Maximum	μ_r at Small Magnetization
Ferromagnetic		
Nickel	50	50
Cobalt	60	60
Cast Iron	90	60
Machine Steel	450	300
Transformer Iron	5500	3000
Silicon Iron	7000	3500
Very Pure Iron	8000	4000
Permaloy	100,000	25,000
Supermalloy	1000,000	100,000
Paramagnetic		
Beryllium	1.00000079	
Aluminum	1.00000065	
Ni-Cl	1.000040	
Diamagnetic		
Silver	0.99999981	
Wood	0.99999950	
Paraffin	0.99999942	
Copper		
Lead		
Mercury		

value of 1. *The relative permeability* of 1 for air *must not be* confused with *the absolute permeability value* of air which is $(4\pi \times 10^{-7})$. The values of relative permeability are often listed in tables in engineering handbooks. (See Table 11-1.)

11-12.4 High Permeability Materials

High permeability materials are materials, such as soft iron, which are very easily magnetized because they have a very high magnetic conductivity. They produce a very high concentration of magnetic lines of flux within the material while a magnetizing force is present.

EXAMPLE

Figure 11-18 shows the flux density between the poles of a magnet with air ($\mu_r = 1$) as the medium. If a piece of soft iron with a μ_r of 50 is inserted between the poles of the same magnet as shown in Fig. 11-19, the flux density in that iron (and thu between the poles)

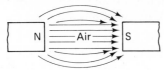

FIGURE 11-18 Note the magnetic flux spreading out between the poles.

FIGURE 11-19 The magnetic flux is concentrated by the soft iron.

will be 50 times greater than the flux density of the magnetic lines of force between the poles in Fig. 11-19. Note the crowding together of the lines of flux of the magnetic field in Fig. 11-19. The permeability of the soft iron increases the flux density. High permeability metals, however, are materials with poor retentivity, so they possess only a small amount of residual magnetism.

It would appear from Fig. 11-20 that magnetic lines of force follow the path of least opposition in the same manner that current takes the

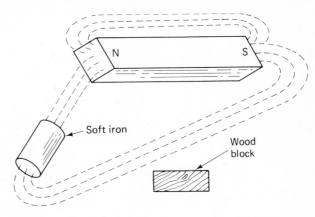

FIGURE 11-20 Demonstration of the effect of permeability. The magnetic lines of force are diverted and concentrated by the high-permeability material. Low-permeability wood has no effect.

path of least resistance in an electric circuit (greatest current through the smallest resistance in parallel). Permeability is thus analogous to conductivity in electric circuits. The magnetic lines of force in Fig. 11-20 are shown passing through the iron rather than through the air around it and the wood. Such metals of high permeability are called ferromagnetic materials.

11-12.5 Types of Materials

FERROMAGNETIC MATERIALS

Ferromagnetic materials (Table 11-1) become strongly magnetized by a weak magnetizing force. They become magnetic in the same direction as the magnetizing force. They include such materials as nickel, cobalt, iron, steel, and alnico alloy. See Fig. 11-21.

PARAMAGNETIC MATERIALS

Materials such as aluminum, platinum, manganese, and chromium have a permeability very slightly better than air (see Table 11-1) and can only be slightly magnetized. They become magnetized in the same direction as the magnetizing force.

DIAMAGNETIC MATERIALS

Diamagnetic materials (Table 11-1) have a lower permeability than air. Magnetic lines of force usually go around them, but those that do go through them tend to spread out (see Figs. 11-20 and 11-22). In other words, diamagnetic materials tend to offer a slight opposition to magnetic lines of force. They magnetize in a direction opposite to the magnetizing force. Zinc, lead, mercury, gold, silver, copper, and its alloys such as brass are all diamagnetic metals.

In practice, paramagnetic and diamagnetic materials (some are gases) are considered nonmagnetic materials because they have virtually no more effect on a magnetic field than does free space.

11-12.6 Magnetic Shielding

High-permeability metals such as soft iron are good materials for shielding coils. A watch, for example, is made antimagnetic (shielded) by making the case of soft iron (Fig. 11-23).

11-12.7 Ferrite Iron

Soft iron, even though it has high μ and low retentivity, is not a suitable material for shielding when the magnetic field must change polarity and magnitude at the speed of radio frequencies. A recently-developed ceramic material called ferrite will do what soft iron is unable to do and is used for loop antennas, cores for antennas, and radio-frequency cores.

Ferrites are ceramic ferromagnetic materials having the properties of

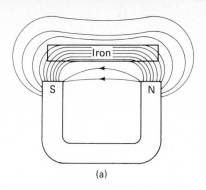
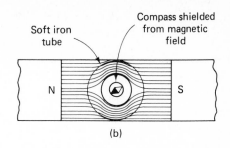

FIGURE 11-21   (a) The ferromagnetic (high-permeability) material, iron, draws flux lines into itself and reduces the magnetic field around it. (b) Lines of flux follow the high-permeability material and shields objects within it.

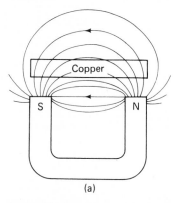
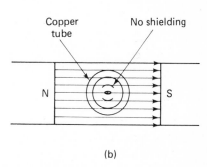

FIGURE 11-22   The diagmagnetic material, copper, has no more influence on flux lines than air. It is useless for magnetic shielding.

very high-permeability iron. The ceramic material is an insulator, and the powdered iron is the conductor. Because of the minute size of the powdered iron, eddy currents are very low ($I^2 R$ losses are low), and ferrite has high efficiency as a shield at radio frequencies.

11-12.8 Low-Permeability Ferromagnetic Materials

Low-permeability ferromagnetic metals such as cast iron and nickel alloys retain their magnetism once the magnetizing force is removed and so have

Table 11-2   Comparison Between SI and CGS Magnetic Units

	Symbol	Formula	SI Units		Obsolete CGS Units	
Flux	ϕ	—	Weber (Wb)	Wb = 10^8 Lines of Flux	Maxwell (Mx)	1 Mx = 1 line
Flux Density	B	$B = \dfrac{\phi}{A}$	Tesla (T)	$T = \dfrac{Wb}{m^2}$	Gauss	Gauss = $\dfrac{Mx}{cm^2}$

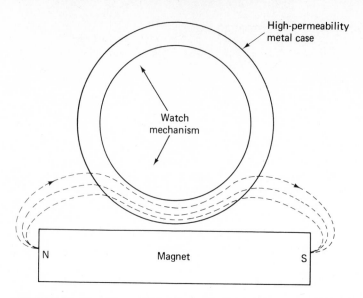

FIGURE 11-23   The effect of a magnetic field on an antimagnetic watch case.

a high *retentivity* and a large amount of residual magnetism. They are materials usually used for permanent magnets.

11-13   Magnetics Terms

11-13.1   Retentivity

The ability of any material to retain the magnetism it has acquired even after the magnetizing force is removed is called *retentivity*.

11-13.2   Residual Magnetism

The amount of magnetism that a material retains after the magnetizing force has been removed is called the *residual magnetism*.

EXERCISES

1. What is an important characteristic of Lodestone?

2. Name practical applications of magnetism.

3. Which are magnetic materials?
   a. Tungsten
   b. Nickel
   c. Copper
   d. Iron

4. Why are some materials magnetic while others are not?

5. Define the term domain.

6. Approximately how many atoms make a domain?

7. Does a domain have magnetic polarity?

8. Name three main classes of artificial magnets.

9. Describe the differences among each of the three classes of artificial magnets.

10. List six characteristics of magnetic lines of force.

11. What is meant by degaussing?

12. State the equation for the magnetic force between two poles.

13. What will be the relative value of the magnetic force between two magnetic poles if the distance between them is tripled?

14. List three practical applications of horseshoe magnets.

15. Describe a toroid electromagnet.

16. List two advantages of a toroid magnet.

17. State two possible causes for a permanent magnet to lose its magnetism.

18. Define
    a. Flux density
    b. Flux
    c. Weber

19. A magnet has a total flux of 1800 μWb at the end of one pole whose cross-sectional area is 3.4×10^{-4} m^2. What is the flux density at the end of that pole?

20. A bar has a cross-sectional area of 5×10^{-6} m^2. How much flux is required to provide a flux density of 6×10^{-2} T?

21. Define permeability.

22. Does the value $4\pi \times 10^{-7}$ represent the relative or absolute permeability of air?

23. How much is the permeability changed (increased or decreased) if the air core between the poles of a magnet is replaced by iron with a permeability of 60?

24. What is the analogy to permeability in electric circuits?

25. Which of the following materials are paramagnetic?
    a. Silver
    b. Nickel
    c. Aluminum
    d. Iron

26. Define
    a. Retentivity
    b. Residual magnetism
    c. Tesla

27. What would be the flux density at the end of a magnetic pole having a cross-sectional area of 0.01 m^2 and possessing 60 μWb of flux lines?

28. What must be the cross-sectional area of a magnetic pole if it contains a total flux of 900 μWb and has a flux density of 7.941 T?

29. How much flux is at the end of a magnetic pole if a flux density of 1.5 T is contained within an area of 10×10^{-6} m^2?

30. State the relationship between permeability of air, relative permeability, and absolute permeability.

chapter **12**

ELECTROMAGNETISM

12-1 Electromagnetism

Electromagnetism is sometimes called a magnetic circuit (circuit meaning circle) because the magnetic lines of flux in the magnetic field form closed loops. The word electromagnetism can be broken into two parts; electro meaning electrical, and magnetism. An electromagnet is a component which becomes a magnet when current flows through it.

The electromagnet, often referred to as a solenoid, has many applications. It is used in relays, buzzers, headphones, motors, generators, to lift scrap iron in scrap yards, in electric brakes, solenoid water valves, on automatic washing machines, and in video and audio tape recorders.

12-2 Magnetism and Current

Around 1820, a Danish physicist, Oersted, discovered that a magnetic field is produced around a wire in which current is flowing. It is quite simple to demonstrate the presence of a magnetic field around a wire when current is flowing in it, by connecting a circuit as shown in Fig. 12-1. The circuit consists of a 3-Ω current-limiting resistor* in series with a 12-V power supply capable of delivering at least 4 A of current. One of the connecting wires is passed through a hole in the center of a sheet of cardboard. Iron filings are sprinkled on the paper around the wire and two navigation compasses are placed close to the wire. When the switch is closed the iron filings and compass will align themselves as shown in Fig. 12-1.

*Without the resistor, the current would be infinite and the wire would burn out.

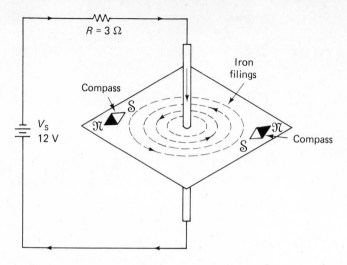

FIGURE 12-1 Magnetic field around a wire in which current is flowing. The dark end of the compass is the N-pole.

The iron filings and compasses in Fig. 12-1 are aligned in a counterclockwise direction. If the battery polarity were reversed, the iron filings would all move until their ends were aligned in a clockwise direction, and the N-poles of the compasses would point in a clockwise direction.

This simple demonstration shows that there is a definite relationship between the direction of current flow and direction of magnetic rotation around a straight wire. A rule which can be used to determine the direction of one of these units when the other is known is called the *left-hand rule* (Fig. 12-2).

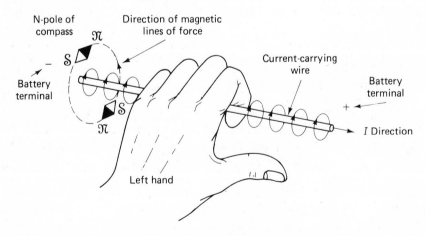

FIGURE 12-2 Left-hand rule for a conductor.

12-3 Left-Hand Rule for Conductors

The direction of spin of the magnetic field around a straight wire can be determined by using the left-hand rule for current flow (Fig. 12-2).

Grasp the wire with the left hand so that the thumb points in the direction of the current flow. (Electrons flow from negative to positive.) The fingers will then circle the conductor in the direction of the magnetic spin (north to south).

Conversely, if the direction of the magnetic field around a wire is identified with a compass, the direction of current flow can be determined by wrapping the fingers around the wire in the direction of the magnetic field. The thumb will then point in the direction of current flow and the negative side of the battery voltage can be identified.

Note: that in some circles, particularly in the electrical industry, current flow is said to flow from the positive side of a dc voltage source to the negative side. This rule is known as the conventional current flow rule. In such cases the right-hand rule must be applied. The same procedure is used as for the left-hand rule except that the right hand is used.

12-4 Flux and Conductors

Because the direction of the magnetic field around a conductor can be determined with the left-hand rule, it is possible to determine the effects of the fields of two parallel conductors which are close together.

Figure 12-3 shows current flowing in opposite directions through two parallel adjacent wires. The lines of force *between* the two conductors

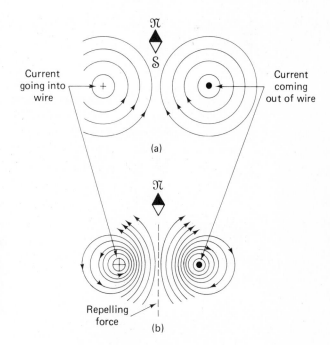

FIGURE 12-3 (a) Direction of magnetic spin around two separated wires in which the current flow is opposite; note direction of compass. (b) Same as (a) but showing the repelling force between the wires when they are close together.

Current going into wire

Current coming out of wire

(a)

Repelling force

(b)

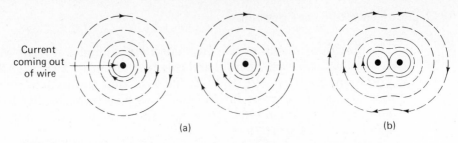

Current
coming out
of wire

(a)

(b)

FIGURE 12-4   (a) Magnetic fields around each of two separated wires carrying current in the same direction. (b) Same as (a), but wires are close together. Their fields reinforce one another, and there is a pulling together.

rotate in the same direction and produce a condition much the same as occurs between two like magnetic poles. The conductors repel each other and move apart. The magnetic-force lines on the outside flow in opposite direction and oppose one another, weakening the fields on the outside of the coil.

Figure 12-4 shows two adjacent conductors carrying current in the same direction. The flux lines *between* the two conductors cancel one another because they flow in opposite directions, while those on the outside flow in the same direction. These aiding magnetic fields combine with one another and pull the two conductors together, reinforcing the two magnetic fields and producing a single strong magnetic field. This is what takes place when a conductor is wound into a coil. The magnetic field around one turn of wire reinforces the magnetic field of the adjacent turn (Figs. 12-4b and 12-5b).

12-5   Flux and Coils

12-5.1   Increasing Magnetic Flux

If the wire in Fig. 12-4 is wound into a coil, the magnetic fields of adjacent turns will add together to form a concentrated field down the axis of the coil, called its core. The lines of force follow the axis of the coil and then spread apart once they leave the north-pole end of the coil. They form a complete path back to the south-pole end of the coil. The field on the outside of the coil is less concentrated, i.e., weaker because the lines of flux spread out (Fig. 12-5).

The strength of an electromagnet can be increased by:

1. Increasing the number of turns of wire in the coil
2. Increasing the amount of current flowing through the wire
3. Replacing the air inside the coil with a material with a high permeability.

For example in Fig. 12-6 a 20-V battery is shown connected in series with a 10-turn air-core coil and a 4-Ω current-limiting resistor. If

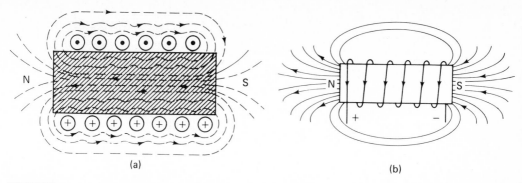

(a)   (b)

FIGURE 12-5 (a) A cross-sectional view of the coil shown in (b). Each turn reinforces the other. (b) The resulting magnetic field of several turns of wire adjacent to one another.

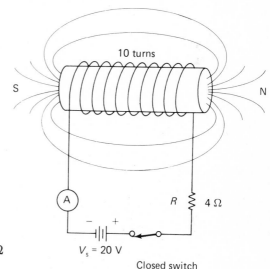

FIGURE 12-6 Ten-turn coil, 4-Ω resistor, 5-A current circuit.

the battery voltage were doubled or the 4-Ω limiting resistor were decreased to 2 Ω, the current would double and the magnetic field would double. If the number of turns of wire were doubled, the magnetic field would double. If a soft iron coil with a μ_r of 5000 were inserted, the magnetic field would increase 5000 times.

12-5.2 Left-Hand Rule for Magnetism

The magnetic polarity of the ends of a *coil* may be determined by using the *left-hand rule for magnetism* which states that if the coil is grasped so that the fingers of the left hand follow the direction of current flow around the circumference of the coil, the thumb will point in the direction of the north pole (Fig. 12-7).

Note that the coils in Fig. 12-7 are both wound clockwise. The mag-

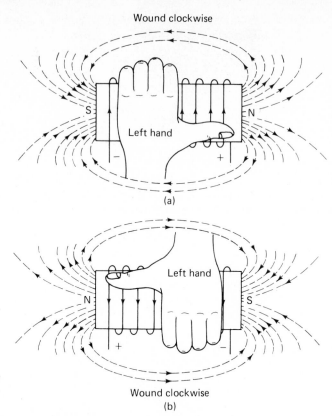

Wound clockwise

(a)

Wound clockwise

(b)

FIGURE 12-7 Left-hand rule for coils (the coils are wound clockwise).

netic polarity at the end of the coil in Fig. 12-7a is opposite that in Fig. 12-7b because the battery polarities have been reversed.

The coils in Fig. 12-8 are wound counterclockwise. The negative polarity of the battery in Fig. 12-8a is connected to the left-hand side of the coil as it is in Fig. 12-7a, but notice that the left-hand end of the coil in Fig. 12-8a now has a N-magnetic pole instead of a S-magnetic pole even

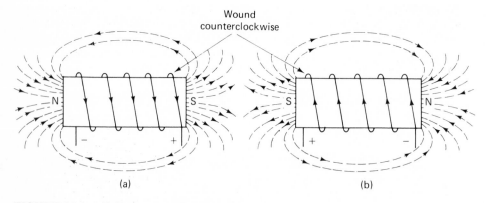

Wound counterclockwise

(a)   (b)

FIGURE 12-8 Coils are wound counterclockwise.

though the battery polarity is the same. This is because the coil turns have been wound in the opposite direction. Observe that the same difference exists between the coils in Figs. 12-7b and 12-8b.

The direction of coil winding and the battery polarity together determine magnetic polarity.

12-5.3   Left-Hand Rule for Current

If the direction of the current flow through a *coil* is to be determined when the magnetic polarity of the coil is known (determined by a navigator's compass), grasp the coil with the left hand so that the thumb points in the direction of the N-pole. The fingers of the left hand will then point in the direction of the current flow. This is the identical rule to the left-hand rule for conductors.

12-5.4   Clarification

It must be emphasized at this time that even though it is said that the magnetic field is around the wire or coil, it is not because of the wire. The wire is merely the pipe carrying the electrons. The magnetic field comes from the spin of the electrons, not from the metal in the copper wire. For example, a narrow beam of electrons travels through the eight-to ten-inch space in the vacuum between the cathode and phosphor screen in the picture tube of a television set. A magnetic field exists around the beam of electrons and makes possible the production of pictures on the screen. There is no wire inside the picture tube. The magnetic field exists around the electron beam only.

12-6   Magnetic Measurements

Unfortunately, several different units have been used in the past to measure magnetism. The SI (International System, also called the MKS; meter, kilogram, second) system is now generally recognized as the official system. Prior to 1976 the CGS (centimeter, gram, second) system was used extensively in Canada and the United States, but some countries also used the English Units. Table 12-3 at the end of this chapter shows the comparison between the MKS and CGS unit systems. More confusion was created by the fact that names of physicists who made certain magnetic discoveries were also used as part of the unit system.

12-6.1   Flux

Flux in electromagnetism has exactly the same meaning as in permanent magnets. It is the total number of lines of force that flow out of the N-pole of an electromagnet and enter the S-pole.

the symbol for flux is the greek letter phi ϕ

the SI unit is the weber (Wb) = 10^8 flux lines

Because of the large number of lines included in the unit 1 Wb, flux is often stated in μWb. One μWb = 1×10^2 lines of flux.

12-6.2   Flux Density (B)

Flux density as discussed in permanent magnetism is the same thing in electromagnetism.

$$\text{flux density} = \frac{\text{flux}}{\text{square area}} \tag{1}$$

$$B = \frac{\phi}{A} \quad (\text{teslas}) \tag{2}$$

where B is teslas (Wb/m)
ϕ is webers
A is area in m^2

the letter symbol for flux density is B

the SI unit for flux density is teslas (T)

12-6.3   Magnetomotive Force (F$_m$)

Flux in an electromagnet is produced by current flowing in a wire wound in the form of a coil. The more turns on the coil, the more concentrated the lines of force in the core area. The more current which flows, the stronger will be the magnetic field around each turn of wire.

The product of the current multiplied by the number of turns of the coil is called the magnetomotive force.

magnetomotive force = number of turns $\times$ current

$$F_m = IN \tag{3}$$

where F_m is magnetomotive force in amperes (A)
N is number of turns
I is current in amperes

the symbol for magnetomotive force is F_m

the SI unit for magnetomotive force is A

PROBLEM 12-1 ────────────────────────

Two amps of current is flowing in a coil which has 80 turns. What magnetomotive force is produced?

SOLUTION   *(cover solution and solve)*

$$F_m = IN$$

$$= 2 \times 80$$

$$= 160 \text{ A}$$

If the coils in the above examples were stretched to twice their length, the intensity of their magnetic fields would be half as great because the turns would be further apart and would have less reinforcing effect; thus there would be less concentration of the lines of force. Magnetomotive force therefore does not measure the true magnetizing force until the length of the coil is included in the formula. Since the strength of the magnetic field decreases with length, the field intensity is inversely proportional to the length of the coil.

$$\text{field intensity } (H) = \frac{\text{current} \times \text{number of turns}}{\text{length of coil}} \tag{4}$$

$$H = \frac{IN}{l} = \frac{F_m}{l} \quad (\text{A/m}) \tag{5}$$

where I is current in
amperes
N is number of turns
l is length of coil in
meters
H is expressed in
A/m = amperes/1
meter length

PROBLEM 12-2 ────────────────────────────────

A current of 750 mA is flowing in a coil with 480 turns. The coil is a toroid having an average length of 8 cm. What is the field intensity (H) of the magnetic field produced?

SOLUTION   *(cover solution and solve)*

$$H = \frac{IN}{l}$$

$$H = \frac{(750 \times 10^{-3}) \times 480}{8 \times 10^{-2}}$$

$$= 750 \times 60 \times 10^{-3} \times 10^{2}$$

$$= 750 \times 60 \times 10^{-1}$$

$$= 4500 \text{ A/m}$$

PROBLEM 12-3 ────────────────────────────────

A coil consisting of 240 turns is wound on a form. The length of the coil (core) is 0.04 m. What current is required to produce a field intensity of 1900 A/m?

SOLUTION   (cover solution and solve)

$$H = \frac{IN}{l}$$

Transpose

from $Hl = IN$

then $I = \dfrac{Hl}{N}$

$$= \frac{1900 \times 0.04}{240}$$

$$= \frac{(1.9 \times 10^3) \times (4 \times 10^{-2})}{240}$$

$$= 0.316 \text{ A}$$

$$= 316 \text{ mA}$$

Methods of calculating average magnetic path-length and areas of a toroid and rectangular-shaped core are shown at the end of this chapter.

12-6.5 Magnetizing Force and Flux Density Relationship

F_m and magnetizing force calculations have been made, but as yet they show no relationship to magnetic flux or flux density. The relationship between flux density and magnetizing force can be shown on a graph called the magnetizing curve, in which B is plotted against H (Fig. 12-10). The curve is plotted in the following manner:

PLOTTING THE B-H CURVE

An iron core electromagnet of a fixed number of turns and length is connected in series with a step-variable resistor (R_1), a switch Sw$_2$, a source of voltage (V_s), and an ammeter. A magnetic flux measuring device, called a fluxometer, is attached to the core so as to measure the flux density within the core (Fig. 12-9).

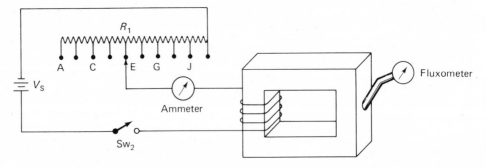

FIGURE 12-9   A crude test/experiment to measure flux density versus field intensity.

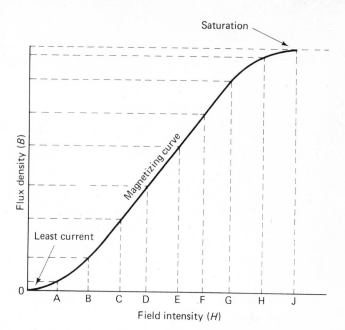

FIGURE 12-10 A magnetizing curve produced by a test/experiment as shown in Fig. 12-9. The flux density, B, is produced by the field intensity, H.

1. The measurements are begun with R_1 at its maximum resistance and Sw_2 open. H is zero and B is zero. Plot these B and H values on the graph paper (see Fig. 12-10).

2. Sw_2 is closed and R_1 is in position A.
   a. The current and flux density are measured.
   b. The field intensity (H) is calculated (current $\times$ turns).
   c. The H and B values are plotted on the graph paper.

3. R_1 is moved to position B which increases the current. Repeat a., b., and c. in 2 above for position B.

4. R_1 is moved to position C and the current is increased still more. Repeat a., b., and c. in 2.

5. The above routine is continued until an increase in current in the coil no longer increases the flux density. This condition is called *magnetic saturation.*

6. Connect all the plots together and the resulting curve is called the magnetizing curve (Fig. 12-10).

12-6.6 B-H Magnetizing Curve

The *B-H* magnetizing curve in Fig. 12-11 shows the relationship between the magnetizing force and the flux density value when permalloy is used as the core of the coil.

Air produces a linear relationship between flux density (B) and field intensity (H); that is, the ratio of increase in flux density to increase in field intensity is one-to-one (Fig. 12-12).

Ferromagnetic materials do not have a linear relationship between

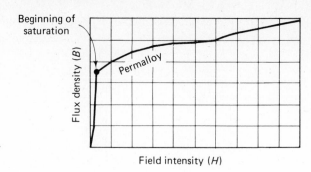

FIGURE 12-11 The *B-H* curve for permalloy.

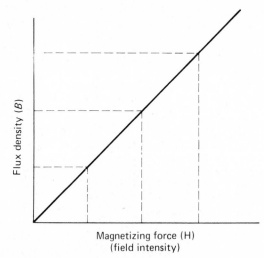

FIGURE 12-12 The magnetizing curve for an air-core coil. The curve is straight, indicating that the flux density is increasing in direct proportion to the field intensity and that there is no saturation value.

field intensity (*H*) and flux density (*B*). That is to say, *B* will not double if *H* is doubled. Each type of material has its own unique characteristic curve shape. See Fig. 12-13.

Notice that the wrought iron and sheet steel curves rise at a linear rate, then begin to flatten off. The flattening of the curves occurs at the

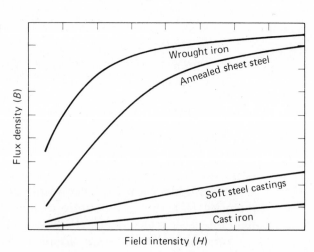

FIGURE 12-13 The magnetizing curves of four different types of metals.

Ch. 12 Electromagnetism

value at which the field intensity (H) begins to bring all the domains into line and to cause the condition called magnetic saturation.

The question should now be asked, why do different metals have characteristic curves of different shapes? The answer is that each material has a different value of permeability.

12-6.7 Permeability (μ)

Absolute permeability, usually referred to as just permeability, is the ability of a metal to permit magnetic lines of force to pass through the material when it is placed in a magnetizing field. In other words, permeability is analogous to conductance in electric circuits. Conductance in electric circuits is the reciprocal of resistance; similarly, permeability is the reciprocal of the opposition to magnetic lines of force, called *reluctance*.

FLUX DENSITY AND PERMEABILITY

The magnetizing curves as shown in Fig. 12-13 would be produced if a coil had its core replaced with the metals plotted in Fig. 12-13. The difference in the permeability values of the different metals accounts for the different shapes of curves.

Permeability of a metal may be determined from *B-H* magnetizing curves, as shown in Fig. 12-13, by dividing the flux density by the value of field intensity which was required to produce that flux.

$$\text{permeability} = \frac{\text{flux density}}{\text{field intensity}} \tag{6}$$

$$\mu = \frac{B}{H} \quad \text{(H/m)} \tag{7}$$

where μ is the permeability of the metal
(absolute permeability)
B is flux density in teslas (Wb/m^2)
H is field intensity in ampere per meter (A/m)

Equation (7) can be interpreted as follows: the less the amount of field intensity required to produce a specific flux density, the greater must be the permeability of the metal.

Rearrange Eq. (7) into Eq. (8).

$$B = \mu H \tag{8}$$

Equation (8) tells us that the flux density of a magnetic field is dependent upon the permeability of the metal being magnetized, as well as upon the strength of the magnetizing force (field intensity).

the letter symbol for permeability is μ

the SI unit for permeability is H/m

PERMEANCE

Permeability is the ability of a metal to conduct magnetic lines of force; permeance is a measure of that conductivity. The value of absolute permeability as obtained from the values of flux density (B) and field intensity (H) is given for a bulk of material with no length or area specified. The value of absolute permeability therefore merely indicates ability of a material to conduct magnetic lines of force. As flux lines, by contrast, are not a true measure of magnetic strength until area is specified, permeability is not a meaningful measure of magnetic conductivity until length and area are taken into account.

$$\text{permeance} = \frac{\text{permeability} \times \text{area}}{\text{length}} \qquad (9)$$

$$P_{\text{m}} = \frac{\mu A}{l} \quad \text{henries (H)} \qquad (10)$$

$$\begin{aligned}
\text{where} \quad & P_{\text{m}} \text{ is henries (H)} \\
& A \quad \text{is area in square meters} \\
& l \quad \text{is length of the magnetic circuit} \\
& \quad\quad \text{in meters} \\
& \mu \quad \text{is permeability}
\end{aligned}$$

Notice that the value of permeance in Eq. (10) is equal to the value of permeability when the units of length and area are each equal to one. Thus permeance per unit length and cross sectional area of a material is called permeability. Although permeability may be expressed in webers per ampere-turn, it is usually expressed as a pure number. Magnetic conductivity of a magnetic material increases as the area of the magnetic circuit increases and the length of the magnetic circuit decreases. The strength of a magnetic field, for the same field intensity, will therefore increase as the area of the material increases, but will decrease as the length increases.

The unit of permeability is used much more often than permeance, as designers are usually comparing the abilities of different magnetic materials to concentrate magnetic lines of force.

Transposing Eq. (10) produces another equation for permeability.

$$\mu = \frac{Pl}{A} \quad \text{henries per meter (H/m)} \qquad (11)$$

Equation (11) gives the permeability (magnetic conductivity) of one unit area of a magnetic material.

the letter symbol for permeance is P

the SI unit for permeance is henry (H)

There are three types of permeability.

1. Absolute permeability (μ)

2. Permeability of air (μ_0)
3. Relative permeability (μ_r)

PERMEABILITY (μ)

Permeability as discussed earlier is the magnetic conductivity of a magnetic material.

PERMEABILITY OF AIR (μ_0)

Air is used as the reference for comparing the permeability of all other materials. Air has a permeability of $4\pi \times 10^{-7}$, which is approximately 1.26×10^{-6}.

RELATIVE PERMEABILITY (μ_r)

Relative permeability is the comparison of a given material to air as a path for magnetic lines of force. In other words, relative permeability expresses the magnetic conductance of a certain material compared to the conductance of air.

$$\text{relative permeability} = \frac{\text{absolute permeability}}{\text{permeability of air}} \tag{12}$$

$$\mu_r = \frac{\mu}{\mu_0} \tag{13}$$

Relative permeability has no unit. It is a ratio of two values.

Table 12-1 shows a list of different ferromagnetic materials and their relative permeability values. Table 12-1 states relative permeability values (μ_r). The B-H magnetizing curves (Fig. 12-14) provide absolute permeability values. There is a distinct difference between the two values, but they are closely related.

Table 12-1   Relative Permeability (μ_r) of Various Materials

μ_r Maximum		μ_r at Small Magnetization
Ferromagnetic		
Nickel	50	50
Cobalt	60	60
Cast Iron	90	60
Machine Steel	450	300
Transformer Iron	5500	3000
Silicon Iron	7000	3500
Very Pure Iron	8000	4000
Permalloy	100,000	25,000
Supermalloy	1000,000	100,000

12-6.8 Interpretation of B-H Curves

B-H magnetization curves are useful for deriving absolute permeability values. A study of *B-H* magnetization curves shows that the amount of flux density created in a magnetic material differs considerably, up to the saturation value, as the strength of the magnetizing force increases. Since flux density is equal to permeability times magnetizing force ($B = \mu H$), the different amounts of change in flux density, for equal changes in field intensity, must be due to a change in the permeability value of the magnetic material. It is the change in permeability value which explains the curvature in *B-H* magnetization curves. A study of the *B-H* magnetization curves in Fig. 12-14 shows the equal changes in the field intensity as ΔH_1, ΔH_2 and ΔH_3. Notice that the amounts of the associated changes in flux density, ΔB_1, ΔB_2 and ΔB_3 are each different in value, and that the greatest permeability, for transformer iron, for example, exists when the magnetizing force is smallest.

INCREMENTAL PERMEABILITY

The magnetic field intensity in some magnetic circuits varies between two limits such as those shown by ΔH_1, ΔH_2, or ΔH_3 in Fig. 12-14. This situation occurs whenever a pulsating direct current flows in a coil, as is the case in many audio transformers. The permeability value calculated for a limited range of magnetizing force and the resultant flux density is called incremental permeability.

Incremental permeability, therefore, is the value reached by dividing the flux density change by the field intensity change which caused the flux density change. Incremental permeability is called *differential permeability* when the variation between two field intensity limits is set by the peak negative voltage and peak positive voltage of an alternating current.

the symbol for incremental permeability is $\Delta \mu$

The following calculations show examples of incremental or differential permeability ($\Delta \mu$) values for transformer iron. The values for the calculations are obtained from the transformer iron curve on the graph in Fig. 12-14.

The incremental permeability ($\Delta \mu$) value for:
ΔB_1, ΔH_1 is equal to

$$\Delta \mu = \frac{\Delta B_1}{\Delta H_1} = \frac{0.43}{800} = 537.5 \times 10^{-6}$$

$\Delta \mu$ for ΔB_2, ΔH_2 is equal to

$$\Delta \mu = \frac{\Delta B_2}{\Delta H_2} = \frac{0.125}{800} = 156.3 \times 10^{-6}$$

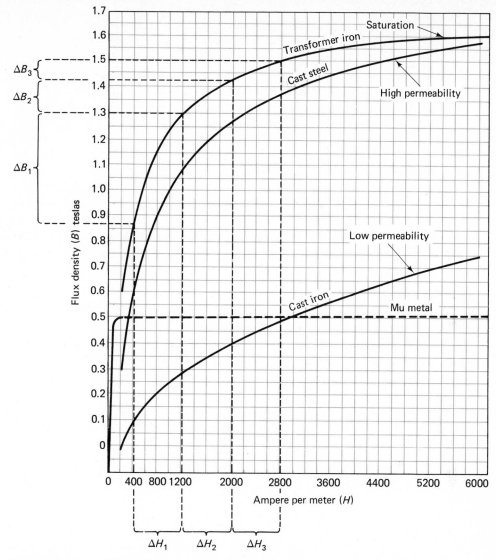

FIGURE 12-14   Magnetizing curves and their values for four different metals.

$\Delta\mu$ for ΔB_3, ΔH_3 is equal to

$$\Delta\mu = \frac{\Delta B_3}{\Delta H_3} = \frac{0.075}{800} = 93.8 \times 10^{-6}$$

Notice that the incremental permeability values decrease sharply for the transformer iron as the strength of the magnetizing force (H) increases. B-H magnetizing curves are useful in design work to determine:

1. The value of magnetizing force required to produce a flux density of a specific value in a magnetic material.

2. The value of flux density which will be produced when a particular magnetic material is subjected to a magnetizing force (H) of a specific strength.

3. The value of absolute permeability of a magnetic material for a particular magnetizing force (H) or flux density (B).

To determine the amount of current which must be supplied to a coil wound with a definite number of turns in order to produce a specific amount of flux, one must know permeability. Permeability in this situation is obtained by dividing a single value of flux density obtained from the *B-H* magnetizing curve (Fig. 12-14) by the value of the field intensity which caused that amount of flux density. The key to these calculations, however, is the conversion of the required number of flux lines into flux density ($B = \phi/A$), which involves the cross-sectional area of the magnetic path.

PROBLEM 12-4

Interpreting the graphs on Fig. 12-14. What value of field intensity is required to produce 1 T of flux density in

1. A piece of cast steel
2. Transformer iron

SOLUTION *(cover solution and solve)*

Field intensity for

1. Cast steel with $B = 1$ T.
   a. Refer to the cast steel curve on the graph in Fig. 12-14. Locate 1 T on the B axis.
   b. Run a horizontal line to the cast steel curve.
   c. The H value immediately below the intersection is 980 A/m of magnetizing force.
2. Transformer iron with B = 1 T.
   Repeat the procedure in 1. but use the transformer iron curve. The field intensity value is 500 A/m of magnetizing force.

PROBLEM 12-5

What value of field intensity (magnetizing force), H is required to produce 3×10^{-4} Wb of flux in a transformer iron core with a cross sectional area of 2×10^{-4} m^2? (See Fig. 12-15.)

SOLUTION *(cover solution and solve)*

$$\text{amount of flux } (\phi) = 3 \times 10^{-4} \text{ Wb}$$

$$\text{area} = 2 \times 10^{-4} \text{ m}^2$$

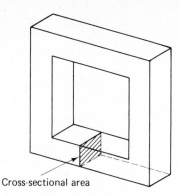

FIGURE 12-15 A core-type core or closed box core.

Cross-sectional area

$$\text{flux density} = \frac{\phi}{A} = \frac{3 \times 10^{-4}}{2 \times 10^{-4}} = 1.5 \text{ T}$$

Refer to the transformer-iron curve on the *B-H* magnetizing graph, Fig. 12-14 and draw a horizontal line to the right from the 1.5-T value until it intersects with the transformer-iron curve. A vertical line from the intersection down to the A/m value produces a magnetizing force of 2800 A/m.

PROBLEM 12-6

1. A field intensity of 2600 A/m is applied to a coil wound on one leg of a closed loop core. What flux density is produced in the core if the core is made of
   a. Transformer iron
   b. Cast steel

2. How much flux will flow in each core if the cross-sectional area of each core is 5×10^{-4} m²?

SOLUTION *(cover solution and solve)*

1. Flux Densities.
   a. Transformer iron.
      Refer to the *B-H* magnetizing curve in Fig. 12-14. Run a vertical line up to the transformer iron curve from the 2600 A/m value. The value to the left of the intersection is approximately 1.48 T of flux density.
   b. Cast steel.
      Run a vertical line from the 2600 A/m value to the cast steel curve. The value of flux density to the left of the intersection is 1.35 T.

2. Amount of flux flow in each core.
   a. Transformer iron.
      The flux density in the transformer iron is 1.48 T. The area of the core is 5×10^{-4} m². The amount of flux flowing in the

core is

$$\phi = BA$$

$$= 1.48 \times 5 \times 10^{-4} = 7.4 \times 10^{-4} \text{ Wb}$$

b. Cast steel.
     The flux density in the cast steel is 1.35 T. The area of the core is 5×10^{-4} m^2. The amount of flux developed in the core is

$$\phi = BA$$

$$= 1.35 \times (5 \times 10^{-4}) = 6.75 \times 10^{-4} \text{ Wb}$$

PROBLEM 12-7

What is the permeability value of transformer iron when it is subjected to a field intensity of 1800 A/m?

SOLUTION   (*cover solution and solve*)

1. Refer to the graph, Fig. 12-14.
2. Locate 1800 A/m on the H axis. The value of B in transformer iron with an 1800 A/m magnetizing force is 1.35 T.

$$\mu = \frac{B}{H} = \frac{1.35}{1800} = 7.5 \times 10^{-4} \text{ H/m}$$

PROBLEM 12-8

What is the permeability of transformer iron when it is subjected to a field intensity of 2800 A/m?

SOLUTION   (*cover solution and solve*)

Refer to the graph in Fig. 12-14 and repeat the procedure used in Problem 12-8.

$$\mu = \frac{B}{H} = \frac{1.5}{2800} = 5.36 \times 10^{-4} \text{ H/m}$$

A comparison of Problems 12-7 and 12-8 shows that there is no constant value of permeability. Rather, the permeability of a particular material decreases with an increase in the strength of the magnetizing force (H). Also observe that steep B-H magnetization curves are produced in magnetic materials with high permeability values and that low gradient curves indicate metals with low permeability values.

Air-core coils or coils with nonmagnetic cores produce linear (straight) B-H magnetization curves, as shown in Fig. 12-16. The linear curve indicates that any increase in field intensity will produce a proportionate increase in flux density (B); thus, any change in field intensity (H) will produce a proportionate amount of change in flux density (B). The

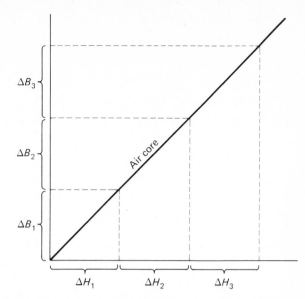

FIGURE 12-16 *B-H* magnetization curve for air. Note the equal changes in *B* for equal changes in *H*.

permeability of air or nonmagnetic material must therefore always be constant, as one divided by one in the equation $\mu = B/H$ will always produce a quotient of one.

Air is always used as a reference against which to compare the magnetic conductivity (permeability) value of other magnetic materials, because air has a constant permeability value (air has an absolute permeability value of $4\pi \times 10^{-7}$ or approximately 1.26×10^{-6}, and it has a low magnetic conductivity compared to most materials.

Relative permeability is equal to $\mu_r = \mu/\mu_0$ [Eq. (13)]. Relative permeability tables such as Table 12-1 usually show two values: maximum value, and small magnetization value. These are the two extreme permeability values, produced by large and small magnetizing forces respectively, in magnetic materials. A study of Table 11-1 shows that materials such as wood and paper with relative permeability values lower than that of free space (a vacuum) are called diamagnetic. Those materials such as aluminum which have a relative permeability value slightly greater than that of free space are paramagnetic. Materials such as steel, nickel, and iron have relative permeability values from 50 to 8000 times the permeability of air and are called ferromagnetic.

The relative permeability values of all paramagnetic and diamagnetic materials, which are approximately equal to free space, are considered to be one in all computations.

PERMEABILITY, RELATIVE PERMEABILITY, AND PERMEABILITY OF AIR

Because the permeability of air is a constant; $4\pi \times 10^{-7}$, the value of either absolute permeability (μ) or relative permeability (μ_r) may be determined if the other is known.

Relative permeability (μ_r) of electromagnetic circuits is usually more meaningful and is the most often-quoted value because it automatically describes how much the flux density of a magnetic field will increase if air is replaced by some specific magnetic material. The following problem shows the method of conversion.

PROBLEM 12-9

Find the relative permeability (μ_r) of the following metals. Use $\mu_0 = 4\pi \times 10^{-7}$.

1. Nickel $(\mu = 6.28 \times 10^{-5})$ H/m
2. Soft (transformer) iron $(\mu = 6.9 \times 10^{-3})$ H/m
3. Permalloy $(\mu = 5.6 \times 10^{-2})$ H/m

SOLUTION   *(cover solution and solve)*

1. Nickel

$$\mu_r = \frac{\mu}{\mu_0}$$

$$= \frac{6.28 \times 10^{-5}}{4\pi \times 10^{-7}}$$

$$= 49.8$$

2. Transformer iron

$$\mu_r = \frac{\mu}{\mu_0}$$

$$= \frac{6.9 \times 10^{-3}}{4\pi \times 10^{-7}}$$

$$= 5491$$

3. Permalloy

$$\mu_r = \frac{\mu}{\mu_0}$$

$$= \frac{5.6 \times 10^{-2}}{4\pi \times 10^{-7}}$$

$$= 44,563$$

12-6.9 Reluctance

Reluctance is the opposition which a magnetic material or magnetic circuit offers to the passage of magnetic lines of force. Reluctance is the reciprocal of permeability and permeance; therefore, reluctance is related to permeability and permeance in much the same way as resistance is related to conductance in electric circuits.

Reluctance is the reciprocal of permeance.

$$\text{reluctance} = \frac{1}{\text{permeance}} \tag{14}$$

and

$$R_m = \frac{1}{P} \quad \text{reciprocal of henries (H}^{-1}) \tag{15}$$

where P is henries (H)

R_m is H^{-1} (note the inversion to P)

RELUCTANCE AND PERMEABILITY

Reluctance is the reciprocal of permeance, but since permeance and permeability are directly related, reluctance is also related to permeability in the following manner.

$$R_m = \frac{1}{P}$$

$$\text{but} \quad P = \frac{\mu A}{l} \tag{10}$$

$$R_m = \frac{1}{\frac{\mu A}{l}} = \frac{l}{\mu A} \text{ (H/m)}$$

$$\text{reluctance} = \frac{\text{length}}{\text{permeability} \times \text{area}} \tag{16}$$

$$R_m = \frac{l}{\mu A} \quad (\text{H}^{-1}) \tag{17}$$

where l is length of magnetic circuit in meters

μ is absolute permeability

A is cross sectional area of the end of one pole in square meters

A study of Eq. (17) shows that reluctance, the opposition to magnetic lines of force, will increase if the length of the magnetic circuit increases, if the area of the end of one pole decreases, or if the permeability of the magnetic material decreases. The relationship between reluctance and permeability, then, includes the length and area of the magnetic circuit; therefore, equations relating relative permeability values and reluctance can only compare the two values under the conditions of unity length and area. Equation (18), for example, can be used only to demonstrate that the greater the relative permeability, the less the relative reluctance.

$$R_{m_r} = \frac{1}{\mu_r} \tag{18}$$

Ferromagnetic materials have a high relative permeability value. Therefore, they have a low relative reluctance value. The permeability of air or a vacuum is low; consequently, the reluctance of air is high.

the symbol for reluctance is R_m

the unit of reluctance is the reciprocal of henry (H^{-1})

Note that the unit of reluctance is the inverse of the unit of permeance.

PROBLEM 12-10

Find the reluctance of a circular bar of nickel ($\mu = 6.28 \times 10^{-5}$ H/m) having a radius of 0.0005 m and a length of 0.01 m.

SOLUTION   *(cover solution and solve)*

$$A = \pi r^2$$
$$= 3.14(5 \times 10^{-4})^2$$
$$= 78.5 \times 10^{-8} \text{ m}^2$$
$$= 7.85 \times 10^{-7} \text{ m}^2$$

$$R_m = \frac{l}{\mu A} \quad \text{(absolute value)}$$

$$= \frac{1 \times 10^{-2}}{(6.28 \times 10^{-5}) \times (7.85 \times 10^{-7})}$$

$$= \frac{10^{-2} \times 10^{12}}{49.298}$$

$$= \frac{10^{10}}{49.298}$$

$$R_m = 2 \times 10^8 \text{ H}^{-1}$$

PROBLEM 12-11

Suppose that the length of nickel bar is the same as in Problem 12-10 but the radius is twice as wide. What would be the reluctance now?

SOLUTION   *(cover solution and solve)*

$$A = \pi r^2$$
$$= 3.14 \times (0.10^{-3})^2$$
$$= (3.14 \times 10^{-6}) \text{ m}^2$$

$$R_m = \frac{l}{\mu A} \quad \text{(absolute value)}$$

$$= \frac{1 \times 10^{-2}}{(6.28 \times 10^{-5}) \times (3.14 \times 10^{-6})}$$

$$= \frac{10^{-2} \times 10^{11}}{19.719}$$

$$= \frac{10^{9}}{19.719}$$

$$= \frac{100 \times 10^{7}}{19.719}$$

$$= 5.07 \times 10^{7} \text{ H}^{-1}$$

$$= 0.507 \times 10^{8} \text{ H}^{-1}$$

Doubling the radius of the rod decreased the reluctance by 1/4 (1/4 of 2×10^{8} = 0.507 $\times 10^{8}$). Reluctance increases directly as the length of the core material increases.

12-6.10 Ohm's Law for Magnetic Circuits

Three magnetic units which have already been discussed are magnetomotive force (F_m), flux (ϕ), and reluctance (R_m). The relationship between them is known as Ohm's Law for magnetic circuits.

$$\textbf{flux} = \frac{\textbf{magnetomotive force}}{\textbf{reluctance}} \tag{19}$$

$$\phi = \frac{F_m}{R_m} \quad \text{(Wb)} \tag{20}$$

where ϕ is number of flux lines in webers
F_m is magnetomotive force (A)
R_m is reluctance (H^{-1})

This relationship is analogous to Ohm's Law, $I = E/R$. If two of the values are known, the third can be calculated.

F_m (*IN*) is comparable to voltage in that it is the pressure which produces lines of force.

A *flux* line in a magnetic circuit is the result of magnetomotive force and is analogous to current.

Reluctance, the inverse of permeability, is the opposition which a material offers to magnetic flux. It is analogous to resistance in an electric circuit.

Ohm's Law for magnetic circuits does not apply to air-core coils because air has a relative permeability of 1, and because reluctance is the inverse of permeability, it too would have a value of only 1. Coils seldom if ever have air for their cores; a high permeability material offers greater efficiency and field strength. The discussion of magnetism from now on will always concern iron cores.

PROBLEM 12-12

What will be the number of flux lines produced at the end of an iron-core coil, if the core material has a reluctance of 4.8×10^3 H^{-1}, and a magnetizing force of 10 A is applied to the coil?

SOLUTION *(cover solution and solve)*

$$\phi = \frac{F_m}{R_m}$$

$$= \frac{10}{4.8 \times 10^3}$$

$$\phi = 2.08 \times 10^{-3} \text{ Wb}$$

$$= 2.08 \text{ mWb}$$

PROBLEM 12-13

What is the reluctance of a magnetic circuit in which a total flux of 1.8×10^{-3} Wb is set up by a 5-A current flowing in a solenoid having 280 turns of wire?

SOLUTION *(cover solution and solve)*

$$\phi = 1.8 \times 10^{-3}$$

$$F_m = 5 \times 280 = 1400 \text{ AT}$$

$$R_m = \frac{F_m}{\phi} = \frac{1400}{1.8 \times 10^{-3}}$$

$$= 778 \times 10^3 \text{ H}^{-1}$$

Now that several necessary magnetic measurements are known, some mathematical relationships can be established among them. For example, the relationship between permeability (μ) and field intensity (H) is

$$\mu = \frac{B}{H}$$

12-6.11 Mathematical Proof

Prove that $\qquad\qquad \mu = \dfrac{B}{H}$

$$F_m = \phi \times R_m \qquad\qquad\qquad (21)$$

$$\text{but} \quad H = \frac{F_m}{l} \tag{22}$$

Substitute Eq. (21) in Eq. (22).

$$H = \frac{\phi \times R_m}{l} \tag{23}$$

$$\text{from} \quad B = \frac{\phi}{A} \tag{24}$$

$$\text{then} \quad \phi = B \times A \tag{25}$$

Substitute Eq. (25) in Eq. (23).

$$H = \frac{BA \times R_m}{l} \tag{26}$$

$$\text{but} \quad R_m = \frac{l}{\mu A} \tag{27}$$

Substitute Eq. (27) in Eq. (26).

$$H = \frac{BA \left(\dfrac{l}{\mu A} \right)}{l} \tag{28}$$

$$= B\!\!\!/A \, \frac{\not{l}}{\mu\!\!\!/A} \times \frac{1}{\not{l}} \tag{29}$$

$$H = \frac{B}{\mu} \tag{30}$$

$$\text{therefore} \quad \mu = \frac{B}{H} \tag{31}$$

12-7 Power Loss in Ferromagnetic Materials

There are three major losses which occur in ferromagnetic materials, or in particular in the iron cores of transformers. These losses are:

1. Hysteresis loss
2. Eddy current loss
3. Leakage flux loss

12-7.1 Hysteresis

Hysteresis is one of the many power losses which can occur in iron core inductances of all types when ac currents flow in them, especially in transformers.

The word hysteresis means to lag behind. Hysteresis in magnetic circuits is the lagging of the magnetic field behind the magnetizing force which creates it. The additional magnetizing force required to overcome molecular friction to bring the magnetic field into step with the magnetizing force produces heat which is dissipated and lost in space. This loss is known as *hysteresis loss*.

A plot on graph paper of the magnitude of the magnetizing force (H) and the associated polarity and magnitude of the resulting magnetism is known as the *hysteresis loop or curve*.

DEVELOPING THE HYSTERESIS CURVE

The magnetizing curve as discussed in Section 12-6.5 was produced by applying current, in increasing amounts, to a coil wound on a magnetic core. Each time the current was increased, the resulting magnetic flux was measured with a fluxometer and the H and B values were recorded. This process was continued until full magnetic saturation was reached. A curve was then plotted on graph paper from the recorded data. The curve showed the relationship between the flux (B) and the magnetizing force (H) (Fig. 12-17).

The plotting of a hysteresis loop (curve) is merely a continuation of plotting the magnetizing curve. The magnetizing current is reversed, then brought to zero until the full cycle of current application has been completed and the magnetizing force is back to its original value.

The following description outlines the procedure which could be used to produce a *hysteresis loop*:

1. (Switch S_3 is in position P.)
   Figure 12-18 is a schematic drawing of a very simple method which may be used to measure hysteresis of a core material. Figure 12-18 is similar to the experiment set up to plot the magnetizing curve of a magnetic material, except that the current through the coil may be reversed by S_3. Reversing the current produces a flux of the reverse polarity ($-B$).

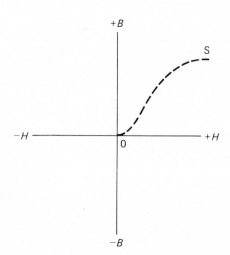

FIGURE 12-17   The dotted curve line is a *B-H* curve similar to that developed in Section 12-6.5. The curve is produced in the first step of the hysteresis test measurement.

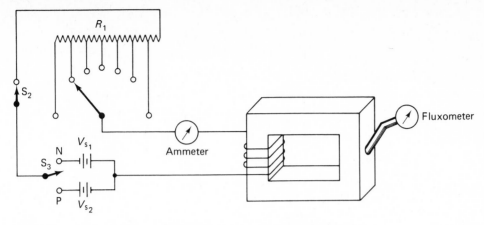

FIGURE 12-18   A crude test/experiment for measuring hysteresis.

2. Fig. 12-19 (Switch S₂ is opened.)
   If the switch S_2 between the coil and the voltage source in Fig. 12-18 were opened when saturation (+S) was reached, the magnetizing force would drop to zero (point R in Fig. 12-19), but the magnetic flux (ϕ) in the core would still be large because the retentivity of the core would hold a certain amount of residual magnetism (R to 0, Fig. 12-19). The amount of residual magnetism remaining will depend upon the type of metal in the core. A magnetic material which retains a large amount of residual magnetism is said to have good retentivity. *Retentivity* is that property of magnetic materials which enables that material to hold magnetism.

3. Fig. 12-20 (Switch S₃ is placed in position N.)
   In order to reduce the residual magnetism to zero, the current would have to be reapplied in the opposite direction. In other

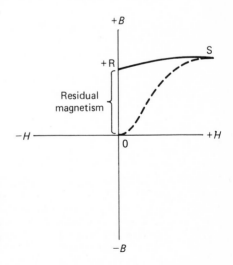

FIGURE 12-19   The curve formed from point +S to point R is the first portion of the overall hysteresis curve. Note the residual magnetism that is the cause of hysteresis.

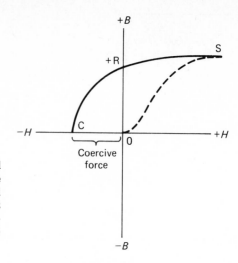

FIGURE 12-20 The curve formed from point +R to point C is the second portion of the overall hysteresis curve. Note the amount of magnetizing force, called *coercive force*, required to reduce the residual magnetism to zero.

words, switch S_3 would have to be changed from P to N. This would result in a reversal of the polarity of the magnetizing force from (+H) to (−H) and a demagnetization of the residual magnetism in the iron core. The amount of demagnetizing force (0 to C, Fig. 12-20) required to bring the residual magnetism to zero is called the *coercive force*.

4. Fig. 12-21 (The current is increased.)
   If the current in the coil were allowed to flow even after the core had been demagnetized, the iron would again become magnetized but in the (−B) polarity. If the current was increased in value until the iron was saturated (−S), the curve from C to −S would be produced (Fig. 12-21).

5. Fig. 12-22 (Switch S_2 is opened.)
   Opening the switch S_2 after the saturation point (−S) had been reached would cause the magnetizing force to collapse to zero

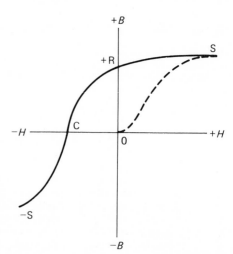

FIGURE 12-21 Note the C to −S portion of the hysteresis curve, which is similar to the *B-H* magnetizing curve to saturation but in the opposite magnetic polarity (−S) to that in Fig. 12-19.

248

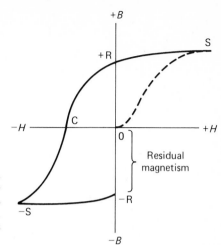

FIGURE 12-22   The curve formed from point −S to point −R is the fourth portion of the hysteresis curve, resulting from the residual magnetism from point 0 to point −R.

(Fig. 12-22), but the residual magnetism (0 to −R) would remain and the −S to −R portion of the hysteresis curve would be formed.

6. Fig. 12-23 (Switch S_3 is placed in position P.)
    The direction of the magnetizing force current would now have to be reversed to (+H) by moving switch S_3 (Fig. 12-18) from N back to P in order to remove the residual magnetism. The magnetizing force would increase as the current was increased until the amount of coercive force (F to 0, Fig. 12-23) brought the residual magnetism in the iron to a zero value.

7. Fig. 12-24 (The current is increased.)
    Applying the magnetizing force (current) until and then after the core is demagnetized, and then increasing the amount of current flow in the coil, would again bring the flux in the coil in the +B direction and finally into + saturation (+S). Curve F to +S is now formed, completing the curve envelope (Fig. 12-24).

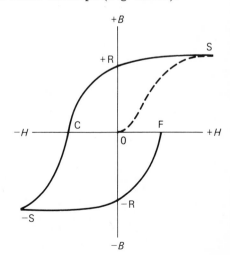

FIGURE 12-23   The   coercive force 0 to F (magnetizing force) bringing the residual magnetism to zero.

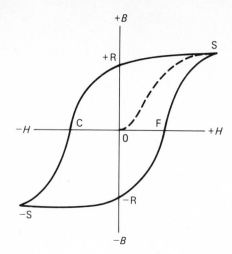

FIGURE 12-24 The magnetizing force is continued until the saturation point (+S) is reached once again as it was in Fig. 12-19. The resulting *B-H* curve formed from point F to point +S completes the hysteresis loop.

The curve enclosed by points +S, +R, C, –S, –R, and F in Fig. 12-24 forms what is called a hysteresis loop or double-valued curve (the magnetizing curve is a single-valued curve). The dotted line is the initial *B-H* curve required to bring the magnetic material to the saturation point (+S) so that the hysteresis test may begin.

SUMMARY:

The word hysteresis means to lag behind. The hysteresis curve or loop comes about because the retentivity of the material causes the magnetic flux to lag behind the magnetizing force in value and polarity. The retentivity causes the following phenomena:

1. Residual magnetism (flux), represented by 0 to R and 0 to –R on the hysteresis loop, remains after the magnetizing force has been reduced to zero.

2. The amount of coercive force, represented by 0 to C and 0 to F (the amount of magnetizing force to bring the residual magnetism to zero), is quite large.

HYSTERESIS LOSS

The area inside the loop represents the amount of work or power used to change the value and polarity of the magnetism in the material. The larger the area enclosed by the hysteresis loop, the greater the energy loss. The smaller the loop, the less the loss and the more efficient the core material. Losses are greatest for steel and hard iron and progressively less for soft iron, permalloy, and superalloy. These power losses can be very high in high-frequency circuits such as are used in electronics. The power losses manifest themselves in the form of heat and are brought about by the energy required to reverse the polarity of the magnetic domains. The heat energy is wastefully dissipated into the surrounding air. Power loss in transformers used to supply electronic equipment would be a serious problem if good quality soft iron or higher permeability metals were not used.

Power loss due to hysteresis losses can be reduced only by using core material of the highest permeability permitted by cost. The losses, however, are insignificant when the magnetizing force changes slowly, but this seldom occurs even in the field of electrical power.

Materials such as Alnico 5, which is used for magnets, have very wide and square hysteresis loops because of their high retentivity characteristics (Fig. 12-25), but power losses are of no concern in permanent magnets because the magnetic field is never changed in polarity.

Different materials produce different shapes of hysteresis loops. Figure 12-26 shows a curve for each of three different materials.

The soft iron curve is typical of high permeability materials which may be used in most electromagnetic applications.

The curve for soft iron shows that soft iron produces the least amount of power loss and is typical of very high quality transformer iron.

Materials with the lowest permeability and highest retentivity have the highest hysteresis losses.

The curve in Fig. 12-27 is the typical shape for ferrite iron required for computer memory cores (used to store data). The curve shows strong residual magnetism after the magnetizing current has been removed, and the almost vertical line indicates a nearly instant reversal of magnetic

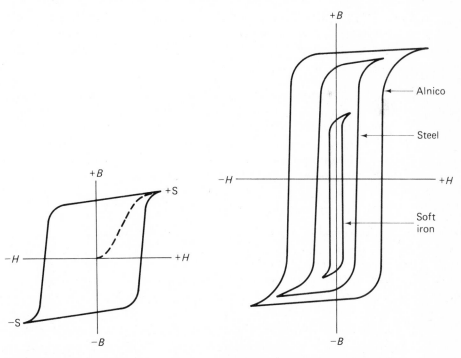

FIGURE 12-25 Hysteresis loop for Alnico-5.

FIGURE 12-26 Comparative hysteresis loops for three different metals.

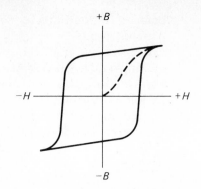

FIGURE 12-27 A rectangular-shaped hysteresis loop for ferrite iron used for computer memory cores.

flux (ϕ) polarity with the application of a reversed current. The rapid change from $+B$ to $-B$ is analogous to a switching action at high speeds; a switching property required of computers.

12-7.2 Eddy Current Loss

When an alternating current flows in a coil to which an ac voltage is applied, the resulting magnetic field changes in polarity in accordance with the changing polarity of the ac current. This changing magnetic field induces a voltage in the iron core of a coil. The induced voltage is larger near the outside surface of the core than in the center of the core. The difference in potential between the outside surface and the center of the coil causes currents to flow in the iron core in a circular path like that of a whirlpool or eddy. These currents are called eddy currents and they travel through the cross section of the coil at right angles to the length of the coil (see Fig. 12-28).

Eddy currents cause a waste in power because they produce I^2R losses in the iron core which are dissipated as heat from the core. The higher the frequency of the applied voltage, the greater the eddy currents (because of rate of change of the magnetic field) and the greater the power loss due to those eddy currents.

The question may be asked, why do eddy current losses occur in iron-

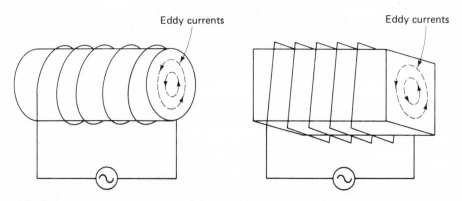

FIGURE 12-28 Iron-core coils showing eddy currents.

core coils but not in air-core coils? The answer lies in two electrical phenomena:

1. Induced currents
2. dc resistance of iron

Energy is drawn from the source voltage to produce the magnetic field in an air-core coil. This energy is stored in the magnetic field of the coil (a hypothetically pure inductor) until the circuit current and magnetic field collapse. The collapsing circuit current causes the magnetic flux to collapse and all the stored energy in the magnetic field to return to the voltage source. On the other hand, when a voltage source is connected to an iron-core coil circuit, the expanding magnetic field induces a voltage and current in the iron core. The induced current flowing in the iron core is not stored energy and therefore is not returned to the voltage source when the circuit current collapses. It flows through the dc resistance (ohmic) of the iron and produces I^2R losses. This is a heat energy loss, whose energy originated in the voltage source.

Eddy current losses are made negligible while high flux density is maintained, by one of two systems.:

1. Laminated-iron core
2. Powdered-iron core

LAMINATED-IRON CORE

Eddy currents can be reduced by replacing a solid iron core by one which is made up of slices of iron (cut parallel to the length of the coil) called laminations (see Fig. 12-29). These laminated sheets of iron are insulated from each other by a very thin coating of iron oxide and varnish. The varnish increases the resistance of the cross-sectional area of the core, and the length of any one eddy-current path is limited to the thickness of each lamination.

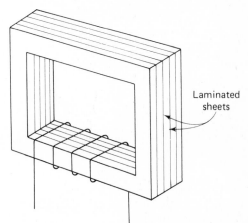

Laminated sheets

FIGURE 12-29 Iron-core laminated sheets.

Laminations are used in transformers operating in the range of audio frequencies. As frequency increases, flux movement increases and eddy current loss increases. Powdered-iron cores are used for higher-frequency applications such as radio frequencies (above 20 kHz).

Powdered-iron cores consist of small particles of finely ground soft iron which are held together by a high-resistance glue. This decreases the eddy current values.

12-7.3 Leakage Flux

Magnetic leakage takes place between the legs or sides of horseshoe- or rectangular-shaped magnetic cores, because magnetic lines of force try to become as short as possible. Air has permeability, as small as it may be, and flux lines attempt to cross the air gap between legs of the magnetic core, pulling away from the ends of the magnetic pole pieces. This diversion or leakage around the pole piece ends reduces the flux density at the ends of the poles. Leakage is thus a complete waste of magnetic field strength (Fig. 12-30).

Leakage flux is usually small enough to be neglected. It is negligible because the magnetic circuit has a much smaller reluctance than the air path between the legs of the magnet. Leakage flux can be kept to a minimum by

1. Increasing the permeance of the magnetic circuit path.
2. Keeping the length of the magnetic circuit path shorter than the air path between the sides of the magnet.

12-8 Magnetic Circuits

Magnetic circuits and electric circuits have many similarities and equivalents. See Table 12-2.

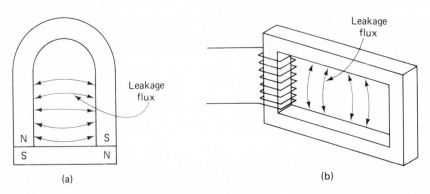

(a) (b)

FIGURE 12-30 Leakage flux in magnetic circuits.

Table 12-2 Comparison of Electric and Magnetic Circuits

Electric Circuits	Magnetic Circuits
Conductance	Permeance or Permeability
Resistance	Reluctance (R_m)
Resistances add in series. $$R_T = R_1 + R_2 + \cdots + R_n$$ (Fig. 12-31b)	Reluctances add in series. $$R_{m_T} = R_{m_1} + R_{m_2} + \cdots + R_{m_n}$$ (Fig. 12-32b)
Voltage	Magnetomotive force (F_m)
Kirchhoff's Voltage Law	Ampere's Circuital Law
Current	Flux ϕ
The sum of the currents arriving at a point is equal to the sum of the currents leaving that point.	The sum of the magnetic flux lines arriving at a point is equal to the sum of the flux lines leaving that point.
Ohm's Current Law $$I = \frac{V}{R}$$	Ohm's Magnetic Law $$\phi = \frac{F_m}{R_m}$$
Electric circuits may be series, parallel or series–parallel	Magnetic circuits may be series, parallel or series–parallel
$$V_T = V_1 + V_2 + V_3$$	$$F_m = F_{m_1} + F_{m_2} + F_{m_3}$$

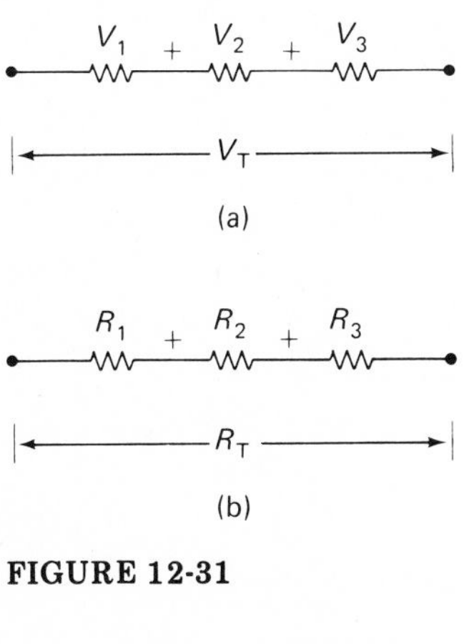

FIGURE 12-31

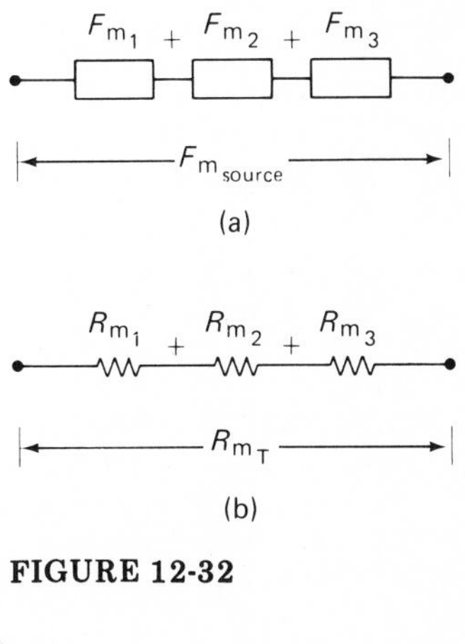

FIGURE 12-32

Most manufactured transformers have a closed-box or square construction as shown in Fig. 12-33. The construction is made of several pieces of laminated iron which are assembled and bolted together after the coil has been put in place. The upper section is constructed in the form of an upside down U while the lower section is a straight I.

A line of magnetic flux follows the loop indicated by the dashed line, passing first through one section and then the other. Thus, the sections of the magnetic circuit are in series.

The *first characteristic* of a series magnetic circuit is that the total reluctance of the series sections is equal to the sum of the individual reluctances.

$$R_{m_T} = R_{m_1} + R_{m_2} + R_{m_3} + \cdots + R_{m_n}$$

The law that the total reluctance of a series circuit is equal to the sum of individual reluctances is very important, because the same material need not be, and often is not, used in all sections of a magnetic circuit. As an example, one section may be of cast steel and another section may be sheet steel. The magnetic characteristics of the two types of steel are much different.

The *second characteristic* of a series magnetic circuit is Ampere's Circuital Law: The total magnetomotive force in a series circuit is equal to the sum of the individual magnetomotive force sources. The series magnetic circuit is similar to the series electric circuit; there is only one path by which flux lines may travel. The rules of series magnetic circuits, therefore, resemble the rules of series electric circuits. This law is similar to Kirchhoff's Voltage Law.

Ampere's Circuital Law states that the sum of the magnetomotive force sources is equal to the sum of the magnetomotive force drops. The magnetomotive drop is equal to field intensity (H) times the length of the magnetic path.

from Eq. (5) $H = \dfrac{F_m}{l}$

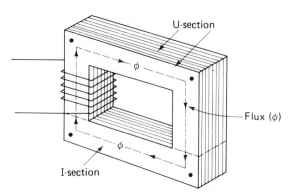

FIGURE 12-33   A core-type core or closed box core. Note the laminated core.

$$\text{magnetomotive force} = \text{field intensity} \times \text{length}$$

$$F_m = Hl \qquad\qquad (32)$$

where F_m is magnetomotive force in A

H is field intensity in A/m

l is length of magnetic path in meters

12-8.2 Average Magnetic Path Length and Areas

The calculation for reluctance, $R_m = l/\mu A$ involves the average length, l of the magnetic path and the cross-sectional area, A of the same magnetic path. Some magnetic circuits are of the square construction and others are of the toroid or round-rod construction. Reluctance calculations for all of these magnetic circuits are the same except that the calculations for the average magnetic path length and the area are different. The following example calculations outline a standard method of calculating average lengths and areas for

1. Rectangular or square circuits
2. Round-rod circuits
3. Toroid circuits

PROBLEM 12-14

RECTANGULAR CLOSED CORE

There are different methods of calculating the average length of the magnetic path in a rectangular closed core. One method is that used in Problem 12-14. The procedure is to first divide the core into a U-section and an I-section. The next step is to calculate the average length of each of the two sections. The final step is to add the two average lengths together.

Calculate

1. The average length of the magnetic series circuit path of the rectangular core in Fig. 12-34
2. The cross-sectional area of the core

SOLUTION *(cover solution and solve)*

1. Average length of magnetic path in U-section

outside length of the U-section = 3 + 3.5 + 3 = 9.5 cm

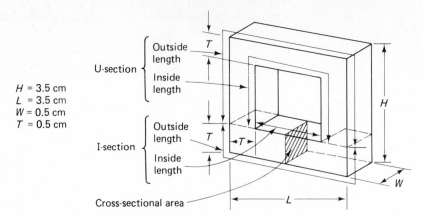

FIGURE 12-34 A rectangular (closed box) core.

$$H = 3.5 \text{ cm}$$
$$L = 3.5 \text{ cm}$$
$$W = 0.5 \text{ cm}$$
$$T = 0.5 \text{ cm}$$

$$\text{inside length of U-section} = 2.5 + 2.5 + 2.5 = 7.5 \text{ cm}$$

$$\text{average length} = \frac{9.5 + 7.5}{2} = 8.5 \text{ cm}$$

Average length of magnetic path in I-section

$$\text{outside length} = 3.5 + 0.5 + 0.5 = 4.5 \text{ cm}$$

$$\text{inside length} = 2.5 \text{ cm}$$

$$\text{average length} = \frac{2.5 + 4.5}{2} = 3.5 \text{ cm}$$

Total average length of magnetic path in the core.

$$l_T = l_U + l_I$$
$$= 8.5 + 3.5$$
$$= 12 \text{ cm}$$

2. Cross-sectional area

$$A = T \times W \text{ (see Fig. 12-34)}$$
$$= (0.5 \times 10^{-2}) \times (0.5 \times 10^{-2}) = 0.25 \text{ cm}^2$$
$$= 0.25 \times 10^{-4} \text{ m}^2$$

ROUND-ROD

1. Average length of magnetic path = length of rod.
2. Cross-sectional area (A) of round rod. (See Fig. 12-35.)

$$A = \pi r^2 = \pi \left(\frac{\text{diameter}}{2}\right)^2$$

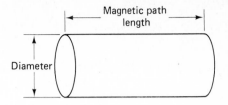

FIGURE 12-35 A round rod core.

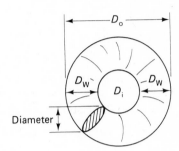

FIGURE 12-36 A toroid core.

TOROID

1. Average length of a toroid coil.

   The average length of a toroid coil and core is its average circumference. (See Fig. 12-36.)

   $$l_{avge} = \text{circumference}_{avge}$$

   $$= \frac{\text{inner circumference} + \text{outer circumference}}{2} \qquad (33)$$

   $$= \frac{\pi D_i + \pi D_o}{2}$$

   $$= \pi \frac{(D_i + D_o)}{2}$$

2. Cross-sectional area of toroid core (in meters).

   $$A = \pi r^2$$

   $$= \pi \left(\frac{D_w}{2}\right)^2 \quad \text{m}^2 \qquad (34)$$

   where D_w is the thickness of the core in meters

   $$D_w \text{ is } \frac{D_o - D_i}{2} \qquad (35)$$

   D_w is derived in the following manner

   $$D_o = D_w + D_i + D_w \quad \text{(see Fig. 12-36)}$$

$$D_o = D_i + 2D_w$$

$$\therefore\ 2D_w = D_o - D_i$$

$$\text{and}\ \ D_w = \frac{D_o - D_i}{2}$$

$$\text{area}\ (A) = \pi \left(\frac{D_o - D_i}{4}\right)^2$$

PROBLEM 12-15

Refer to Fig. 12-36. A toroid core has an inner diameter (D_i) of 4 cm and an outside diameter (D_o) of 6 cm.

1. What is its average length in meters?
2. What is the circular cross-sectional area?

SOLUTION   *(cover solution and solve)*

1.
$$l_{avge} = \frac{\text{inner circum}\,(C_1) + \text{outer circum}\,(C_o)}{2}$$

$$\text{where}\ \ C_1 = \pi D_i$$
$$= 3.14 \times (4 \times 10^{-2})$$
$$= 12.57 \times 10^{-2}\ m$$

$$\text{and}\ \ C_o = \pi D_o$$
$$= 3.14 \times (6 \times 10^{-2})$$
$$= 18.85 \times 10^{-2}\ m$$

$$l_{avge} = \frac{(12.57 \times 10^{-2}) + (18.85 \times 10^{-2})}{2}$$

$$= \frac{31.42 \times 10^{-2}}{2}\ m$$

$$= 15.71 \times 10^{-2}\ m$$

2. Circular cross-sectional area

$$A = \pi \left(\frac{D_w}{2}\right)^2$$

$$\text{but}\ \ D_w = \frac{D_o - D_i}{2}$$

$$= \frac{(6 \times 10^{-2}) - (4 \times 10^{-2})}{2}$$

$$= 1 \times 10^{-2}\ m$$

$$A = 3.14 \left(\frac{1 \times 10^{-2}}{2} \right)^2$$

$$= 3.14 \, (0.05 \times 10^{-2})^2$$

$$= 3.14 \times (0.25 \times 10^{-4})$$

$$= 0.785 \times 10^{-4} \; \text{m}^2$$

PROBLEM 12-16

Determine the total reluctance for the series magnetic circuit with two types of laminated steel, as shown in Fig. 12-37. The steel in the U-section has an absolute permeability value of 9.6×10^{-4} H/m. The I-section has a permeability of 1.48×10^{-3} H/m. The thickness of the varnish accounts for 5% of the thickness dimensions.

SOLUTION   *(cover solution and solve)*

1. Reluctance of U-shaped section.
    - a. Find the average length of the magnetic path.

        length of outside path = 3 + 3.5 + 3 = 9.5 cm

        length of inside path = 2.5 + 2.5 + 2.5 = 7.5 cm

        average length of magnetic path = $l_{\text{avge}} = \dfrac{9.5 + 7.5}{2} = 8.5$ cm

    - b. Find the cross-sectional area of the U-shaped section.

        Area $(A) = (0.5 \times 10^{-2}) \times (0.5 \times 10^{-2}) = 0.25 \; \text{cm}^2$

        $= 0.25 \times 10^{-4} \; \text{m}^2$

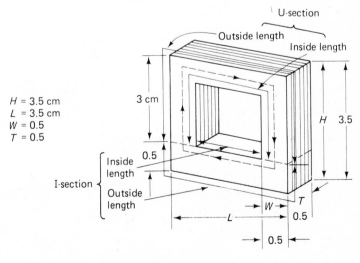

FIGURE 12-37   A rectangular (closed box) core.

The area must be reduced by 5%, because varnish accounts for 5% of the thickness dimensions.

$$\text{actual area } (A) = (0.25 \times 10^{-4}) \times 0.95$$

$$= 0.2375 \times 10^{-4} \text{ m}^2$$

c.   Find the reluctance of the U-section.

$$R_{m_U} = \frac{l}{\mu A}$$

$$= \frac{8.5 \times 10^{-2}}{(9.6 \times 10^{-4})(0.2375 \times 10^{-4})}$$

$$= 372.8 \times 10^4$$

$$= 3.728 \times 10^6 \text{ H}^{-1}$$

2.   Reluctance of the I-section.
  a.   Find the average length of the magnetic path in the I-section.

$$\text{outside length} = 3.5 + 0.5 + 0.5 = 4.5$$

$$\text{inside length} = 2.5$$

$$l_{\text{avge}} = \frac{4.5 + 2.5}{2} = 3.5 \text{ cm}$$

  b.   Find the cross-sectional area of the I-section.

$$\text{area } (A) = 0.5 \times 0.5 = 0.25 \text{ cm}^2$$

$$= 0.25 \times 10^{-4} \text{ m}^2$$

Varnish accounts for 5% of the thickness. Therefore, actual steel area is 95% of total area.

$$A = (0.25 \times 10^{-4}) \times 0.95$$

$$= 0.2375 \times 10^{-4} \text{ m}^2$$

  c.   Find the reluctance of the I section.

$$R_{m_I} = \frac{l}{\mu A}$$

$$= \frac{3.5 \times 10^{-2}}{1.48 \times 10^{-3} \times 0.2375 \times 10^{-4}}$$

$$= 9.957 \times 10^5$$

$$= 0.9957 \times 10^6 \text{ H}^{-1}$$

3.   Total reluctance of the circuit, R_{m_T}.

$$R_{m_T} = R_{m_U} + R_{m_I}$$

$$= (3.728 \times 10^6) + (0.9957 \times 10^6)$$

$$= 4.7237 \times 10^6 \text{ H}^{-1}$$

PROBLEM 12-17

Assume the following values exist for Fig. 12-37:

$$R_{m_U} = 3.728 \times 10^6 \text{ H}^{-1}$$

$$R_{m_I} = 0.9957 \times 10^6 \text{ H}^{-1}$$

$$R_{m_T} = 4.7237 \times 10^6 \text{ H}^{-1} \text{ (T = total)}$$

The total average length of the magnetic path is (8.5 + 3.5) = 12 cm. The total flux required to be generated in the laminated steel is 800 μWb.
Calculate:

1. The total magnetomotive force required to produce 800 μWb of flux.
2. The magnetomotive drop in the U-section.
3. The magnetomotive drop in the I-section.
4. The value of source magnetomotive force.
5. The amount of field intensity to produce the magnetomotive force.

SOLUTION   (cover solution and solve)

1. F_m required.
    Ohm's Law for magnetism states

$$F_m = \phi \times R_{m_T}$$

    If $\phi = 800 \ \mu\text{Wb}$

    and $R_m = 4.7237 \times 10^6 \text{ H}^{-1}$

    then $F_m = (800 \times 10^{-6}) \times (4.7237 \times 10^6)$

$$= 3778.96 \text{ A}$$

2. The number of flux lines is the same in every part of a series circuit. Therefore the magnetomotive drop in each part of the circuit will be $F_m = \phi \times R_m$.

$$F_{m_U} = \phi \times R_m$$

    If $\phi = 800 \ \mu\text{Wb}$

    and $R_m = 3.728 \times 10^6 \text{ H}^{-1}$

$$F_{m_U} = (800 \times 10^{-6}) \times (3.728 \times 10^6)$$

$$= 2982.4 \text{ A}$$

3. Magnetomotive force in the I section.

$$F_{m_I} = \phi \times R_m$$

    If $\phi = 800 \ \mu\text{Wb}$

$$\text{and} \quad R_{m_I} = 0.9957 \times 10^6 \ \text{H}^{-1}$$

$$F_{m_I} = (800 \times 10^{-6}) \times (0.9957 \times 10^6)$$

$$= 796.56 \ \text{A}$$

4. The sum of the magnetomotive force drops equals the source magnetomotive force, F_{m_s} (s = sum).

$$F_{m_s} = F_{m_U} + F_{m_I}$$

$$= 2982.4 + 796.56$$

$$F_{m_s} = 3778.96 \ \text{A}$$

5. The field intensity required to produce a magnetomotive force of 3778.96 A is equal to

$$\text{field intensity} \ (H) = \frac{\text{magnetomotive force} \ (F_m)}{\text{length} \ (l_T)}$$

$$H = \frac{F_{m_s}}{l}$$

$$\text{but} \quad F_{m_s} = 3778.96$$

$$\text{and} \quad l_T = 8.5 + 3.5 = 12 \ \text{cm}$$

$$= 12 \times 10^{-2} \ \text{m}$$

$$\therefore \ H = \frac{3778.96}{12 \times 10^{-2}}$$

$$= 314.9 \times 10^2 \ \text{A/m}$$

12-8.3 Air Gap

The air space between the poles of a magnet is its air gap. Air increases the total reluctance of a magnetic circuit because air has high reluctance. The air gap between the poles of a magnet may be filled with a nonmagnetic material such as glass, brass, or fiberboard, but these materials do not change the effects or characteristics of the air gap. The air gap is always in series with the total magnetic circuit since all the magnetic flux must pass through the air gap in order to pass from one pole to the other.

Magnetic lines of force tend to repel one another, and therefore they spread or bulge out in a nonmagnetic area such as in an air gap. This spreading or bulging out of flux lines is called *fringing* (Fig. 12-38). Fringing increases the area which the magnetic flux exits and enters; it thus decreases the flux density of the magnetic field of the poles.

The shorter the air gap between poles, the stronger the field in the gap for a given pole strength. A decrease in gap length decreases the total reluctance to the magnetic flux ($R_m = l/\mu A$). The purpose of a short air gap in a magnet, for example in a tape recorder head, is to concentrate the

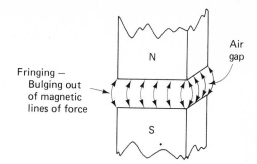

Fringing —
Bulging out
of magnetic
lines of force

N

Air
gap

S

FIGURE 12-38   The phenomenon
of fringing.

magnetic field so as to produce maximum induction in the magnetic
tape which passes over the air gap. The air gap is filled with a nonmagnetic insert and the complete unit is then buffed to a polished surface for
a smooth passage of the plastic tape which passes over it.

The effect of fringing is insignificant when the air gap is very short.
Correction for fringing in magnetic circuit calculations, in situations where
the air gap is long and the cross-sectional area of the poles is small, is made
by adding an arbitrary value of up to 10% of the cross-sectional area to
the initial cross-sectional area calculation.

An air gap is present inside the E-shaped permanent magnet in which
the voice coil of a speaker operates, between the field pole pieces and the
armature of motors and generators, and between the pole piece and the
bobbin of meter movements, so that the armature and bobbin may have
space to rotate.

12-8.4   Air Gaps and Magnetic Circuits

Air gaps interrupt the high permeability path provided by iron and increase the total reluctance of the magnetic circuit. The air gap is in
series with the iron magnetic path since all flux lines must pass through the
air gap to complete the magnetic loop. The increased reluctance of the
effectively increased cross-sectional area created by the air gap must be
added to the other reluctances in the magnetic circuit in exactly the same
manner as the reluctance value of different types of metals is added. The
magnetic circuit in Problems 12-16 and 12-17 had two different metals in
its path. An air gap in the path of the same circuit would be treated in
the same manner as a third different metal. Nonmagnetic materials such
as brass are considered the same as air gaps.

PROBLEM 12-18

Calculate the total reluctance of the magnetic circuit in Problem 12-16
if two pieces of brass, each 2 mm thick, are inserted between each leg
of the U-section and the I-section. Assume that varnish accounts for
5% of the thickness of the core. Add 2.5% to the initial cross-sectional
area for fringing.

SOLUTION   *(cover solution and solve)*

1. The total reluctance of the U-section and I-section will be the same as already calculated in Problem 12-16. The value is equal to

$$R_{m_U} + R_{m_I} = 4.7237 \times 10^6 \ H^{-1}$$

2. Reluctance of the air gap.
    a.   Cross-sectional area of air gap $(A) = T \times W$

    But reduce thickness of core by 5%

    thickness $(T) = 0.5 \times 95\% = 0.475$ cm

    actual area $(A) = (0.475 \times 10^{-2}) \times (0.5 \times 10^{-2}) = 0.2375$ cm^2

$$= 0.2375 \times 10^{-4} \ m^2$$

The air gaps in this problem are reasonably small; only 2 mm thick. Therefore, allow 2.5% to the cross-sectional area for fringing.

   final area $(A) = 0.2375 \times 10^{-4} \ m^2$

$$2.5\% \text{ of } 0.2375 \times 10^{-4} = 0.0059 \times 10^{-4}$$

total square area $= 0.2375 \times 10^{-4} + 0.0059 \times 10^{-4}$

$$= 0.2434 \times 10^{-4} \ m^2$$

   b. Reluctance of air gap

$$R_{m(ag)} = \frac{l}{\mu A}$$

$$= \frac{4 \times 10^{-3}}{(1.26 \times 10^{-6}) \times (0.2434 \times 10^{-4})} \quad \text{(2 air gaps)}$$

$$= \frac{4 \times 10^{-3}}{30668 \times 10^{-10}}$$

$$= 130.429 \times 10^6 \ H^{-1}$$

   c. Total reluctance of magnetic circuit

$$R_{mT} = R_{m(U \& I)} + R_{m(ag)}$$

$$= (4.72 \times 10^6) + (130.43 \times 10^6)$$

$$= 135.15 \times 10^6 \ H^{-1}$$

PROBLEM 12-19

What current is required for a 1000-turn coil to produce the necessary field intensity to create 2×10^{-3} Wb of flux in a 0.25 cm-wide air gap in a laminated cast steel core (Fig. 12-39)? The core has an average

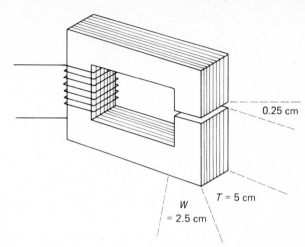

FIGURE 12-39 Cast steel core with a 0.25-cm air-gap and a 1000-turn coil.

0.25 cm

$T = 5$ cm

$W = 2.5$ cm

magnetic path length of 30 cm and a permeability value of 9×10^{-4} H/m. Allow 5% for varnish, but ignore fringing; it is negligible because of the small air gap.

SOLUTION   *(cover solution and solve)*

1. Cross-sectional area.

$$A_{(ag)} = T \times W$$

Reduce thickness of core by 5% for varnish.

$$\text{thickness} = T = 5 \times 0.95 = 4.75$$

Area of air gap

$$A_{(ag)} = (4.75 \times 10^{-2}) \times (2.5 \times 10^{-2})$$
$$= 11.875 \text{ cm}^2$$
$$= 11.875 \times 10^{-4} \text{ m}^2$$

2. Reluctance of air gap.

$$R_{m\,(ag)} = \frac{l}{\mu A}$$

$$= \frac{0.25 \times 10^{-2}}{(1.26 \times 10^{-6})(11.875 \times 10^{-4})}$$

$$= \frac{0.25 \times 10^{8}}{14.9625}$$

$$= 1.67 \times 10^{6} \text{ H}^{-1}$$

3. Reluctance of core material.

$$R_{m \text{ (steel)}} = \frac{l}{\mu A}$$

$$= \frac{29.75 \times 10^{-2}}{(9.0 \times 10^{-4})(11.875 \times 10^{-4})}$$

$$= \frac{29.75 \times 10^{6}}{106.875}$$

$$= 0.278 \times 10^{6} \ \text{H}^{-1}$$

4. Total reluctance of circuit.

$$R_{m_T} = R_{m \text{ (steel)}} + R_{m \text{ (ag)}}$$

$$= (0.278 \times 10^{6}) + (1.67 \times 10^{6})$$

$$= 1.948 \times 10^{6} \ \text{H}^{-1}$$

5. Flux required (2×10^{-3} Wb).
   From Ohm's Magnetic Law

$$\phi = \frac{F_m}{R_m}$$

$$F_m = \phi \times R_m$$

$$= (2 \times 10^{-3})(1.948 \times 10^{6}) \ \text{A}$$

$$= 3.896 \times 10^{3} \ \text{A}$$

6. Amount of magnetomotive force required to produce 2×10^{-3} Wb of flux in 1.948×10^{6} AT/Wb of reluctance.

$$F_m = \phi \times R_m$$

$$= (2 \times 10^{-3}) \times (1.948 \times 10^{6})$$

$$= 3.896 \times 10^{3} \ \text{A}$$

7. Value of current which must be supplied to the 1000-turn coil to produce 2×10^{-3} Wb of flux.

$$\text{from} \quad F_m = IN$$

$$\text{then} \quad I = \frac{F_m}{N}$$

$$= \frac{3.896 \times 10^{3}}{1000}$$

$$= 3.896 \ \text{A}$$

12-8.5 Parallel Magnetic Circuits

Many transformers are constructed with a shell-type core as shown in Fig. 12-40. The advantage of this type of core is its appreciable reduction of

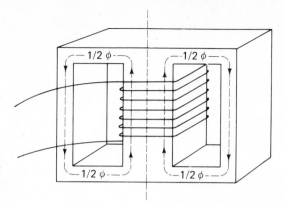

FIGURE 12-40 A shell-type core, showing the flux in the center leg dividing, causing half to go into each half of the core.

leakage. The winding for a shell-type core is placed on the center leg, which creates two separate paths of magnetic flux. The two paths make the core a parallel magnetic circuit, and usually the two paths are equal; therefore, the core can usually be divided into two equal sections. The first step in calculating the value of flux which will flow in each half of the core is to divide the core in half. The next step is to determine the required magnetomotive force in a way similar to calculating voltage in an electric circuit.

The total flux of a parallel circuit is the sum of the flux in each branch, which is analogous to current flow in parallel circuits. The magnetomotive force is common to each branch, just as voltage is common to all branches in parallel electric circuits. It is important to realize that there is no truly parallel magnetic circuit but only a close approximation of such a circuit.

PROBLEM 12-20

What value of current must be supplied to a coil containing 800 turns of wire which is wound on the center leg of the core as shown in Fig. 12-40, and which is required to produce 4×10^{-3} Wb of flux in the core? The core is a shell-type construction made of laminated transformer iron. The average magnetic path length in the core is 25 cm. The transformer iron has a permeability value of 11.4×10^{-4}. Varnish accounts for 5% of the thickness of the core.

SOLUTION   *(cover solution and solve)*

1. Divide the magnetic circuit down the center line and find the area of one-half of the core.

$$A = (2 \times 10^{-2}) \times (2 \times 10^{-2}) \times 95\% = 3.8 \text{ cm}^2$$

$$= 3.8 \times 10^{-4} \text{ m}^2$$

2. Reluctance of 1/2 of the core.

$$R_{\mathrm{m}} = \frac{l}{\mu A}$$

$$= \frac{25 \times 10^{-2}}{(11.4 \times 10^{-4})(3.8 \times 10^{-4})}$$

$$= 0.577 \times 10^6 \ H^{-1}$$

3. The total flux to be produced is 4×10^{-3} Wb; therefore, each half of the core must contain 2×10^{-3} Wb of flux lines, and the magnetomotive force (F_m) equals

$$F_m = \phi R_m$$

$$= (2 \times 10^{-3})(0.577 \times 10^6)$$

$$= 1.154 \times 10^3 \ A$$

Magnetomotive force is also equal to

$$F_m = IN$$

4. There are 800 turns on the coil; therefore, the current flow in the coil must be

$$I = \frac{F_m}{N} = \frac{1154}{800} \ A$$

$$= 1.44 \ A$$

The magnetomotive force of 1.154×10^3 A will cause the other half of the total flux of 4×10^3 Wb to flow in the second half of the shell-type core.

12-8.6 Relays

Relays are selected for applications according to

1. Current (current rating) which their contacts must carry
2. The voltage of the circuit which they are to control
3. The current available for the relay coil, or the resistance of the relay coil winding

Resistance of the relay coil is also important when the relay is the only load in the primary circuit, as the coil resistance is also the current limiter for the circuit.

The current rating of the coil determines the number of turns that the relay coil must have in order to produce the required flux density to close the contacts. The current rating also determines the size of the wire which is wound. The larger the current rating, the larger must be the AWG (American Wire Gauge) size of wire. Relay coils are wound with transformer wire, which is a copper wire that has been dipped in a high-voltage varnish.

Table 12-3   Magnetic Units—Summary Note

Term	Symbol	mks or SI Unit	Abbreviation	(Obsolete) CGS Units	Some Alternate Equations
Flux	ϕ	Webers = 10^8 lines	Wb	Maxwells = 1 line	BA $\dfrac{F_m}{R_m}$
Flux Density	B	teslas = $\dfrac{\text{Webers}}{\text{m}^2}$	T	Gauss = $\dfrac{\text{Mx}}{\text{cm}^2}$	$\dfrac{\phi}{A}$ μH
Magnetomotive Force	F_m	ampere	A	0.796 AT = 1 Gilbert	IN $\dfrac{R_m \phi}{l}$ Hl
Field Intensity (Magnetizing Force)	H	ampere/meter	A/m	Oersteds = $\dfrac{\text{Gilbert}}{\text{cm}}$	$\dfrac{F_m}{l}$ $\dfrac{R_m \phi}{l}$ $\dfrac{B}{\mu}$
Permeability (Absolute)	μ	henries/meter	H/m	$\dfrac{\text{Gauss}}{\text{Oersted}}$	$\dfrac{B}{H}$ $\dfrac{Pl}{A}$ $\dfrac{l}{R_m A}$ $(1.26 \times 10^{-6}) \times (\mu_r)$
Permeance	P	henry	H	—	$\dfrac{1}{R_m}$ $\dfrac{\mu A}{l}$
Permeability of Air	μ_0	$4\pi \times 10^{-7}$ or $\cong 1.26 \times 10^{-6}$	—	1.26×10^{-6}	
Relative Permeability	μ_r	—	—	—	$\dfrac{\mu}{\mu_0}$
Reluctance	R_m	reciprocal henry	H^{-1}	$\dfrac{\text{Gilberts}}{\text{Maxwell}}$ = Rels	$\dfrac{1}{P}$ $\dfrac{F_m}{\phi}$ $\dfrac{l}{\mu A}$
Number of Turns on a Solenoid	N or T	$\dfrac{F_m}{\text{Current}}$	—	—	$\dfrac{F_m}{I}$
Current through a Solenoid	I or A	$\dfrac{F_m}{\text{Turns}}$	—	—	$\dfrac{F_m}{N}$

1. State the relationship between the direction of current flow and the direction of the magnetic field around a straight conductor.

2. List three ways of increasing the strength of the magnetic field of an electromagnet.

3. State the left-hand rule for magnetism in a coil.

4. What relationship is there between
   a. Magnetic polarity and direction of coil winding?
   b. Direction of magnetic field and battery polarity?

5. State the left-hand rule for direction of current flow in a coil when the magnetic polarity at the end of the coil is known.

6. Is the magnetic field around a wire due to the presence of the wire or due to the electrons which flow in that wire?

7. Define the terms
   a. Flux
   b. Flux density
   c. Magnetomotive force

8. What is the unit of
   a. Flux?
   b. Flux density?
   c. Magnetomotive force?

9. What two conditions must be present to produce magnetomotive force?

10. A 6-V battery is connected to a solenoid consisting of 160 turns of wire. What magnetomotive force will be produced if 4 A of current flow in the coil?

11. Which of the two units, magnetomotive force or field intensity measures the true strength of a magnetizing force?

12. A 12-V battery causes 250 mA of current to flow in a 1200-turn coil 12 cm long. Calculate the strength of the magnetizing force (field intensity).

13. What does a fluxometer measure?

14. a. Do ferromagnetic materials produce linear *B-H* curves?
    b. Explain the reason for your answer in part a.

15. Define the terms
    a. Permeability
    b. Permeance

16. What is the property of permeability analogous to in electric circuits?

17. What is the reciprocal of the property of permeability?

18. What is the relationship between permeability, flux density, and field intensity?

19. What is the unit sometimes used for permeability?

20. Describe the difference between the term permeance and permeability.

21. State the symbolic equation for permeance and indicate what each symbol represents.

22. When are permeance and permeability values equal to one another?

23. What is the permeability value of air?

24. Write the equation showing the relationship between absolute permeability, relative permeability, and permeability of air.

25. State the difference between permeability, incremental permeability, and differential permeability, how they are calculated and why.

26. List three things which *B-H* magnetizing curves help to determine in design work.

27. Refer to the *B-H* magnetizing curves in Fig. 12-14 and determine the flux density which would be produced in transformer iron if a magnetizing force of 2400 A were applied to a coil wound on one leg of a core-type core of a transformer.

28. The flux density in a laminated transformer iron core is 1.46 T. The area of the core is 4×10^{-4} m^2. What is the number of flux lines produced in the core?

29. A piece of transformer iron is in a magnetizing force of 800 A. What is its permeability? (Refer to the *B-H* magnetizing curves in Fig. 12-14.)

30. Draw the magnetization curve for air.

31. A piece of nickel has a relative permeability value of 60. What is its absolute permeability value?

32. State the reciprocal of the property of permeance.

33. Write the equations showing the relationship between reluctance and
    a. Permeance
    b. Permeability

34. Calculate the reluctance of a circular bar of cast iron ($\mu = 17.857 \times 10^{-5}$ H/m) which has a radius of 4×10^{-4} m and a length of 1×10^{-2} m.

35. Write in symbolic form the equation for Ohm's Law for Magnetic Circuits. Explain what each symbol represents.

36. How much flux is produced in an iron-core coil if the iron core material has a reluctance of 5×10^3 H^{-1} and the field intensity is 9 A?

37. Define the terms:
    a. Hysteresis
    b. Coercive force
    c. Residual magnetism
    d. Retentivity

38. List three types of losses which may occur in magnetic materials.

39. What may be done to reduce each of the three types of losses?

40. Explain the term leakage flux.

41. List at least six equivalents or similarities between electric circuits and magnetic circuits.

42. State Ampere's Circuital Law.

43. Write the equation relating magnetomotive force, field intensity, and length of magnetic path.

44. How does an air gap affect the total reluctance of a magnetic circuit?

chapter 13

INDUCTANCE

13-1 Inductance

When used in applications other than electromagnets, coils are called inductors. Inductors are used in many operations. In dc operations they may be used for timing circuits, referred to technically as time-constant circuits. Computers at one time used dozens of L/R time-constant circuits. Filter chokes, R.F. chokes, and transformers are all applications of inductors in ac circuits. The coil in an automobile ignition system is an inductor and makes possible the increase of the 12-V battery voltage to 20,000 to 30,000 V.

The word inductor is a noun. A coil is called an inductor, but the property (characteristic) of an inductor is called inductance.

13-2 Induce

Inductance! What does the word mean? The word inductance is derived from the word "induce," meaning to persuade. Inductance is the property of an electrical circuit by which a magnet can induce (produce) electric current within it. Figure 13-1a is a pictorial demonstration of voltage being induced into a conductor when the conductor is moved at right angles to the lines of magnetic flux in which it is present. There is no voltage induced into the conductor in Fig. 13-1b because it is being moved parallel to the lines of flux in which it is present and therefore is not being cut by those lines of flux.

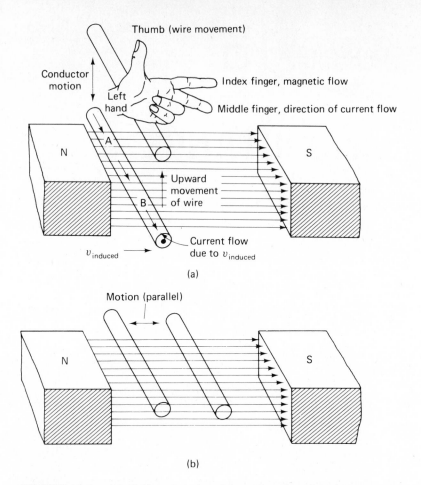

Thumb (wire movement)

Conductor
motion

Left
hand

Index finger, magnetic flow

Middle finger, direction of current flow

N

A

S

Upward
movement
of wire

B

$v_{induced}$

Current flow
due to $v_{induced}$

(a)

Motion (parallel)

N

S

(b)

FIGURE 13-1   (a) Voltage is induced into the wire when it is moved at right angles to the lines of force (magnetic induction).   (b) There is no voltage induced when the wire is moved parallel to the lines of force (no cutting).

13-3 Electromagnetic Induction

13-3.1 Magnetism and Voltage

At about the same time that Oersted noticed that a magnetic field sur-rounds a wire through which current is flowing, two physicists—Michael Faraday of England and Joseph Henry of the United States—discovered that voltage is induced across the terminals of a conductor when the con-ductor cuts or is cut by the magnetic field of a magnet. They noticed that the voltage ceased as soon as the motion was stopped. They also observed that the magnitude of the voltage depended directly upon the speed at which the wire was cut by the magnetic field. The direction of resulting current flow in the conductor can be determined by the left-hand genera-tor rule.

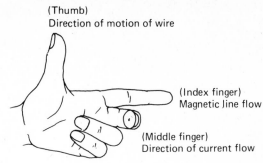

(Thumb)
Direction of motion of wire

(Index finger)
Magnetic line flow

(Middle finger)
Direction of current flow

FIGURE 13-2 Left-hand generator rule.

13-3.2 Left-Hand Generator Rule for Direction of Induced Current in a Wire

The *left-hand generator rule* is as follows (see Fig. 13-2):

1. Extend the thumb, index finger and middle finger of the left hand at mutually right angles as shown in Fig. 13-2.
2. Point the thumb in the direction of the wire's motion, or in the outward or inward direction of magnetic motion of build-up or collapse if a coil is being studied.
3. Point the index finger in the direction of the magnetic flow N to S (S to N inside coils).
4. The middle finger will point in the direction of the current flow (in coils, in the top half of the winding).

13-3.3 Effect of Inserting a Magnet into a Coil

As current can be induced into a wire when it is cut by a magnetic field [Fig. 13-1], so can current be induced into a coil by cutting its turns with the magnetic field of a bar magnet [by moving it into and out of the coil (Fig. 13-3)]. There cannot be current without voltage, so there must

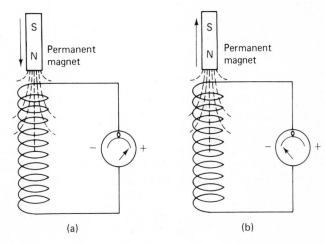

FIGURE 13-3 Effect of moving a permanent magnet into and out of a coil. (a) N-pole moving down; pointer moves right of zero. (b) N-pole moving up; pointer moves left of zero.

(a)

(b)

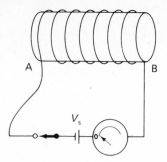

FIGURE 13-4   First instant that switch is closed.   Note the zero current.

also be a voltage induced which can actually be measured across the terminals of the coil (Fig. 13-3).

There will be voltage across the coil as long as the magnet is moving. If the magnet is held still, there will be no voltage. Moving the magnet down into the coil will produce a voltage of one polarity; moving the magnet up will produce the opposite polarity of voltage (Fig. 13-3a). Reversing the polarity of the magnet or the direction of the winding of the coil will also produce a voltage of opposite polarity.

13-4 Counter Voltage (CEMF)

If time could be sensed in milliseconds or microseconds, one would notice that there would be no current flow in the coil in Fig. 13-4 until a short time after the switch was closed. The reason is explained in one of the definitions for inductance which states that:

> an inductance is that characteristic of a component called an inductor which opposes any change in the current (direction or magnitude) which flows through it

The inductor opposes current attempting to flow into or out of it because a voltage is produced within it and across its turns when the magnetic field of the current from the outside source changes in magnitude. This induced voltage is called counter voltage (CEMF), self-induced voltage or simply induced voltage (v_L). The change in magnitude of the magnetic field of the outside-source current cuts the adjacent turns and induces voltage into the turns in the same way that the magnetic field of a moving bar magnet cuts the turns of a coil and induces voltage into the coil (Fig. 13-3).

The continually changing ac voltage of ac circuits always has a counter voltage opposing it in inductive circuits. This explains why very low resistance transformer primary windings have very little current flow in them until the secondary winding is loaded.

13-5 Production of Induced Voltage

The following steps are an instructional and pictorial explanation of how voltage is induced into an inductor. Inductors have no effect on dc cur-

Ch. 13 Inductance

rent and are not used to control it, but a battery circuit is nevertheless used in this explanation of induced voltage for simplicity.

Two separate conditions will be used in the following explanation of induced voltage.

Condition 1: A source voltage is connected to a coil. With this system we will examine how induced (CEMF) voltage is produced and how it opposes source current.

Condition 2: The source voltage is disconnected from the coil and the current energy stored in the coil in the form of a magnetic field attempts to collapse. With this system we will learn how an induced voltage of opposite polarity is produced by the collapse of the current energy stored in the coil and how the induced voltage attempts to keep the current flowing in the open-circuited coil. This induced voltage is the cause of arcing at switch contacts when they are opened.

CONDITION 1: A COIL IS CONNECTED TO A SOURCE VOLTAGE

1. See Fig. 13-5. The switch is open. The battery current is zero and there is no magnetic field.

2. The switch is closed to the battery, and the battery current attempts to flow and produce a magnetic field. The magnetic field moves outward as shown in Fig. 13-6.

FIGURE 13-5   The switch is open, battery current is zero, and there is no magnetic field.

FIGURE 13-6   *Below:* (a) Cutaway view of coil. The battery current attempts to flow into the top of the coil after the switch is closed. (b) Note the direction of the battery current and the resultant magnetic field movement after the switch is closed.

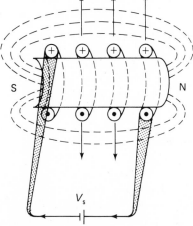

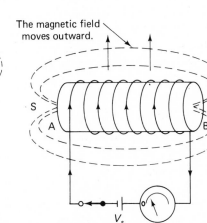

(a)   (b)

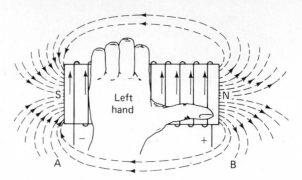

FIGURE 13-7 Polarity of the magnetic field produced by the battery.

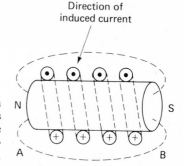

FIGURE 13-8 Note the direction of the induced current in the coil as the result of the changing magnetic field produced by the battery current attempting to increase in value.

3. The left-hand rule is used to determine the polarity of the magnetic field produced at the ends of the coil by the battery current. The B end of the coil is the N-magnetic pole. See Fig. 13-7.

4. The magnetic field moving outward and sideways cuts adjacent turns of the coil and induces a current and voltage into the coil turns, as shown in Fig. 13-8.

5. The left-hand generator rule (Fig. 13-9) establishes the direction of the induced current shown in Fig. 13-8. Note that the induced current is flowing out of the top half of the coil winding (Fig. 13-8) whereas the direction of the battery current was flowing into the top half of the coil.

6. There cannot be current flow without voltage. The direction of the induced current is caused by the polarity of the induced voltage. The induced voltage is equivalent to a battery connected in the middle of the coil with a polarity as shown in Fig. 13-10. Notice that the induced voltage is series-opposing the battery voltage and is equal in value to it. The induced voltage is, therefore, counter to the battery voltage and is often called *counter voltage* (CEMF).

SUMMARY:

There is no current flow in an inductive circuit in the first instant that the switch is closed between a coil and a source of voltage, because of induced voltage. The induced voltage produces an equal and counter-

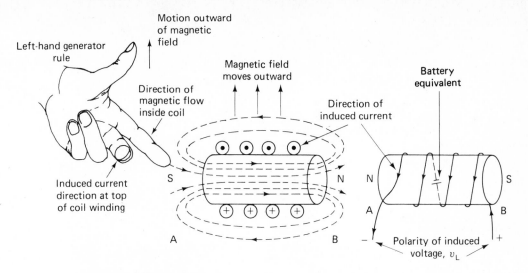

FIGURE 13-9 The direction of the induced current is determined by the left-hand generator rule.

FIGURE 13-10 The polarity of the induced voltage is determined from the direction of the induced current.

ing voltage to the attempted immediate rise of the battery current from zero to a steady state value; this restricts the flow of the battery current. The battery current magnetic field decreases in its rate of change as soon as the initial increase in battery current has been halted, and a smaller induced voltage (v_L) is produced. A smaller induced voltage results in less opposition to the source voltage, and the battery current slowly increases until it reaches its steady-state (maximum) value. The value of the steady-state current is controlled by the circuit resistance. The magnetic field is large when the battery current has reached its steady-state value, but it is constant in value; therefore, no counter (induced) voltage is produced when the battery current has reached its maximum value.

The production of an induced countering voltage to a source voltage may be summarized as follows:

1. Induced current opposes source current (battery). Compare Fig. 13-8 with Fig. 13-6a.
2. Induced voltage opposes source voltage (battery) (Fig. 13-11).
3. The induced magnetic field (produced by the induced current) opposes the magnetic field produced by the source current. Compare Fig. 13-9 with Fig. 13-7.

The circuits in Fig. 13-11 represent induced voltage and source voltage and show the relative polarity of the induced voltage with respect to the source voltage. Notice that the induced voltage is counter to the source voltage when the source current is attempting to increase in value.

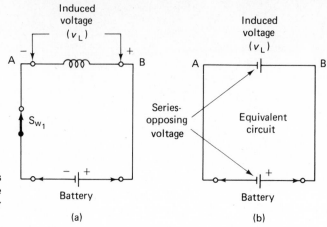

FIGURE 13-11 Note that (b) is the equivalent circuit for (a). The induced voltage opposes the battery voltage when the switch is closed.

(a) (b)

CONDITION 2: THE SWITCH IS OPENED BETWEEN THE COIL AND VOLTAGE SOURCE

The presence of a CEMF voltage to a source voltage when the switch between a coil and a voltage source is closed was discussed in Condition 1. What happens when the switch S_{w_1} between the coil and battery is opened?

1. Assume that the switch S_{w_1} in Condition 1 is closed and the source current has reached its steady value. The battery current and the magnetic field will be large but of a constant value; therefore, there will be no CEMF produced in the coil and the battery current will flow in the direction as shown in Fig. 13-12(a) and (b).

2. The switch S_{w_1} is opened but the current energy stored in the magnetic field is still present as illustrated in Fig. 13-13.

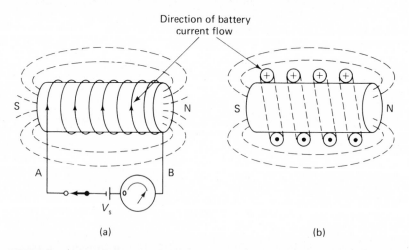

(a) (b)

FIGURE 13-12 Note that: (1) the current flow is from the voltage source; (2) the current is maximum and constant in value; and (3) a strong magnetic field is present but steady.

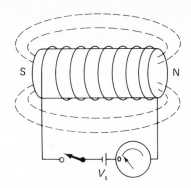

FIGURE 13-13 The magnetic field is present but the battery current is zero because the switch is open.

V_s

3. The magnetic field in the coil does not remain constant once the switch S_{w_1} is opened but attempts to collapse. The collapsing magnetic field moves inward, cuts adjacent turns and induces current and voltage in the coil of opposite polarity to the original induced voltage in Condition 1. Compare Fig. 13-10 and Fig. 13-15.

4. The direction of the induced current is determined by the left-hand rule as shown in Fig. 13-14. Current can flow only when

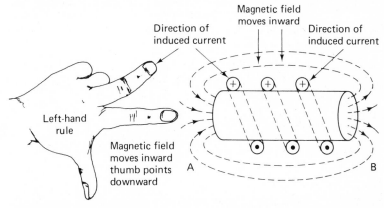

Direction of induced current

Magnetic field moves inward

Direction of induced current

Left-hand rule

Magnetic field moves inward thumb points downward

A

B

FIGURE 13-14 The direction of the induced current is determined with the left-hand generator rule.

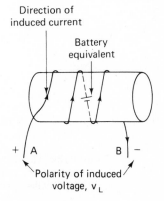

Direction of induced current

Battery equivalent

FIGURE 13-15 The direction of the induced voltage is determined from the direction of the induced current.

+ A

B −

Polarity of induced voltage, v_L

there is voltage; therefore, the polarity of the induced voltage can be determined from the direction of the induced current. The direction of the induced current indicates an induced voltage equivalent to a battery located in the middle of the coil with the polarity as shown in Fig. 13-15. Compare the polarity of this induced voltage with the polarity of the source voltage (Fig. 13-16a) and notice that they are series-aiding. The induced voltage—depending upon the speed at which the switch is opened— may be greater in value than the source voltage before the induced voltage arcs across the open switch contacts. The arc completes the circuit and allows the induced current to flow in the same direction as the disconnected battery current.

SUMMARY:

The explanation of the production of a counter voltage to an open circuit may be summarized as follows:

1. The induced current attempts to keep the source current flowing even after the circuit switch is opened. Compare the direction of the induced current in Fig. 13-14 with the direction of the source current in Fig. 13-12(b).

2. The induced voltage series-aids the source voltage. Compare the polarity of the induced voltage in Fig. 13-16 with the polarity of the source voltage.

3. The induced current flows in the same direction as the source current; therefore, the magnetic field produced by the induced current series-aids the magnetic field produced by the source current.

The relative polarity of the induced voltage with respect to the disconnected source voltage is shown in the circuits shown in Fig. 13-16.

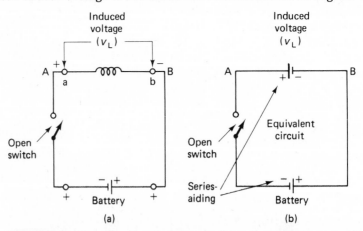

FIGURE 13-16   The circuit shown in (b) is the equivalent circuit of (a). Note that the batteries are actually disconnected, but the polarity of the induced voltage is such that it would aid the battery voltage.

13-6 Lenz's Law

A German physicist, Lenz, performed an experiment in the middle 1800s similar to the one just described. His observations were put into a law called Lenz's Law. There are three ways of stating this law. Choose the one which you understand the best.

1. The direction of current induced in a conductor will be such that its magnetic field will oppose the action that produced the induced current.

2. The direction of i_L is such that the magnetic field which it produces will oppose the magnetic field which produced i_L.

3. The direction of the i_L is such that any flux resulting from it will develop a current which tends to oppose any change in the original flux or current.

13-7 Flux Linkage

The magnetic field from each conductor may cut several conductors, and the magnetic field from serveral conductors may cut one conductor. This process is called flux linkage.

13-8 Self-Inductance

The phenomenon of induced voltage which has just been described comes about by self-induced current and voltage. Self-inductance is actually taking place, but the shorter expression "inductance" is generally used in its place.

13-9 Magnitude of Counter Voltage

The effect of changing the polarity of a magnet which was inserted into a coil, or changing the direction of its motion has been discussed, but nothing has been said about the rate of the motion.

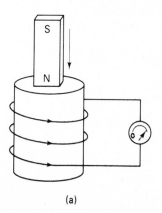

(a)

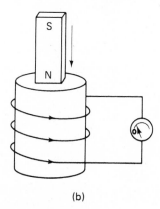

(b)

FIGURE 13-17 (a) Bar magnet dropped into coil; large volt. (b) Bar magnet slowly inserted into coil; small volt.

If the bar magnet were dropped into a coil (Fig. 13-17a), a large induced voltage would register on the voltmeter. Inserting the bar slowly (Fig. 13-17b) will produce only a small voltage registration. The more rapidly that magnetic lines of force cut a conductor, the greater will be the induced voltage.

Conclusion: Induced voltage is proportional to the rate at which a magnetic field cuts a coil. This fact was also observed by Michael Faraday when he discovered that electricity could be produced by cutting a conductor with magnetic flux. The rule establishing this fact is called Faraday's Law.

13-10 Faraday's Law

Faraday's Law states that the value of the voltage induced (v_L) into a coil is directly proportional to the rate of change of the magnetic flux cutting the turns of that coil.

$$\text{induced voltage} \propto \text{rate of change of flux}$$

$$v_L \propto \frac{d\phi}{dt} \frac{\text{(Wb)}}{\text{(s)}}$$

When the magnetic flux is due to a changing current, and this flux in turn cuts adjacent turns, Faraday's Law may be restated thus:

The value of the induced voltage (v_L) in a coil is directly proportional to the rate of change of current flow in that coil.

$$\text{induced voltage} \propto \text{rate of change of current}$$

$$v_L \propto \frac{di}{dt}$$

13-11 Unit of Inductance

Induced voltage cannot be computed with rate of change of current alone. The number of turns must also be included. The relationship between (di/dt) and number of turns expresses the value of inductance.

A coil has an *inductance of 1 henry* if 1 V is induced into the coil when the current in it is changing at the rate of 1 A/s.

$$\text{inductance} = \frac{\text{induced voltage}}{\text{rate of current change/second}}$$

$$L = \frac{-v_L}{\frac{di}{dt}}$$

from which induced voltage

$$v_L = \frac{di}{dt} L$$

where v_L is induced voltage

L is inductance in henries

$\dfrac{di}{dt}$ is $\dfrac{A}{s}$

The negative sign indicates that the induced voltage is in a direction opposite to the voltage that produced the current change. It is not used in calculations.

the symbol for inductance is L

the unit of inductance is H (henry)

PROBLEM 13-1

The current in an inductor changes from 12 A to 16 A in 2 s. How much is the inductance of the coil if the induced voltage is 40 V?

SOLUTION *(cover solution and solve)*

$$L = \frac{v_L}{\dfrac{di}{dt}}$$

$$= \frac{40}{\dfrac{16 - 12}{2}}$$

$$= \frac{40}{2}$$

$$= 20 \text{ H}$$

PROBLEM 13-2

What is the v_L across a coil of 10 H when the current through it is changing from 6 to 14 mA in 8 μs?

SOLUTION *(cover solution and solve)*

$$L = \frac{v_L}{\dfrac{di}{dt}}$$

$$v_L = \frac{di}{dt} L$$

$$v_L = \frac{(14 - 6) \times 10^{-3} \times 10}{8 \times 10^{-6}}$$

$$= \frac{80 \times 10^{-3}}{8 \times 10^{-6}}$$

$$= 10,000 \text{ V}$$

$$= 10 \text{ kV}$$

$X_L = \dfrac{1}{2\pi f L}$

PROBLEM 13-3

What value of inductance is necessary to induce a voltage of 250 V if the current changes at the rate of 5 A in 10 ms?

SOLUTION   *(cover solution and solve)*

$$L = \frac{v_L}{\dfrac{di}{dt}} = \frac{250}{\dfrac{5}{10 \times 10^{-3}}}$$

$$= \frac{250}{\dfrac{5}{10^{-2}}}$$

$$= 50 \times 10^{-2}$$

$$= 0.5 \text{ H}$$

13-12 Physical Determinants of Inductance

The calculations just made for inductance involved electrical units, and could be used to determine, for example, what inductance is required if a certain induced voltage were to be produced when the current in the coil was changing at so many amperes per second. If a coil of such an inductance were to be made, one would have to know the physical aspects of that coil such as length and number of turns. The inductance of a coil depends upon the following physical parameters.

$$L = \frac{N^2 \mu A}{l}$$

where L is inductance in henries
N is number of turns
μ is permeability of the core material
l is length of the core in meters
A is area in square meters (m^2)

Winding coils is a specialized job done in a coil factory; it is important, however, that one realizes what factors are important and how they affect the inductance of a coil. It is not uncommon to have to change the value of a bought coil slightly. Observe these facts:

1. Doubling the μ value or square area of the coil doubles the inductance.
2. Doubling the length decreases the inductance value by half.
3. Doubling the number of turns quadruples the inductance value.

PROBLEM 13-4

A coil is wound on an air-core circular cardboard form which has a diameter of 2 cm and is 4 cm long. What will be the inductance of the coil if it is wound with 150 turns?

SOLUTION   *(cover solution and solve)*

Given:

$N = 150$

μ of air is 1.26×10^{-6}

$l = 4 \text{ cm} = 4 \times 10^{-2} \text{ m}$

$D = 2 \text{ cm} = 2 \times 10^{-2} \text{ m}$

$r = \dfrac{\text{diameter}}{2}$

$A = \pi r^2 = 3.14 \times (1 \times 10^{-2})^2$

$\qquad\quad = 3.14 \times (1 \times 10^{-4})$

$$L = \frac{N^2 \mu A}{l}$$

$$= \frac{(150)^2 \times (1.26 \times 10^{-6}) \times (3.14 \times 10^{-4})}{4 \times 10^{-2}}$$

$$= \frac{222.5 \times 10^2 \times 10^{-10}}{4 \times 10^{-2}}$$

$$= 222.5 \times 10^{-6}$$

$$= 222.5 \; \mu\text{H}$$

PROBLEM 13-5

What is the inductance of a 200-turn coil which has a magnetic core with a relative permeability of 2500? The core is 2 cm long and the area of the ends is 3 cm^2.

SOLUTION   *(cover solution and solve)*

Given:

$N = 200$

$l = 2 \text{ cm} = 2 \times 10^{-2} \text{ m}$

$A = 3 \text{ cm}^2 = 3 \times 10^{-4} \text{ m}$

$\mu_r = 2500$

$$\mu_r = \frac{\mu}{\mu_0}$$

$$\mu = \mu_r \times \mu_0$$
$$= 2500 \times (1.26 \times 10^{-6})$$

$$L = \frac{N^2 \mu A}{l}$$

$$= \frac{200^2 \times (25 \times 10^2 \times 1.26 \times 10^{-6}) \times (3 \times 10^{-4})}{2 \times 10^{-2}}$$

$$= \frac{(4 \times 10^4) \times 94.5 \times 10^2 \times 10^{-10}}{2 \times 10^{-2}}$$

$$= 189 \times 10^{-2}$$

$$= 1.89 \text{ H}$$

13-13 Practical Considerations

13-13.1 Dc Resistance of Inductor

The dc resistance value of coils may be from a fraction of an ohm to a few hundred ohms depending upon the current rating, the inductance of the coil (number of turns) and the application in which the coil is used.

13-13.2 Radio-Frequency Resistance

Radio-frequency resistance is called *skin effect*. Radio-frequency currents have a tendency to flow near the surface of a conductor instead of through it. The use of fine multistrand conductors, called Litz wire, for coils increases the total surface area of the conductor in the coil and reduces the RF resistance in the same way that a large diameter wire reduces dc resistance.

Inductors used in radio-frequency applications such as antenna coils and radio-frequency transformers are wound of Litz wire which is very small in diameter and thus has a relatively large dc resistance per foot but a low RF resistance. Litz wire consists of a number of very fine strands of wire, each insulated by a thin fabric material, twisted together to form one conductor. The advantage of Litz wire is its low resistance to radio-frequency currents.

13-13.3 Connecting Inductors Electrically (Not Magnetically)

There are two types of coupling between inductors:

1. Electrical coupling
2. Magnetic (L_M) coupling

The total inductance of inductors connected together electrically is calculated in the same manner as is resistance, if the inductors are placed so that there is no magnetic coupling between them.

Series connection:

$$L_T = L_1 + L_2 + L_3 + \cdots + L_n$$

Parallel connection:

$$L_T = \frac{L_1 \times L_2}{L_1 + L_2}$$

To compute the total inductance of coils connected together electrically and positioned so that there is magnetic coupling between them one must also take the magnetic coupling into account. The magnetic coupling is called mutual inductance.

13-14 Mutual Inductance

We have seen that self-inductance, commonly known as inductance, is the phenomenon which occurs when current changes in a coil; a resulting magnetic field moves in and out of adjacent turns and induces a second current in the adjacent turns of the coil.

If a second coil were placed close enough to the first coil so that the changing magnetic lines of force in the first also cut the second (Fig. 13-18) a voltage would be induced into the second coil. The effect which induces a voltage into a second coil when the magnetic field in the first changes is called mutual inductance.

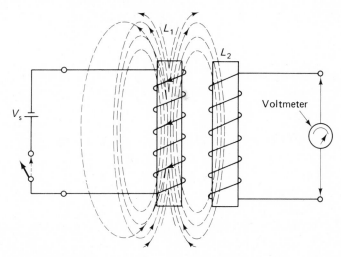

FIGURE 13-18 Mutual inductance: magnetic field from coil 1 cuts the turns of coil 2.

the symbol for mutual inductance is L_M

the unit of mutual inductance is the henry, H

In other words, mutual inductance is a term used to describe the relationship that exists between coils when a change of current in one induces a voltage and current into the other, and the flux produced in each cuts through the other.

Two coils are said to have a mutual inductance of 1 H when a rate of change of 1 A/s in one coil induces a voltage of 1 V into the second coil.

Mutual inductance, however, comes about and can be calculated only if a certain condition is met. That condition is called *coefficient of coupling*.

13-15 Coefficient of Coupling

The degree or amount of magnetic coupling between two coils is known as the *coefficient of coupling* (K). The maximum coefficient of coupling occurs when every line of flux in the first coil cuts the turns of the second coil. This degree of closeness of the coils is known as *tight coupling*.

One-hundred percent coupling (theoretical only) is expressed as 1. Coupling of less than 100% is always expressed in decimal values; for example, 75% coupling is expressed as 0.75. Loose coupling produces a small value of coupling; for example, if only 1% of the lines of force from coil 1 were cutting coil 2, then the coefficient of coupling would be 0.01.

Coefficient of coupling is dependent not only upon the closeness of one coil to another but also upon the angle between the two. Coils placed parallel to one another (Figs. 13-19 and 13-20) produce the highest value of coupling for a specific distance apart. At right angles (perpendicular), zero coupling occurs regardless of the closeness (Fig. 13-21). This is because the expanding magnetic lines of flux in the first coil move parallel to the turns in the second coil and no cutting can take place.

Coefficient of coupling is expressed by the letter K. K is a ratio. It has no units.

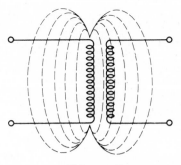

FIGURE 13-19

Tight coupling
K is high

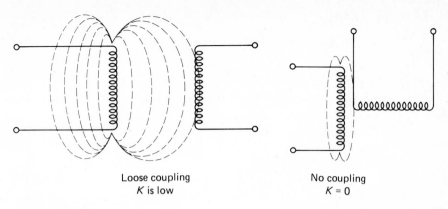

Loose coupling
K is low

No coupling
$K = 0$

FIGURE 13-20   **FIGURE 13-21**

$$K = \frac{\text{lines of flux in coil 2}}{\text{lines of flux in coil 1}}$$

$$K = 1 = 100\% \text{ coupling}$$

$$K = 0 = \text{no coupling}$$

The relationship between two coils which are coupled, and their coefficient of coupling, is

$$L_M = K\sqrt{L_1 \times L_2}$$

where L_M is the mutual inductance of the two coils in henries (H)

L_1 is the inductance of the first coil in henries (H)

L_2 is the inductance of the second coil in henries (H)

K is the coefficient of coupling, in decimal values

PROBLEM 13-6

Calculate the mutual inductance of a 2-H coil coupled to a 6-H coil. The coefficient of coupling is 0.8.

SOLUTION   *(cover solution and solve)*

Given:

$L_1 = 2$ H

$L_2 = 8$ H

$K = 0.8$

$$L_M = K\sqrt{L_1 \times L_2}$$
$$= 0.8\sqrt{2 \times 6}$$
$$= 2.77 \text{ H}$$

PROBLEM 13-7

What is the coefficient of coupling between two coils if one has an inductance of 12 mH, the other has an inductance of 36 mH, and the mutual inductance of the two is 18 mH?

SOLUTION   (cover solution and solve)

Given:

$L_1 = 12 \times 10^{-3}$
$L_2 = 36 \times 10^{-3}$
$L_M = 18 \times 10^{-3}$

from $L_M = K\sqrt{L_1 \times L_2}$

$$K = \frac{L_M}{\sqrt{L_1 \times L_2}}$$

$$= \frac{18 \times 10^{-3}}{\sqrt{(12 \times 10^{-3}) \times (36 \times 10^{-3})}}$$

$$= \frac{18 \times 10^{-3}}{\sqrt{432 \times 10^{-6}}}$$

$$= \frac{18 \times 10^{-3}}{20.78 \times 10^{-3}}$$

$$= 0.86$$

PROBLEM 13-8

What must be the inductance of a coil, L_2, which can be used with a 4-H coil, L_1, to produce a mutual inductance of 3 H, if the coefficient of coupling between the two coils is 0.5?

SOLUTION   (cover solution and solve)

Given:

$L_1 = 4 \text{ H}$
$L_M = 3 \text{ H}$
$K = 0.5$

from $L_M = K\sqrt{L_1 \times L_2}$

$$\text{then} \quad L_{M^2} = K^2 \times L_1 \times L_2$$

$$\therefore L_2 = \frac{(L_M)^2}{K^2 \times L_1}$$

$$= \frac{3^2}{(0.5)^2 \times 4}$$

$$= \frac{9}{0.25 \times 4}$$

$$= 9 \text{ H}$$

13-16   Inductors and L_M

Coils are connected in series if there is need for an inductance larger than that available from a single inductor. Mutual inductance affects the total value of inductance when coils are connected close enough that there is coupling between them. Mutual inductance adds to the total inductance, or it can subtract from it depending upon whether the coils are in a series-aiding or series-opposing arrangement.

13-16.1   Inductors in Series with no Coupling

The inductance of coils connected in series and placed at right angles to one another or far enough apart so there is no mutual inductance is added like resistors in series.

$$L_T = L_1 + L_2 + \cdots + L_n$$

13-16.2   Series-Aiding Coils and Mutual Inductance

Two inductors are said to be connected series-aiding (Fig. 13-22) when the leads of the two coils are so connected that the magnetic fields of the two coils add to one another when the coils are in close proximity (mutual inductance).

The total inductance of two coils connected series-aiding magnetically (Fig. 13-22) is expressed by

$$L_T = L_1 + L_2 + 2L_M$$

13-16.3   Series-Opposing and Mutual Inductance

Two coils are connected series-opposing when they are connected electrically in series so that the magnetic field of one coil is opposite to (and opposes) the magnetic field of the second coil (Fig. 13-23). These magnetic fields will link and will wholly or partially cancel each other depending upon their distance from one another (mutual inductance).

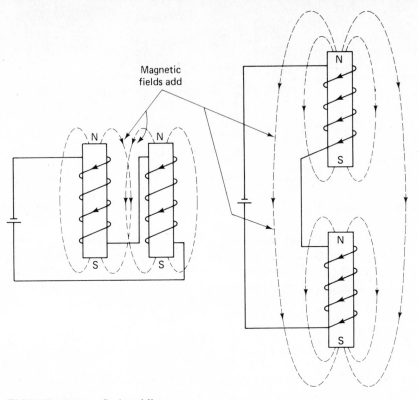

Magnetic
fields add

FIGURE 13-22   Series-aiding.

The equation for the total inductance of two magnetically series-opposing coils is:

$$L_T = L_1 + L_2 - 2L_M$$

PROBLEM 13-9

A 12-H coil and an 8-H coil are connected together but placed at right angles to one another. What is the total inductance?

SOLUTION   *(cover solution and solve)*

(There is no magnetic coupling when the coils are at right angles.)

Given:

$L_1 = 12 \text{ H}$
$L_2 = 8 \text{ H}$

$$L_T = L_1 + L_2$$
$$L_T = 12 \text{ H} + 8 \text{ H}$$
$$= 20 \text{ H}$$

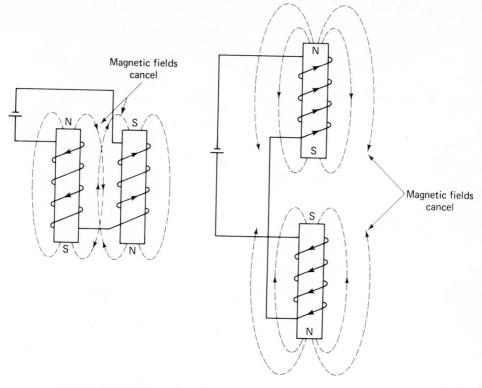

FIGURE 13-23 Series-opposing magnetically.

PROBLEM 13-10

A 10-H and a 12-H coil are connected series-aiding magnetically and have a mutual inductance of 2 H. What is the total inductance?

SOLUTION *(cover solution and solve)*

Given:

L_1 = 10 H

L_2 = 12 H

L_M = 2 H

$$L_T = L_1 + L_2 + 2L_M$$
$$= 10 + 12 + (2 \times 2)$$
$$= 26 \text{ H}$$

If the above coils were connected series-opposing, what would be the total inductance?

$$L_T = L_1 + L_2 - 2L_M$$
$$= 10 + 12 - (2 \times 2)$$
$$= 18 \text{ H}$$

PROBLEM 13-11

A 50-mH coil is connected series-aiding magnetically to a 75-mH coil. The coefficient of coupling (K) is 0.8. What is the total inductance?

SOLUTION   *(cover solution and solve)*

Given:

$L_1 = 50 \times 10^{-3}$

$L_2 = 75 \times 10^{-3}$

$K = 0.8$

$$L_M = K\sqrt{L_1 \times L_2}$$

$$= 0.8\sqrt{(50 \times 10^{-3}) \times (75 \times 10^{-3})}$$

$$= 0.8\sqrt{3750 \times 10^{-6}}$$

$$= 0.8 \times (61.24 \times 10^{-3})$$

$$= 49 \times 10^{-3} \text{ H}$$

$$= 49 \text{ mH}$$

$$L_T = L_1 + L_2 + 2L_M$$

$$= (50 \times 10^{-3}) + (75 \times 10^{-3}) + 2(49 \times 10^{-3})$$

$$= (50 \times 10^{-3}) + (75 \times 10^{-3}) + 98 \times 10^{-3}$$

$$= 223 \times 10^{-3} \text{ H}$$

$$= 223 \text{ mH}$$

PROBLEM 13-12

A 4-H coil is connected to an 8-H coil so that the two are magnetically series-opposing. What is the coefficient of coupling (K) if the total inductance is 2 H?

SOLUTION   *(cover solution and solve)*

Given:

$L_1 = 4 \text{ H}$

$L_2 = 8 \text{ H}$

$L_T = 2 \text{ H}$

$$\text{From} \quad L_T = L_1 + L_2 - 2L_M$$

$$\text{then} \quad 2L_M = L_1 + L_2 - L_T$$

$$\therefore L_M = \frac{L_1 + L_2 - L_T}{2}$$

$$= \frac{4 + 8 - 2}{2}$$

$$= 5 \text{ H}$$

$$\text{From} \quad L_M = K\sqrt{L_1 \times L_2}$$

$$\text{then} \quad K = \frac{L_M}{\sqrt{L_1 \times L_2}}$$

$$K = \frac{5}{\sqrt{4 \times 8}}$$

$$= 0.88$$

13-16.4 Inductance in Parallel and Mutual Inductance

Inductances may be connected in parallel if the current rating of a single inductance is not large enough to carry the current in a circuit.

Coils connected in parallel will have the same voltage (v_L) induced across each of them (voltage is common in parallel circuits).

13-16.5 Parallel Coils with Aiding Fields (Fig. 13-24)

$$L_T = \frac{L_1 \times L_2 - (L_M)^2}{L_1 + L_2 - 2L_M}$$

↑_____ Note the negative sign when aiding.
See Fig. 13-24.

13-16.6 Parallel Coils with Opposing Fields (Fig. 13-25)

$$L_T = \frac{L_1 \times L_2 - (L_M)^2}{L_1 + L_2 + 2L_M}$$

↑_____ Note the plus sign when opposing.
See Fig. 13-25.

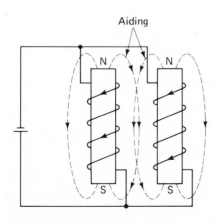

FIGURE 13-24   Parallel-aiding magnetically.

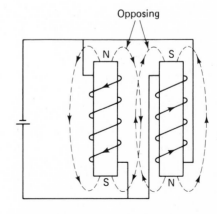

FIGURE 13-25   Parallel-opposing magnetically.

Coils connected in parallel and spaced or angled so there is no magnetic coupling between them are added together in the same that resistors in parallel are added.

$$L_T = \frac{L_1 \times L_2}{L_1 + L_2}$$

PROBLEM 13-13

A 6-H and a 4-H coil are connected in a magnetically parallel-aiding arrangement. What is the total inductance if the coefficient of coupling (K) between them is 0.4?

SOLUTION   *(cover solution and solve)*

Given:

L_1 = 6 H
L_2 = 4 H
K = 0.4

$$L_M = K\sqrt{L_1 \times L^2}$$
$$= 0.4\sqrt{6 \times 4}$$
$$= 1.96$$

Parallel Aiding

$$L_T = \frac{L_1 \times L_2 - (L_M)^2}{L_1 + L_2 - 2L_M}$$
$$= \frac{(6 \times 4) - (1.96)^2}{6 + 4 - 2(1.96)}$$
$$= \frac{20.15}{6.08}$$
$$= 3.314 \text{ H}$$

PROBLEM 13-14

What would be the total inductance if the coils in Prob. 13-13 had been connected parallel-opposing?

SOLUTION   *(cover solution and solve)*

$$L_T = \frac{L_1 \times L_2 - (L_M)^2}{L_1 + L_2 + 2L_M}$$

$$= \frac{(6 \times 4) - (1.96)^2}{6 + 4 + 2(1.96)}$$

$$= \frac{20.15}{13.92}$$

$$= 1.45 \text{ H}$$

13-17 Transformers

One of the best working examples of mutual inductance is a transformer. A transformer will have L_2 actually wound over the top of L_1 so that the coefficient of coupling will be as tight as possible (Fig. 13-26).

The subject of transformers is extensive, but it is dealt with lightly at this moment to illustrate the principle of mutual inductance as it applies to power transformers.

There are many different classifications of transformers. For example, there are audio transformers, radio-frequency transformers, output transformers, filament transformers, and power transformers. All must be operated on ac voltage. Transformers cannot be operated on dc voltage because the magnetic field in the primary circuit must be moving in order to induce a voltage into the secondary circuit.

13-17.1 Construction

Transformers used at power line frequencies and audio frequencies are constructed with laminated high-permeability soft iron cores. The coefficient of coupling is as high as possible, approaching the value of 1. Radio frequency transformers on the other hand are wound on cores of low permeability and low coefficient of coupling, from as low as 0.002 to 0.75. Radio-frequency transformers are used in tuned circuit applications.

13-17.2 Primary and Secondary

The input side of a transformer (L_1) is called the *primary winding* and the output winding (L_2), which is cut by the magnetic lines of force from the primary winding, is called the *secondary winding*.

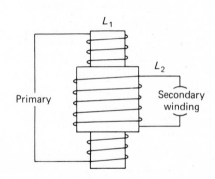

FIGURE 13-26 A transformer (pictorial view).

There can never be more than one primary winding on a transformer, although some transformers may have several taps on the one primary winding. There can, however, be several secondary windings. The primary winding is always the first winding on a core, and all secondary windings are wound in succession on top of and over the primary winding.

The amount of voltage which will be induced into the secondary winding depends on three main factors:

1. Coefficient of coupling ($K \cong 1$ in power transformers)
2. Primary voltage
3. Turns ratio of secondary to primary winding

13-17.3 Principle of Operation

The principle of operation is that a changing current in the primary sets up a changing flux that cuts and induces a current and voltage in the secondary winding. If the number of turns in the secondary winding is the same as the number of turns in the primary, and there is unity coupling between them, then the voltage induced in each turn of the secondary will be the same as the self-induced voltage of each turn in the primary. Thus there will be the same voltage across the secondary winding as is applied to the primary winding (from definition of inductance).

13-17.4 Isolation Transformer

The transformer, as described above, can neither step up nor step down the voltage. Its only function is to prevent equipment connected on the secondary winding from being connected directly to the ground on the primary side. It is used in power line applications to protect persons from shock and to protect transformerless equipment from shorting against other transformerless equipment.

13-18 Voltage Ratio

The ratio of the voltage obtained from the secondary winding of a transformer to the voltage applied to the primary winding is dependent upon the ratio of the number of turns in the secondary winding to the number of turns in the primary winding.

$$\frac{\text{number of turns in the secondary}}{\text{number of turns in the primary}} = \frac{\text{voltage in secondary}}{\text{voltage in primary}}$$

$$\frac{N_s}{N_p} = \frac{V_s}{V_p}$$

13-18.1 Step-Up Transformer

If the secondary winding has more turns than the primary winding, the secondary voltage will be larger than the primary voltage and the trans-

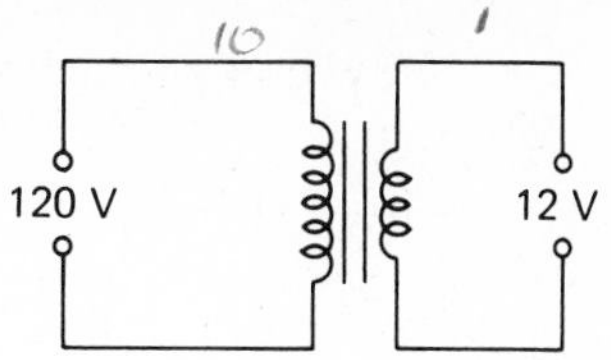

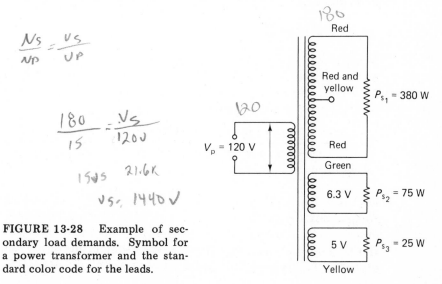

10 *1*

120 V   12 V

FIGURE 13-27   Step-down transformer.   The secondary voltage is
less than the primary voltage.

former is called a step-up transformer. This type of power transformer
is often used to provide large voltages for tube-operated equipment.

13-18.2 Step-Down Transformer

If the secondary winding has fewer turns than the primary winding, there
will be less voltage in the secondary than in the primary and the transformer is known as a step-down transformer (Fig. 13-27). Such a transformer is always a part of the ac adapter circuit used to operate calculators and transistor radios. All transistor equipment operated on power
line voltage is powered by step-down transformers.

13-18.3 Power Transformers

Some types of equipment require different values of voltage for various
circuits. A transformer used in such equipment is often a combination
of a step-down and a step-up transformer and is called a power transformer. The transformer as illustrated in Fig. 13-28 is such a type and is
used in many types of equipment.

PROBLEM 13-15

Find the voltage induced into a 180-turn secondary winding if 120 V
is applied to the primary which is wound with 15 turns of wire.

$$\frac{N_S}{N_P} = \frac{V_S}{V_P}$$

$$\frac{180}{15} = \frac{V_S}{120 V}$$

$15 V_S$ $21.6 K$

$V_S = 1440 V$

180

Red

Red and
yellow P_{s_1} = 380 W

120

V_p = 120 V   Red

Green

6.3 V P_{s_2} = 75 W

5 V P_{s_3} = 25 W

Yellow

FIGURE 13-28   Example of secondary load demands. Symbol for
a power transformer and the standard color code for the leads.

SOLUTION   *(cover solution and solve)*

Given:

$V_p = 120$ V
$N_s = 180$
$N_p = 15$

$$\text{from} \quad \frac{N_s}{N_p} = \frac{V_s}{V_p}$$

$$V_s = \frac{N_s \times V_p}{N_p}$$

$$= \frac{180 \times 120}{15}$$

$$= 1440 \text{ V}$$

PROBLEM 13-16

A transformer primary has 120 V applied to it. How many turns must the secondary winding have, to have 600 V induced into it when the primary winding has 25 turns?

SOLUTION   *(cover solution and solve)*

Given:

$V_p = 120$
$V_s = 600$
$N_p = 25$

$$\text{from} \quad \frac{N_s}{N_p} = \frac{V_s}{V_p}$$

$$N_s = \frac{N_p \times V_s}{V_p}$$

$$= \frac{25 \times 600}{120}$$

$$= 125 \text{ turns}$$

13-19   Power in Primary and Secondary Windings

An undeniable law in physics states that energy (power) can be neither created nor destroyed but can only be converted into one form or another. Such is the case with wattage (power). If the load across a secondary winding requires 240 W of power, that power must come from the primary winding (and in turn from the power company).

$$P_{\mathrm{p}} = P_{\mathrm{s}}$$

$$\therefore\ I_{\mathrm{p}} V_{\mathrm{p}} = I_{\mathrm{s}} V_{\mathrm{s}}$$

If a transformer is listed in a transformer or electronic catalogue as having a primary rating of 120 V and 5 A, this does not mean that 5 A will always be flowing in the primary. It only means that the wire with which the primary is wound can carry 5 A without overheating and burning out. If the secondary winding load draws 240 W of power, the primary will have only 2 A of current flowing in it.

EXAMPLE

$$P_{\mathrm{p}} = P_{\mathrm{s}}$$

$$\therefore\ I_{\mathrm{p}} V_{\mathrm{p}} = P_{\mathrm{s}}$$

$$I_{\mathrm{p}} = \frac{P_{\mathrm{s}}}{V_{\mathrm{p}}}$$

$$I_{\mathrm{p}} = \frac{240}{120} = 2\ \mathrm{A}$$

An increase or decrease in power demand in the secondary will demand more or less current, respectively, from the primary.

Take note that the current drawn from the primary of a transformer is determined completely [disregarding heat (power) losses] by the power demand of the secondary or secondaries.

PROBLEM 13-17

A transformer has 100 turns in the primary and 600 turns in the secondary. If 120 V is applied to the primary and a 480-Ω resistance is connected across the secondary, what is the primary current? (Assume no losses.)

SOLUTION   (cover solution and solve)

Given:

$N_{\mathrm{p}} = 100$

$N_{\mathrm{s}} = 600$

$V_{\mathrm{p}} = 120$

$R_{\mathrm{s}} = 480$

$$\text{from}\quad \frac{V_{\mathrm{s}}}{V_{\mathrm{p}}} = \frac{N_{\mathrm{s}}}{N_{\mathrm{p}}}$$

$$V_{\mathrm{s}} = \frac{V_{\mathrm{p}} N_{\mathrm{s}}}{N_{\mathrm{p}}}$$

$$V_{\mathrm{s}} = \frac{120 \times 600}{100} \qquad = 720\ \mathrm{V}$$

$$\text{but} \quad P_s = \frac{V_s^2}{R_s} = \frac{720 \times 720}{480} = 1080 \text{ W}$$

$$\text{and} \quad P_p = P_s$$

$$\therefore \ I_p V_p = 1080$$

$$I_p = \frac{1080}{120} \qquad = 9 \text{ A}$$

PROBLEM 13-18

A power transformer has a 1-to-6 turns ratio. The voltage applied to the primary is 120 V and the load connected across the secondary is 2880 Ω. Find I_p. (Assume no losses.)

SOLUTION   *(cover solution and solve)*

Given:

$$\frac{N_s}{N_p} = \frac{6}{1}$$

$$V_p = 120 \text{ V}$$

$$R_s = 2880$$

$$\text{from} \quad \frac{V_s}{V_p} = \frac{N_s}{N_p}$$

$$V_s = \frac{V_p N_s}{N_p} = 120 \times \frac{6}{1} = 720 \text{ V}$$

$$\text{but} \quad P_s = \frac{V_s^2}{R_s} = \frac{720 \times 720}{2880} = 180 \text{ W}$$

$$\text{and} \quad P_p = P_s$$

$$\therefore I_p V_p = 180$$

$$\text{and} \quad I_p = \frac{180}{V_p} = \frac{180}{120}$$

$$= 1.5 \text{ A}$$

13-20 Current Ratio

As stated above, the power demanded by the load in the secondary comes from, and is equal to, the power supplied by the primary (assuming no losses).

Therefore

$$P_p = P_s \tag{1}$$

$$\text{and} \quad V_p I_p = V_s I_s \tag{2}$$

$$\therefore \frac{V_p}{V_s} = \frac{I_s}{I_p} \qquad (3)$$

$$\text{but} \quad \frac{N_p}{N_s} = \frac{V_p}{V_s} \left\{ \begin{array}{l} \text{voltage/turns} \\ \text{ratio formula} \end{array} \right. \qquad (4)$$

Substitute V for I in (3)

$$\text{and} \quad \frac{N_p}{N_s} = \frac{I_s}{I_p} \left\{ \begin{array}{l} \text{current/turns} \\ \text{ratio formula} \end{array} \right. \qquad (5)$$

The voltage/turns ratio equation [Eq. (4)] shows the units in direct proportion to one another, but the current and turns in the current/turns ratio equation [Eq. (5)] are inversely proportional to one another. In other words, when the turns ratio steps up the voltage, the current supply will be stepped down for the same power demanded.

PROBLEM 13-19

A step-down transformer with a 20/1 ratio is connected to a 6-Ω load. What are the secondary and primary currents if 120 V is applied to the primary?

SOLUTION   *(cover solution and solve)*   20 / 1

Given:

$$\frac{N_s}{N_p} = \frac{1}{20}$$
$$R_s = 6 \ \Omega$$
$$V_p = 120 \ \text{V}$$

$$\text{from} \quad \frac{V_s}{V_p} = \frac{N_s}{N_p}$$

$$\text{then} \quad V_s = \frac{V_p N_s}{N_p}$$

$$V_s = \frac{120 \times 1}{20} = 6 \ \text{V}$$

$$I_s = \frac{V_s}{R_s} = \frac{6}{6} = 1 \ \text{A}$$

$$\text{from} \quad \frac{N_p}{N_s} = \frac{I_s}{I_p}$$

$$I_p = \frac{V_s I_s}{V_p}$$

$$\text{then} \quad I_p = \frac{N_s I_s}{N_p} = \frac{1 \times 1}{20} \times 10^3$$

$$= 50 \ \text{mA}$$

If the transformer in the previous problem had been a step-up transformer of a 1/20 ratio, what would be the secondary and primary currents if a 6-Ω resistor were connected across the secondary?

SOLUTION   (cover solution and solve)

Given:

$$\frac{N_s}{N_p} = \frac{20}{1}$$
$$R_s = 6\ \Omega$$
$$V_p = 120\ \text{V}$$

$$\text{from}\quad \frac{V_s}{V_p} = \frac{N_s}{N_p}$$

$$\text{then}\quad V_s = \frac{V_p N_s}{N_p}$$

$$\text{and}\quad V_s = \frac{120 \times 20}{1} = 2400\ \text{V}$$

$$\text{but}\quad I_s = \frac{V_s}{R_s} = \frac{2400}{6} = 400\ \text{A}$$

$$\text{from}\quad \frac{N_p}{N_s} = \frac{I_s}{I_p}$$

$$I_p = \frac{N_s I_s}{N_p} = \frac{20 \times 400}{1}$$

$$= 8000\ \text{A}$$

Note that the current in the secondary winding of the step-down transformer is larger than the primary current, but that the voltage has been stepped down. In the step-up transformer, the converse is true.

13-21   Total Power in Power Transformers

The power (wattage) which each secondary winding demands must come from the primary winding. The total power (wattage) supplied by the primary is the sum of the power demand (wattage) of the secondary loads.

The current that will flow in the primary is calculated by dividing the primary voltage into the total of the secondary wattages.

EXAMPLE (SEE FIG. 13-28)

$$\text{From} \quad P_p = I_p V_p = P_{s_1} + P_{s_2} + P_{s_3}$$

$$\text{then} \quad I_p = \frac{P_{s_1} + P_{s_2} + P_{s_3}}{V_p}$$

$$I_p = \frac{380 + 75 + 25}{120} = 4 \text{ A}$$

13-22   Transformers and Power Transfer

Maximum power can be transferred from a source of voltage to the load only when the load resistance is equal to the resistance of the source of voltage. This fact was discussed and demonstrated in Sec. 5-7.1. Impedance has properties very similar to resistance, but impedance exists only where there is ac voltage. Transformers can operate only on ac voltage. Therefore, the term impedance will be used interchangeably with resistance in this discussion about maximum power transfer and impedance matching.

The load offered by speakers and transistors is usually referred to as impedance. It is obvious from the discussion on maximum power transfer in Sec. 5-7.1 that nearly all the power would be dissipated in the internal impedance of a transistor if it had a value of 50 Ω and it were connected in series with a 4-Ω speaker impedance. In such a case, very little power would be transferred to the speaker and the speaker would produce almost no signal output. Connection of a 4-Ω speaker to a 4000-Ω impedance would make the condition even worse.

One of the most important features of a transformer in an electronic circuit is its ability to make a small load impedance appear to be large and equal to the large internal impedance of a voltage source, or to make the small impedance of a voltage source appear large and equal to a large load impedance. This transformation is called impedance matching. Impedance matching is one of the most common uses of transformers other than providing power for power supplies. Some examples where transformers are used for impedance matching are:

1. Matching loudspeakers to audio amplifiers or stereo equipment
2. Matching deflection coils to power amplifiers or transistors in television receivers, television cameras, and radar receivers.
3. Matching any low-resistance device to a high-resistance device, and vice versa.

The transformer, therefore, has the ability to effectively transform impedance; that is, to step up or step down impedance as demonstrated in the following development of the impedance matching equation for transformers.

13-22.1 Deriving the Impedance Transformation Equation

The voltage ratio equation is

$$\frac{N_p}{N_s} = \frac{V_p}{V_s} \tag{6}$$

The current ratio equation is

$$\frac{N_p}{N_s} = \frac{I_s}{I_p} \tag{7}$$

Multiply equation (6) by (7) and

$$\left(\frac{N_p}{N_s}\right)^2 = \frac{V_p}{V_s} \times \frac{I_s}{I_p} \tag{8}$$

Interchange the denominators on the right half of Eq. (8) and

$$\left(\frac{N_p}{N_s}\right)^2 = \frac{V_p}{I_p} \times \frac{I_s}{V_s} \tag{9}$$

$$\text{but} \quad \frac{V_p}{I_p} = Z_p \tag{10}$$

$$\text{and} \quad \frac{I_s}{V_s} = \frac{1}{Z_s} \tag{11}$$

Substitute Eq. (10) and Eq. (11) in Eq. (9) and

$$\left(\frac{N_p}{N_s}\right)^2 = Z_p \times \frac{1}{Z_s} \tag{12}$$

Therefore

$$\left(\frac{N_p}{N_s}\right)^2 = \frac{Z_p}{Z_s} \tag{13}$$

$$\text{and} \quad \frac{N_p}{N_s} = \sqrt{\frac{Z_p}{Z_s}} \tag{14}$$

$$\frac{\text{number of turns in the primary}}{\text{number of turns in the secondary}} = \sqrt{\frac{\text{impedance in the primary side}}{\text{impedance in the secondary side}}}$$

$$\frac{N_p}{N_s} = \sqrt{\frac{Z_p}{Z_s}}$$

The transformer can, therefore, transform impedance values in the same way that it transforms voltage values.

The following problem will prove that a transformer will match a small impedance to a large impedance.

PROBLEM 13-21

A voltage source has an internal impedance of 3600 Ω. The load is 100 Ω. What transformer turns ratio will make the two impedances appear equal?

SOLUTION   *(cover solution and solve)*

Given:

Primary impedance = 3600 Ω

Secondary impedance = 100 Ω

$$\frac{N_p}{N_s} = \sqrt{\frac{Z_p}{Z_s}}$$

$$\frac{N_p}{N_s} = \sqrt{\frac{3600}{100}}$$

$$= \frac{6}{1}$$

The two impedances will appear equal if the primary winding has 6 turns for every 1 turn in the secondary winding.

PROBLEM 13-22

An audio transformer is required to match the 4-Ω impedance of a speaker to the 64-Ω output impedance of a power transistor. Calculate the required turns ratio of the transformer.

SOLUTION   *(cover solution and solve)*

Given:

Primary impedance = 64 Ω

Secondary impedance = 4 Ω

$$\frac{N_p}{N_s} = \sqrt{\frac{Z_p}{Z_s}}$$

$$\frac{N_p}{N_s} = \sqrt{\frac{64}{4}}$$

$$= 4/1 \text{ turns ratio}$$

The transformer must have four turns in the primary for every turn in the secondary winding in order to match the two impedances and permit maximum power transfer.

13-23   Transformer Power Losses

Most operating power transformers feel warm to the touch. Underrated or poor-quality transformer feel hot to the touch. The heat is power loss which is being dissipated and lost into the air. The power losses are due to three factors:

1. Hysteresis losses (see Section 12-7.1)
2. Copper losses (I^2R losses in the resistance of the copper wire windings)
3. Eddy current losses

13-23.1   Hysteresis Loss

Hysteresis loss is reduced by using an iron core with the highest permeability cost will permit (see Sec. 12-7.1).

13-23.2   Copper Loss

Copper loss is the I^2R heat loss which occurs in the wire of transformers. Copper loss can be reduced by operating the primary and secondaries with a wire size (AWG size) which is rated at or above the size required to carry the current.

13-23.3   Eddy Current Loss

Magnetic fields building up in the primary and secondary windings induce currents into the core material. These currents are called eddy currents and could become quite large if the core material were left as a solid piece of iron, because the currents would circulate through the cross-section of the core. The currents produce an I^2R power loss with the resistance of the core material. To reduce eddy current losses in power transformers, the core is made of very thin sheets of iron called laminations which have been dipped in high-voltage insulating enamel and bolted together.

13-24   Practical Considerations

The leads of multiple winding (secondaries) transformers are color coded so they may be identified one from the other as to what voltages they supply. See Fig. 13-28.

Power transformers are selected from the electronic or transformer catalogue according to the minimum voltage and current rating required of each of the secondary windings.

13-25   Autotransformers

Some transformers such as those used in the high voltage circuits of television sets to produce 20 kV for the picture tube, or those used in auto-

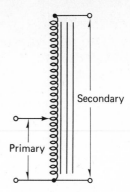

FIGURE 13-29   Autotransformer.

mobile ignition systems (coils) to step up the 12-V battery voltage to 20 kV for the spark plugs, are a type which have a tap on the secondary winding.   The tap serves as the primary winding; such transformers are called autotransformers (Fig. 13-29).

13-26   *L/R* Time-Constants (Rise)

A timing circuit can be made by connecting a resistor in series with an inductance and a source of voltage (see Fig. 13-30).   The time necessary for the voltage to increase across the resistor to some specific value can be determined—or changed—by the value of the resistor and the inductance.   Hundreds of *L/R* circuits were used in early computers as time-delay circuits, primarily to prevent a problem known as "race."

13-26.1   *L/R* Time Constants

L/R time constants, contrary to what their name implies, are not used today for timing circuits (RC and flip-flop *IC* circuits are more convenient for this), nor are these circuits used for shaping voltage waveforms (because of cost and space).   The knowledge of *L/R* time constants is useful, however, in the design of energy storage circuits such as deflection circuits in radar receivers, television receivers, and in automobile ignition systems.

The value of the inductance, in henries, divided by the resistance in ohms gives the time in seconds which it takes for the current in a circuit, or the voltage across the resistor, to rise to 63.2% of maximum value.

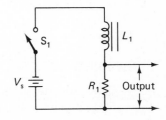

FIGURE 13-30   *L/R* time-constant circuit.

DEFINITION #1

One time constant is the time, in seconds, which it takes for current in an L/R circuit to rise to 63.2% of the steady-state current value.

$$1 \text{ time constant} = \frac{\text{inductance}}{\text{resistance}}$$

$$1 \ TC = \frac{L}{R}$$

where L is in henries
       R is in ohms
       TC is in seconds

the symbol for TC is the greek letter τ (tau)

DERIVING MATHEMATICALLY THE RELATIONSHIP BETWEEN TIME, RESISTANCE, AND INDUCTANCE

From the induced voltage equation,

$$v_L = L \frac{di}{dt} \quad \text{(Sec. 13-11)} \tag{15}$$

comes

$$L = \frac{v_L}{\dfrac{di}{dt}} = v_L \frac{dt}{di} \tag{16}$$

Divide both sides by R and

$$\frac{L}{R} = \frac{v_L}{R} \times \frac{dt}{di} \tag{17}$$

From Ohm's Law

$$\frac{v_L}{R} = I \tag{18}$$

Substitute I for v_L/R in Eq. (17) and

$$\frac{L}{R} = \cancel{I} \frac{dt}{d\cancel{i}}$$

$$\frac{L}{R} = t$$

$$1 \ TC = \frac{L}{R}$$

PROBLEM 13-23

Calculate the time constant of a 10-H coil connected in series with a 500-Ω resistor and a 20-V battery.

$$1 \ TC = \frac{L}{R}$$

$$= \frac{10}{500} \ s$$

$$= \frac{10}{500} \times 10^3$$

$$= 20 \ ms$$

PROBLEM 13-24

How much time, in seconds, will it take for current to rise to 63.2% of its steady-state (maximum) value in a circuit in which a 16-V source of voltage is connected in series with a 4-kΩ resistor and a 12-mH coil?

SOLUTION   *(cover solution and solve)*

Current rises to 63.2% of its maximum value in 1 time constant (τ).

$$1 \ TC \ (\tau) = \frac{L}{R}$$

$$= \frac{12 \times 10^{-3}}{4 \times 10^3} \ s$$

$$= 3 \times 10^{-6} \ s$$

$$\tau = 3 \ \mu s$$

13-26.2   *Explanation for the Exponential Curve*

The time constant determines the time, in seconds, which it takes for the current in an inductive-resistive circuit to increase in magnitude from zero to 63.2% of the maximum (steady-state) value. The increase in the magni-

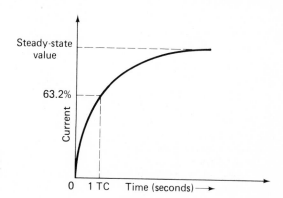

FIGURE 13-31   An   exponential curve showing the percentage of steady-state current after 1 TC.

tude of the current is almost a linear increase up to the 63.2% value; then the increase slows down in its rate of change and completes its increase to the steady-state value at a rate which graphs as a curved line (see Fig. 13-31). The total curve is called an exponential curve. The voltage developed across the resistor is a product of circuit current and the resistance; therefore, the voltage across the resistor in a resistive-inductive circuit increases exponentially and not at a steady linear rate. To better understand the time constant formula and the exponential increase in the voltage across the resistor, the behavior of each separate component operating in a dc circuit must first be considered.

RISE OF CURRENT IN A PURE RESISTANCE CIRCUIT

According to Kirchhoff's Law, the sum of the voltage drops across the loads in a series circuit must always equal the source voltage. This statement means that the current in the circuit rises to the maximum value the instant the switch is closed, and must remain at that value until the switch is opened. Figure 13-32 shows the waveform which would be produced across the resistor if the switch were opened 4 s after it was closed.

The most important fact to remember about a pure resistive circuit is that the maximum current value is referred to as the steady-state current value. In summary—the current in a pure resistive circuit reaches steady-state current value instantly.

RISE OF CURRENT IN A HYPOTHETICAL PURE INDUCTIVE CIRCUIT

It is not possible to make a coil which has pure inductance, because the coil is wound with wire and wire has resistance. However, it is possible to theorize about the events in a hypothetical pure inductance circuit.

In the study of the production of induced voltage, it was seen that when the dc voltage was first applied to a coil, a countervoltage was instantly produced. This induced voltage ($-v_L$) is always equal in absolute value to the source voltage (V_s). For example, suppose the source voltage is 100 V. The induced voltage (v_L) will be 100 V.

$$V_s = v_L = 100 \text{ V}$$

$$= L\frac{di}{dt} = 100 \text{ V}$$

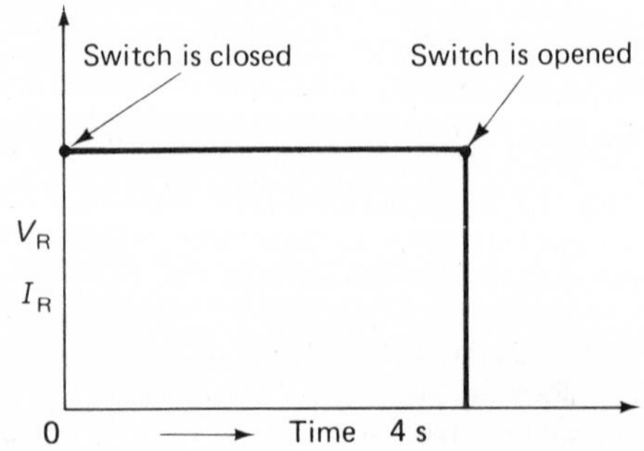

FIGURE 13-32 Current in a resistive circuit increases to maximum instantly. Notice that the waveform is rectangular or square in shape.

The induced voltage, v_L, in a hypothetical pure inductive circuit will not change in value since it is equal to V_s; therefore, it has a constant value. The value of the inductance, L, is also constant.

If both the induced voltage v_L and the inductance L are constant in value, then the ratio of v_L/L is also constant in value.

The equation for induced voltage is

$$v_L = L \frac{di}{dt}$$

from which the equation

$$\frac{v_L}{L} = \frac{di}{dt}$$

can be derived. Thus the rate of change of current is equal to the ratio of the induced voltage over the inductance value.

Since v_L/L is a constant value in a pure inductive circuit,

$$\textbf{rate of change of the current} = \frac{di}{dt}$$

must be constant (fixed) in value.

Time is never constant. Because time is always changing in value, the value of the current has to change (di) because the ratio of di/dt is a constant value

$$\frac{di}{dt} = \frac{v_L}{L}$$

In summary—the value of the current in a pure inductive circuit never reaches a steady-state value but keeps increasing in value as time changes;

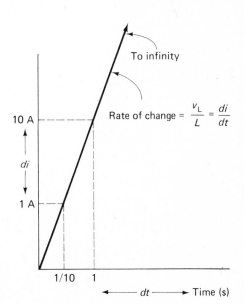

10 A — — — — — Rate of change $= \dfrac{v_L}{L} = \dfrac{di}{dt}$

To infinity

di

1 A

1/10   1

dt Time (s)

FIGURE 13-33 Current in a pure inductive circuit continues to increase toward infinity at a di/dt rate. Note that the rate of change is a linear rise.

at infinite time the current theoretically will reach infinite value (see Fig. 13-33).

EXAMPLE OF CURRENT RATE OF CHANGE (ROC)

If a 10-H coil is connected in series with a voltage source (V_s) of 100 V, the initial induced voltage v_L is equal to V_s; therefore, $v_L = 100$ V. The initial rate of change of current (ROC) from the transposed equation for induced voltage is

$$\frac{v_L}{L} = \frac{di}{dt}$$

$$= \frac{100 \text{ V}}{10 \text{ H}} = 10 \text{ A/1 s}$$

$$\text{therefore} \quad \frac{di}{dt} = 10 \text{ A/1 s}$$

$$\text{or} \quad = 1 \text{ A/(1/10) s}$$

In other words as time passes

Table 13-1 Rate of Change of Current

In	Current will be
1/10 s	1 A
1 s	10 A
2 s	20 A
3 s	30 A
4 s	40 A
Infinity s	Infinity A

From Table 13-1, it can be seen that the current in a hypothetical pure inductive circuit would continue to increase in value at a linear rate until at infinite time the value of the current would be infinite in value. This can be plotted on a graph as shown in Fig. 13-33.

In summary—three important points must be noted about a pure inductive circuit:

1. There is no steady-state current value.
2. The current value in a pure inductive circuit increases in value at the initial rate of change, (ROC) as determined by di/dt.
3. The value of di/dt is determined by the ratio of v_L/L.

RISE OF CURRENT IN A PRACTICAL INDUCTOR CIRCUIT

In practice, all coils do have dc resistance and most circuits have resistance added to the coil circuit; therefore, there will be a steady-state

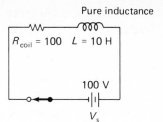

$R_{coil} = 100$ $L = 10$ H

100 V

V_s

FIGURE 13-34 *L/R* circuit showing the resistance of the coil as removed and separate from the coil.

current value in a practical inductive circuit, as set by the resistance of the circuit.

In the discussion and calculation of *L/R* circuits, the coil resistance is always thought of as being removed from the coil and added to the circuit resistance, and the coil itself is considered to be a pure inductance (see Fig. 13-34). Therefore, a practical inductive circuit will have an initial rate of change of current as well as a steady-state current value.

As shown in the graph of Fig. 13-33, the current in a hypothetical pure inductive circuit will keep increasing to infinity because there is no current-limiting resistor. Of course, infinity current is not possible because no source of voltage could supply such a value. There is always, of course, resistance in practical circuits.

Notice that the current rate of change formula

$$\frac{di}{dt} = \frac{v_L}{L} \quad \text{shows that}$$

1. The rate of change of the current varies *inversely* with L (a small L = a large rate of change).
2. The rate of change varies *directly* with the source voltage (a large V_s = a large rate of change because v_L always equals V_s).

DEVELOPMENT OF THE EXPONENTIAL CURVE IN A PRACTICAL *L* CIRCUIT

The circuit shown in Fig. 13-35 is a typical resistive-inductive circuit and will now be used along with the curves shown in Fig. 13-36 to explain the development of exponential curves (curves A and B). An exponential curve is actually a combination of the square wave curve (Fig. 13-32) developed in a pure resistive circuit and a linear rise curve (Fig. 13-33) produced in a hypothetical pure inductive circuit.

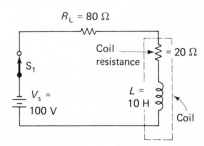

$R_L = 80$ Ω

Coil resistance $= 20$ Ω

S_1

$V_s = 100$ V

$L = 10$ H

Coil

FIGURE 13-35 An *L/R* time-constant circuit. The coil resistance plus the load resistance $= R_T =$ 100 Ω. The voltage V_R is the voltage across R_T.

Sec. 13-26 L/R Time-Constants (Rise)

Two facts must be kept in mind during the explanation of the development of an exponential curve:

1. The sum of all voltage drops, in a series circuit, whether they are inductive or resistive voltage drops, must be equal to the source voltage.
2. The sum of all voltage drops opposes the source voltage and controls the value of current flow.

The switch is closed (Fig. 13-35) to the source voltage, and circuit current attempts to flow and to reach its steady-state value instantly, but the equal and opposing voltage induced (v_L) in the coil (see curve B, Fig. 13-36) opposes this initial attempted instant current rise. Since no current can flow the moment that the switch is closed, there can be no voltage developed across the circuit resistance (R_T) and the value of v_R is zero (curve A, Fig. 13-36).

the letter symbol for a changing voltage is v

di is the letter symbol for a changing current

The initial current which does flow changes at a rate of di/dt. A study of Table 13-1 shows that in the example circuit,

$$\frac{di}{dt} \text{ in } \frac{1}{100} \text{ s} = \frac{1}{10} \text{ A}$$

$$\text{and in } \frac{1}{10} \text{ s} = 1 \text{ A}$$

The important fact to observe is that a very small current will begin to flow a fraction of a microsecond after the switch is closed, and then will increase in value at a di/dt rate. The result, initially, will be a very small voltage, then an increasing voltage developed across the circuit resistance

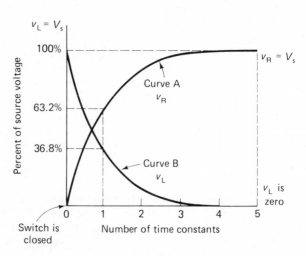

FIGURE 13-36   The increase in the voltage across the resistor and the decrease in the voltage across the coil at an exponential rate. Note the number of time constants, not total time involved.

(R_T) (curve A, Fig. 13-36), and a subsequent decreasing induced voltage in the coil (curve B, Fig. 13-36).

$$\text{Note} \quad v_R + v_L = V_s \quad \text{(KVL)}$$

The decrease in the induced voltage causes a decrease in the di/dt. A reduced di/dt produces a smaller opposing induced voltage because

$$v_L = L\frac{di}{dt}$$

and the circuit current increases, producing a larger v_R. The amount by which the voltage across the resistor (v_R) increases is the amount that the induced voltage v_L decreases at that instant. (The sum of curve B plus curve A at any instant is always equal to the supply voltage.)

The interdependent process of the increasing voltage across the resistor (v_R) and the decreasing induced voltage (v_L) forms a closed loop until the voltage across the resistor (v_R) is equal to the supply voltage V_s, and the induced voltage (v_L) is zero. Exponential curves are produced because the circuit current increases by the same percentage of the remaining steady-state current value during each given period in time (called a time constant) until the circuit current rise reaches the steady-state value.

In summary—the amount of *change* in current decreases with each change in time. Another definition for one time constant can be derived from the above explanation.

13-26.3 *Another Definition for One Time Constant*

DEFINITION #2

One time constant is the time necessary for the current in a circuit to reach its steady-state (V_s/R_T) value if the current were to rise continuously at its initial rate of change.

$$\frac{di}{dt} = \frac{v_L}{L}$$

PROOF FOR DEFINITION

Experimentation with different inductance values, resistance values, and voltage source values would show that the value of the inductance in henries divided by the value of the resistance in ohms will always produce a period of time in seconds, in which the circuit current would reach its steady-state value (V_s/R) if the initial rate of change of the current (di/dt) = v_L/L were not impeded by the resistance of the circuit. The ROC curve (straight curve) shown in Fig. 13-37 is derived from the values of the components in the circuit in Fig. 13-35. The circuit has the following values.

$$R_T = 100\ \Omega$$

$$L = 10 \text{ H}$$

$$V_s = 100 \text{ V}$$

$$\text{time} = \frac{L}{R} = \frac{10}{100} = \frac{1}{10} \text{ s (1 time constant)}$$

$$\text{steady-state current value } (I_{ss}) = \frac{V_s}{R} = \frac{100}{100} = 1 \text{ A}$$

If there were no resistance in the circuit, the current rate of change would be:

$$\frac{v_L}{L} = \frac{di}{dt} = \frac{V_s}{L} \quad \text{(the first instant that the switch is closed)}$$

$$= \frac{100}{10} = 10 \text{ A/1 s}$$

$$\text{or} \quad = 1 \text{ A in } \frac{1}{10} \text{ s}$$

From the di/dt calculation, the current would be 1 A in 1/10 s if there were no resistance to impede the current flow.

One amp is the steady-state current value; 1/10 s is the value obtained from di/dt. The rate of change calculation verifies the statement about the circuit components' relationship.

The above definition for one time constant is a theoretical definition useful only for explaining the relationship between the circuit components of an L/R circuit. The first definition (in Sec. 13-26.1) is a working definition and is used to calculate current, voltage, and time values in L/R timing circuits. Note: the source voltage is not part of the time constant equation.

13-26.4 Time Constants and Current Progression Percentage

The time constant shows the relationship among the circuit components which causes the circuit current to rise from 0 to 63.2% of the steady-state current value in one time constant. The percentage value of current will differ from 63.2% at different periods in time, such as at 2 TC, 3 TC, 4 TC and 5 TC. The progressive percentage method, Table 13-2, proves this fact.

The progressive percentage method is not a practical solution for calculating current percentages for a given number of time constants, but it illustrates in steps how current values do change with a given number of time constants. It, therefore, explains the shape of the exponential curve.

PROGRESSIVE PERCENTAGE METHOD FOR THE EXPONENTIAL CURVE

The interpretation of the first definition for a time constant indicates that:

Table 13-2   Progressive Percentage Method

EXAMPLE:

Refer to Fig. 13-35.

$$I_{max} = \frac{V_s}{R} = \frac{100}{1} = 100 \text{ A}$$

TC 1:

current increase = 63.2% $\times$ 100 = 63.2 A

Remaining current to increase 100 − 63.2 = 36.8 A.

TC 2:

current increase = 63.2% of 36.8 = 23.26 A

Remaining current to increase 100 − (63.2 + 23.26) = 13.54 A.

TC 3:

current increase = 63.2% of 13.54 = 8.56 A

Remaining current to increase 100 − (63.2 + 23.26 + 8.56) = 4.98 A.

TC 4:

current increase = 63.2% of 4.98 = 3.14 A

Remaining current to increase 100 − (63.2 + 23.26 + 8.56 + 3.14) = 1.84 A.

TC 5:

current increase = 63.2% of 1.84 = 1.16 A

Remaining current to increase 100 − (63.2 + 23.26 + 8.56 + 3.14 + 1.16) = 0.993 A.

The current is considered to have reached its steady-state value after 5 TC.

In the first time constant, the current increases to 63.2% of the steady-state (maximum) current value.

In the second time constant, the current increases 63.2% of the remaining current value, and so on, until the current has reached its maximum value. The total time is 5 TC.

The example used in Table 13-2 shows in detail the progressive percentage increase in current that accounts for the exponential curve.

13-26.5 *Graphical Method of Obtaining Current Percentage*

The progression method described above is almost useless when calculating percentage of current flow for fractions of time constants, and at best, it is a tedious and cumbersome method. A rapid but less accurate method is to estimate the current percentage values from standard exponential graphs such as the curve in Fig. 13-37.

The exponential graph in Fig. 13-37 is also called a universal time-constant chart because it can be used for charge conditions in resistive-capacitive circuits as well as for rise of current conditions in resistive-inductive circuits.

The expression "percentage of current flow" is often referred to as the value of instantaneous current. The word "instantaneous" means the value of current at any unit (or parts thereof) of time constants; the value of instantaneous current obtained is expressed as a percentage of the maximum (steady-state) circuit current.

The universal time constant chart may be used to derive values of instantaneous current or voltage for a specific number of time constants in the following manner:

1. Suppose one wishes to determine the *percentage value* of maximum circuit current in an inductive-resistive circuit, two time constants in time after the switch is closed to the supply voltage.

2. Refer to the universal time constant circuit (Fig. 13-37) and locate the number 2 on the horizontal axis which is calibrated in time constants.

3. Draw a vertical line from the number 2 on the horizontal axis until it intersects the curve.

4. Draw a horizontal line to the left from the point where the vertical line intersects the curve, until the horizontal line intersects the vertical (Y) axis which is calibrated in percentages.

5. The percentage value (86.5%), intersected on the Y-axis by the

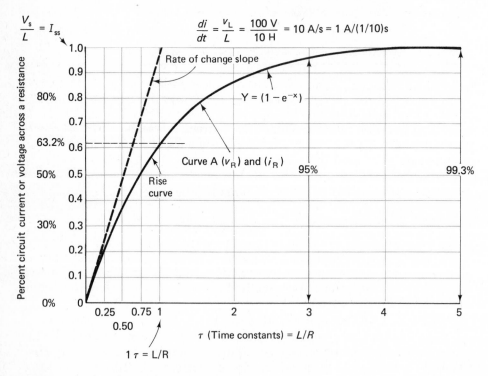

FIGURE 13-37   Time-constant chart.

horizontal line, is the *percentage of* maximum current or supply voltage for the specific instant in time equal to two time constants.

Use this procedure and the universal time-constant chart in Fig. 13-37 to determine the *percentage of* maximum current or supply voltage value for the following time-constant problems.

PROBLEM 13-25

What is the percentage current flow at the following time constants (from the graph in Fig. 13-37)?

1. 0.5 TC
2. 0.75 TC
3. 1.5 TC
4. 3.0 TC

SOLUTION   *(cover solution and solve)*

1. 0.5 TC $\simeq$ 37%
2. 0.75 TC $\simeq$ 51%
3. 1.5 TC $\simeq$ 80%
4. 3 TC $\simeq$ 95%

13-26.6 *Percent, Instantaneous Current and Voltage*

The value of current which is less than steady-state value is referred to as instantaneous current, because it is the value of current at any instant in time, or at any value of time constant.

Once the percentage of steady-state current is determined, the value of instantaneous current may be obtained by multiplying the percentage value by the steady-state current value.

$$i = \% \times I_{max}$$

the letter i is the symbol used for instantaneous current

the letter symbol for instantaneous voltage across a resistor is v_R

The value of voltage across a resistor at any instant in time, or at any value of time constant, is obtained by multiplying the instantaneous current value by the resistance.

$$v_R = i \times R$$

Another way of calculating v_R is

$$v_R = \% \times V_s$$

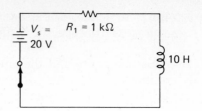

FIGURE 13-38

PROBLEM 13-26

1. What will be the percentage of the steady-state current if the switch is closed to a 10-H coil in series with a 1-kΩ resistor and a 20-V source voltage for 3 TC?

2. What will be the value of the instantaneous current after 3 TC?

3. What value of voltage will be measured across the resistor at the end of the third TC?

SOLUTION   *(cover solution and solve)*

1. Find the percent of current flow at 3 TC by referring to the universal time-constant chart and follow Steps 1 to 5 in Sec. 13-26.5. The current will be 95% of I_{max} in 3 TC.

2. The value of instantaneous current after 3 TC is

$$i = \% \times I_{ss}$$

$$\text{but} \quad I_{ss} = \frac{V_s}{R} = \frac{20}{1 \times 10^3} \times 10^3$$

$$= 20 \text{ mA}$$

$$i = 95 \times (20 \times 10^{-3})$$

$$= 19 \text{ mA}$$

3. The voltage (v_R) across the resistor after 3 TC is

$$v_R = \% \times V_s$$

$$v_R = 95\% \text{ of } 20$$

$$= 19 \text{ V}$$

13-26.7   Use of Epsilon to Solve Exponential Curve Problems

Both the progressive percentage method and the graphical method of calculating percentage current for specific time constant values have disadvantages. The most accurate and rapid method of making these calculations is with the use of epsilon, ϵ, the natural log, because exponential curves are functions of the natural log. The Greek letter epsilon is used to represent the natural log.

FINDING THE NATURAL LOG (ln x) OF A NUMBER

Natural logs are used in much the same way as are common logs. The difference between them is that with common logs the number 10 is used as the base, whereas with natural logs the number 2.71828 is used as the base. The symbol ln x is used to indicate $\log_{2.71828}$ (ln is pronounced "lon"). The letter ϵ is used in place of both the ln x symbol and $\log_{2.71828}$ notation.

In other words, when one sees the symbol $e^x = 7.389$, x represents the natural log of 7.389. The value of the exponent x may be found from the natural log tables or with the use of a scientific calculator. (Punch 7.389; then the ln x button.) The value of the exponent x for the number 7.389 is 2.

The natural log (ln x) of 20.0855 is 3. (Punch 20.0855; then ln x.)

$$\ln 20.0855 = 3 \quad \therefore e^3 = 20.0855$$
$$\ln 2.71828 = 1 \quad \therefore e^1 = 2.71828$$
$$\ln 54.59815 = 4 \quad \therefore e^4 = 54.59815$$
$$\ln 148.413 = 5 \quad \therefore e^5 = 148.413$$

FINDING ln x OF A NUMBER WHEN THE EXPONENT IS NEGATIVE (e^{-x}) (RECIPROCAL OF e^x)

The procedure is similar to that of taking ln x of a number, except that the reciprocal of the number must be taken before the log is taken.

Recall the law of exponents for a negative exponent.

$$a^{-b} = \frac{1}{a^b} \tag{19}$$

$$\text{If} \quad a^{-b} = C \tag{20}$$

Substitute (19) in (20)

$$\text{then} \quad \frac{1}{a^b} = C \tag{21}$$

$$\text{or} \quad a^b = \frac{1}{C} \tag{22}$$

EXAMPLES

Find the value of the exponent x for the following numbers:

1. e^{-x} of 0.30

SOLUTION:

One cannot compute a negative natural log.

$$\text{Reciprocal of } 0.30 = \frac{1}{0.30} = 3.33 = e^x.$$

Take the natural log (ln x) of 3.33 = 1.2

$$\therefore x = 1.2$$

Note that the log was taken of a number greater than 1. Thus the characteristic (1) of the natural log was positive.

2. e^{-x} of 50

SOLUTION:

One cannot compute a negative natural log.

Reciprocal of 50 = $\dfrac{1}{50}$ = 0.02 = e^x.

Take the log (ln x) of 0.02 = -3.912

$$\therefore x = -3.912$$

Note that the log (e^x) was taken of the number 0.02, which is less than 1. The log answer on the calculator was negative. The negative sign can be represented by the bar sign over the characteristic ($-3 = \bar{3}$) for all numbers less than 1. Recall that the mantissa is always positive and that the bar sign refers only to the characteristic. The bar sign is ignored when the exponent x represents the number of time constants.

3. e^{-x} of 6

SOLUTION:

One cannot compute a negative natural log.

Reciprocal of 6 = $\dfrac{1}{6}$ = 1.6667 = e^x.

Take the log (ln x) of 1.6667 = -1.791

$$\therefore x = -1.791$$

FINDING THE ANTI-LOG (e^x) OF THE NATURAL LOG

The anti-log of a natural log is the mathematical operation which is performed to find the value of a number when the exponent is known. The answer may be obtained by the use of the anti-log tables for natural logs, or by using the calculator (punch the value of x; then the (e^x) button). (Note that we are now using "e" to represent the natural log.

EXAMPLES

Find the antilog of "e^x" for the following exponents:

1. e^2

[punch 2; then (e^x)] e^x = 7.389

2. $e^{1.4} = 4.055$

3. $e^4 = 54.598$

4. $e^1 = 2.71828$

5. $e^5 = 148.413$

FINDING THE ANTI-LOG WHEN THE EXPONENT IS NEGATIVE

It is not possible to directly find the anti-log (e^x) of a negative exponent. The procedure is to first take the anti-log of the absolute value of
the exponent, then take the reciprocal of the anti-log. The answer is the
anti-log of the negative exponent.

EXAMPLES

Find the value of the number for the following exponents x of e.

1. e^{-2}

SOLUTION:

One can take only the anti-log of e^x.

$$e^2 = \text{anti-log } (e^x) \text{ of } 2 = 7.389$$

Reciprocal of $7.389 = \dfrac{1}{7.389} = 0.13533 = e^{-2}$.

$$\therefore \ e^{-2} = 0.13533$$

2. e^{-5}

SOLUTION:

One can take only the anti-log of e^x.

$$e^5 = \text{anti-log } (e^x) \text{ of } 5 = 148.413$$

Reciprocal of $148.413 = \dfrac{1}{148.413} = 0.0067379 = e^{-5}$.

$$\therefore \ e^{-5} = 0.0067379$$

3. $e^{-0.6}$

SOLUTION:

One can take only the anti-log of e^x.

$$e^{0.6} = \text{anti-log } (e^x) \text{ of } 0.6 = 1.8221$$

Reciprocal of $1.8221 = \dfrac{1}{1.8221} = 0.5488 = e^{-0.6}$.

$$\therefore \ e^{-0.6} = 0.5488$$

SUMMARY OF PROCEDURES:

1. To find x of e^{-x} for a number
   a. Take the reciprocal of the number
   b. Take the log ($\ln x$) of the reciprocal
2. To find the number for e^{-x}
   a. Take the anti-log (e^x)
   b. Take the reciprocal of the resultant number

13-26.8 Natural Log Method of Solving Time Constant Problems

The percentage of current flow at any one specified time is a function of the natural log, e^x. Percentages are based on decimal values (values less than one), so the negative exponent of the natural log (e^{-x}) is used.

The expression e^{-x} represents the decimal value of current yet to flow; therefore, the value of e^{-x} must be subtracted from 1 to determine the decimal value of current which is actually flowing.

$$1 - e^{-x} = \text{current flowing}$$

The letter Y is used to indicate percentage. Percentage is obtained by multiplying the decimal value by 100.

The formula for the percentage of steady-state current (I_{max}) flowing for any one time constant or part thereof is:

$$\% I_{max} = Y = [(1 - e^{-x}) \times 100] \%$$

where Y is % of I_{max}
e is natural log (base = 2.71828)
x is number of time constants
1 is maximum value (100%)

PROBLEM 13-27

Calculate the percentage of current flowing or % of voltage across a resistor in an L/R circuit after

1. 1 TC
2. 1.4 TC
3. 2.5 TC
4. 5 TC
5. 4.5 TC
6. 0.8 TC

SOLUTION   *(cover solution and solve)*

1. 1 TC

$$Y = 1 - e^{-x} \times 100 = \% \text{ of } I_{max}$$
$$\text{but } x = 1$$
$$\therefore \ Y = 1 - e^{-1}$$

One can only take the anti-log of e^x.

$$e^1 = \text{anti-log } (e^x) \text{ of } 1 = 2.71828$$

$$\text{Reciprocal of } 2.71828 = \frac{1}{2.71828} = 0.3678 = e^{-1}.$$

$$e^{-1} = 0.3678$$

$$\therefore Y = 1 - 0.368$$

$$= 0.632$$

$$Y = 63.2\% \text{ of steady-state current } (I_{ss})$$

2. 1.4 TC

$$Y = 1 - e^{-1.4}$$

$$= 1 - 0.2465 = 0.753 = 75.3\% \text{ of } I_{ss}$$

3. 2.5 TC

$$Y = 1 - e^{-2.5}$$

$$= 1 - 0.082 = 0.9179 = 91.8\% \text{ of } I_{ss}$$

4. 5 TC

$$Y = 1 - e^{-5}$$

$$= 1 - 0.0067 = 0.993 = 99.3\% \text{ of } I_{ss}$$

5. 4.5 TC

$$Y = 1 - e^{-4.5}$$

$$= 1 - 0.0111 = 0.9889 = 98.89\% \text{ of } I_{ss}$$

6. 0.8 TC

$$Y = 1 - e^{-0.8}$$

$$= 1 - 0.4493 = 0.5507 = 55.07\% \text{ of } I_{ss}$$

13-26.9 Percent, Instantaneous Current, and Voltage (Recall from Sec. 13-26.6)

The value of current which is less than steady-state value is referred to as instantaneous current, because it is the value of current at any instant in time, or value of time constant.

Once the percentage of steady-state current is determined, the value of instantaneous current may be obtained by multiplying the percent value times the steady-state current value.

$$i = \% \times I_{max}$$

the letter i is the symbol used for instantaneous current

the letter symbol for instantaneous voltage across a resistor is v_R

The value of voltage across a resistor at any instant in time, or at any value of time constant, is obtained by multiplying the instantaneous current value by the resistance.

$$v_R = i \times R$$

Another way of calculating v_R is

$$v_R = \% \times V_s$$

PROBLEM 13-28

1. What will be the percentage of the steady-state current if the switch is closed to a 10-H coil in series with a 1-kΩ resistor and a 20-V source voltage for 3 TC?
2. What is the value of the voltage across the resistor after 3 TC?
3. What is the value of the instantaneous current in the circuit after 3 TC?

SOLUTION   (cover solution and solve)

1. $$Y = 1 - e^{-x} = \% \text{ of } I_{max}$$

$$\text{but } x = 3$$

$$\therefore \ Y = 1 - e^{-3}$$

One can take only the anti-log of e^x.

$$e^3 = \text{anti-log } (e^x) \text{ of } 3 = 20$$

$$\text{Reciprocal of } 20 = \frac{1}{20} = 0.04978 = e^{-3}.$$

$$e^{-3} = 0.04978$$

$$\therefore Y = 1 - 0.04978$$

$$= 0.95$$

$$Y = 95\% \text{ of steady-state current}$$

Current will reach 95% of I_{max} in 3 TC.

2. The voltage (v_R) across the resistor after 3 TC is

$$v_R = \% \times V_s$$

$$v_R = 95\% \text{ of } 20$$

$$= 19 \text{ V}$$

3. The value of instantaneous current after 3 TC is

$$i = \% \times I_{ss}$$

$$\text{but} \quad I_{ss} = \frac{V_s}{R} = \frac{20}{1 \times 10^3} \times 10^3$$

$$= 20 \text{ mA}$$

$$i = 95 \times (20 \times 10^{-3})$$

$$= 19 \text{ mA}$$

13-26.10 Methods of Determining Number of Time Constants

The number of time constants (x) were given in the preceding example problems. How does one determine the number of time constants if they are not given?

There are two ways of determining the number of time constants.

METHOD 1:

When the percentage of circuit current is known, the number of time constants, x is determined by

$$e^{-x} = 1 - Y \quad (\text{from } Y = 1 - e^{-x})$$

> where e is natural log
> x is number of time constants
> Y is % of maximum current flow

METHOD 2:

When the percentage of current is unknown the number of time constants is calculated by

$$x = \frac{t}{\tau}$$

> where x is number of time
> constants
> t is total time involved
> in seconds
> τ is time of one time
> constant in seconds

EXAMPLE OF METHOD #1 ─────────────

How many time constants will it take for current in a series circuit to reach 70% of its steady-state value?

SOLUTION   *(cover solution and solve)*

From $Y = 1 - e^{-x}$ $(= \% \text{ of } I_{max})$

then $Y - 1 = -e^{-x}$

$$\text{and} \quad e^{-x} = 1 - Y$$

$$\text{but} \quad Y = 70\%$$

$$\therefore e^{-x} = 1 - 0.70 = 0.30$$

$$= 0.30$$

One cannot compute a negative natural log.

Reciprocal of $0.30 = \dfrac{1}{0.30} = 3.333 = e^x$.

Take the log $(\ln x)$ of $3.333 = 1.2$.

$$\therefore x = 1.2 \text{ TC}$$

PROBLEM 13-29

How many time constants will it take for the current to reach 4.5 A after the switch in a series circuit is closed to an inductance of 4 H, a 2-Ω resistance and a 12-V source voltage (V_s)?

SOLUTION   *(cover solution and solve)*

Number of time constants for the instantaneous current to reach a value of 4.5 A:

1. Calculate I_{max}.

$$I_{max} = \frac{V_s}{R} = \frac{12}{2} = 6 \text{ A}$$

2. Calculate percentage which the instantaneous current is of I_{max}.

$$\% = \frac{4.5}{6} = 0.75 = 75\%$$

$$\therefore Y = 75\%$$

3. Determine number of time constants, x.

$$\text{from} \quad Y = 1 - e^{-x}$$

$$e^{-x} = 1 - Y$$

$$= 1 - 0.75$$

$$= 0.25$$

One cannot compute a negative natural log.

Reciprocal of $0.25 = \dfrac{1}{0.25} = 4 = e^x$.

Take the log $(\ln x)$ of $4 = 1.386$.

$$\therefore x = 1.386 \text{ TC}$$

METHOD 2:

When the percentage of current is unknown, the number of time constants is calculated by

$$\text{number of time constants} = \frac{\text{total time (in seconds)}}{\text{number of seconds in 1 TC}}$$

$$x = \frac{t}{\tau} = \frac{t}{\dfrac{L}{R}}$$

$$\text{or} \quad x = \frac{tR}{L}$$

ANALOGY FOR THE EQUATION FOR NUMBER OF TIME CONSTANTS

What is the number of cookie boxes if there are a total of 120 cookies and each box holds 10 cookies?

Answer:

$$\text{number of boxes} = \frac{\text{total cookies}}{\text{number of cookies in 1 box}}$$

$$\text{number of boxes} = \frac{120}{10} = 12 \text{ boxes}$$

EXAMPLE FOR METHOD #2

The switch to a 20-mH coil in series with a 2-kΩ resistor and a 12-V battery is left closed for 40 μs. For how many time constants is the switch left closed?

SOLUTION   *(cover solution and solve)*

$$1 \text{ TC } (\tau) = \frac{L}{R} = \frac{20 \times 10^{-3}}{2 \times 10^{3}} = 10 \ \mu\text{s}$$

Time involved is 40 μs.

$$\text{number of time constants } x = \frac{t}{\dfrac{L}{R}} = \frac{t}{\tau} = \frac{40 \ \mu\text{s}}{10 \ \mu\text{s}}$$

$$= 4 \text{ TC}$$

PROBLEM 13-30

For how many time constants has a switch been left closed to a 15-H coil in series with a 100-Ω resistor and a 25-V source of voltage, if it is left closed for 90 ms before it is opened?

SOLUTION   (cover solution and solve)

The number of TC during which the switch has been closed is

$$x = \frac{t}{\frac{L}{R}} = \frac{tR}{L}$$

$$= \frac{(90 \times 10^{-3}) \times 100}{15}$$

$$= 0.6 \text{ TC}$$

PROBLEM 13-31

The switch to a series circuit is closed to a 200-mH coil, a 20-Ω resistor and 12-V source of voltage for 6 TC. How long in seconds is the switch closed?

SOLUTION   (cover solution and solve)

$$\text{from} \quad x = \frac{t}{\frac{L}{R}}$$

$$\text{then} \quad t = \frac{xL}{R} \text{ s}$$

$$= \frac{6 \times (200 \times 10^{-3})}{20} \text{ s}$$

$$= 60 \text{ ms}$$

PROBLEM 13-32

What value of coil resistance must a 200-mH coil possess to permit current to rise in a circuit for 2 TC, if the circuit is closed for 4 ms?

SOLUTION   (cover solution and solve)

$$\text{from} \quad x = \frac{t}{\frac{L}{R}}$$

$$\text{then} \quad \frac{xL}{R} = t$$

$$\text{and} \quad R = \frac{xL}{t}$$

$$= \frac{2 \times (200 \times 10^{-3})}{4 \times 10^{-3}}$$

$$= 100 \ \Omega$$

PROBLEM 13-33

1. For how many time constants does the voltage increase across a 600-Ω resistor after the switch connects it to a 12-H coil and a 15-V battery for 10 ms?
2. Calculate the percentage of I_{ss} current flow after 10 ms.
3. What is the instantaneous current value after 10 ms?
4. What is the voltage across the resistor after 10 ms?

SOLUTION   (cover solution and solve)

1. Number of time constants during which the voltage increases

$$1 \text{ TC} = \frac{L}{R} = \frac{12}{600} \text{ s}$$

$$= 20 \text{ ms}$$

The switch is closed for 10 ms, which is equivalent to x time constants.

$$x = \frac{t}{\tau} = \frac{10 \times 10^{-3}}{20 \times 10^{-3}}$$

$$\therefore x = 0.5 \text{ TC}$$

2. The percentage of steady-state current after 10 ms

$$Y = 1 - e^{-x}$$

$$\text{but} \quad x = 0.5$$

One can take only the anti-log of e^x.

$$e^{0.5} = \text{anti-log } (e^x) \text{ of } 0.5 = 1.6487$$

Reciprocal of $1.6487 = 0.6065 = e^{-0.5}$.

$$Y = 1 - 0.6065$$

$$= 0.3935 \times 100 = 39.35\%$$

3. The value of instantaneous current after 10 ms

$$i = \% \times I_{ss}$$

$$\text{but} \quad I_{ss} = \frac{V_s}{R} = \frac{15}{600} \times 10^3$$

$$= 25 \text{ mA}$$

$$i = 0.3935 \times (25 \times 10^{-3})$$

$$= 9.83 \text{ mA}$$

4. Voltage across the resistor after 10 ms

$$v_R = i \times R$$
$$= (9.83 \times 10^{-3}) \times 600$$
$$= 5.9 \text{ V}$$

PROBLEM 13-34

An L/R circuit has a 4.8-kΩ resistor in series with a 3-mH inductance and a 9.6-V power source. Calculate the following values for the circuit, 1.125 μs after the circuit switch is closed:

1. Instantaneous current
2. Voltage across the resistor
3. Voltage across the inductor
4. Time for the current to reach its maximum value

SOLUTION   *(cover solution and solve)*

1.
$$\tau = \frac{L}{R} = \frac{3 \times 10^{-3}}{4.8 \times 10^3}$$
$$= 0.625 \ \mu s$$

Number of TC in 1.125 μs is

$$x = \frac{\text{time involved}}{\text{time in 1 TC}}$$

$$x = \frac{1.125 \ \mu s}{0.625 \ \mu s} = 1.8 \text{ TC}$$

Percentage of I_{ss} flowing after 1.125 μs is

$$Y = 1 - e^{-x}$$

$$\text{but} \quad x = 1.8$$

One can find only the anti-log of e^x.

$$e^{1.8} = \text{anti-log } (e^x) \text{ of } 1.8 = 6.0496$$

$$\text{Reciprocal of } 6.0496 = \frac{1}{6.0496} = 0.16529 = e^{-1.8}.$$

$$Y = 1 - 0.16529 = 0.8347$$
$$= 83.47\% \text{ of } I_{ss}$$

$$\text{but} \quad I_{ss} = \frac{V_s}{R} = \frac{9.6}{4.8 \times 10^3} = 2 \text{ mA}$$

Instantaneous current at 1.125 μs is

$$i = 0.835\% \times (2 \times 10^{-3})$$

$$= 1.67 \text{ mA}$$

2. Voltage across R_1 after 1.125 μs is

$$v_R = (1.67 \times 10^{-3}) \times (4.8 \times 10^3)$$

$$= 8.016 \text{ V}$$

3. Induced voltage

$$\text{from} \quad v_L + v_R = V_s$$

$$v_L = V_s - v_R$$

$$= 9.6 - 8.016$$

$$= 1.584 \text{ V}$$

4. Current reaches maximum value after

$$1 \text{ TC} \times 5$$

$$= 0.625 \ \mu s \times 5 = 3.125 \ \mu s$$

PROBLEM 13-35

A relay which has an inductance of 6 H and a resistance of 480 Ω has an ampere rating (contacts closed) of 40 mA. How quickly will the contacts close after the switch is closed to a 24-V source of voltage?

SOLUTION   *(cover solution and solve)*

1. Steady-state current value I_{ss} is

$$I_{ss} = \frac{V_s}{R} = \frac{24}{480} = 50 \text{ mA}$$

2. Time of 1 TC is

$$1 \text{ TC} = \frac{L}{R} = \frac{6}{480} \text{ s}$$

$$= \frac{6}{480} \times 10^3 = 12.5 \text{ ms}$$

3. Find the number of time constants it takes before the relay contacts close.
   a. Percentage of I_{ss} required to close relay contacts is

$$\% = \frac{I_{relay}}{I_{ss}} \times 100$$

$$= \frac{40 \times 10^{-3}}{50 \times 10^{-3}} \times 100 = 80\%$$

b. From $Y = 1 - e^{-x}$

then $e^{-x} = 1 - Y$

but $Y = 80\%$

$\therefore e^{-x} = 1 - 0.80$

$= 0.20$

Reciprocal of $0.20 = \dfrac{1}{0.20} = 5 = e^x$.

The log $(\ln x)$ of $5 = 1.609 = x$.
The contacts close in 1.609 TC. Total time to close relay contacts is

time = (time in 1 TC) $\times$ (number of TC)

$= (12.5 \times 10^{-3}) \times 1.609$

$= 20.11$ ms

13-27 *L/R* Decay Curves

The previous discussion on L/R circuits concerned the condition in which the switch is closed and the circuit current increases to a steady-state value. What happens when the switch is opened and the circuit is broken?

The voltage across a resistor exists only as long as current is present ($V = I \times R$). Voltage disappears the instant that the switch is opened. The current which flows in a resistor produces an $I^2 R$ power loss which is dissipated into space.

In a pure inductive circuit, the current is stored in the magnetic field surrounding the coil in the form of magnetic energy. A voltage (v_L) is induced into the coil by the collapsing stored magnetic energy when the switch is opened, and this produces a new current which flows in the same direction as did the previous battery current. The effect is that the v_L attempts to keep the current, which was flowing before the circuit was opened, from collapsing. It is this characteristic of an inductance which produces the exponential decay curve shown in Fig. 13-40.

13-27.1 *Explanation for the Exponential Decay Curve*

The decay curve is produced in the following manner:
If the switch S_1 were closed to A in the circuit shown in Fig. 13-39 for 5 TC or more, the circuit current would be $V_s/R = 100$ V/100 = 1 A. The 1 A is steady-state current flowing in the coil. There will be 1 A of energy* stored in the coil in the form of a magnetic field, which will attempt to collapse through the short and R_1 when the switch S_1 is

*Energy (W) = $1/2 L I^2$ J.

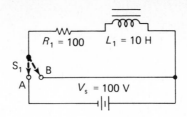

FIGURE 13-39

opened at A and closed at B. The resulting voltage (v_L) induced in the coil acts as a battery and causes the current to attempt to keep flowing in the circuit in the same direction as did the previous battery current. An IR voltage drop is produced across R_1, equal in value to the induced voltage v_L. The current gradually decreases in value in an exponential curve until it reaches zero. The current energy is dissipated in the air in the I^2R power loss of R_1.

13-27.2 *Time Constants and Percentage of I_{ss} on Decay*

The circuit current decay curve is shown in the time constant chart in Fig. 13-40. The curve represents

1. Decay of the circuit current
2. Decay of the induced voltage
3. Voltage drop across the resistor

Notice that the relationship between the percentage of current decay and the number of time constants indicates the percentage of current yet to decay (i.e., still flowing). This description differs from the rise L/R time constant which stated the percentage of maximum current to which the current had risen (i.e., the amount of current actually flowing).

The percentage of current yet to decay (still flowing) in x number of time constants on decay may be determined by either one of two methods:

METHOD 1:

Consulting the decay curve on the universal time constant chart, Fig. 13-40.

METHOD 2:

Using epsilon, e.

METHOD 1: DECAY CURVE ON UNIVERSAL CHART

The percentage of current that is still flowing in a resistive inductive circuit at a specific instant in time after the circuit switch is opened is determined in the same manner as the rise current was determined in Sec. 13-26.5. The only difference is that the answer obtained in the decay condition represents the value to which the current or voltage has de-

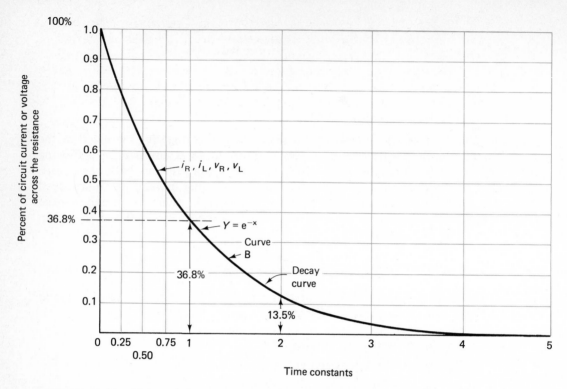

FIGURE 13-40   Universal time-constant graph.

cayed after the switch is opened, instead of the value to which the current has increased in the case of the rise or charge curve conditions.

Use the procedure outlined in Sec. 13-26.5 and the decay curve in Fig. 13-40 to determine the amount of current yet to decay (still flowing) for the following example time constants.

EXAMPLE

What is the percentage of current flow at the following time constants (refer to the graph in Fig. 13-40)?

Time Constant	Answer
1. 0.5 TC	60.6%
2. 0.75 TC	47 %
3. 1.0 TC	36.8%
4. 1.5 TC	22.3%
5. 2.0 TC	13.5%
6. 2.5 TC	8.2%
7. 3.0 TC	5 %

Ch. 13 Inductance

The value of instantaneous current for any particular time constant may be determined, once the percentage of the I_{max} has been determined, by multiplying the percentage by the maximum current value.

The value of the instantaneous voltage across a resistor is obtained by multiplying the instantaneous current value by the resistance of the resistor. The following problem outlines the sample solution.

PROBLEM 13-36

A circuit has a 10-H coil connected in series with a 100-Ω resistor and a 100-V battery. Draw the circuit, and calculate the value of current yet to decay 1 TC after the switch is closed to position B. Assume that the switch S_1 has been closed for 8 TC.

SOLUTION *(cover solution and solve)*

   a. Refer to the universal time constant chart and follow the instructions as previously outlined to find the percentage decay.

   Answer:

$$36.8\% \text{ of } I_{ss} \text{ yet to decay}$$

   b. Find I_{max}

$$I_{ss} = \frac{V_s}{R} = \frac{100}{100} = 1 \text{ A}$$

   c. Find the value of the current yet to decay.

$$i = \% \times I_{ss}$$
$$= 36.8\% \times 1 \text{ A}$$
$$= 0.368 \text{ A}$$

There is still 0.368 A of current flowing after 1 TC of decay.

METHOD 2: THE USE OF EPSILON, e

A decaying exponential curve is also a function of epsilon, e, because the instantaneous current is approaching 0% on decay in an exponential curve. The percentage of current yet to decay, or still flowing, in time constants on decay is determined by:

percent (Y) of steady-state current (I_{ss}) on decay,

$$Y = e^{-x}$$

> where Y is % current yet to decay
> x is number of time constants
> e is natural log (base 2.71828)

The percentage of I_{ss} current yet to decay after one time constant is

$$Y = e^{-1}$$

$$= 36.8\%$$

Instantaneous current and voltage values are determined in exactly the same manner as they were determined in Sec. 13-26.9.

PROBLEM 13-37 (METHOD 2)

Calculate the value of current yet to decay in the circuit in Fig. 13-41 after the switch, S_1, has been placed in position B for one time constant. (Assume that the switch, S_1, was closed in position A for 8 TC.)

SOLUTION (cover solution and solve)

a. Find the percentage of current yet to decay after 1 TC.

$$Y = e^{-x}$$

but $x = 1$

One can take only the anti-log of e^x.

e^1 = Take the A.L. (e^x) of 1 = 2.71828

Reciprocal of $2.71828 = \dfrac{1}{2.71828} = 0.368 = e^{-1}$.

$$\therefore Y = 36.8\% \text{ of } I_{ss}$$

b. Find I_{max}.

$$I_{ss} = \frac{V_s}{R} = \frac{100}{100} = 1 \text{ A}$$

c. Find the value of the current yet to decay.

$$i = \% \times I_{ss}$$

$$= 36.8\% \times 1 \text{ A}$$

$$= 0.368 \text{ A}$$

There is still 0.368 A of current flowing after 1 TC of decay.

13-27.3 *Circuit Resistance and Decay Time*

The time, in seconds, contained in 1 TC is inversely proportional to the value of the total circuit resistance ($\tau = L/R$). In other words, the larger the circuit resistance, the less the time, in seconds, contained in 1 TC for rise or decay. The value of a time constant for a circuit is calculated with the same equation whether the circuit is in a decay or a rise condition; thus, it is not the condition which will change the value of the time constant, but the value of the resistance or inductance of the circuit.

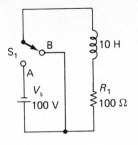

FIGURE 13-41

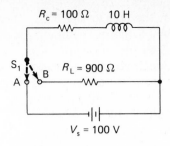

FIGURE 13-42

Compare the time contained in 1 TC for the circuits shown in Figs. 13-41 and 13-42 when the switch is in position B.

The circuit in Fig. 13-41 when the switch is in position B has a resistance of 100 Ω and an inductance of 10 H. The time of 1 TC in position B is

$$\tau = \frac{L}{R} = \frac{10}{100} \times 10^3$$

$$= 100 \text{ ms}$$

The current has decayed to 36.8% of I_{ss} in 100 ms; that is, 62.8% has already flowed.

The circuit in Fig. 13-42 in position B has a resistance of (100 + 900) = 1000 Ω and an inductance of 10 H. The time of 1 TC in position B is

$$\tau = \frac{L}{R} = \frac{10}{1000} \times 10^3$$

$$= 10 \text{ ms}$$

The current has decayed to 36.8% of I_{ss} in only 10 ms in Fig. 13-42, whereas it took 100 ms to decay in the circuit in Fig. 13-41.

SUMMARY:

The larger the resistance of the circuit the more rapid the decay of the induced voltage and induced current. In other words, the time constant is shorter.

13-27.4 Resistance and Instantaneous Voltage on Decay

The circuit shown in Fig. 13-42 has an added load resistance (R_L) when the switch is placed in position B.

The total resistance of the circuit when the switch is placed in position B will be 1000 Ω and during the decay the value of the instantaneous voltage developed across the coil will be larger than the source voltage for the following reason:

The 1 A of energy ($V_s/R = 100/100 = 1$ A) which flows in the coil in the circuit in Fig. 13-42 when the switch is placed in position A is stored in the coil in the form of magnetic energy. This 1 A of magnetic energy will attempt to collapse the instant that the switch is moved from position A to the decay position, B. The attempted instant decay produces an induced voltage (v_L) in the coil which acts as a source of voltage across resistors R_c and R_L. The voltage developed across the total resistance is equal to (and opposes) the induced voltage which is acting as the source voltage during decay.

$$v_{R_T} = v_L \text{ (acting } V_s)$$

The voltage developed across the total resistance the first instant that the switch is placed in the decay position B is equal to

$$v_{R_T} = 1 \text{ A} \times 1000 \text{ }\Omega$$
$$= 1000 \text{ V}$$

Because the v_{R_T} is 1000 V, the induced voltage (v_L) must also be 1000 V ($v_R = v_L$).

The value of $v_{L(max)}$ the first instant of decay is many times larger than the 100-V source voltage during the rise period.

13-27.5 Value of Instantaneous Current

The value of the instantaneous current for any one time constant or part of during rise time is dependent upon the value of the steady-state current. The steady-state current is dependent upon the source voltage (V_s) and circuit resistance (R_T) during charge time. The difference in circuit resistance values between the rise and decay conditions determines the time, in seconds, of the time constants for rise and decay. (Problem 13-44 reveals these facts.)

PROBLEM 13-38 _____

The switch in the circuit in Fig. 13-43 is in position A for 1.4 TC before it is placed in the decay position B. Calculate the value of current for the circuit 12 μs after the switch is placed in the decay position B.

SOLUTION *(cover solution and solve)*

Switch is closed to Position A.

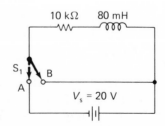

FIGURE 13-43

1. Current flow after 1.4 TC rise time (in position A)

$$Y = 1 - e^{-1.4}$$

$$= 75.34\%$$

(Steady-state current Position A) $I_{ss} = \dfrac{V_s}{R} = \dfrac{20}{10 \times 10^3} = 2 \text{ mA}$

(Instantaneous current Position A) $i = \% \times I_{ss}$

$$= .7534 \times 2 \times 10^{-3} = 1.5 \text{ mA}$$

The 1.5 mA is the current energy stored in the coil in the form of a magnetic field. It will produce the induced voltage (v_L) on decay when the switch is in position B.

Switch is in Position B (decay position).

2. Find the time in 1 TC.

$$\tau = \frac{L}{R} = \frac{80 \times 10^{-3}}{10 \times 10^3}$$

$$= 8 \times 10^{-6} \text{ s}$$

$$= 8 \text{ } \mu s$$

3. Find the number of TC (x).

$$x = \frac{t}{\tau} = \frac{12 \times 10^{-6}}{8 \times 10^{-6}}$$

$$= 1.5 \text{ TC}$$

4. Find the percent of current yet to decay.

$$Y = e^{-x}$$

$$\text{but} \quad x = 1.5$$

One can take only the anti-log of e^x.

$$e^{1.5} = \text{(take the A.L. (e^x) of 1.5)} = 4.4817$$

Reciprocal of $4.4817 = \dfrac{1}{4.4817} = 0.2231.$

The current still flowing is 22.31% of $I_{\text{position A}}$.

5. Find the value of current still flowing after 12 μs in position B.

(Instantaneous current in Position B) $\quad i = \% \times I_{\text{position A}}$

$$= (22.31\%) \times (1.5 \times 10^{-3})$$

$$= 0.3346 \text{ mA} = 334.6 \text{ } \mu A$$

The value of 334.6 μA is still flowing after 12 μs of decay.

Suppose that the switch in the circuit in Fig. 13-44 is in position A for 180 ms before it is put into position B. Calculate:

1. The number of time constants during which the switch is left at A
2. The value of the instantaneous current when the switch is put to B
3. The magnetic field energy stored when the switch is placed at B
4. The amount of maximum v_L when the switch is placed at B
5. The value of instantaneous current 0.7 TC after the switch is placed at B

SOLUTION   *(cover solution and solve)*

Switch in Position A.

1. $1 \text{ TC} = \dfrac{L}{R} = \left(\dfrac{5}{50}\right) \times 1000 = 100 \text{ ms}$

    time involved = 180 ms

    number of TC = $\dfrac{\text{time involved}}{\text{time in 1 TC}} = \dfrac{180}{100} = 1.8 \text{ TC}$

2. Instantaneous current (i_A)

    $$i_A = (I_{max}) \times Y$$

    $$Y = 1 - e^{-1.8}$$

    $$= 1 - 0.0165$$

    $$= 0.8347$$

    and $I_{max} = \dfrac{V}{R} = \dfrac{100}{50} = 2 \text{ A}$

    $\therefore \ i_A = 2 \times 0.8347$

    $= 1.669 \text{ A}$

3. Magnetic field energy stored = 1.669 A. ·

Switch in Position B.

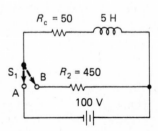

FIGURE 13-44

4. Induced voltage (v_L) at switch position B

$$v_L = i \times (R_1 + R_2)$$
$$= 1.669 \times (50 + 450)$$
$$= 834.5 \text{ V}$$

5. Instantaneous current 0.7 TC after switch is placed in position B.

$$i_B = i_A \times Y$$
$$Y = e^{-x}$$
$$= e^{-0.7}$$
$$= 0.4966$$
$$i_B = 1.669 \times 0.4966$$
$$= 0.829 \text{ A}$$

13-28 Practical Considerations

The very large induced voltage, observed above, which occurs on the opening of switch contacts, is the cause of many problems. The arcing at the brushes of motors is due to the v_L. These arcs cause burning of brush contacts as well as interference on radio and television, because they travel through space sometimes up two or three city blocks. The problem is usually cured by the use of capacitors which absorb the v_L, or by solid state devices which short out or dissipate the energy.

The points in the distributor of a car would be burned out very quickly, because of the large v_L which is created when the points open, if a capacitor (condenser) were not connected across them.

The above problems are only two of many which are caused by the large voltage induced in a circuit when it is opened.

13-29 Rise and Decay Curves

Figure 13-45 shows the rise and decay curves on the same graph. Curve A is the curve formed by the rise of current in the circuit and the consequent voltage developed across the resistor during the time that the voltage is applied. Curve B is the curve formed by the decay of the v_L across the coil during the time that the switch is in the discharge position.

13-30 Time Constants and Frequency

There are many applications of L/R time constant circuits other than timing circuits; waveshaping is one example. Waveshaping depends upon the relationship between the time contained in one time constant and the duration of one period of the applied voltage pulse of some particular

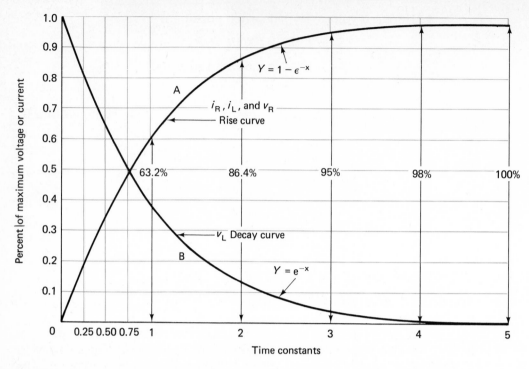

FIGURE 13-45 Rise and decay curves.

shape. In such applications, time constants are called waveshaping circuits or networks and are classified as one of three types:

1. Long time constant
2. Short time constant
3. Medium time constant

A period is the time duration, or length of time, which one voltage pulse or cycle lasts. It is the inverse of the frequency and is generally stated in milliseconds or microseconds: $T = 1/f$ seconds.

An L/R circuit cannot be said to have a long time constant nor a short time-constant until the time contained within the time constant is compared to one period of the frequency of the signal voltage which is to be applied to the L/R circuit. The L/R time constant may be long compared to one frequency but very short when compared with another frequency.

13-30.1 The Length of Time Constants Versus Period

Suppose a 1-mH coil is connected in a waveshaping circuit with a 1 kΩ resistor, and a square pulse is applied to the waveshaping circuit (Figs. 13-46 and 13-47).

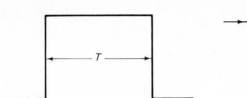

FIGURE 13-46   A positive square pulse. T = period = $1/f$ (seconds).

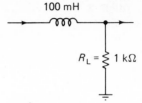

FIGURE 13-47   An L/R time-constant circuit.

TIME CONSTANT OF THE L/R CIRCUIT

The time in the L/R time constant is

$$\tau = \frac{L}{R} = \frac{(100 \times 10^{-3})}{10^3}$$

$$= 100 \ \mu s$$

1. Is the above L/R time constant a short or a long time constant?
2. What will be the shape of the waveform at the output of the L/R time constant circuit?

The answer to these questions depends upon the frequency of the applied square wave voltage as demonstrated by the following three comparison conditions.

COMPARISON 1 (FIG. 13-48)

Suppose that a 1-MHz square wave voltage pulse is applied to the L/R time constant circuit. The period of the 1-MHz positive square wave volt-

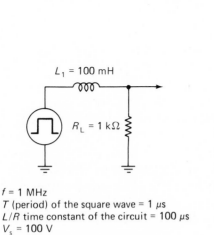

f = 1 MHz
T (period) of the square wave = 1 μs
L/R time constant of the circuit = 100 μs
V_s = 100 V

(a)

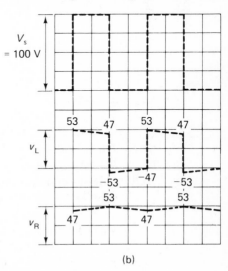

(b)

FIGURE 13-48   A long time constant and the waveforms produced. (A high-frequency pulse is applied.)

age pulse is

$$\text{Frequency} = 1,000,000 \text{ cycles/s}$$

$$T = \frac{1}{f} = \frac{1}{10^6}$$

$$= 1 \ \mu s$$

1. The length of the time constant is 100 times longer than the time which the square wave positive pulse lasts. The L/R circuit is said to have a long time constant compared to the frequency of the square wave signal.

2. Notice that the output voltage waveform across the resistor in the graph in Fig. 13-48b for the long time constant is a near flat output voltage even though the input voltage is a square wave. Long L/R time constant circuits are useful for filter circuits. Notice that the sum of the voltage values of the two waveforms is always equal to the value of the pulse voltage. For example

    when   voltage across R_L = 47 V

    voltage across coil = 53 V

    sum   100 V = pulse voltage

COMPARISON 2 (FIG. 13-49)

1. Suppose that a 50-cycle square wave positive voltage pulse is applied to the same L/R circuit as in Comparison 1, and the circuit has a time constant of 100 μs (Fig. 13-49). The period of the 50-cycle positive square wave voltage pulse is computed as follows:

    frequency = 50 cycles/s

$$T = \frac{1}{f} = \frac{1}{50} \ s$$

$$= 20,000 \ \mu s$$

The time contained (100 μs) in the time constant is only 1/200 as long as that of the period of a 50-cycle square positive voltage pulse that has a time constant of 20,000 μs. The L/R circuit thus has a short time constant.

2. Notice that the shape of the output waveform across the inductance of the L/R time constant circuit when a 50-cycle (20,000 μs) square wave voltage pulse is applied is a spiked waveform, and the wave shape across the resistor is much like a square wave.

Short time constant L/R circuits are used as differentiation circuits when a spike output voltage waveshape is required. The output voltage is taken from the inductance.

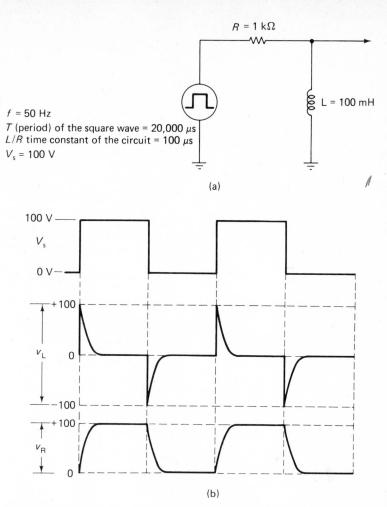

$R = 1\ \text{k}\Omega$

$f = 50$ Hz

T (period) of the square wave = 20,000 μs
L/R time constant of the circuit = 100 μs
$V_s = 100$ V

$L = 100$ mH

(a)

100 V

V_s

0 V

+100

v_L 0

−100

+100

v_R

0

(b)

FIGURE 13-49   A short time-constant condition.   (a) The same circuit as Fig. 13-47 but with a low-frequency square-wave pulse. (b) The effect of a short L/R time constant on a square-wave pulse. Note the shape of the pulse across each component.

SUMMARY:

In comparison 1, the time constant was called a long time constant because 1 TC was much longer than 1 period of the square voltage pulse.

In comparison 2, the time constant was a short time constant because 1 TC was much shorter than the period of the square voltage pulse. The actual values are not important.

COMPARISON 3 (FIG. 13-50)

1. A medium time constant is one which is approximately equal to the length of the period of a voltage pulse of the L/R circuit (Fig. 13-50). For example, an L/R circuit with a time constant of 100

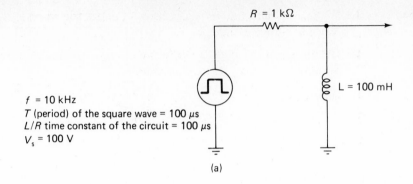

$R = 1\ k\Omega$

$f = 10\ kHz$
T (period) of the square wave = 100 μs
L/R time constant of the circuit = 100 μs
$V_s = 100$ V

$L = 100$ mH

(a)

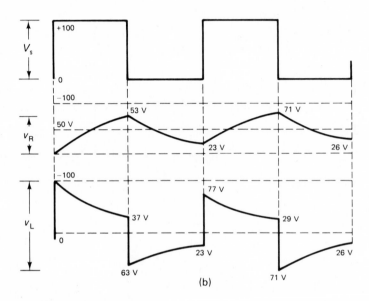

(b)

FIGURE 13-50   A medium time-constant condition. (a) The same circuit as Fig. 13-47 but with a medium-frequency square-wave pulse. (b) The effect of a medium L/R time constant on a square-wave pulse. Note the shape of the pulse across each component.

μs would be considered a medium time constant if the applied voltage pulse had a period of 100 μs. A voltage with a period of 100 μs would be occurring at a frequency of 10,000 cycles per second.

2. Medium time constant L/R circuits are used for changing waveshapes.

EXERCISES

1. Define the term inductance.

2. What two conditions must be satisfied to induce a voltage into a conductor?

3. Why is voltage not induced into a conductor when it is moved parallel to the lines of force?

4. What determines the polarity of voltage or direction of current flow which is induced into a conductor?

5. State the left-hand generator rule for voltage.

6. What is the effect of moving a magnet into and out of a coil?

7. What is the effect of reversing the poles of a magnet which is moved into and out of a coil?

8. An inductance is described as the characteristic of a component. What is the characteristic?

9. A voltage is induced into a coil when a battery is first connected to the coil. Is the voltage series-aiding or series-opposing the battery voltage?

10. Does the induced voltage in a coil series-aid or series-oppose the battery voltage when the switch is opened between the coil and battery?

11. a. Is there current flow in a coil the first instant that the switch is closed to a battery?
    b. Explain your answer to a.

12. State Lenz's Law.

13. Define:
    a. Flux linkage
    b. Self-inductance
    c. Unit of inductance of one henry

14. What does the magnitude of induced voltage depend upon?

15. State Faraday's two Laws.

16. Write the equation for inductance, showing how it is related to induced voltage.

17. Write the equation for induced voltage showing its relationship to inductance.

18. The current in an inductor changes from 18 A to 14 A in 2 s. What is the inductance of the coil if 20 V is induced into the coil?

19. What is the magnitude of the induced voltage in a coil of 8 H if the current flow in it changes from 2 mA to 8 mA in 10 μs?

20. What value of inductance is necessary to induce a voltage of 125 V when the current changes at the rate of 10 A in 5 ms?

21. State the formula for inductance which one may use to design a coil of some specific inductance value.

22. What would be the inductance of a 200-turn air-core coil if it had the following characteristic values:

$$\text{Length} = 4 \text{ cm}$$

$$\text{Area} = 2 \times 10^{-6} \text{ m}^2$$

23. What would be the inductance of the air-core coil in question 22 if the air core were replaced by a magnetic core having a relative permeability value of 5000?

24. Briefly describe a practical method of quickly changing the inductance of a coil.

25. Define skin effect.

26. Two inductors, 150-mH and 300-mH, are connected in series so that there is no magnetic coupling between them. What is their total inductance?

27. Two inductors, 40-mH and 60-mH, are connected in parallel so that there is no magnetic coupling between them. What is their total inductance?

28. State the difference between self-inductance and mutual inductance.

29. Write the symbol for mutual inductance.

30. State the definition for the unit of one henry of mutual inductance.

31. Define the term coefficient of coupling.

32. Is the value of coefficient of coupling less than or greater than one?

33. What is the mutual inductance of a 4-mH coil and a 3-mH coil, which have a coefficient of coupling of 0.4?

34. What is the coefficient of coupling between two coils magnetically coupled, if one coil has an inductance of 24 H, the second coil has an inductance of 9 H and the mutual inductance is 8.18 H?

35. What must be the inductance of a coil, L_1 which can be used with an 8-H coil, L_2 to produce a mutual inductance of 4 H if the coefficient of coupling between the two coils is 0.5?

36. A 100-mH coil is connected magnetically and electrically series-aiding to a 150-mH coil. The coefficient of coupling is 0.8. What is their total inductance?

37. A 12-mH coil is connected in parallel with an 8-mH coil and there is no magnetic coupling between them. What is the total inductance?

38. Explain why transformers cannot be operated on dc voltage.

39. What is the approximate value of the coefficient of coupling between the primary and secondary winding of a power transformer?

40. What is the secondary voltage of a transformer having 36 turns of wire in the primary and 1512 turns of wire in the secondary winding when 120 V of ac is applied to the primary winding?

41. What is the secondary voltage of a transformer having a 1/16 turns ratio when 100 V of dc voltage is applied to the primary?

42. What current is drawn from a primary winding which has 120 V of ac voltage applied to it, when the secondary winding is drawing 600 W of power (assume 100% efficiency)?

43. An impedance-matching transformer is connected between 4200 Ω in the primary and 8 Ω in the secondary. What must be the primary-to-secondary turns ratio to properly match the two impedances?

44. List three losses which occur in power transformers and which may cause transformers to get hot.

45. Draw an autotransformer and label the primary and secondary winding.

46. What methods are used to reduce the following losses in transformers:
    a. Copper loss
    b. Eddy current loss
    c. Hysteresis loss

47. Write the Greek letter symbol for a time constant.

48. Calculate the time constant of a 5-H coil connected in series with a 1200-Ω resistor and a 9-V battery.

49. How many seconds will it take for the current to rise to 63.2% of the maximum value, in a circuit consisting of a 16-mH coil and a 4-kΩ resistor connected in series with a 9-V battery?

50. Write the initial rate of change of current formula as it relates to induced voltage, inductance, time, and current.

51. A 5-H inductor of pure inductance (zero dc resistance) is connected in series with a 50-V battery. What, theoretically, would be the final current flow after 4 s?

52. State one of the definitions for L/R time constant.

53. For how many time constants has the switch been closed if the current in a closed circuit is at 70% of its maximum value?

54. A switch is closed to a 5-H inductor connected in series with a 2-kΩ resistor and a 12-V battery.
    a. What will be the percentage of steady-state current flow after 3 TC?
    b. What value of voltage will be measured across the resistor at the end of the third time constant?
    c. What will be the value of instantaneous current after 3 TC?

55. A switch is closed to a 2-H coil, a 4-Ω resistor, and a 24-V battery. How many time constants must elapse before the current reaches 2.25 A?

56. How many seconds is a switch closed to an L/R circuit if it is closed for 12 TC? The circuit consists of a 100-mH coil, a 40-Ω resistor, and a 6-V battery.

57. What happens to the current which flows into an inductor of pure inductance?

58. The switch to a 50-V battery in series with a 50-Ω resistor and a 5-H inductor has been closed for 14 TC before the switch is opened. Determine, at 2.5 TC after the switch is opened:

   a. The value of current yet to decay
   b. The amount of current actually flowing in the circuit
   c. The voltage across the resistor
   d. The value of induced voltage in the inductor

59. The switch in the circuit in Fig. 13-51 is closed to position A for 150 ms before it is placed at position B. Calculate:

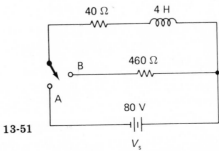

FIGURE 13-51

   a. The number of time constants during which the switch is left in position A.
   b. The instantaneous current flow when the switch is placed at position B.
   c. The value of induced voltage in the coil when the switch is placed at position B.
   d. The amount of time in seconds necessary for the current to decay to zero after the switch has been placed at position B.

60. Does the circuit shown in Fig. 13-52 have a short or a long time constant, compared to a 1-MHz positive pulse?

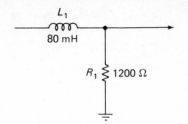

FIGURE 13-52 An *L/R* time-
constant circuit.

61. Draw the waveshape of voltage output which would be obtained
from Fig. 13-52 if a 60-Hz square wave voltage were applied to it.

chapter **14**

CAPACITANCE

14-1 Introduction

A capacitor is formed when two metal plates (conductors), separated by an electrical insulator (dielectric), are close enough together that when a voltage is connected across them, the electric charge on one side of the insulator influences the electric charge on the other side of the insulator.

14-2 Capacitor

The basic capacitor consists of metal plates facing one another and separated by an insulating material called the dielectric (Fig. 14-1).

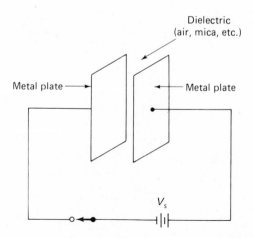

FIGURE 14-1   Basic capacitor.

The two plates of a capacitor must always be metal because they are the conductors which carry the electrons to and from the dielectric. The material used for the dielectric determines the type of capacitor. If the dielectric is air, the capacitor is called an air capacitor; these capacitors are usually variable in value and are used in tuning circuits, e.g., in radios. If the dielectric is mica, the capacitor is called a mica capacitor, and so on.

14-3 Purpose of Capacitors

The purpose of a capacitor in any circuit in which it is used is to act as a storage tank for electrons. As a water tank stores water to be used when large quantities are demanded at one time, so a capacitor stores electrons until a large quantity is demanded at a specific period of time by the rest of the circuit.

There are many capacitors in most electronic equipment. Capacitors are used in filter circuits to remove unwanted ac voltages or interfering signals. They are used in tuning circuits in all electronic equipment which picks signals out of the air, such as CB radios, AM radios, FM radios, and televisions. Capacitors are used as coupling devices when it is necessary to pass on ac voltages but block dc voltages. These are only a few of dozens of capacitor applications.

14-4 Capacitors and Conductance

Capacitors do not pass current, because the dielectric is an electrical insulator. They merely charge and discharge in ac electrical circuits, and the effect is the same as if they passed ac voltages and currents.

14-5 Principle of Operation

Electrons flow from the negative side of the battery into the top plate of the capacitor (Fig. 14-2) when a voltage is applied to it. Negative charges repel, so the electrons adding to the top side of the capacitor repel the electrons in the fixed atoms of the dielectric. The electrons in the dielectric atoms are not free to move away from their atoms, but being repelled, they stretch and distort their orbits toward the plate connected to the positive side of the battery voltage. The close proximity of these fixed electrons to the capacitor bottom plate forces electrons off the bottom conductive plate of the capacitor. These electrons leave behind positively charged ions (atoms lacking an electron are called cations) on the bottom plate. This process of electrons flowing into the top side of the capacitor and out of the bottom plate (flowing into and out of the capacitor) continues until the charge difference between the negatively charged top plate and the positively charged bottom plate produces a voltage equal in value to the battery voltage. Current now stops flowing into the capacitor

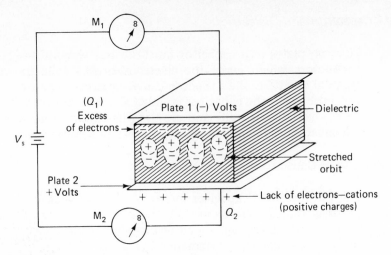

FIGURE 14-2 Charging condition. Both meters read the same value. Note the stretched orbits and positive ions on the bottom plate caused by the repelling force of electrons on the top plate. Note polarity of charging voltage.

and the capacitor is said to be charged (see Fig. 14-3). The voltage remains across the capacitor terminals even after the capacitor has been removed from the charging circuit (Fig. 14-4), indicating that the capacitor has stored the electrons which have charged it. While the capacitor is charging, the same number of electrons flow into plate 1 as flow out of plate 2. In other words, Meter 1 will read the same value as Meter 2 (Fig. 14-3). Note that the current is not flowing through the dielectric but merely is charging the top side of the capacitor negatively and the bottom side positively. The difference in charges across the dielectric produces a voltage between the two plates and an electric (electrostatic) field across the dielectric.

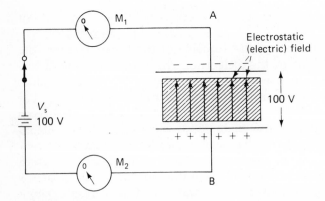

FIGURE 14-3 The electric field across the charged capacitor. Note that the current is zero.

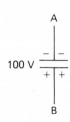

FIGURE 14-4 A disconnected capacitor—still charged.

14-6 Clarification

Most texts state that an electric field and electrostatic charge exist between the plates. This is true only in the sense that the plates touch each side of the insulator, and therefore the plates may be used as a reference. In fact, however, the plates are only the conductors. It is the dielectric between the plates that has the field and charge exerted across it, regardless of whether the dielectric is air, mica, polystyrene or another material.

14-7 Electrostatic Field Measurements

An electrostatic or electric field is the sum of all of the electrostatic lines of force which exist between the plates of a capacitor when it is connected across a source of voltage. The term *electrostatic* pertains to static electricity or an electric charge at rest. An electrostatic line of force is that line of force which exists between an electron on the negative plate and its associated positive ion on the positive plate (Fig. 14-3).

14-7.1 Electric Field Strength (E)

Electric field strength is the magnitude (in V/m) of the electrostatic (electric) field which exists between the plates, or across the dielectric of a capacitor, due to the pulling force on the electrons by the positive ions on the positively charged side of the dielectric. The pulling force (field strength) on the electrons is inversely proportional to the thickness of the dielectric (distance between the plates). The greater the distance, the less the pull. The force is also directly proportional to the voltage developed across the plates by the difference in charges.

Electric field strength (E) is comparable to field intensity (*H*) in magnetic circuits. For this reason, it sometimes is called *electric field intensity*.

$$\text{electric field strength} = \frac{\text{voltage between the plates}}{\text{thickness of dielectric } (d)}$$

$$E = \frac{V}{d} \ \text{V/m}$$

where E is in volts/meter
V is voltage source
d is distance between the
plates (in meters)

14-7.2 Electric Flux Density (D)

An electrostatic line of force is an imaginary line which connects every electron in the negatively charged side of the dielectric to a positive ion (an atom deficient of an electron) in the positively charged side of the dielectric. Each line of force is called an electric flux line. Since there is an electric flux line for every electron, electric flux is measured in terms of coulombs (charge). However, as in magnetism where the number of magnetic lines tell nothing about the strength of the magnetic field until it is stated in unit area, the magnitude of charge has no meaning in capacitance unless the magnitude of the charge is stated in unit area.

$$\text{electric flux density} = \frac{\text{coulombs (of charge)}}{\text{area of one plate}}$$

$$D = \frac{Q}{A} \ \text{C/m}^2 \quad (\text{coulombs per m}^2)$$

where D is electric flux density
in coulombs/m^2
Q is in coulombs
A is area of one plate
(in m^2)

14-7.3 Permittivity (ϵ) (Relationship Between E, D, and ϵ)

Permittivity (ϵ) is the ease with which an electrostatic (electric) field can be established in a dielectric.

Electric field strength (E) is the magnitude, in volts/meter, of the force of the electrostatic field on the charge in a capacitor.

Electric field density is the quantity of the charge (electrons) per unit area in a charged capacitor.

Dividing flux density (D) by field strength (E) produces permittivity—a measure of the ease with which an electric field will be established in the dielectric.

$$\text{permittivity} = \frac{\text{flux density}}{\text{field strength}}$$

$$\epsilon = \frac{D}{E} \quad (\text{farads per meter, F/m})$$

where ϵ (epsilon) is permittivity (in farads/meter)
D is electric flux density [in coulombs/(m^2)]
E is electric field strength (in volts/m)

Every dielectric has its own permittivity value. The permittivity value of a vacuum is used as the absolute or reference value.

The permittivity of a vacuum is measured by placing two plates opposite one another in a vacuum. A voltage is then connected to the terminals of the plates, and the electric flux density and field strength are measured. The ratio of the two parameters is the permittivity of the vacuum (free space). This value is a constant (ϵ_0).

$$\epsilon_0 = \frac{D}{E} = 8.85 \times 10^{-12}$$

The value $\epsilon_0 = 8.85 \times 10^{-12}$ is important because it is used in the calculations for the capacitance value of all capacitors.

14-8 Factors Affecting Capacitance

The three factors affecting capacitance are permittivity, square area of the plates, and the distance between the plates.

14-8.1 Permittivity (ε)

Permittivity (ϵ) is one of the three factors which affect the capacitance value of a capacitor. It is determined by the type of dielectric material used. The dielectric of a capacitor is the most important component of the capacitor. The type of dielectric material can make the capacitor a reliable component or an undependable one; it determines whether the capacitor will be bulky physically for a specific capacity or have a low or high working voltage. (Working voltage is the maximum voltage which may be placed across the capacitor before its dielectric will break down and the capacitor will become shorted. See Sec. 14-13.)

Permittivity (ϵ) is the measure of the ease with which a flux field is established in a capacitor.

As higher permeability materials increase the number of flux lines in magnetic circuits, higher permittivity dielectric materials increase the number of electric flux lines in dielectrics; because there is one electron for each electric flux line, this increases the charge (coulombs).

If two capacitors have the same plate area and distance between the plates, but the permittivity (ϵ) of one is three times greater than the permittivity (ϵ) of the other, the capacity of the first capacitor will be three times greater than the second one.

$$\text{capacity of a capacitor, } C = \frac{\epsilon_0 K A}{d}$$

where K is the dielectric constant (see Sec. 14-8.2).

If ϵ_0 increases three times, the capacity will increase three times.

The permittivity of a vacuum is the value with which the permittivity of all other materials is compared, and its value is called the absolute permittivity value, (ϵ_0) (see Sec. 14-7.3).

$$\epsilon_0 = 8.85 \times 10^{-12} \text{ farads/meter}$$

14-8.2 Dielectric Constant (K)

Dielectric constant is the commonly used name for relative permittivity. Dielectric constant is the ratio of the ease with which a particular dielectric will establish an electrostatic flux field to the ease with which air will establish a flux field. Each type of dielectric has its own permittivity value (ϵ). The calculation of capacity values uses the relative permittivity value (K) called dielectric constant.

$$\text{dielectric constant} = \frac{\text{permittivity of a dielectric material}}{\text{permittivity of free space}}$$

$$K = \frac{\epsilon}{8.85 \times 10^{-12}}$$

where K is dielectric constant (a ratio)

ϵ is permittivity of the material

8.85×10^{-12} is permittivity value of a vacuum

It is conventional practice in calculating capacitance values to use the dielectric constant value, which is then multiplied by 8.85×10^{-12}, to obtain the permittivity value of the dielectric used.

Table 14-1 Dielectric Constants of Various Dielectrics

Dielectric	K
Vacuum	1.0
Air	1.0
Polystyrene	2–3
Mica	5–7.5
Ceramics	500–7500
Paper—oiled or waxed	2.5–3
Glass	6–8
Aluminum oxide	10
Tantalum oxide	11
Transformer oil	4
Porcelain ceramic	6
Bakelite	7

Note that the capacity of a capacitor will increase seven times if its air dielectric is replaced by bakelite.

14-8.3 Plate Area

The dielectric constant (relative permittivity), K affects the capacitance value of capacitors, but the capacity of a capacitor is also directly dependent upon the square area of the plates. The greater the plate area, the greater the area on each side of the dielectric, and the greater the space to store electrons.

capacity is proportional to plate area

$$C \propto A$$

The third factor that affects the capacity is the distance between the plates, which is controlled by the thickness of the dielectric. The thicker the dielectric material, the greater the distance between plates and the less the storage capabilities, because the field strength is less.

Capacity is inversely proportional to the space between the plates

$$C \, \alpha \, \frac{1}{d}$$

14-9 Calculating Capacity

The parameters affecting the capacity of a capacitor can now be combined into a formula for calculating the value of a capacitor.

$$\text{Capacity } (C) = \frac{\epsilon_0 KA}{d}$$

$$= \frac{(8.85 \times 10^{-12}) KA}{d}$$

where ϵ_0 is absolute permittivity of air = (8.854×10^{-12})
K is dielectric constant of dielectric material
A is area (m^2) of one plate
d is distance between plates (meters)
C is the capacity in farads

PROBLEM 14-1

What is the capacity of a capacitor which is made of two strips of aluminum foil each 4 cm wide and 10 m long, separated by a 24-μm thick mylar dielectric? K of mylar = 3.

SOLUTION   *(cover solution and solve)*

$$C = \frac{\epsilon_0 KA}{d}$$

$$C = \frac{8.85 \times 10^{-12} \times K \times A}{d} \quad \text{farads}$$

$$A = (4 \times 10^{-2}) \times 10 = 40 \times 10^{-2} \text{ m}$$

$$= \frac{(8.85 \times 10^{-12}) \times 3 \times (40 \times 10^{-2})}{24 \times 10^{-6}}$$

$$= \frac{1062 \times 10^{-14}}{24 \times 10^{-6}}$$

$$= 44.25 \times 10^{-8}$$

$$= 0.4425 \ \mu\text{F}$$

14-10 Unit of Capacitance—Farad

If a capacitor stores 1 coulomb (6.24×10^{18} electrons) of charge when 1 V is applied across it, the capacitor is said to have a capacity of one farad.

$$C = \frac{Q}{V}$$

where C is in farads
             Q is 1 coulomb
             V is 1 volt

A capacitor of one farad would measure several meters in width, height, and length. Common sense tells us that such large physical sizes could never be used in electronic equipment; thus, one farad is never a unit of capacity actually used in electronics. The value of capacitors commonly used in electronic equipment is in either microfarads, nanofarads or micromicrofarads, the latter value usually being expressed in picofarads (pF).

14-10.1 Symbols

Microfarads = μF = 1×10^{-6} F

Nanofarads = nF = 1×10^{-9} F

Micromicrofarads = picofarads = pF = 1×10^{-12} F

14-10.2 Value Applications

Capacitance values of less than 1000 pF are generally used in radio frequency applications.

Values from 1000 pF (1 nF) to 10 μF are generally used in audio circuits.

Values greater than 10 μF are commonly used in power supply filter circuits.

14-10.3 Conversion of pF to nF or μF and Vice Versa

Capacitors are sometimes quoted in electronic catalogues in pF or nF values but on circuit schematics (drawings) they are shown in μF values, or vice versa. It is necessary to convert from one unit to the other in such cases. The conversion is made as follows:

Recall that there are:

10^6 microparts in a whole

10^9 nanoparts in a whole

10^{12} micromicroparts or picoparts in a whole

Refer to Chap. 2, Sec. 2-7.1 for the principle of conversion.

PROBLEM 14-2

1. Convert 2500 pF to nF.
2. Convert 2500 pF to μF.

SOLUTION   *(cover solution and solve)*

1. $(2500 \times 10^{-12}) \times 10^9 = 2.5$ nF
2. $(2500 \times 10^{-12}) \times 10^6 = 0.0025$ μF

PROBLEM 14-3

1. Convert 0.0047 μF to nF
2. Convert 0.0047 μF to pF.

SOLUTION   *(cover solution and solve)*

1. $(0.0047 \times 10^{-6}) \times 10^9 = 4.7$ nF
2. $(0.0047 \times 10^{-6}) \times 10^{12} = 4700$ pF

EXAMPLE

Convert:	Answer
1. 1000 nF to μF	1 μF
2. 0.05 μF to pF	50,000 pF
3. 400,000 pF to nF	400 nF
4. 1,000,000 pF to nF	1000 nF
5. 0.2 μF to pF	200,000 pF
6. 400 nF to μF	0.4 μF
7. 1 μF to nF	1000 nF

14-11 Dielectric Strength

Take note that dielectric strength is completely different from field strength. Any type of dielectric material will break down and conduct if the voltage applied across it exceeds a certain value. The voltage value beyond which a dielectric will break down and conduct is known as its dielectric strength. Dielectric strength varies with the type of material and

Table 14-2   Dielectric Strength of Materials

Dielectric	Volt/μm
Air	2.95
Ceramic	3.9–11.81
Porcelain	11.81
Transformer oil	15.75
Bakelite	15.75–19.69
Paper	19.65
Polystyrene	39.37
Glass	78.74
Mica	196.85

its thickness. A list of dielectric materials and their dielectric strengths is given in Table 14-2.

$$\text{breakdown voltage} = \text{dielectric strength} \times d$$

where d is the distance (space between the plates)

PROBLEM 14-4

What is the breakdown voltage (voltage rating) of an air dielectric capacitor if the spacing (d) between the plates is 0.01 mm?

SOLUTION   *(cover solution and solve)*

$$\text{dielectric strength of air} = 2.95 \text{ V}/\mu\text{m}$$

$$\text{voltage rating} = \text{space} \times \text{dielectric strength}$$
$$= (1 \times 10^{-5}) \times (2.95 \times 10^{6})$$
$$= 29.5 \text{ V}$$

PROBLEM 14-5

What is the voltage rating of a paper capacitor if the dielectric is 2 $\times$ 10^{-5} m thick?

SOLUTION   *(cover solution and solve)*

$$\text{dielectric strength of paper} = 19.65 \text{ V}/\mu\text{m} = 19.65 \times 10^{6} \text{ V/m}$$

$$\text{voltage rating} = \text{space} \times \text{dielectric strength}$$
$$= (2 \times 10^{-5}) \times (19.65 \times 10^{6})$$
$$= 393 \text{ V}$$

14-12   Effect on Capacity of Dielectric Constant (K) and Dielectric Strength

Electronic equipment manufacturers usually want the smallest physical sized capacitor which will meet voltage and capacity requirements. Capacitor manufacturers therefore seek dielectric materials of the highest permittivity and dielectric strength possible. Comparison of ceramic and paper as dielectric materials shows the preference for ceramic in the above problems.

A ceramic capacitor which has a dielectric constant (K) of 600 will have two hundred times the capacity for the same physical size as a paper capacitor which has a K of 3. In other words, for the same capacity, the ceramic will be much smaller physically than the paper capacitor. The ceramic material for the same dielectric thickness will not withstand quite as much voltage as the paper, however.

A ceramic capacitor with a 0.03 mm dielectric thickness will withstand:

$$\text{voltage breakdown} = (3 \times 10^{-5}) \times (11.81 \times 10^6) = 564 \text{ V}$$

A paper capacitor with a 0.03 mm dielectric thickness will withstand:

$$\text{voltage breakdown} = (3 \times 10^{-5}) \times (19.65 \times 10^6) = 589.5 \text{ V}$$

14-13 Working Voltage (W.V.D.C.)

The working voltage of a capacitor is determined by the dielectric strength and the thickness of the dielectric used in the manufacture of the capacitor. It is that value of voltage which, if exceeded, will break down the dielectric of the capacitor and cause it to become shorted. Often the working voltage (W.V.D.C.) value is written on the side of the capacitor, but if not, it is always specified by the manufacturer and listed in the electronic catalogues.

Electrolytic and tantulum capacitors have a chemical which forms an oxide. The oxide becomes the dielectric for the capacitor. The terminals of these types of capacitors are marked with a negative (-) and a positive (+) sign. A voltage applied to the capacitor in polarities opposite to those marked on the capacitor causes the oxide dielectric to break down and conduct current, and the capacitor is said to be shorted. It is for this reason that these capacitors must always be operated in circuits in which the dc voltage is larger than the ac voltage. Specially constructed electrolytic capacitors called nonpolarized capacitors are used in crossover networks in speaker systems.

The maximum working voltage of electrolytic and tantulum capacitors is 450 V dc. Mica capacitors are available which have a working voltage of several thousands of volts. These have been used in high-voltage applications such as the 20 kV high-voltage supply of television sets.

The working voltages of most capacitors used in electronic equipment range from 25 to 600 V.

14-14 Capacity, Voltage, and Charge

The definition of a capacitor shows the relationship between the capacity of a capacitor, the voltage developed across it, and the amount of electrons stored in it. The definition states:

A capacitor has a capacity of one farad if one volt is developed across it when it stores 1 coulomb (6.24×10^{18}) of electrons. In other words, the relationship between the three units is:

$$\text{capacity (of capacitor)} = \frac{\text{coulomb}}{\text{volts}}$$

$$C = \frac{Q}{V}$$

or, transposing

$$\text{charge in coulombs} = \text{capacity} \times \text{voltage}$$

$$Q = CV$$

transposing again

$$\text{voltage (developed across a capacitor)} = \frac{\text{coulombs}}{\text{capacity}}$$

$$V = \frac{Q}{C}$$

where Q is charge in
coulombs
C is capacity
in farads
V is voltage
in volts

The following problems, using the relationship between capacity, voltage, and charge help to explain why:

1. The smallest capacitor in series capacitor circuits always has the largest voltage developed across it.
2. The largest capacitor in a parallel arrangement will store the greatest amount of potential current for use by other circuits, but the voltage will be common.
3. The total capacity of two capacitors connected in series is always less than the smaller of the two.
4. The total capacity of capacitors connected in parallel is the sum of the capacities.

PROBLEM 14-6

How many coulombs of electrons are stored in a charged 3300-pF capacitor across which is 600 V?

SOLUTION   (cover solution and solve)

$$Q = CV$$
$$= (3300 \times 10^{-12}) \times 600$$
$$= 198 \times 10^{4} \times 10^{-12}$$
$$= 198 \times 10^{-8}$$
$$\doteq 1.98 \ \mu\text{C}$$

PROBLEM 14-7

What voltage is developed across a 0.01-μF capacitor if it contains 0.5 μC of charge?

SOLUTION   *(cover solution and solve)*

$$V = \frac{Q}{C}$$

$$= \frac{0.5 \times 10^{-6}}{0.01 \times 10^{-6}}$$

$$= 50 \text{ V}$$

PROBLEM 14-8

What is the capacity of a capacitor that charges to 1.45×10^{-9} C (coulombs) when a voltage of 150 V is applied to it?

SOLUTION   *(cover solution and solve)*

Given:

$Q = 145 \times 10^{-11}$
$V = 150$ V

$$C = \frac{Q}{V}$$

$$= \frac{(145 \times 10^{-11})}{150} \times 10^{12}$$

$$= 9.67 \text{ pF}$$

14-15 Connecting Capacitors

14-15.1 Capacitors in Series

Connecting capacitors in series effectively increases the distance between the plates; therefore, the capacity is decreased (Fig. 14-5). The total capacity of capacitors in series is calculated in the same way as resistance is calculated for resistors in parallel as shown in the following derivation.

PROOF:

From Kirchhoff's Law

$$V_{C_1} + V_{C_2} = V_T \tag{1}$$

$$\text{but} \quad V = \frac{Q}{C} \tag{2}$$

FIGURE 14-5 A 1-mm space between the plates of 2 capacitors
connected in series is equivalent to
1 capacitor with a 2-mm dielectric
thickness (space).

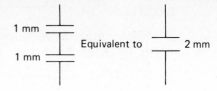

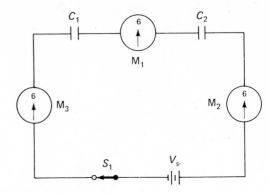

FIGURE 14-6 The 3 current meters read the same value of current;
therefore, the charge is the same in
all capacitors.

Substitute Eq. (2) in Eq. (1) and

$$\frac{Q_1}{C_1} + \frac{Q_2}{C_2} = \frac{Q_T}{C_T} \tag{3}$$

But in a series circuit, the charge, Q is the same as in every part of the
circuit (Fig. 14-6).

$$Q_1 = Q_2 = Q_T \tag{4}$$

Therefore the Q values cancel out and (3) becomes

$$\frac{1}{C_2} + \frac{1}{C_1} = \frac{1}{C_T} \tag{5}$$

$$C_T = \frac{C_1 \times C_2}{C_1 + C_2}$$

14-15.2 Capacitors in Parallel

Connecting capacitors in parallel effectively increases the square area of
the plates; therefore, the capacity is increased (Fig. 14-7).

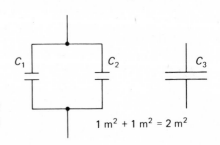

FIGURE 14-7 The square area of
the plates of the two capacitors is
equal because $1 \text{ m}^2 + 1 \text{ m}^2 = 2 \text{ m}^2$.

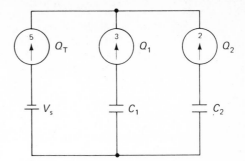

FIGURE 14-8 Parallel capacitors. The current is a different value in each capacitor because each capacitor has a different capacity. The total current is the sum of Q_1 and Q_2.

The total capacity of capacitors connected in parallel is calculated in the same way as is the resistance of resistors in series.

$$C_T = C_1 + C_2 + \cdots + C_n$$

DERIVING $C_T = C_1 + C_2$

In parallel circuits the total charge $= Q_T$.

$$Q_T = Q_1 + Q_2 \quad \text{(see Fig. 14-8)} \tag{6}$$

$$\text{but} \quad Q = CV \tag{7}$$

Substitute Eq. (7) in Eq. (6) and

$$C_T V_T = C_1 V_1 + C_2 V_2$$

In parallel circuits, $V_T = V_1 = V_2$. The voltages are common; therefore, they cancel.

$$\therefore \quad C_T = C_1 + C_2$$

14-16 Capacity and Developed Voltage

14-16.1 Series Circuits

The largest voltage is always developed across the smallest capacitor when capacitors are connected in series. Capacitors connected in series store the same amount of charge, Q. If two capacitors store the same charge but the capacity of one is 1/4 that of the other, the smaller capacitor will have four times more voltage developed across it than will the larger capacitor as shown in the following proof:

PROOF

Refer to the values in the circuit in Fig. 14-9.

$$C_T = \frac{C_1 \times C_2}{C_1 + C_2}$$

$$C_T = \frac{1 \times 4}{5} = 0.8 \ \mu\text{F}$$

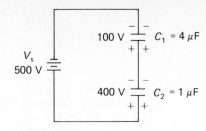

FIGURE 14-9 The smallest voltage is across the largest capacitor.

Charge flowing in each part of the circuit is

$$Q = VC_T$$

$$= 500 \times (0.8 \times 10^{-6}) = 400 \times 10^{-6}$$

$$= 400 \ \mu C$$

$$(4 \ \mu F) \ V_{C_1} = \frac{Q_T}{C_1} = \frac{400 \times 10^{-6}}{4 \times 10^{-6}} = 100 \ V$$

$$(1 \ \mu F) \ V_{C_2} = \frac{Q_T}{C_2} = \frac{400 \times 10^{-6}}{1 \times 10^{-6}} = 400 \ V$$

There are occasions when electrolytic capacitors must be connected in series, such as in a filter circuit, because the working voltage of the available capacitor is not large enough to withstand the voltage of the circuit in which the capacitor is to be used.

Exceeding the capacitor working voltage will cause a breakdown of the capacitor dielectric and will result in a shorted capacitor. The capacity of the capacitors in series arrangement must be equal in order to divide the voltage equally among them. The disadvantage of this arrangement is that the total capacity of two capacitors in series is only half the capacity of one of the capacitors.

14-16.2 Parallel Circuits

The voltage developed across each capacitor in a parallel circuit will be the same as that across every other capacitor, but the smaller capacitor will store the lesser amount of charge (Fig. 14-10).

PROBLEM 14-9

A 5-μF capacitor (C_2) is connected in series with a 10-μF capacitor (C_1) and a 400-V source of voltage. Calculate:

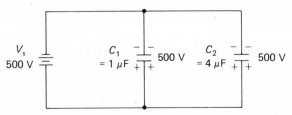

FIGURE 14-10 The same voltage is across all the capacitors in a parallel circuit.

1. Total capacity
2. The charge on each capacitor
3. The voltage developed across each capacitor
4. The minimum working voltage required of each capacitor

SOLUTION   *(cover solution and solve)*

1. Series capacitance

$$C_T = \frac{C_1 \times C_2}{C_1 + C_2}$$

$$= \frac{(10 \times 10^{-6}) \times (5 \times 10^{-6})}{15 \times 10^{-6}} = 3.33 \ \mu F$$

2. Charge

$$Q_T = CV$$

$$= (3.33 \times 10^{-6}) \times 400 = 1332 \times 10^{-6} \ C$$

The capacitors are in series, so the charge (electron flow) is the same in and out of both capacitors.

3. Voltage

$$V_{C_1} = \frac{Q_1}{C_1} = \frac{1332 \times 10^{-6}}{5 \times 10^{-6}} = 266.4 \ V$$

$$V_{C_2} = \frac{Q_2}{C_2} = \frac{1332 \times 10^{-6}}{10 \times 10^{-6}} = 133.2 \ V$$

total voltage = 399.6 V

4. Minimum working voltage for C_1 is 267 V. The practical value would be 400 V because manufacturers make only 200- W.V.D.C. and 400-W.V.D.C. capacitors in these higher voltages.

PROBLEM 14-10

Three capacitors: a 10-μF (C_1), a 4-μF (C_2), and a 2-μF (C_3) are connected in parallel with a 250-V voltage source.
Calculate

1. The total capacity
2. The charge stored in each capacitor
3. The voltage developed across each capacitor
4. The minimum working voltage (W.V.D.C.)

SOLUTION   *(cover solution and solve)*

1.

$$C_T = C_1 + C_2 + C_3$$
$$= 10 \ \mu F + 4 \ \mu F + 2 \ \mu F = 16 \ \mu F$$

2. Charge

$$Q_1 = C_1 V_1$$
$$= (10 \times 10^{-6}) \times (250) = 2500 \ \mu C$$
$$Q_2 = C_2 V_2$$
$$= (4 \times 10^{-6}) \times 250 \quad = 1000 \ \mu C$$
$$Q_3 = C_3 V_3$$
$$= (2 \times 10^{-6}) \times 250 \quad = 500 \ \mu C$$

3. The supply voltage is developed across each capacitor.

4. Minimum W.V.D.C. is the supply voltage.
   Practical value would be 400 V.

14-16.3 Leakage Resistance

The high dc resistance of dielectric materials does permit a very small current to flow through the dielectric when a dc voltage is applied across it. This current is called leakage current. It allows charges which have been acquired to leak away once the capacitor is removed from the circuit, and prevents indefinite electron storage. Leakage current increases exponentially with temperature increase. At room temperature, leakage current may be negligible. It will be of appreciable value at high temperatures.

The time in which the charge will decay to 36.8% (1 TC) of its initial value is expressed by time constant RC. The R is the leakage resistance value and the C is the capacity value of the capacitor.

Table 14-3   Typical Time Constants

Material	Length of 1 TC
Polystyrene	several days
Impregnated paper	several hours
Tantalum	1 to 2 hours
Ceramic	several minutes
Ordinary Electrolytic	several seconds

14-17 Capacitor Construction

There are many different types of capacitors, each one classified according to the dielectric used between the plates.

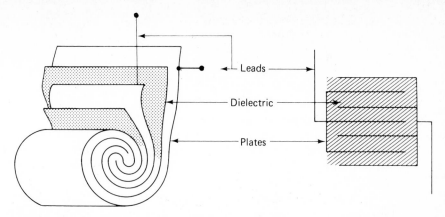

FIGURE 14-11 Tubular constructions. **FIGURE 14-12** Layer construction.

There are two basic types of construction:

1. Tubular construction, as shown in Fig. 14-11. This construction can be used only with pliable dielectric materials such as plastic or paper.

    Tubular construction consists of two long narrow strips of aluminum foil between which are sandwiched a slightly wider and longer piece of dielectric. The assembly is rolled in a tight cylinder and then inserted into a cylindrical case after leads have been attached to each foil. The case is then hermetically sealed against humidity.

2. Layer construction, as shown in Fig. 14-12, is used with rigid materials such as mica or ceramic. In this type of construction, metal foil is alternately layered with dielectric. Then all of the odd sheets are connected together to a lead and the same is done with all of the even sheets. The assembly is then encapsulated in a plastic covering.

14-18 Capacitor Types

The following is a list of the most common types of capacitors.

1. Mica
2. Ceramic
3. Electrolytic (aluminum—dry)
4. Tantulum
5. Air
6. Plastic—Polystyrene
          —Mylar
          —Metallized film

—Metallized polycarbonite

—Metallized polyester

7. Paper

14-18.1 Mica Capacitors

These are made with the layer type construction in which thin sheets of tinfoil are alternated with layers of mica. All of the odd-numbered tin foil sheets are connected to a common lead, and the same is done with the even numbered plates. The finished unit is then encapsulated in a case of bakelite to make a small flat rectangular package.

Silver mica capacitors are constructed by firing a thin layer of silver on each side of the mica dielectric, and then attaching a lead to each side of silver and encapsulating the capacitor in a plastic case.

Mica capacitors are used in applications requiring very narrow tolerance capacitances. They have a high stability value in severe environmental conditions and have a high operating temperature. They have a zero temperature coefficient. The values range from 100 to 50,000 pF.

14-18.2 Ceramic Capacitors

Ceramic dielectrics have about the highest dielectric constant among dielectric materials. Ceramic material is made by baking barium-strontium titantate in a kiln at an extremely high temperature. Each side of the ceramic is then silver-coated. The silver coats become the plates to which leads are attached, and then the complete unit is encapsulated by dipping it into molten plastic.

14-18.3 Electrolytic Capacitors (Dry-Aluminum)

An electrolytic capacitor, regardless of type, consists of two conducting electrodes (plates) separated by an electrically conductive chemical called electrolyte. The electrolyte forms a nonconducting oxide (dielectric) on the electrode to which the positive terminal of a dc voltage is connected. The process of connecting voltage across the terminals of an electrolytic capacitor so as to form the dielectric (non-conducting oxide) on the positive terminal is called *forming the capacitor*. Forming is part of the process in the manufacture of all electrolytic capacitors, because the resulting oxide is the only dielectric separating the two plates of the capacitor.

The oxide dielectric generally breaks down if a dc voltage is connected across the electrolytic capacitor in a polarity opposite that which was connected across it when it was formed. The consequence is a shorted capacitor which soon will overheat and in a short time will explode as a result of the buildup of steam in the sealed unit. The oxide dielectric may also break down if the capacitor is left unused for a few months or years.

Regardless of the cause of dielectric breakdown, electrolytic capacitors can generally be restored by repeating the process of forming. The process is then called *reforming* because the capacitor has previously undergone the same process in the factory.

The capacities of capacitors range from 1 μF to 100,000 μF, and the working voltage ranges from 4 V.W.D.C. to 600 V.W.D.C. Capacitors larger than 50 μF are generally used for filtering circuits in power supplies. Capacitors smaller than 50 μF can be found in coupling circuits in transistor equipment.

14-18.4 Tantalum Capacitors

Tantalum capacitors are used whenever high capacity and small physical size are the most important design considerations. They also have low leakage characteristics and low inductance, it should be noted.

Tantalum capacitors are the latest development in electrolytic capacitors. They are the tubular roll-up construction type. They differ from the standard aluminum electrolytic capacitors in that instead of aluminum, a thin sheet of tantalum is used for the plates. A thin layer of oxide on one side of the plate serves as the dielectric. The dielectric constant of the tantalum oxide is higher than that of aluminum oxide. These capacitors must be operated with a specific polarity as marked on the capacitor.

Tantalum-pellet capacitors do not need reforming and have an expected shelf life of more than 10 years. The advantages which they offer over aluminum electrolytics are a greater capacity with less bulk, and such an extremely small leakage current (micro-amps) that they may be used in multivibrators (electronic switches). Their disadvantage is a low working voltage range of from 6 V to about 50 V. They are expensive, but are capable of operating over a temperature range of $-55°C$ to $+125°C$ with a negligible change in capacitance.

Tantalums range in capacity from 0.1 μF to 330 μF. They are smaller physically than aluminum electrolytics.

14-18.5 Air-Dielectric Capacitors

These capacitors are usually the variable capacitors used to tune radio receivers and transmitters, such as AM, FM, CB, and police radios. They consist of one set of fixed plates called the stator, intermeshed with a set of rotating plates called the rotor. The stator and rotor plates are separated by a narrow air space which is the dielectric. The capacity is maximum when the plates are completely meshed (greatest plate area) and minimum when the rotor is completely out of mesh with the stator (least plate area).

14-18.6 Plastic Dielectric Capacitors

There are many types of capacitors which have a plastic dielectric. The most common of these capacitors listed in catalogues are:

1. Polystyrene film type
2. Polyester film type
3. Mylar capacitors (PVC)
4. Polystyrene capacitors
5. Metallized polyester
6. Metallized polycarbonate
7. Metallized film type

The above capacitors differ in two ways:

1. The type of polymer (plastic) used as the dielectric
2. The method by which the dielectric is installed between the plates.

The dielectric of the above capacitors consists of thin films of polymer material. The prime characteristic of these plastics is their high electrical insulation resistance (which affects leakage current) at room temperatures. The most common polymer films used for capacitor dielectrics are:

1. Mylar (Dupont's trade name for polyethylene terephthalate)
2. Polycarbonate
3. Polystyrene
4. Polyester

MYLAR CAPACITOR

The mylar capacitor is a member of the polyethylene terephthalate polymer family. It is a tough white polymer with high tensile strength, free of pinholes. It has good insulating properties over a reasonably wide range of temperatures. It is available in films as thin as 3.5 μm. It is used alone or in combination with paper as a dielectric.

These capacitors are made by sandwiching the mylar film between two long strips of aluminum or tin foil. The plastic film is slightly wider and longer than the metal plates so that the plates cannot touch and short together. The assembly is then rolled into a tubular form, leads are connected to each plate and the assembly is encased in a plastic round covering. The lead connected to the outer plate is marked by a black ring at the end of the capacitor. This lead should be connected to the least positive voltage, as it is the outside strip and shields the capacitor from outside magnetic fields.

POLYCARBONATE CAPACITORS

Polycarbonate is a polyester of carbonic acid and bisphonics. It has the good physical properties of mylar but has a lower power loss. Its temperature coefficient characteristics are closer to zero than mylar and it is available in film form as thin as 2 μm.

These capacitors are generally made by coating both sides of the plastic film with a metallic conductive material which acts as a plate. Leads are attached to each plate. The assembly is then formed into a rectangular shape and encapsulated by a moistureproof case. The capacitor constructed in this fashion is called a *metallized polycarbonate* capacitor. Metallization reduces the physical size for a specific capacity and working voltage.

Applications are the same as for polyester capacitors, but a polycarbonate capacitor would be used if a higher quality capacitor was desirable. Polycarbonate capacitors are particularly suited for high frequencies and for steep pulse applications. They are self-healing. Their capacity range is from 0.01 to 2.2 μF.

POLYSTYRENE CAPACITORS

Polystyrene is a hydrocarbon clear thermoplastic material. It has a lower permittivity value than mylar and polycarbonate materials but it has a higher dissipation factor. Its tensile strength for winding is lower than mylar; therefore, it is not available in less than an 8 μm thickness. For this reason, it is not normally used in metallized capacitors, and it is generally of tubular construction.

Polystyrene capacitors are suitable for tuned circuits in filter devices such as television and carrier telephone equipment. They have a small negative temperature coefficient and thus have high stability. They also have low loss at high frequencies.

They range in capacity from 0.001 μF to 2 μF and in working voltages from 100 V to 1600 V.

POLYESTER CAPACITORS

There are three types of polyester capacitors available:

1. Polyester
2. Metallized polyester
3. Polyester film foil

The first two are of the tubular construction. The third type is formed into a rectangular shape after the plastic film has been metallized.

1. Polyester Capacitor Applications

This is a general-purpose capacitor used mostly for coupling and decoupling. It is commonly used in both domestic and professional equipment. Its capacity range is from 0.01 μF to 2.2 μF.

2. Metallized Polyester Applications

Applications are the same as for polyester, but it is a better quality capacitor and is self-healing. Its capacity ranges from 0.01 μF to 5 μF.

3. Polyester Film Foil Applications

Applications are the same as for polyester capacitors, but a metallized polyester capacitor would be used if a higher quality capacitor were desirable. Polyester film foil capacitors are particularly suited for high frequencies and steep pulse applications. They are self-healing. Their capacity range is from 0.01 μF to 2.2 μF.

14-19 Practical Considerations

Capacitors, with the exception of aluminum electrolytics, have a shelf life of tens of years. Aluminum electrolytic capacitors need reforming periodically if they are stored for more than one to two years, because the oxide dielectric breaks down and the capacitor, if not reformed, will act more like a resistor than a capacitor. Reforming is done by applying the working voltage to the capacitor through approximately a 1000-Ω resistor for one hour. Successful reforming is indicated by a normal ohmmeter test reading for the capacity and working voltage of the capacitor.

Defective capacitors are not always easily detected. An open capacitor may be detected by the resumption of normal circuit operation when a known good capacitor of equal value is shunted across the suspected capacitor. Other than this test, the best and fastest method to test a suspected capacitor is to replace it immediately with one of its kind (in capacity and working voltage). Shorted capacitors are often detected with an ohmmeter check, but this check is reliable only if the capacitor is nearly or totally shorted. Capacity checkers are useful but often slow.

Color-coded values on capacitors are generally read in the same way as values are read on resistors, with the values being in picofarads. The fifth band, however, states the working voltage of the capacitor. For example, a yellow fifth band means that the working voltage of that capacitor is 400 V.

14-20 Capacitive-Resistive (C-R) Time-Constant Circuits

14-20.1 Charging a Capacitor

As discussed in Sec. 14-14, the expression "charging a capacitor" refers to a charge of so many coulombs flowing into, or being stored in a capacitor.

$$Q = CV \tag{8}$$

For a given capacitor, the charge is directly proportional to the voltage source applied across the capacitor. It must be noted that voltage cannot be developed across a capacitor until it takes a charge (electrons flow into it).

$$V = \frac{Q}{C} \tag{2}$$

All circuits have resistance; therefore, capacitors cannot charge instantly when a voltage source is connected in series with them, because the resistance limits the value of current that can flow into the capacitor.

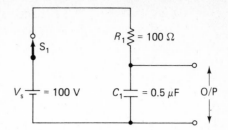

FIGURE 14-13 A capacitive-resistive time-constant circuit.

Figure 14-13 shows a 0.5-μF capacitor connected in series with a 100-Ω resistor and a 100-V source of voltage. The instant that the switch is closed in the circuit, the voltage across the uncharged capacitor is zero because there is no electron difference between the plates; thus there is no electric field and voltage across it. At that same instant the voltage across R_1 will be equal to the applied voltage.

$$\text{because} \quad V_s = v_R + v_C$$

$$\text{and} \quad v_C = 0$$

$$\text{thus} \quad v_R = V_s$$

The voltage $v_R = i \times R$; therefore, the current at the instant that the switch is closed must be

$$i = \frac{100 \text{ V}}{100 \text{ }\Omega} = 1 \text{ A}$$

The 1 A current (1 coulomb/1 second) flows into the capacitor and starts to charge it (Fig. 14-13). As soon as electrons flow into a capacitor, a voltage is developed across it.

$$V = \frac{Q}{C}$$

The voltage developed across the capacitor is series-opposing the V_s; thus, the effective voltage ($v_{eff} = V_s - v_C$) causing current to flow into C_1 decreases.

$$v_{eff} = V_s - v_C$$

The current flowing in the circuit is

$$i = \frac{v_{eff}}{R}$$

The above equation shows that i decreases as v_{eff} decreases. Time, in seconds, has been passing while this current cycle has been taking place. The current change can be plotted against time and will produce the decaying exponential curve as shown in Fig. 14-14.

The voltage across a capacitor is the difference between the source voltage and resistor voltage ($v_C = V_s - v_R$). If the current is decreasing, then the voltage across the resistor is also decreasing (Fig. 14-15), and the resistor voltage becomes zero when the capacitor voltage has reached the supply voltage value (Fig. 14-15).

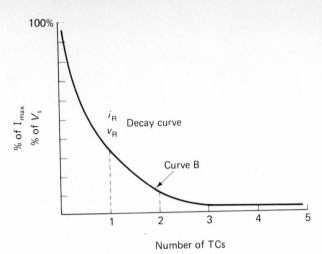

FIGURE 14-14 The circuit current and resistor voltage decay curve.

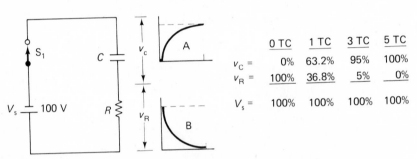

FIGURE 14-15 The voltage rise across the capacitor and the voltage decay across the resistor curves during charge time. Note: The sum of v_C and v_R always equals V_s.

It takes time for the capacitor to charge through the resistor. The time, in seconds, involved to fully charge the capacitor is five times the product of the resistance value and the capacitance value (Figs. 14-16 and 14-17). The product of the resistance and capacitance is called a time constant, one definition for which is as follows:

14-20.2 Definition of C-R Time Constant

One *C-R* time constant is the time, in seconds, which it takes for the voltage across a capacitor to reach 63.2% of the applied voltage (V_s) (Figs. 14-15 and 14-17).

The time constant is the product of the values of the resistance and the capacitance.

1 time constant (τ) = resistance $\times$ capacitance

$$1 \text{ TC } (\tau) = RC$$

where τ is in seconds
R is in ohms
C is in farads

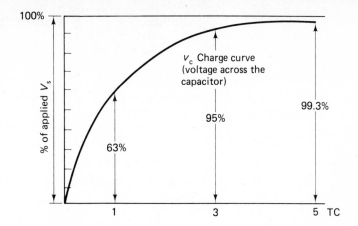

FIGURE 14-16   The capacitor charge (voltage) curve.   Notice that the voltage across the capacitor increases.

A second definition for a *C-R* time constant is that one time constant is the time which it would take for the voltage across a capacitor to rise to the value of the source voltage if the capacitor charge were to continue to rise at its initial rate of change $dv/dt = V_s/CR$ (see Fig. 14-17) during the charging time.

but when v_C is charged, $v_C = V_s$

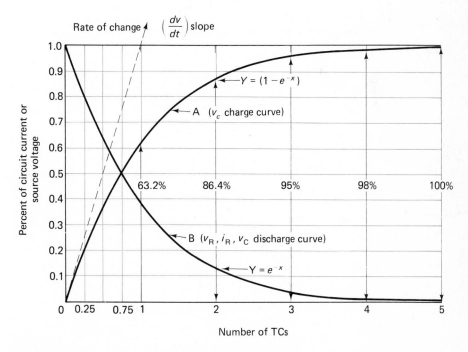

FIGURE 14-17   Universal time-constant chart.   Curve A is representative of Fig. 14-16; Curve B is representative of Fig. 14-14.

therefore $\dfrac{dv_C}{dt} = \dfrac{V_s}{CR}$ (rate of change)

and $t = CR$

where t is time

The capacitor voltage does not rise at a constant rate but continually decreases as the capacitor becomes charged because the v_C opposes V_s. It can be seen that increasing the C-R time constant results in an increase in the time required for the voltage to rise. This means that the rate at which the capacitor charges decreases as the capacitor becomes charged.

the letter symbol for one time constant is the Greek letter tau (τ)

PROBLEM 14-11

A 2-μF capacitor (C_1) is connected in series with a 100-kΩ resistor (R_1) and a 100-V voltage source. How long will it take, in seconds, to charge the capacitor to 63.2% of its full charge (100 V)?

SOLUTION   *(cover solution and solve)*

$$1 \text{ TC } (\tau) = RC$$
$$= (10^5) \times (2 \times 10^{-6})$$
$$= 2 \times 10^{-1}$$
$$= 0.2 \text{ s}$$
$$= 200 \text{ ms}$$

14-20.3 Comparison of Capacity to a Large Resistance

If C_1 were 2 μF and R_1 were increased to 1 MΩ, 1 TC would be

$$1 \text{ TC } (\tau) = RC$$
$$= 10^6 \times (2 \times 10^{-6})$$
$$= 2 \text{ s}$$

It would take the capacitor 2 s to charge to 63.2% of its full charge, instead of 200 ms as in Problem 14-11. In other words, the larger the resistor, the less rapid the charge of the capacitor.

14-20.4 Comparison of Resistance to a Large Capacity

If the resistor were left at 100 kΩ but the value of the capacitor were increased to 4 μF, 1 TC would be

$$1 \text{ TC } (\tau) = RC$$
$$= (10^5) \times (4 \times 10^{-6})$$
$$= 0.4 \text{ s}$$
$$= 400 \text{ ms}$$

A comparison of the above three examples shows that the time required to charge a capacitor to 63.2% of its full charge is directly proportional to the value of either the resistance or the capacitance, or both.

14-20.5 Proof for Definition

The current which flows the instant that the switch is closed in a C-R circuit is the value of the maximum circuit current. It is calculated by $i = V_s/R$. The maximum current is also referred to as the initial current, but it is not the steady-state current value. The steady-state current value in C-R circuits is zero.

PROOF FOR 1 TIME CONSTANT $\tau = C \times R$:

Initial current

$$i = \frac{V_s}{R} \tag{9}$$

$$\text{also} \quad i = \frac{Q}{t}$$

$$Q = i \times t \tag{10}$$

$$\text{also} \quad Q = CV_C \tag{11}$$

therefore

$$Q = i \times t = CV_C \tag{12}$$

$$\text{and} \quad i \times t = CV_C \tag{13}$$

Substitute Eq. (9) in Eq. (13) for i and

$$\frac{V_s}{R} \times t = CV_C \tag{14}$$

$$\text{and} \quad \cancel{V_s} \times t = CR\cancel{V_C} \tag{15}$$

$$V_s \text{ cancels } V_C. \text{ Therefore,} \tag{16}$$

$$t = CR \tag{17}$$

The symbol t is the time, in seconds, of one time constant.

Another proof that $t = C \times R$ begins by assuming that the product of capacity and resistance equals some other variable.

$$RC = \frac{\cancel{V}}{I} \times \frac{Q}{\cancel{V}} \quad \text{because} \quad R = \frac{V}{I} \text{ and } C = \frac{Q}{V} \tag{18}$$

$$\therefore \quad RC = \frac{Q}{I} \tag{19}$$

$$\text{but} \quad \frac{Q}{I} = t \quad (\text{Sec. 2-1.2}) \tag{20}$$

Substitute Eq. (20) in Eq. (19) and

$$RC = t \qquad\qquad (21)$$

PROBLEM 14-12

A series circuit has the following two components:

1. How long will it take capacitor C_1 to charge to 63.2% of its full charge if $C_1 = 1\ \mu F$ and $R_1 = 470\ k\Omega$?
2. How long will it take the voltage across the resistor to fall to 36.8% of the applied voltage if $C_1 = 1\ \mu F$ and $R_1 = 470\ k\Omega$?

SOLUTION (cover solution and solve)

1.
$$1\ TC = CR$$
$$= (1 \times 10^{-6}) \times (47 \times 10^4)\ s$$
$$= (47 \times 10^{-2}) \times 10^3\ ms$$
$$= 470\ ms$$

It would take 470 ms for the capacitor to charge to 63.2% of its full charge.

2. It would take 470 ms for the voltage across the resistor to fall to 36.8% of the applied voltage.

14-20.6 Time and Time Constants

The formula used to calculate the amount of time (in seconds) contained in one time constant is

$$1\ time\ constant = capacity \times resistance$$

$$1\ TC = CR$$

The number of time constants, x, is calculated by dividing the amount of *time* a voltage is applied to an R-C circuit by the time contained in 1 TC.

$$number\ of\ time\ constants = \frac{total\ time\ (in\ seconds)}{number\ of\ seconds\ in\ 1\ TC}$$

$$x = \frac{t}{R\text{-}C} = \frac{t}{\tau}$$

PROBLEM 14-13

The switch to a 300-$k\Omega$ resistor in series with a 0.1-μF capacitor and 12-V voltage source is closed for 60 ms.

1. For how many time constants, x, is the switch closed?

2. For how many times constants, x, is the switch closed if it closed for 24 ms?

SOLUTION (cover solution and solve)

1.

$$\text{time in 1 TC } (\tau) = CR$$
$$= (0.1 \times 10^{-6}) \times (3 \times 10^5) \text{ s}$$
$$= (3 \times 10^{-2}) \text{ s}$$
$$= (3 \times 10^{-2}) \times 10^3 \text{ ms}$$
$$= 30 \text{ ms}$$

The switch is closed for 60 ms.

Switch is closed $\dfrac{t}{R\text{-}C} = \dfrac{60}{30} = 2$ TC

2.

$$\text{1 TC } (\tau) = 30 \text{ ms.}$$

Time is 24 ms.

$$x = \frac{24}{30} = 0.8 \text{ TC}$$

14-20.7 Time Constants and Capacitor Charge

Definition 1 for a time constant (Sec. 14-20.2) states the relationship between the circuit components and time. The value of capacitor voltage will be different than 63.2% of the source voltage at different periods in time, such as at 2 TC, 3 TC, 4 TC, and 5 TC. The progressive percentage solution shown in Table 14-4 proves this statement.

The progressive percentage method, Table 14-4, is not a practical solution for calculating capacitor voltage percentages for a given number of time constants, but it illustrates in steps how capacitor voltage values do change in a given number of time constants. It also demonstrates why the charge curve is an exponential curve.

Definition 1 for a *C-R* time constant indicates that:

In the first time constant, the voltage across the capacitor rises to 63.2% of the source voltage.

In the second time constant, the voltage increases by 63.2% of the voltage yet to be developed across the capacitor, and so on until the capacitor voltage has reached the source voltage value. The total time involved is 5 TC.

The example in Table 14-4 shows in detail the progressive percentage solution for a standard exponential curve.

Table 14-4 Progressive Percentage Method

A resistor and a capacitor are connected in series with a 100-V voltage source. The following calculations show how the voltage is developed across the capacitor for each of the five time constants. (Compare with Fig. 14-17 for each TC.)

1. The capacitor voltage increases to 63.2% in the first time constant.

$$63.2\% \times 100 = 63.2 \text{ V}$$

Remaining voltage to increase is $100 - 63.2 = 36.8$ V.

2. Second time constant

$$\text{voltage increase} = 63.2\% \text{ of } 36.8 = 23.26 \text{ V}$$

Remaining voltage to increase is $100 - (63.2 + 23.26) = 13.54$ V.

3. Third time constant

$$\text{voltage increase} = 63.2\% \text{ of } 13.54 = 8.56 \text{ V}$$

Remaining voltage to increase is $100 - (63.2 + 23.26 + 8.56) = 4.98$ V.

4. Fourth time constant

$$\text{voltage increase} = 63.2\% \text{ of } 4.98 = 3.14 \text{ V}$$

Remaining voltage to increase is $100 - (63.2 + 23.26 + 8.56 + 3.14) = 1.84$ V.

5. Fifth time constant

$$\text{voltage increase} = 63.2\% \text{ of } 1.84 = 1.16 \text{ V}$$

Remaining voltage to increase is $100 - (63.2 + 23.26 + 8.56 + 3.14 + 1.16) = 0.993$ V.

The capacitor is considered to have reached its full charge after 5 TC (Fig. 14-17).

14-20.8 Determining Percentage of Capacitor Charge

The progressive percentage method of calculating the voltage developed across a capacitor for x number of time constants is almost useless for fractions of time constants. Its use is limited almost entirely to instructional purposes. The same two methods as were used in L/R time constant problems are also used for C-R time constant solutions. The two methods are:

Method 1. Graphical method with the universal TC chart.

Method 2. Use of the natural log, epsilon.

14-20.9 Method 1—Graphical Method of Determining Percentage of Voltage or Current

This method is exactly the same as was used in the L/R time constant circuit, Sec. 13-26.5. Refer to steps 1 to 5 in Sec. 13-26.5 for a review of the graphical method. The charge curve on the TC chart, Fig. 14-17 should also be used with the instructions. Do the following time constant example problems for practice.

PROBLEM 14-14

What is the percentage of current flow at the following time constants (from the charge curve in Fig. 14-17)?

	Time Constant	Answer
1.	At 0.50 TC	39.35%
2.	0.75 TC	47.24%
3.	1.0 TC	63.2 %
4.	1.5 TC	77.68%
5.	2.0 TC	86.50%
6.	2.5 TC	91.80%
7.	3.0 TC	95.00%

14-20.10 Percentage, Instantaneous Current, and Voltage

The value of the voltage developed across the capacitor at any charge time may be obtained by multiplying the supply voltage value by the percentage of charge.

$$v_C = \% \times V_s$$

The value of the instantaneous voltage developed across the resistor during charge of the capacitor may be obtained by subtracting the voltage developed across the capacitor from the supply voltage.

$$v_R = V_s - v_C$$

The value of the voltage developed across the resistor during discharge of the capacitor is equal to the capacitor voltage.

$$v_R = v_C$$

The value of instantaneous current is obtained by dividing the voltage developed across the resistor by the resistance value.

$$i = \frac{v_R}{R}$$

14-20.11 Method 2—Use of the Natural Log to Solve Capacitor Percentage Charge

The natural log is the most precise method of calculating the percentage of voltage developed across a capacitor at x number of time constants. The method is exactly the same as was used with L/R time constant circuits in Sec. 13-26.7. Refer to Sec. 13-26.7 for review of the procedure using the natural log.

The method for finding the value of instantaneous voltage and current was just outlined in Sec. 14-20.10.

The percentage of capacitor charge at any one specified time is a function of e^x. Percentages are based on decimal values, so the reciprocal (e^{-x}) of the expression e^x is used.

Since e^{-x} states the decimal value of current yet to flow, the value of e^{-x} must be subtracted from 1 to determine the decimal value of current which is actually flowing.

$$Y = 1 - e^{-x}$$

The letter symbol Y is used to indicate percentage. Percentage is obtained by multiplying the decimal value by 100.

The formula showing the percentage of the source voltage to which the capacitor is charged for any one time constant or part thereof is:

$$\% \, V_s = Y = (1 - e^{-x}) \times 100$$

where Y is % of V_s
e is natural log (base 2.71828)
x is number of time constants
1 is maximum value (100%)

PROBLEM 14-15

Calculate the percentage of source voltage developed across a capacitor after:

1. 1   TC
2. 1.4 TC
3. 2.5 TC
4. 5   TC   *Answer* = 99.30%
5. 4.5 TC   = 98.89%
6. 0.8 TC   = 55.07%

SOLUTION   (cover solution and solve)

1. $Y = 1 - e^{-x} \times 100 = \%$ of $V_s = v_C$

but $x = 1$

One can take only the anti-log of e^x.

$$e^1 = \text{anti-log } (e^x) \text{ of } 1 = 2.71828$$

Reciprocal of $2.71828 = \dfrac{1}{2.71828} = 0.3678$.

$$\therefore Y = 1 - 0.3678$$

$$Y = 0.6322 = 63.2\%$$

2. 1.4 TC

$$Y = 1 - e^{-1.4}$$

$$= 1 - 0.2465 = 0.753 = 75.3\% \text{ of } V_s = v_C$$

3. 2.5 TC

$$Y = 1 - e^{-2.5}$$

$$= 1 - 0.082 = 0.9179 = 91.8\% \text{ of } V_s = v_C$$

PROBLEM 14-16

A 0.02-μF capacitor (C_1) is connected in series with a 10-kΩ resistor (R_1), a switch, and a 200-V battery.
Draw the circuit and calculate:

1. The time of 1 TC
2. The number of time constants in 280 μs
3. The voltage across the capacitor 280 μs after the switch is closed
4. The voltage across the resistor 280 μs after the switch is closed
5. The current flow in the circuit 280 μs after the switch is closed

SOLUTION *(cover solution and solve)*

1.

$$1 \text{ TC} = CR$$

$$= (0.02 \times 10^{-6}) \times (1 \times 10^{4}) \text{ s}$$

$$= (2 \times 10^{-8}) \times (1 \times 10^{4}) \times 10^{6} \ \mu s$$

$$= 200 \ \mu s$$

2. Number of TC (x)

$$x = \frac{\text{time}}{\text{time in 1 TC}} = \frac{280 \ \mu s}{200 \ \mu s}$$

$$= 1.4 \text{ TC}$$

3. Capacitor voltage

$$Y = 1 - e^{-x}$$

$$\text{but } x = 1.4$$

One can take only the anti-log of e^{x}.

$$e^{1.4} = \text{anti-log } (e^{x}) \text{ of } 1.4 = 4.055$$

Reciprocal of $4.055 = \dfrac{1}{4.055} = 0.2466.$

$$Y = 1 - 0.2466$$

$$\therefore Y = 0.7534 = 75.3\ \%$$

$$v_\text{C} = 75.3\% \text{ of } V_\text{s}$$

Capacitor voltage is $0.753 \times 200 = 150.6$ V.

4. Voltage across R_1, 280 μs after the switch is closed

Method 1.

$$\text{from} \quad V_\text{s} = v_\text{R} + v_\text{c}$$

$$\therefore \quad v_\text{R} = V_\text{s} - v_\text{c}$$

$$= 200 - 150.6 = 49.4 \text{ V}$$

Method 2. Resistor voltage from decay formula

$$\% \text{ of } V_\text{s} = Y = e^{-x} \quad \text{but} \quad x = 1.4$$

$$\text{then} \quad Y = 24.66\%$$

$$\text{and} \quad v_\text{R} = 24.66\% \text{ of } V_\text{s}$$

$$= 24.66\% \text{ of } 200$$

$$= 49.3 \text{ V}$$

5. Amount of current flow after 280 μs

$$I_\text{ss} \frac{V_\text{s}}{R} = \frac{200}{10 \times 10^3} = 20 \text{ mA}$$

$$\text{and} \quad Y = e^{-1.4} = 24.66\%$$

$$\therefore \quad i = 0.2466 \times (20 \times 10^{-3}) = 4.932 \text{ mA}$$

14-20.12 Methods of Determining Number of Time Constants

There are two methods of determining the number of time constants. Method 1 has already been discussed.

Method 1. When the percentage of current is to be determined, the number of time constants, x, is calculated by

$$x = \frac{t}{\tau}$$

Method 2. When the percentage of capacitor voltage is known, the number of time constants, x, is determined by

$$e^{-x} = 1 - Y \quad \text{(from } Y = 1 - e^{-x}\text{)}$$

PROBLEM 14-17

How many time constants are required for the voltage across a capacitor to charge to 70% of the source voltage?

SOLUTION *(cover solution and solve)*

$$\text{from} \quad Y = 1 - e^{-x} = \% \text{ of } V_s$$

$$\text{then} \quad Y - 1 = -e^{-x}$$

$$\text{and} \quad e^{-x} = 1 - Y$$

$$\text{but} \quad Y = 70\%$$

$$\therefore e^{-x} = 1 - 0.70 = 0.30$$

One cannot compute a negative natural log.

The reciprocal of $0.30 = \dfrac{1}{0.30} = 3.333$.

Take the log, ($\ln x$) of $3.333 = 1.2 = e^x$.

$$x = 1.2 \text{ TC}$$

Calculator operation of the natural log:

1. Press 0.30; then the $\dfrac{1}{x}$ button.

   Result is 3.333.
2. Press 3.33; then the $\ln x$ button.

$$x = 1.2 \text{ TC}$$

PROBLEM 14-18

A 200-kΩ resistor (R_1) is connected in series with a 0.001-μF capacitor (C_1) and a 300-V battery. Draw the circuit and:

1. Determine the time necessary for the capacitor to charge to 25% of the applied voltage.
2. Determine the time required for the capacitor to charge to 180 V.
3. Determine the time necessary for the capacitor voltage to reach 75% of full charge.
4. Calculate the voltage across the resistor when the capacitor voltage is at 75% of full charge.
5. Determine the time required for the voltage across the resistor to decay to 120 V.
6. Calculate how much time will elapse after the switch is closed before the instantaneous current is 15% of the initial current value.

SOLUTION *(cover solution and solve)*

1. Time to charge the capacitor to 25% of V_s

$$\text{from} \quad Y = 1 - e^{-x}$$

$$\text{then} \quad e^{-x} = 1 - Y$$

$$\text{but} \quad Y = 0.25$$

$$\therefore e^{-x} = 1 - 0.25 = 0.75.$$

One can find only the log of e^x.

$$\text{Reciprocal of } 0.75 = \frac{1}{0.75} = 1.33333 = e^x$$

Take the log ($\ln x$) of 1.3333.

$$\therefore x = 0.287 \text{ TC} \quad \text{(number of time constants)}$$

$$1 \text{ TC} = RC$$

$$= (2 \times 10^5) \times (1 \times 10^{-3} \times 10^{-6}) \text{ s}$$

$$= (2 \times 10^{-4}) \times 10^6 \text{ } \mu\text{s}$$

$$= 200 \text{ } \mu\text{s}$$

$$\text{total time} = (\text{time of 1 TC}) \times (\text{number of TC})$$

$$= 200 \text{ } \mu\text{s} \times 0.287$$

$$= 57.6 \text{ } \mu\text{s}$$

2. Time to charge capacitor to 180 V

$$180 \text{ V} = \frac{180}{300} = 0.6 = 60\%$$

$$\text{from} \quad Y = 1 - e^{-x}$$

$$\text{then} \quad e^{-x} = 1 - Y$$

$$\text{but} \quad Y = 0.60$$

$$\therefore e^{-x} = 1 - 0.60 = 0.40$$

One cannot compute a negative natural log.

$$\text{Reciprocal of } 0.40 = \frac{1}{0.40} = 2.5 = e^x.$$

Take the log ($\ln x$) of 2.5 = 0.916

$$\therefore x = 0.916 \text{ TC}$$

$$\text{but 1 TC} = RC = 200 \text{ } \mu\text{s}$$

$$\text{total time} = (\text{time of 1 TC}) \times (\text{number of TC})$$

$$= 200 \text{ } \mu\text{s} \times 0.916$$

$$= 183.3 \text{ } \mu\text{s}$$

3. Time for the capacitor to charge to 75% of V_s

$$\text{from} \quad Y = 1 - e^{-x}$$

$$\text{then} \quad e^{-x} = 1 - Y$$

$$\text{but} \quad Y = 0.75$$

$$\therefore e^{-x} = 1 - 0.75 = 0.25$$

One cannot compute a negative natural log.

$$\text{Reciprocal of } 0.25 = \frac{1}{0.25} = 4.0 = e^x.$$

Take the log ($\ln x$) of 4.0 = 1.386.

$$x = 1.386 \text{ TC}$$

$$\text{but} \quad 1 \text{ TC} = 200 \; \mu s \text{ (see part 1. of solution)}$$

$$\text{total time} = (\text{time of 1 TC}) \times (\text{number of TC})$$

$$= 200 \; \mu s \times 1.386$$

$$= 277 \; \mu s$$

4. Resistor voltage when capacitor voltage is 75%
When v_C is at 75% charge the v_R is at 25% charge, because $V_s = v_R + v_C$.

$$V_s = 300 \text{ V}$$

$$v_R = 25\% \text{ of } 300$$

$$= 75 \text{ V}$$

5. Time for the resistor voltage to decay to 120 V

$$\% \text{ of } V_s = \frac{120}{300} = 40\%$$

$$\text{from} \quad Y = e^{-x}$$

$$\text{then} \quad e^{-x} = Y$$

$$\text{but} \quad Y = 0.40$$

$$\therefore e^{-x} = 0.40$$

One cannot compute a negative natural log.

$$\text{Reciprocal of } 0.40 = \frac{1}{0.40} = 2.5 = e^x.$$

Take the log ($\ln x$) of 2.5 = 0.916.

$$x = 0.916 \text{ TC}$$

$$1 \text{ TC} = 200 \ \mu s \text{ (see part 1 of solution)}$$

$$\text{total time} = (\text{time of 1 TC}) \times (\text{number of TC})$$

$$= 200 \ \mu s \times 0.916$$

$$= 183.3 \ \mu s$$

6. Time for the current to rise to 15% of initial current
The current decays in an *R-C* circuit therefore.

$$\text{from} \quad Y = e^{-x}$$

$$\text{then} \quad e^{-x} = Y$$

$$= 0.15$$

$$x = 1.897 \text{ TC}$$

$$1 \text{ TC} = 200 \ \mu s$$

$$\text{total time} = (\text{time of 1 TC}) \times (\text{number of TC})$$

$$= 200 \ \mu s \times 1.897$$

$$= 379.4 \ \mu s$$

PROBLEM 14-19

A 150-V battery is connected in series with a 200-kΩ resistor (R_1) and a 2000-μF capacitor (C_1). Draw the circuit and calculate:

1. Time of 1 TC
2. Initial charging current
3. Rate of voltage change
4. Time it would take the capacitor to charge to 84 V
5. Time it would take the voltage across the resistor to fall to 600 V
6. The voltage across the capacitor after 240 s
7. The voltage across the resistor after 240 s

SOLUTION   *(cover solution and solve)*

1.

$$1 \text{ TC} = C \times R$$

$$= (2000 \times 10^{-6}) \times (2 \times 10^{5})$$

$$= (2 \times 10^{-3}) \times (2 \times 10^{5})$$

$$= 400 \text{ s}$$

2.

$$\text{initial current} = \text{maximum current}$$

$$= \frac{V_s}{R} = \frac{150}{200,000} = A$$

$$= \frac{(150 \times 10^{-5})}{2} \times 10^6$$

$$= 750 \ \mu A$$

3.

$$\text{rate of change} = \frac{V_s}{CR} \quad \text{(see Sec. 14-20.2)}$$

$$= \frac{150}{(2 \times 10^{-3}) \times (2 \times 10^5)}$$

$$= \frac{150}{4 \times 10^2} = 0.375 \ V/s$$

4. Time to charge to 84 V

$$\% \text{ of } V_s = \frac{84}{150} \times 100 = 56\%$$

from $\quad Y = 1 - e^{-x}$

then $\quad e^{-x} = 1 - Y$

but $\quad Y = 0.56$

$$\therefore e^{-x} = 1 - 0.56 = 0.44$$

One cannot compute a negative natural log.

Reciprocal of $0.44 = \dfrac{1}{0.44} = 2.272 = e^x$.

Take the log ($\ln x$) of $2.272 = 0.8209$.

$$\therefore x = 0.82 \ TC$$

but $\quad 1 \ TC = 400 \ s$

$$\text{total time} = (\text{time of 1 TC}) \times (\text{number of TC})$$

$$= 400 \ s \times 0.82$$

$$= 328 \ s$$

5. Time for the resistor voltage to decay to 60 V

$$\% \text{ of } V_s = \frac{60}{150} \times 100 = 40\%$$

from $\quad Y = e^{-x}$

then $\quad e^{-x} = Y$

$$\text{but} \quad Y = 0.40$$

$$\therefore e^{-x} = 0.40$$

One cannot compute a negative natural log.

$$\text{Reciprocal of } 0.40 = \frac{1}{0.40} = 2.5 = e^x.$$

Take the log ($\ln x$) of $2.5 = 0.916$.

$$x = 0.916 \text{ TC}$$

$$1 \text{ TC} = 400 \text{ s}$$

$$\text{total time} = (\text{time of 1 TC}) \times (\text{number of TC})$$

$$= 400 \text{ s} \times 0.916$$

$$= 366.4 \text{ s}$$

6. Capacitor voltage after 240 s

$$\text{number of TC} = \frac{t}{\tau} = \frac{240}{400} = 0.6 \text{ TC}$$

$$v_C = \% \text{ of } V_s$$

$$= Y = 1 - e^{-x}$$

$$\text{but} \quad x = 0.6$$

One can take only the anti-log (e^x) of e^x.
Take the anti-log (e^x) of $0.6 = 1.822$.

$$\text{Reciprocal of } 1.822 = \frac{1}{1.822} = 0.5488 = e^x.$$

$$Y = 1 - 0.5488 = 0.45 = 45\%$$

$$v_C = 45\% \text{ of } 150 \text{ V}$$

$$= 67.5 \text{ V}$$

7.

$$v_R = 180 - 67.5 = 112.5 \text{ V}$$

14-21 C-R Discharge

The discharge waveform produced by a discharging capacitor is as important as the charge waveform. Sawtooth waveforms may be obtained from across the capacitor as it charges and discharges. Other shapes of waveforms such as a differentiated waveform (Fig. 14-18c and d) are removed from the resistor as the capacitor charges and discharges.

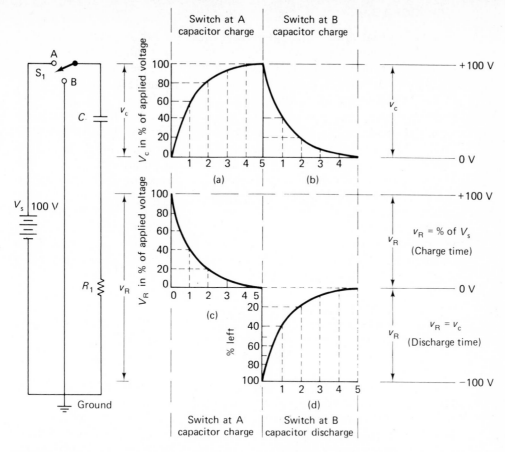

FIGURE 14-18 The combined waveshapes and voltage polarities across the resistor and capacitor during charge and discharge time.

14-21.1 Operation

CAPACITOR VOLTAGE

If the switch S_1 in Fig. 14-19 is closed to position A for 5 TC, the capacitor will become fully charged. When the switch is placed to position B, the capacitor will discharge through the short and through R_1, and it will produce the capacitor discharge curve (Fig. 14-18b).

RESISTOR VOLTAGE AND CHARGE TIME

The first instant that the switch S_1 is closed to position A, the total supply voltage is developed across the resistor with the top of the resistor being positive in polarity with respect to ground (Fig. 14-18c). The voltage decays across the resistor as the capacitor charges, until the resistor voltage becomes zero, at which time the capacitor is fully charged. (Note the voltage polarity of the charged capacitor—the top is positive with respect to ground.)

Suppose that after 5 TC the switch S_1 is placed to position B in Fig. 14-19. The capacitor now acts like a battery and the electrons (charge) in the bottom of the capacitor discharge through resistor R_1, making the top of R_1 negative in polarity with respect to ground (Fig. 14-18d).

The capacitor was charged to 100 V in position A, and when the switch is placed to position B, the current flow through R_1 is the same value as the initial charging current but it flows in the opposite direction. The current flow out of the charged capacitor, therefore, produces 100 V negative voltage across the resistor with respect to ground. The supply of current from the capacitor decreases as the capacitor discharges, and so does the current which is flowing through R_1, resulting in a progressively smaller voltage across R_1 until it is zero when the capacitor is discharged.

Note that the waveform developed across R_1 in Fig. 14-18c is a positive spike of voltage followed by an equal-sized negative spike of voltage in Fig. 14-18d. Voltage of this shape is called a differentiated waveform.

14-21.2 Use of Formulas

The *time* which is necessary for the capacitor to discharge is calculated with the same time constant formula as when considering charge time.

$$1 \text{ TC} = CR$$

The difference is that on discharge the time constant expresses the time in which the voltage or charge across the capacitor will decay; therefore, the decay percentage formula, $Y = e^{-x}$, must be used to calculate the value of voltage across the capacitor or resistor at any point in time while the capacitor is discharging.

14-21.3 Clarification

The universal time constant curves as shown in Figs. 14-17 and 14-18 are graphs of the capacitor and resistor voltages, in percentages, during charge and discharge. Figure 14-18a and b shows exactly the same thing as does Fig. 14-17, but Fig. 14-18a and b also includes the polarity of the voltages on charge and discharge. See Table 14-5.

Table 14-5   Comparison of Curves, Figs. 14-17 and 14-18

Curve in Fig. 14-18	Type of Curve	Curve in Fig. 14-17
(a)	v_C charge curve	A
(b)	v_C discharge curve	B
(c)	v_R decay curve	B
(d)	v_R decay curve from 100 V to 0 V but (d) also shows polarity (negative voltage)	B

14-21.4 Time Constants and Percentage of Capacitor Discharge

The percentage of capacitor discharge in x number of time constants may be determined by one of either of two methods:

Method 1. The discharge curve on the universal time constant chart (Fig. 14-17)

Method 2. The use of the natural log, $Y = e^{-x}$

14-21.5 Method 1—The Universal Time Constant Chart

Exactly the same method is used as was used to obtain the percentage of current decay in L/R time constants. The only difference is that the capacitor discharge curve (Fig. 14-17) is used instead of the current decay curve (Fig. 13-45). Refer to Sec. 13-26.5 for review of the procedure, but use the discharge curve.

14-21.6 Method 2—The Use of the Natural Log

A capacitor discharges in an exponential manner and thus produces an exponential curve on discharge. The discharge of a capacitor is therefore a function of the natural log. The percentage of voltage discharge for x number of time constants may be determined by

$$Y = e^{-x}$$

where Y is % of maximum voltage still
       remaining across the capacitor
x is number of time constants
e is natural log (base 2.71828)

PROBLEM 14-20

A 0.1-μF capacitor is connected in series with a 150-kΩ resistor and a 150-V battery for 8 TC (Fig. 14-19).

1. How many seconds will it take C_1 to completely charge?
2. How many seconds will it take to completely discharge the capacitor in position B?

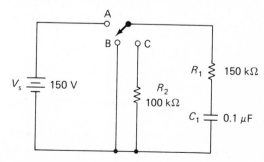

FIGURE 14-19

3. How long will it take capacitor C_1 to completely discharge if the switch is in position C?

SOLUTION   *(cover solution and solve)*

1. Charge

$$1 \text{ TC} = RC$$
$$= (150 \times 10^3) \times (0.1 \times 10^{-6})$$
$$= 15 \times 10^{-3} \text{ s}$$
$$= 15 \text{ ms}$$

$$\text{Complete charge} = \tau \times 5 \text{ TC}$$
$$= (15 \times 10^{-3}) \times 5 \text{ TC}$$
$$= 75 \text{ ms}$$

2. Discharge time

$$1 \text{ TC} = R_1 C_1$$
$$= (150 \times 10^3) \times (0.1 \times 10^{-6}) \text{ s}$$
$$= 15 \text{ ms}$$

$$\text{complete discharge} = 5 \text{ TC}$$
$$= 5 \times (\text{time in 1 TC})$$
$$= (15 \times 10^{-3}) \times 5$$
$$= 75 \text{ ms}$$

Note that it takes the capacitor the same time to discharge as it does to charge if the time constants are the same.

3. Discharge time

$$1 \text{ TC} = (R_1 + R_2) \times C_1$$
$$= (250 \times 10^3) \times (0.1 \times 10^{-6})$$
$$= 25 \text{ ms}$$

$$\text{complete discharge} = 5 \text{ TC}$$
$$= (RC) \times 5 \text{ TC}$$
$$= (25 \times 10^{-3}) \times 5$$
$$= 125 \text{ ms}$$

PROBLEM 14-21

Suppose the switch is closed to position A for 6 TC and then it is placed in position B (Fig. 14-20).

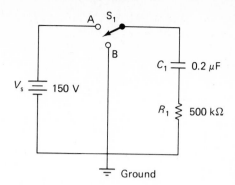

FIGURE 14-20

1. How many seconds will it take for the resistor voltage to decay to 42 V in position B?

2. What will be the polarity of the 42 V across the resistor with respect to ground in position B?

3. What will be the capacitor voltage when the resistor voltage is 42 V while the switch is in position B?

4. How many seconds will it take for the capacitor voltage to fall to 42 V while the switch is in position B?

5. What will be the value of the current flow through the resistor when the capacitor voltage has fallen to 42 V while the switch is in position B?

6. What will be the charge in the capacitor when its voltage has fallen to 42 V (charge = Q)?

7. Draw the shape of the waveform across the resistor and the capacitor. Show the voltage polarities and the value of voltages the instant that the switch is closed to position B.

Capacitor is fully charged when the switch is placed in position B.

SOLUTION *(cover solution and solve)*

1. Time for v_R to decay to 42 V

$$\% \text{ of full charge} = \frac{42}{150} \times 100 = 28\%$$

$$Y = e^{-x}$$

$$0.28 = e^{-x}$$

$$x = 1.27 \text{ TC}$$

$$\text{but} \quad 1 \text{ TC} = RC$$

$$= (500 \times 10^3) \times (0.2 \times 10^{-6})$$

$$= 100 \text{ ms}$$

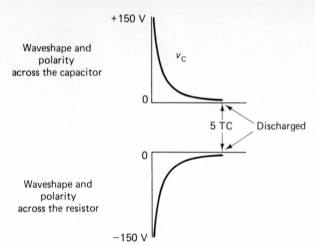

Waveshape and polarity across the capacitor

+150 V

v_C

0

5 TC   Discharged

0

Waveshape and polarity across the resistor

FIGURE 14-21 The waveforms produced across the resistor and capacitor.

−150 V

$$\text{total time} = (\text{time in 1 TC}) \times (\text{number of TC})$$

$$= (100 \times 10^{-3}) \times 1.27$$

$$= 127 \text{ ms}$$

2. Negative

3. Capacitor voltage always equals resistor voltage on discharge.

4. The capacitor voltage and resistor voltage are the same voltage on discharge. The resistor voltage decayed to 42 V in 127 ms. The capacitor voltage, then, decayed to 42 V in 127 ms.

5. Value of current when $v_c = 42$ V

$$v_c = v_R = 42 \text{ V}$$

$$i_R = \frac{v_R}{R} = \frac{42}{500 \times 10^3} \text{ A}$$

$$= \frac{42 \times 10^{-3}}{500} \times 10^6 \text{ } \mu\text{A}$$

$$= 84 \text{ } \mu\text{A}$$

6. Charge in capacitor when $v_c = 42$

$$\text{charge} = Q = CV$$

$$= (0.2 \times 10^{-6}) \times 42$$

$$= 8.4 \text{ } \mu\text{C}$$

7. See Fig. 14-21.

PROBLEM 14-22

Refer to Fig. 14-22.

The switch is left in position A for 40 ms and then it is placed in position B (Fig. 14-22).

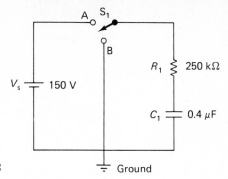

FIGURE 14-22

Ground

Calculate:

1. The voltage across the capacitor at the end of 40 ms of charge.
2. The voltage across the resistor after 40 ms of charge.
3. While the capacitor is charging, what is the polarity with respect to ground of the voltage across:
   a. The resistor
   b. The capacitor
4. At 25 ms after the switch is put in position B, what is the value of voltage across
   a. The resistor
   b. The capacitor
5. While the capacitor is discharging, what is the polarity of the voltage with respect to ground across
   a. The resistor
   b. The capacitor

SOLUTION   *(cover solution and solve)*

In Position A:

1. v_C at end of 40 ms of charge

$$1 \text{ TC} = RC$$

$$= (250 \times 10^3) \times (0.4 \times 10^{-6}) \text{ s}$$

$$= (100 \times 10^{-3}) \times 10^3 \text{ ms}$$

$$= 100 \text{ ms}$$

$$40 \text{ milliseconds} = \frac{\text{total time}}{1 \text{ TC}} = \frac{40 \text{ ms}}{100 \text{ ms}} = 0.40 \text{ TC}$$

On a charge

$$Y = 1 - e^{-x}$$

$$\text{but} \quad x = 0.4$$

One can take only the anti-log of e^x.

$$e^{0.4} = \text{anti-log } (e^x) \text{ of } 0.4 = 1.4918$$

$$\text{Reciprocal of } 1.4918 = \frac{1}{1.4918} = 0.67.$$

$$\therefore Y = 1 - 0.67 = 0.3296 = 32.96\%$$

$$\text{or} \quad Y = 33\%$$

$$v_C = \% \times V_s$$
$$= 0.33 \times 150$$
$$= 49.5 \text{ V}$$

2. v_R after 40 ms of charge

$$\text{On charge} \quad v_R + v_C = V_s$$
$$\text{if} \quad v_C = 49.5 \text{ V}$$
$$\text{and} \quad V_s = 150 \text{ V}$$
$$\text{then} \quad v_R = V_s - v_C$$
$$= 150 - 49.5$$
$$= 100.5 \text{ V}$$

3.

$$v_C \text{ is } +$$
$$v_R \text{ is } +$$

In Position B:

4. v_C 25 ms after position B
   The charge on the capacitor at the end of 40 ms was 49.5 V. This voltage value is the voltage source on discharge.

1 TC on discharge (same as charge) = RC

$$= (250 \times 10^3) \times (0.4 \times 10^{-6}) \text{ s}$$
$$= (100 \times 10^{-3}) \times 10^3 \text{ ms}$$
$$= 100 \text{ ms}$$

$$\text{number of TC of discharge} = \frac{\text{time}}{1 \text{ TC}}$$

$$= \frac{25 \text{ ms}}{100 \text{ ms}}$$

$$= 0.25 \text{ TC}$$

On discharge

$$v_C = (\% \text{ of } V_{chge}) \times V_{chge}$$

where $\%$ of V_{chge} is $Y = e^{-x}$
V_{chge} is charge across the
capacitor before it began
to discharge

$$Y = e^{-0.25}$$

$$= 0.78 = 78\%$$

$$V_{chge} = 49.5 \text{ V}$$

After 25 ms discharge

a.

$$v_C = (\% \text{ of } V_{chge}) \times V_{chge}$$

$$v_C = 0.78 \times 49.5$$

$$= 38.6 \text{ V}$$

b.

$$v_R = v_C \text{ on discharge}$$

$$= 38.6 \text{ V}$$

5. Capacitor voltage is a positive voltage with respect to ground.
Resistor voltage is a positive voltage with respect to ground.

14-22 Time Constant and Frequency

C-R time constants are commonly used in timing circuits. For example, burglar alarms use *C-R* time constant circuits to keep the alarm from ringing until the operator can leave the room Circuits used to turn off the car headlights after the garage doors are closed employ *C-R* timing circuits. These are only a few of the many applications for which time constants are used. Radar, oscilloscopes, computers, and television receivers use many *C-R* time constant circuits.

C-R time constant circuits are commonly used in waveshaping applications. For example, the horizontal sweep circuit which produces the straight horizontal line (horizontal trace) on the oscilloscope or television set employs two *C-R* time constant circuits. One time constant circuit is in the output of the horizontal oscillator and produces a sawtooth voltage to move the electron beam across the cathode ray tube at a constant rate of speed. The other *C-R* circuit is in the input of the horizontal oscillator where it provides the correct timing for the rate of speed of the trace.

14-22.1 *Time Constant Classifications*

Waveshaping time constants are of three categories.

1. Long time constant

2. Short time constant

3. Medium time constant

These classifications, as in L/R time constant circuits, can be made only when the time of 1 TC is compared with the frequency of the applied voltage. It should be remembered that a small capacitor will charge quickly, and that a small resistor connected in series with it as part of the time constant circuit will permit a large current to flow into the capacitor. The two conditions facilitate the rapid charge of the capacitor. Generally speaking, a small capacitor and a small resistor constitute a short time constant.

Large C-R values produce longer time constants. A larger capacitor requires more current to charge it completely, while a larger resistor limits the current more, and thereby increases the time required to charge the capacitor.

It is conventional practice to consider that a circuit has a *long* time constant when the product of the resistor and the capacitor is greater than ten times the period ($T = 1/f$ s) of the applied voltage.

A *short* time constant is a C-R circuit in which the product of the resistor and the capacitor is less than 1/10 of the period of the applied voltage.

A *medium* time constant has a C-R time between the long time constant and the short time constant values.

14-22.2 Square Waves, Frequency, Period, and Time Constants

Most C-R time constants in radar circuits are associated with high-speed timing devices, so the voltages applied are rapid, repeating square wave voltages rather than dc voltages. Many other types of equipment also use square waves.

SQUARE WAVES

Very simply stated, a square wave voltage may be created by connecting a resistor in series with a battery and switch. The voltage across the resistor is equal to the battery voltage when the switch is closed and is zero when the switch is open. The amplitude of the square wave voltage is equal to the battery voltage (Fig. 14-23).

FREQUENCY

The number of times that the switch to the battery is opened and closed determines the frequency of the square wave.

PERIOD

A period is another word for time. A period is the length of time, in seconds, of one cycle. The length of time that the switch to the battery is closed and opened determines the period.

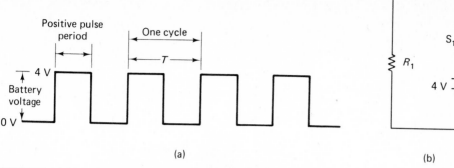

(a)   (b)

FIGURE 14-23   Opening and closing the switch between the battery and resistor R_1 in (b) will produce the square wave shown in (a), which shows the positive pulse and one cycle.

$$\textbf{period} = \frac{1}{\textbf{frequency}}$$

$$T = \frac{1}{f}$$

where T is time in seconds
f is cycles per second
(Hz)

A square wave with a frequency of 100 Hz has a period (T) of

$$T = \frac{1}{f} = \frac{1}{100} \times 10^3 = 10 \text{ ms}$$

COMPARISON OF TIME IN 1 TC AND FREQUENCY

If 1 TC C-R were 2 ms and the period of a square wave were 10 ms, one would say that the voltage was being applied for

$$x = \frac{\text{period}}{RC} = \frac{T}{1 \text{ TC}} = \frac{10}{2} = 5 \text{ TC}$$

14-22.3   Effect of a Short TC in C-R Circuits

The square wave shown in Fig. 14-24 has a frequency of 50 Hz which is equal to

$$\text{period } (T) = \frac{1}{f} = \frac{1}{50} \text{ s}$$

$$= \frac{1}{50} \times 10^3 \text{ ms}$$

$$= 20 \text{ ms}$$

$$\text{or } = (20 \times 10^3) \times 10^6 = 20,000 \text{ } \mu\text{s}$$

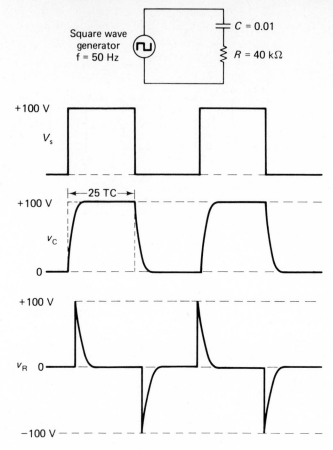

FIGURE 14-24 Waveforms produced from a square wave by a short time constant.

The positive voltage period (1/2 cycle) lasts for 10,000 μs. The TC of the C-R circuit is

$$1 \text{ TC} = C\text{-}R$$

$$= (40 \times 10^3) \times (0.01 \times 10^{-6}) \text{ s}$$

$$= (0.4 \text{ ms})$$

$$= (0.4 \times 10^{-3}) \times 10^6$$

$$= 400 \text{ } \mu\text{s}$$

The positive volt pulse is applied for

$$x = \frac{\text{time}}{CR} = \frac{T}{1 \text{ TC}} = \frac{10,000 \text{ } \mu\text{s}}{400 \text{ } \mu\text{s}} = 25 \text{ TC}$$

The positive voltage is applied to the C-R circuit for 25 times longer than the time of 1 time-constant. Therefore, the duration of the time constant of the C-R circuit is short compared to the duration of the applied voltage.

Compared to the duration of the applied voltage, the capacitor will charge quickly and will remain charged until the voltage is removed. The capacitor will also discharge quickly compared to the time when no voltage is applied. The waveform produced across the capacitor is shown as v_C in Fig. 14-24 and the voltage across the resistor is shown as v_R. Note the negative and positive spikes of voltage developed across the resistor as the result of the short time duration of current flow through R_1. Note the similarity in waveform shapes to those in Fig. 14-18c and d. Most short time constant circuits are used for waveshaping with the output taken from across the capacitor.

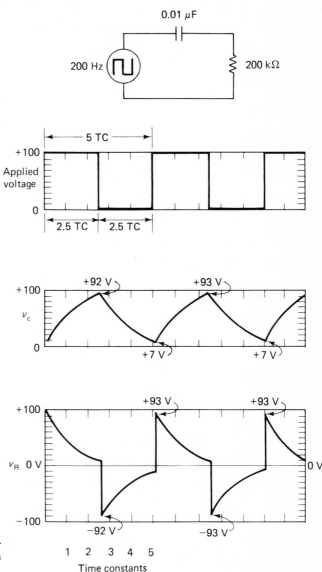

FIGURE 14-25 Waveforms produced from a square wave by a medium time constant.

Figure 14-25 shows a medium time constant circuit. The time constant of the C-R circuit is

$$1 \text{ TC} = RC$$
$$= (200 \times 10^3) \times (0.01 \times 10^{-6})$$
$$= 2 \text{ ms}$$

The frequency of the square wave is 200 Hz.

The period (T) is $\dfrac{1}{f} = \dfrac{1}{200} \times 10^3 = 5$ ms

The positive voltage is applied for 1/2 cycle.

$$\text{period} = 2.5 \text{ ms} \quad (1/2 \text{ of } 5)$$

Therefore, the positive pulse voltage is applied to the C-R circuit for 2.5 TC.

$$X = \frac{\text{time}}{RC} = \frac{T}{1 \text{ TC}} = \frac{2.5 \text{ ms}}{2 \text{ ms}} = 1.25 \text{ TC}$$

The period of the applied signal is 1.25 times longer than the time constant of the circuit; therefore they are almost equal to one another.

Medium time constants produce varying waveshapes across both the resistor and capacitor; therefore, they are used for waveshaping as well as for timing purposes.

14-22.5   *Effect of a Long Time Constant*

Figure 14-26 shows the effect on the waveform when the time constant of the C-R circuit is long compared to the frequency (period) of the applied voltage. The time constant in Fig. 14-26 is 10 times longer than the duration of the period of the voltage applied to it.

$$1 \text{ TC} = CR = (0.1 \times 10^{-6}) \times (1 \times 10^6) = .1 \text{ s}$$
$$= .1 \times 10^3 = 100 \text{ ms}$$

$$\text{frequency} = 100 \text{ Hz} = T = \frac{1}{f} = \frac{1}{100} \times 10^3 = 10 \text{ ms}$$

The positive voltage is applied for 1/2 of the cycle, which is 5 ms (1/2 of 10).

$$X = \frac{\text{time}}{RC} = \frac{T}{1 \text{ TC}} = \frac{5 \text{ ms}}{100 \text{ ms}} = \frac{1}{20} \text{ TC}$$

In other words the time constant is 20 times longer than the duration of 1 period of the applied signal.

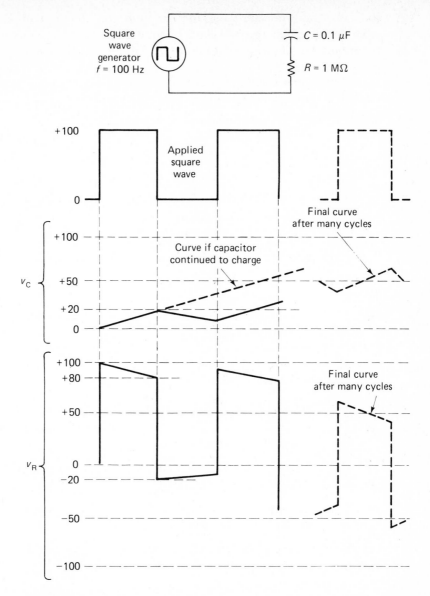

FIGURE 14-26 Waveforms produced from a square wave by a long time constant.

The positive voltage is applied for (1/20) TC. Referring to the universal charge curve, one can see that the capacitor will be charged to only 10% of its full charge in (1/20) TC. The first 10% of the charge is almost a straight line on the curve (linear). During discharge a similar linear curve is produced. Notice that the capacitor never completely charges nor discharges, so the voltage across the capacitor finally becomes nearly constant in value; that is, it approaches a dc voltage. Notice that most of the square wave voltage appears across the resistor.

Long time constant circuits are used for filter circuits in which it is desirable to obtain a dc voltage output from an input which has a large ac component. The output voltage of filter circuits is always taken from across the capacitor.

A long time constant circuit is also used for coupling in audio amplifiers. In this application, the audio signal output is always taken from across the resistor.

EXERCISES

1. What constitutes a capacitor?

2. What is the general purpose of a capacitor?

3. When and why does current stop flowing into a capacitor when it is connected in a closed circuit to a source of voltage?

4. Define:
    a. Electrostatic line of force
    b. Electric field strength
    c. Electric flux density
    d. Permittivity
    e. Dielectric constant
    f. Dielectric strength

5. State two factors which increase electric field strength.

6. Use symbols to write the equation for electric field strength. Explain what each symbol represents.

7. Use symbols to write the equation for electric flux density. Explain what each symbol represents.

8. Write the equation showing the relationship between permittivity, flux density, and field strength.

9. What is the permittivity value of free space or a vacuum?

10. Write an equation showing the relationship between dielectric constant, permittivity of a dielectric material, and permittivity of free space.

11. Write the formula for capacitance. Incorporate all the parameters which affect the capacity of a capacitor. Explain what each symbol represents.

12. Calculate the capacity of a 2-plate capacitor if the plates are separated by a polystyrene dielectric 32 μm thick. The dielectric constant of the polystyrene is 2.6. The area of each plate is 0.5 m^2.

13. State in what applications the following ranges of capacitors would most likely be found:

a. 10 μF and greater _____
b. 1000 pF and smaller _____
c. 1 nf to 10 μF _____

14. Complete the following table of conversions.

pF	nF	μF
2500		
		0.0037
	4.7	
4700		
	39	
	400	
		1
		4

15. Which has the largest dielectric strength value: air, ceramic, or paper?

16. What is the breakdown voltage of a polystyrene capacitor if the polystyrene is 1×10^{-5} m thick?

17. How does the dielectric constant and dielectric strength affect the capacity value of capacitors?

18. What two factors determine the working voltage of a capacitor?

19. Explain what is meant by the term working voltage.

20. What is the dielectric material in tantulum and electrolytic capacitors?

21. What is the maximum working voltage of tantulum and electrolytic capacitors?

22. Write an equation showing the relationship between the voltage developed across a capacitor, the capacity of the capacitor, and the charge in the capacitor.

23. Why does the smallest capacitor in a series circuit have the largest voltage developed across it?

24. Two capacitors, one twice as large as the other, are connected in parallel. Which capacitor will be able to deliver the greatest amount of current?

25. What voltage will be developed across a 300-pF capacitor if it is storing 164 nC of charge?

26. How many coulombs of charge are stored in a 12-μF capacitor which has 500 V developed across it?

27. A 40-μF and a 10-μF capacitor (C_1 and C_2, respectively) are connected in series across a 500-V source of voltage. Determine:
   a. The total capacity
   b. The voltage developed across each capacitor

28. A 20-μF capacitor and 30-μF capacitor are connected in parallel across a 250-V source. Calculate:
   a. The total capacity
   b. The voltage developed across each capacitor

29. What, in the construction of capacitors, determines whether the capacitor will be constructed in a tubular or layer construction?

30. List four classifications of capacitors other than the plastic type.

31. List four types of plastic capacitors.

32. How many time constants are required for a capacitor to charge to 63.2% of the applied voltage?

33. What determines the time in seconds for a capacitor to charge to 63.2% of the applied voltage?

34. How many time constants does it take a capacitor to fully charge?

35. State both definitions for a capacitor time constant.

36. A 750-kΩ resistor is connected in series with a 500-pF capacitor and a 150-V battery. How long will it take, in seconds, to fully charge the capacitor?

37. What will be the value of voltage developed across a resistor after 5 TC if it is connected in series with a capacitor and a 90-V source of voltage?

38. The switch in Fig. 14-27 is closed for 75 ms.
   a. For how many time constants is the switch closed?
   b. For how many time constants is the switch closed in the circuit in Fig. 14-27 if it is kept closed for only 25 ms?

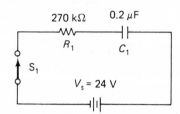

270 kΩ 0.2 μF
R_1 C_1
S_1
V_s = 24 V

FIGURE 14-27

39. Determine the following values for the circuit in Fig. 14-28 after the switch S_1 is closed:

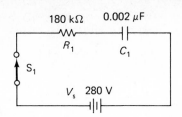

FIGURE 14-28

a. The time it will take for the capacitor C, to charge to 25% of the applied voltage.

b. The time in seconds required for the capacitor to charge to 180 V.

c. The time which will elapse after the switch is closed before the capacitor voltage reaches 70% of its full charge.

d. The value of voltage across the resistor when the capacitor voltage has reached 70% of full charge.

e. The time necessary for the voltage across the resistor to decay to 120 V.

f. The time which will elapse after the switch is closed before the instantaneous current will be 20% of the initial current value.

40. Draw the waveform shape developed across a capacitor during

a. Charge

b. Discharge

41. Draw the waveform shape developed across the resistor in series with a capacitor during

a. Charge of the capacitor

b. Discharge of the capacitor

42.

a. A 0.1-μF capacitor is connected in series with a 250-kΩ resistor and a 160-V battery (Fig. 14-29).

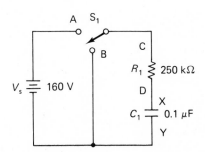

FIGURE 14-29

(1) Determine the voltage developed across capacitor C_1 after the switch S_1 is closed to position A for 2 TC.

(2) How many seconds is the switch left closed to position A if it is closed for 2 TC?

b. The switch, S_1, is opened from position A and closed to position B after 8 TC. Calculate the following values after the switch is closed to position B from position A:

(1) How long, in seconds, will it take for the voltage across the resistor to decay to 38 V after the switch is placed in position B?

(2) What will be the polarity of the 38 V across the resistor?

(3) What will be the capacitor voltage when the voltage across the resistor has reached 38 V?

(4) What will be the polarity of the voltage at point X with respect to Y across the capacitor?

(5) What will be the value of the current flow in the circuit when the voltage across the capacitor has reached 38 V?

(6) What will be the charge in the capacitor when the voltage across it is 38 V?

43. What determines whether a *C-R* time constant is long, short, or medium?

44. Does a large or a small *C-R* value produce the longest time constant?

45. Draw the waveforms produced across a capacitor for a square wave applied to:

a. A short *C-R* time constant

b. A long *C-R* time constant

chapter 15

ALTERNATING CURRENT

15-1 Introduction

Alternating current is a general expression which describes a voltage which is continually changing in polarity, and a current which is continually changing in direction. Remember, dc circuits involve voltages and currents which are of constant magnitude, polarity, and direction. Dc voltages are generally produced chemically by batteries (dc generators are the exception). All ac voltages are produced by mechanical devices called ac generators or alternators which are driven by steam power, water power, or some other form of motor.

the symbol for alternating current or voltage is ac

15-2 Rules of Magnetism

15-2.1 Induced Voltage

The generation of ac voltage is dependent upon a conductor cutting, or being cut by a magnetic field. Voltage is induced into a conductor when the conductor is cut by a magnetic field (Fig. 15-1a), but there will be zero volts when the conductor is moving parallel to the lines of force (Fig. 15-1b).

As current can be induced into a wire when it is cut by a magnetic field (Fig. 15-1a), current can be induced into a coil by cutting its turns with the magnetic field of a bar magnet as it is moved into and out of the coil (Fig. 15-2). There cannot be current without voltage, so there must also be a voltage induced which can actually be measured across the terminals of the coil (Fig. 15-2).

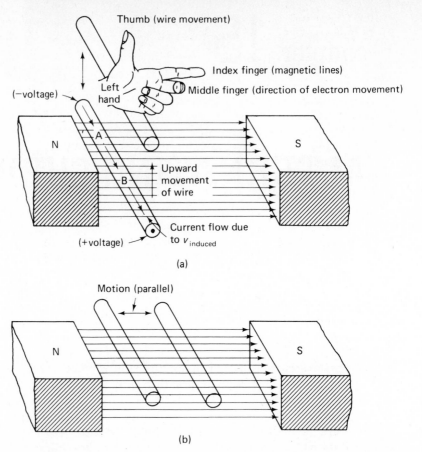

Thumb (wire movement)

Index finger (magnetic lines)

Left
hand

Middle finger (direction of electron movement)

(−voltage)

N

A

B

S

Upward
movement
of wire

(+voltage)

Current flow due
to $v_{induced}$

(a)

Motion (parallel)

N

S

(b)

FIGURE 15-1   (a) Voltage is induced into the wire when it is moved at
right angles to the lines of force (electromagnetic induction). (b) There is
no voltage induced when the wire is moved parallel to the lines of force
(no cutting).

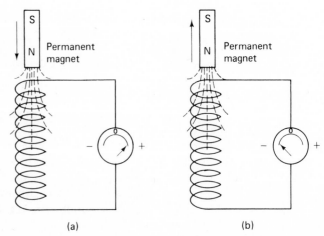

S

N

Permanent
magnet

S

N

Permanent
magnet

FIGURE 15-2   Effect of moving a
permanent magnet into and out of
a coil.   (a) N-pole moving down
moves the pointer to the positive
side. (b) N-pole moving up causes
the meter to read negative.

(a)

(b)

There will be voltage across the coil as long as the magnet is moving. If the magnet is held still, there will be no voltage.

15-2.2 Magnitude of Induced Voltage

The principle underlying the generation of ac voltage was discovered by Michael Faraday of England and Joseph Henry of the United States. The two men also discovered that the magnitude of the voltage depends directly upon the speed at which the conductor is being cut by the magnetic field (Faraday's Law). For example, if the bar magnet in Fig. 15-3a were dropped into the coil, a large induced voltage would register on the voltmeter. If the bar were inserted slowly (Fig. 15-3b), only a small voltage would register. The more rapidly magnetic lines of force cut a conductor, the greater will be the value of the induced voltage.

Conclusion: Induced voltage is proportional to the rate of cut of the coil by a magnetic field. This fact was observed by Michael Faraday when he discovered that electricity could be produced by cutting a conductor with magnetic flux. The rule establishing this fact is called Faraday's Law.

Faraday's Law states that the value of voltage induced into a coil (v_L) is directly proportional to the rate of change of the magnetic flux cutting the turns of that coil.

induced voltage is proportional to the rate of change of flux

$$v_L \alpha \frac{d\Phi}{dt} \frac{Wb}{s}$$

Magnitude is also dependent upon the angle of cut. Maximum voltage is induced into a conductor when it is cut at right angles (90°) by a magnetic field (Fig. 15-1a), but 0 V occurs when the conductor is moved parallel to the lines of flux (Fig. 15-1b). In other words, no voltage is produced when the wire is not being cut.

There are 90 angles of 1° increments between 0° and 90°. The magni-

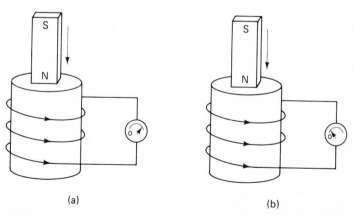

(a)   (b)

FIGURE 15-3 (a) Bar magnet dropped into coil produces a large volt. (b) Bar magnet inserted slowly into coil produces a small voltage.

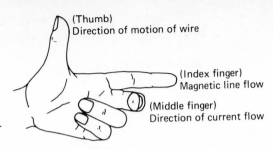

(Thumb)
Direction of motion of wire

(Index finger)
Magnetic line flow

(Middle finger)
Direction of current flow

FIGURE 15-4   Left-hand generator rule.

tude of the voltage induced, up to the maximum value, will increase in proportion to the number of degrees in the angle of cut.

15-2.3 Voltage Polarity and Current Direction

The polarity of the voltage induced into a conductor or coil is dependent upon the direction of motion. This fact is demonstrated with a coil and a magnet as in Fig. 15-2. Moving the magnet down into the coil will produce a voltage of one polarity; moving the magnet up will produce the opposite polarity of voltage. Reversing the polarity of the magnet or the direction of the winding of the coil will also produce a voltage of opposite polarity. The polarity of the voltage induced into a wire and the direction of the resulting current when the circuit is complete are determined by the left-hand generator rule (Fig. 15-4).

15-2.4 Left-Hand Generator Rule

The left-hand generator rule is as follows:

1. Extend the thumb, index finger, and middle finger of the left hand at right angles to one another as shown in Fig. 15-4.
2. Point the thumb in the direction of motion of the wire. (Fig. 15-1a).
3. Point the index finger in the direction of the magnetic flow, N to S between the magnets.
4. The middle finger will point in the direction of the current flow (in coils, in the top half of the winding).
5. Current flows from the negative polarity to the positive polarity. This fact can be used to identify the polarity of the induced voltage.

15-3 Generation of ac Voltage

Now that the rules of the relationship between current and magnetism have been established, the generation of ac voltage and current can be explained.

Figure 15-5a shows a looped conductor shaped into the form of a rectangular coil and placed parallel to the lines of force between the N and S poles of a magnet. No voltage will be induced into the conductor in Fig. 15-5a when it is moved, because the conductor is not being cut by the magnetic flux.

15-3.1 Angle of Cut and Voltage

If the conductor in Fig. 15-5b were moved, voltage would be induced into the looped conductor because it would be moving at an angle of 90° to the flux lines. If the conductor in Fig. 15-5a were moved through all the degrees until it reached the 90° cut in Fig. 15-5b, the value of voltage would gradually increase from zero value to the maximum value as shown by a sine wave. The voltage sine wave curve in Fig. 15-5 is produced by plotting the value of the voltage for each degree of cut against the angle of cut on a piece of graph paper.

15-3.2 Simple Generator

Figure 15-5 shows the principle of operation for the simplest type of a single-phase ac generator. In practice, many loops often are wound together to obtain a larger voltage. The N and S magnetic poles are referred to as the field. In a practical generator, the magnets are electromagnets which are magnetized by a dc voltage.

15-3.3 Voltage Value and Polarity

Figure 15-5 shows five positions of the looped conductor between the field magnets, and the relative voltage value and voltage polarity in each position.

Position (a): Side B of the loop is parallel to the lines of flux and the output is zero.

Position (b): The conductor has moved clockwise through every degree to the right angle (90°) position and the voltage has increased to the peak (maximum) value. The motion of side B has been to the right, so the polarity of the induced voltage is positive as shown by the left-hand rule (Fig. 15-4).

Position (c): The conductor has moved from the right angle cut position in (b) through progressively smaller angles of cut, until the conductor is again moving parallel to the lines of force. The output voltage is reduced in value until it is now zero.

Position (d): Side B of the conductor has moved to the left from the parallel position (c) to position (d), through a continuously increasing angle of cut, until it is now at right angles (90°) to the flux lines. The induced voltage is again at its peak value. Side B in its movement from position (c) to position (d) has travelled to the left, whereas it moved to the right from (a) to (b). The left-hand rule shows that the polarity of the voltage has reversed, and the output of side B will have a negative voltage with respect to the output of side A.

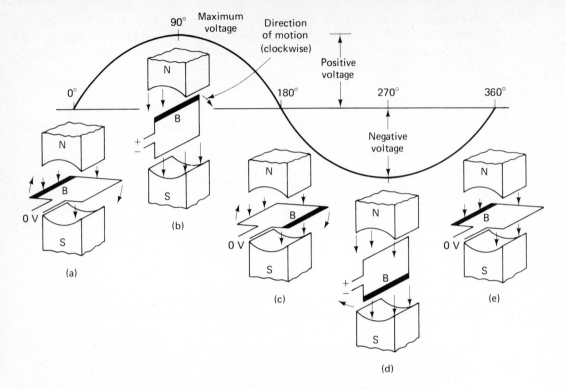

FIGURE 15-5 The principle of the generation of an ac voltage. The amplitude of the sine-wave represents the magnitude of the voltage produced at different angles of cut of the loop by the magnetic field.

Position (e): Side B has returned to its original parallel position, but has been cut by a decreasing angle of cut as it moved from position (d) to (e). The magnitude of the voltage has steadily decreased until it is now 0 V.

The generator loop has travelled one complete mechanical cycle of 360° from position (a) to position (e). The loop has also been cut four times by angles from 0° to 90°. Therefore, it is said that the loop has produced a full 360° electrical cycle, and the loop conductor has been cut by every angle from 0° to 360°.

From position (a) to (b) the loop is said to have been cut by angles from 0° to 90°.

From position (b) to (c) the loop is said to have been cut by angles from 90° to 180°.

From position (c) to (d) the loop has been cut by angles from 180° to 270°.

From position (d) to (e) the loop has been cut by angles from 270° to 360°. The respective magnitude of voltage for each 90° angle is shown in Figs. 15-5 and 15-11.

If either the positive or negative peak value is known, the value of the voltage (instantaneous voltage) at each angle of cut can be determined by taking the sine of that angle and multiplying it by the peak value. This is why this type of ac voltage is called a sine wave or sinusoidal voltage.

SUMMARY:

The magnitude of generated ac voltage is determined by three factors:

1. Angle of cut
2. Rate of cut
3. Number of conductors side by side revolving together

The polarity of generated ac voltages is determined by:

1. Direction of conductor movement with respect to the magnetic field in which it is moving
2. The polarity of the magnetic field in which the conductor is moving

15-4 Sine of the Angle

15-4.1 Sine of the Angle and Phasors

A loop rotating in an ac generator passes through 360° within each cycle, and when the value of the voltage induced into the conductor for each degree is plotted against each degree of the 360°, a sine wave waveform is produced as shown in Figs. 15-5 and 15-11.

The sine wave may also be represented by a vector diagram (Fig. 15-6). The horizontal line (Fig. 15-6) is always the reference arm, and the arm which rotates counterclockwise from it (positive direction) is called the phasor. The angle formed between the X-axis and the phasor arm is called the phase angle. The phasor gives the angle of cut of the loop by the number of degrees of displacement counterclockwise from the reference arm. The length of the phasor is always drawn proportional to the magnitude of the peak (maximum) voltage; a line (O) drawn from the end (point) of the phasor to the horizontal axis (perpendicular to the X-axis) represents, and is proportional to the value of the instantaneous voltage for any angle of cut (Figs. 15-8, 15-9, and 15-10).

15-4.2 Sine of the Angle

The sine of an angle is a trigonometric function of right-angled triangles. The sine of an angle is simply the ratio of the side opposite to the angle of concern, to the hypotenuse. ($Sin\ \theta = O/H$). If the value of the hypotenuse (which is V_{max}) is known and the ratio is known, the value of the instantaneous voltage, O, may be found by:

$$O = \sin \theta \times H$$

$$\text{since} \quad \sin \theta = \frac{O}{H}$$

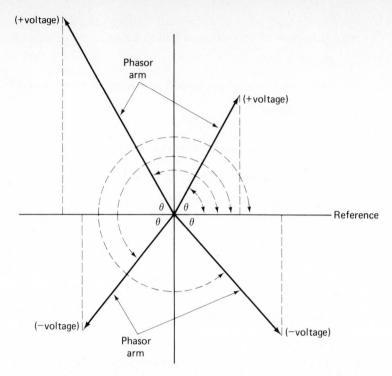

FIGURE 15-6   Vector diagram. Phasor arm in four different positions and forming four different angles.

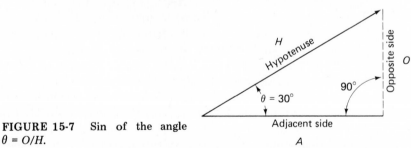

FIGURE 15-7   Sin of the angle $\theta = O/H$.

For example, in Fig. 15-7 the angle θ is 30° and

$$\sin 30° = 0.5$$

If the hypotenuse represents a peak voltage of 100 V, the value of the instantaneous voltage at the opposite side (O) to 30° is

$$v = \sin 30° \times 100 \text{ V}$$
$$= 0.5 \times 100 \text{ V} = 50 \text{ V}$$

15-4.3 Angles Greater than 90°

Most calculators automatically compute the sine of angles greater than 90°. There may be some students, however, who must know the mathe-

Ch. 15 Alternating Current

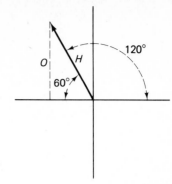

FIGURE 15-8 Angle greater than 90° but less than 180°.

matical mechanics of computing the sine of angles greater than 90°, 180°, or 270°.

15-4.4 Angles Greater than 90° but Less than 180°

The sine of an angle greater than 90° but less than 180° is obtained by first subtracting the angle of travel from 180° and then taking the sine of that angle. (See Fig. 15-8.)

EXAMPLE ─────────────────────────────────

$$\sin 120°$$
$$= \sin (180 - 120) = 60°$$
$$= 0.866$$
$$\sin 170°$$
$$= \sin (180 - 170) = 10°$$
$$= 0.17$$

───

15-4.5 Angles Greater than 180° but Less than 270°

Angles greater than 180° place the phasor arm below the X-axis. All values below the X-axis are negative, so the ratio will be a negative value which, when multiplied by the peak voltage, produces a negative voltage value. The angle is obtained by subtracting 180° from the angle of travel. (See Fig. 15-9.)

EXAMPLE ─────────────────────────────────

$$\sin 190°$$
$$= \sin (190 - 180) = 10°$$
$$= -0.17$$

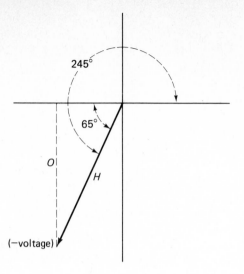

FIGURE 15-9   Angle greater than 180° but less than 270°.

$$\sin 245°$$
$$= \sin (245 - 180) = 65°$$
$$= -0.906$$

15-4.6 Angles Greater than 270° but Less than 360°

The phasor arm is below the X-axis; therefore, the ratio is a negative voltage. The angle is obtained by subtracting the angle of travel from 360°. (See Fig. 15-10.)

EXAMPLE

$$\sin 305°$$
$$= \sin (360 - 305) = 55°$$
$$= -0.819$$
$$\sin 345°$$
$$= \sin (360 - 345) = 15°$$
$$= -0.259$$

15-5 Instantaneous Values

There are three instantaneous values in a generated ac voltage:

1. Instantaneous voltage (v)
2. Instantaneous current (i)
3. Instantaneous power (p)

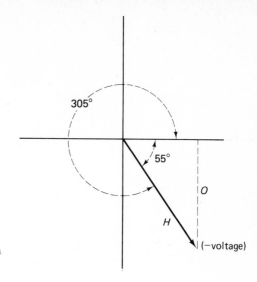

FIGURE 15-10 A phasor arm in the 305° position.

15-5.1 Instantaneous Voltage

Instantaneous voltage is the value of voltage at any specific instant in time. It is never constant, but has a different value for every angle of cut. Its value is directly proportional to the sine of the angle of cut.

the symbol for instantaneous voltage is the small letter v

instantaneous voltage = maximum voltage times sine of the angle

$$v = V_\mathrm{p} \times \sin \theta$$

where v is value of instantaneous voltage

V_p is maximum voltage (or peak voltage)

$\sin \theta$ is the sine of the angle being cut

In other words, the sine of the angle gives the decimal part or percentage of the peak (maximum) voltage at that particular angle of cut. The voltage value can then be obtained by multiplying the value of peak voltage by the percentage.

PROBLEM 15-1 ─────────────────────────────────────

The peak voltage in Fig. 15-11 is 100 V. What is the instantaneous voltage (v) when the conductor is being cut by:

1. 30°
2. 90°
3. 135°
4. 315°

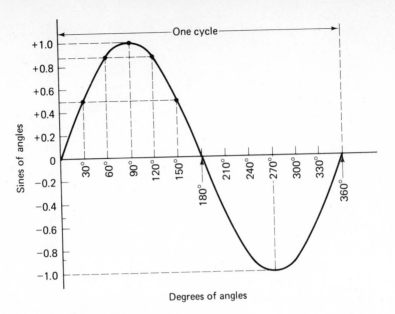

FIGURE 15-11 A sine-wave representation of voltage and the percent of the peak voltage at each angle.

SOLUTION   (cover solution and solve)

$$v = V_p \times \sin \theta$$

1. $$v = 100 \text{ V} \times \sin 30°$$
    $$= 100 \text{ V} \times 0.5 = 50 \text{ V}$$

2. $$v = 100 \text{ V} \times \sin 90°$$
    $$= 100 \text{ V} \times 1.0 = 100 \text{ V}$$

3. $$v = 100 \text{ V} \times \sin 135°$$
    $$= 100 \text{ V} \times 0.707 = 70.7 \text{ V}$$

4. $$v = 100 \text{ V} \times \sin 315°$$
    $$= 100 \text{ V} \times -0.707 = -70.7 \text{ V}$$

15.5.2 Instantaneous Current (i)

the symbol for instantaneous current is the small letter i

Current flows whenever a load is connected to a voltage source. The value of this current is in direct proportion to the voltage applied to the load; therefore, the value of ac current will be directly dependent upon the instantaneous voltage. The value of instantaneous current can be determined by

$$\text{instantaneous current} = \frac{\text{instantaneous voltage}}{\text{load}}$$

$$i = \frac{v}{R_{\mathrm{L}}}$$

The value of instantaneous current can also be determined at any degree of cut by multiplying the peak current by the sine of the angle of that degree.

$$\text{instantaneous current} = \text{sine of } \theta \times \text{peak current}$$

$$i = \sin \theta \times I_{\mathrm{p}}$$

where I_{p} is $\dfrac{V_{\mathrm{p}}}{R_{\mathrm{L}}}$

θ is angle of cut

PROBLEM 15-2

A 40-Ω load is connected across an ac generator which produces a peak voltage of 160 V. What is the value of the instantaneous current (i) at

1. 45°
2. 90°

SOLUTION   (cover solution and solve)

$$I_{\mathrm{p}} = \frac{160}{40} = 4 \text{ A}$$

and $i = I_{\mathrm{p}} \times \sin \theta$

1.

$$i = 4 \text{ A} \times \sin 45°$$

$$= 4 \text{ A} \times 0.707 = 2.828 \text{ A}$$

2.

$$i = 4 \text{ A} \times \sin 90°$$

$$= 4 \text{ A} \times 1 = 4 \text{ A}$$

15-5.3 Instantaneous Power (Resistive Circuits) (p)

the symbol for instantaneous power is the small letter p

Instantaneous power (p) is calculated with the same formula as is used in dc circuits, except that instantaneous values are used in place of dc values.

$$p = v \times i = i^2 \times R = \frac{v^2}{R}$$

SUMMARY:

There are two ways of determining instantaneous voltage:

$$v = \sin \theta \times V_\mathrm{p}$$

$$\text{or} \quad v = i \times R_\mathrm{L}$$

There are two ways of determining instantaneous current:

$$i = \sin \theta \times I_\mathrm{p}$$

$$\text{where} \quad I_\mathrm{p} \text{ is } \frac{V_\mathrm{p}}{R_\mathrm{L}}$$

$$\text{or} \quad i = \frac{v}{R_\mathrm{L}}$$

There are three ways of determining instantaneous power:

$$p = i \times v$$

$$\text{or} \quad p = i^2 \times R_\mathrm{L}$$

$$\text{or} \quad p = \frac{v^2}{R_\mathrm{L}}$$

Table 15-1 summarizes the instantaneous values of voltage, current, and power obtained at four different instants in time for Fig. 15-12. Figure 15-13 is a phasor diagram representation of each of the angles shown in Table 15-1.

Table 15-1   Some Instantaneous Voltage, Current, and Power Values

Degrees from zero	v	i	p
$39°$	113.27 V	2.83 A	320.57 W
$120°$	155.88 V	3.897 A	607.46 W
$228°$	-133.8 V	-3.345 A	447.56 W
$300°$	-155.88 V	-3.897 A	607.46 W

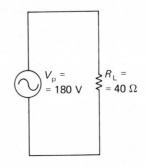

FIGURE 15-12   A pure resistive circuit.

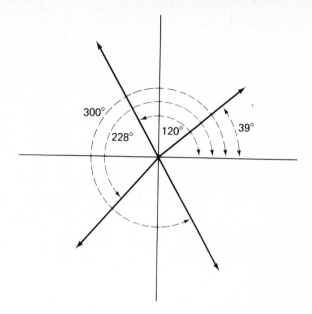

FIGURE 15-13 A phasor diagram representation of the voltage or current at different degrees of the full cycle. Compare with Table 15-1.

15-6 Sine Wave Values

The magnitude of sine wave voltages can be expressed in:

1. Peak to peak voltage
2. Peak voltage (maximum)
3. Effective voltage or rms
4. Average value

15-6.1 Maximum or Peak Value and Peak to Peak Voltage

The sine wave voltage as displayed on the oscilloscope is a picture of the actual ac voltage and shows the maximum (peak) positive value as well as the peak (maximum) negative value. In other words, the peak voltage as well as the peak to peak voltage is shown on the scope (Fig. 15-14).

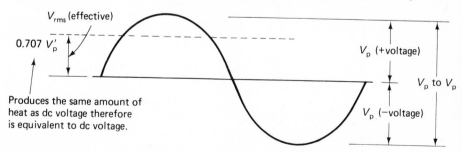

FIGURE 15-14 A sine wave showing the various values of ac voltage.

15-6.2 The Effective Voltage (rms)

Most 60-cycle ac voltage is measured with an ac voltmeter. The voltmeter measures rms, or effective voltage, which is a value equivalent to the same power or heat which a dc voltage would supply.

Note that an ac voltage is never constant in value; it changes from zero to a peak value, whereas a dc voltage is always the same value. Common sense therefore tells us that a 100-V peak ac voltage cannot provide the same heat nor supply the same power to an electric motor as a 100-V dc voltage. In fact, a 100-V peak ac voltage is equivalent to only 70.7 V of dc voltage in its heating or power capabilities. To convert peak ac voltage to rms (effective) voltage, use the following equation:

$$V_{rms} = 0.707 \; V_p$$

To convert peak to peak ac voltage to rms voltage, the peak to peak voltage must first be divided by 2 to obtain the peak value.

$$V_{rms} = 0.707 \; \frac{V_{p\text{-}p}}{2}$$

$$= 0.707 \; V_p$$

DERIVING 0.707

If all of the instantaneous values of the peak voltage in a complete electrical cycle are squared, summed, and then averaged, the square root of the mean (average) will be 0.707 of the peak value. Thus rms (root, mean, square).

The value of 0.707 comes from the inverse of the square root of

$$2 = \frac{1}{\sqrt{2}} = \frac{1}{1.414} = 0.707.$$

15-6.3 Converting rms and Peak to Peak Voltage

It is often necessary to know the peak to peak value of a measured ac voltage.

$$\text{from} \quad V_{rms} = 0.707 \; V_p$$

$$V_p = \frac{V_{rms}}{0.707}$$

$$\text{but} \quad \frac{1}{0.707} = 1.414$$

$$\therefore \quad V_p = 1.414 \; V_{rms}$$

$$\text{but} \quad V_{p\text{-}p} = 2 \times V_p$$

Substitute 1.414 V_{rms} for V_p

$$\text{and} \quad V_{p\text{-}p} = 2.828 \times V_{rms}$$

$$\text{or} \quad V_{\text{p-p}} = 2.828 \ V_{\text{rms}}$$

These are values applicable only to sine wave voltages or currents. Square wave, sawtooth, and complex waveforms use other values.

15-6.4 Average Value of Sine Waves

Average values of voltages and currents are not used as often as are rms values, but they are of importance in, for example, rectifiers, which may be either half-wave or full-wave. A rectifier is an electronic component which will pass on to other circuits only the positive half of the ac cycle in half-wave rectification (Fig. 15-17), or will pass on both halves of the ac cycle in full-wave rectification (see Fig. 15-18). In full-wave rectification the negative half of the cycle is transferred to the positive side of the zero line with the positive half-cycle (Fig. 15-16).

An unrectified sinusoidal ac voltage is symmetrical; that is, the negative half-cycle is exactly the same as the positive half-cycle (Fig. 15-14). If all the instantaneous positive values were added to all the instantaneous negative values, the average would be zero. This is why a dc voltmeter will read zero on an unrectified voltage.

HALF-WAVE AVERAGE (RECTIFIED)

To obtain the average value of an ac voltage, the instantaneous value for every degree from $0°$ to $180°$ in a half-cycle is summed, then averaged. The average value is always 31.8% of the peak voltage value. If the peak voltage was 100 V, for example, the average would be 31.8 V (half-cycle).

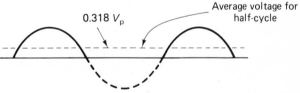

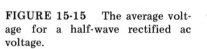

FIGURE 15-15 The average voltage for a half-wave rectified ac voltage.

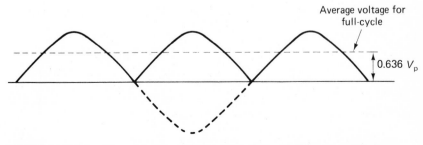

FIGURE 15-16 The average voltage value for a full-wave rectified ac voltage.

FULL-WAVE AVERAGE (RECTIFIED)

The value 0.318 is the average for only a half-cycle. Because each half-cycle is the same as every other half-cycle in a sine wave voltage, the average value of the peak voltage for a complete cycle is

$$0.318 \times 2 = 0.636 \ V_p \ \text{(full wave)}$$

$$V_{avge} = 0.636 \ V_p$$

SUMMARY:

The average current or voltage of an unrectified sine wave is zero.

$$V_{avge}, I_{avge} = 0 \ \text{(sine wave)}$$

The average current or voltage for a rectified half-wave sine wave is 0.318 of peak voltage.

$$V_{avge}, I_{avge} = 0.318 \ V_p \ \text{(half-wave)}$$

The average voltage or current value for a rectified full-wave sine wave is 0.636 of peak voltage.

$$V_{avge}, I_{avge} = 0.636 \ V_p \ \text{(full-wave)}$$

Traditionally the rectified full-wave average value of voltage or current is quoted for all sine wave voltages.

$$V_{avge}, I_{avge} = 0.636 \ V_p$$

15-6.5 Current Values

The same formulas are used for current as are used for voltage.

$$\text{rms current} = 0.707 \ I_p$$

$$\text{peak current} = 1.414 \ \text{rms current}$$

$$\text{peak to peak current} = 2.828 \ \text{rms current}$$

$$\text{average current} = 0.636 \ I_p \ \text{(full-wave)}$$

Table 15-2 Computation of Peak, rms, or Average Value

Given Value	Peak	rms	Average (Full Wave)
rms	1.41 × rms		0.9 rms
Peak		0.707 × Peak	0.636 × Peak
Average (Full Wave)	1.57 × Average	1.11 × Average	
For peak to peak value multiply peak value by 2			

PROBLEM 15-3

What is the peak voltage of a sine wave that causes the same power as a 120-V dc voltage?

SOLUTION   (cover solution and solve)

Dc voltage is equivalent to rms (effective) voltage. Therefore 120 V dc = 120 V_{rms}.

$$V_p = 1.414 \times V_{rms}$$

$$= 1.414 \times 120 = 169.68 \text{ V}$$

PROBLEM 15-4

The instantaneous value of current of a 60-Hz voltage is 12 mA, 30° after zero angle. What is the

1. Peak current
2. Peak to peak current
3. rms current
4. I_{avge} (full-wave)

SOLUTION   (cover solution and solve)

1.   from $i = I_{max} \times \sin \theta$

$$I_{max} = \frac{i}{\sin \theta} = \frac{(12 \times 10^{-3})}{\sin 30} = 24 \text{ mA (peak)}$$

2. $I_{p\text{-}p} = 2 \times I_p$

$$= 2 \times 24 \text{ mA} = 48 \text{ mA (peak to peak)}$$

3. $I_{rms} = 0.707\, I_p$

$$= 0.707 \times 24 \text{ mA} = 16.97 \text{ mA (rms)}$$

4. $I_{avge}(\text{FW}) = 0.636 \times I_p$

$$= 0.636 \times 24 \text{ mA} = 15.26 \text{ mA (average)}$$

15-7 Frequency and Period

15-7.1 Frequency (f)

We have seen that one complete mechanical rotation is also one complete electrical cycle (360°) in the simple two-pole ac generator. If the conductor rotated 60 times in one second, it would be said that the frequency of rotation was 60 cycles per second, or in the SI units, 60 Hz. Since the electrical cycle is the same as the mechanical cycle, the frequency of the electrical ac voltage and current will also be 60 Hz. Frequency is the number of complete cycles occurring each second. The frequency of 60 Hz is the frequency for all North American power sources. Europe has a power line frequency of 50 Hz.

the symbol for frequency is f

the symbol for cycles per second is Hz

The unit *hertz* (Hz) was accepted as the SI unit (international system) for cycles per second in 1957 to honor the German physicist H. R. Hertz.

1 Hz = 1 cycle per second (c/s)

1 kHz = 1 kilohertz = 1000 c/s

1 MHz = 1 megahertz = 10^6 c/s

1 GHz = 1 gigahertz = 10^9 c/s

1 THz = 1 terahertz = 10^{12} c/s

15-7.2 Period (T)

The term "period" is another word for time. A period is the amount of time in seconds required to produce one electrical cycle. A frequency of 60 Hz means that there are 60 full electrical cycles (from $0°$ to $360°$) completed every second. If 60 cycles are made every second, then how long will it take to make one cycle? The answer is 1/60 of a second. This answer is the period for a 60-Hz frequency.

$$\text{period} = \frac{1}{\text{frequency}}$$

$$T = \frac{1}{f}$$

where f is the number of cycles
occurring in 1 s
T is the time in seconds to
make 1 cycle

Table 15-3 Frequency versus Time

Frequency	Period of 1 cycle			
	$ms\ (10^{-3})$	$\mu s\ (10^{-6})$	$ns\ (10^{-9})$	$ps\ (10^{-12})$
25 kHz	0.04	40	40,000	
25 MHz		0.04	40	40,000
25 GHz			0.04	40
25 THz				0.04

PROBLEM 15-5

A radio station operates at a frequency of 1050 kHz. What is the period of one cycle?

SOLUTION   *(cover solution and solve)*

$$T = \frac{1}{f}$$

$$= \frac{1}{1050 \times 10^3} \times 10^6$$

$$= 0.95 \ \mu s$$

PROBLEM 15-6

An audio signal has a period (the time of one cycle) of 250 μs. What is the frequency of this signal?

SOLUTION   *(cover solution and solve)*

$$f = \frac{1}{T}$$

$$= \frac{1}{250 \times 10^{-6}} = 4000 \ \text{Hz}$$

15-8 Time, Angle of Cut, and v

The value of instantaneous voltage can be determined by $v = V_{max} \times \sin \theta$ when one complete cycle is involved. Generators do not produce one cycle and then stop, however. Generators operate for hours or years without stopping.

Theoretically, it may be desirable to determine the value of instantaneous voltage after a specific number of minutes or seconds after a generator has started. Calculations are always based on the assumption that the generator starts with 0 V.

PROBLEM 15-7

What is the value of the instantaneous voltage 12 ms after a 60-Hz two-pole generator producing 260 V_p has started?

SOLUTION   *(cover solution and solve)*

$V_p = 260 \ \text{V}$

$T \ \ = 12 \ \text{ms}$

$f \ \ = 60 \ \text{Hz}$

It takes time for a loop to rotate around the 360° in a complete mechanical cycle and produce the 360° electrical cycle. If the frequency

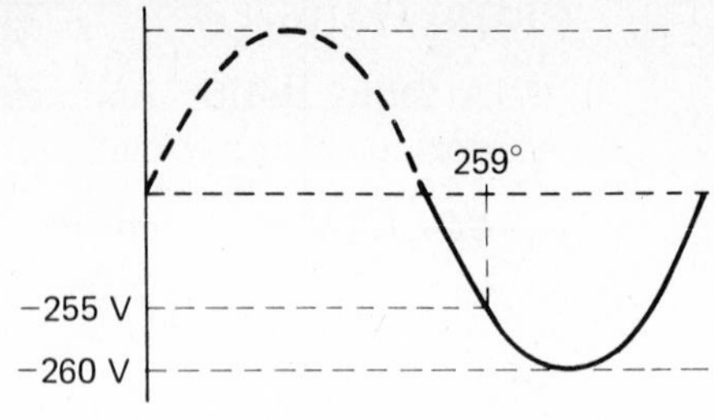

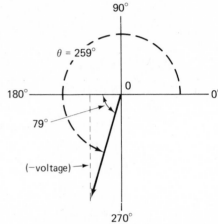

FIGURE 15-17   The angle of cut after 12 ms and the ac voltage produced.

−255 V

−260 V

259°

90°

$\theta = 259°$

180°

79°

0

0°

(−voltage)

270°

FIGURE 15-18   A phasor diagram representation of the angle produced after 12 ms.

of rotation is 60 Hz, then the time of 1 cycle (period) is

$$T = \frac{1}{60} \text{ s for 1 cycle}$$

$$= 16.67 \text{ ms/cycle}$$

$$16.67 \text{ ms} = 1 \text{ cycle} = 360°$$

$$\therefore \ 1 \text{ ms} = \left(\frac{360}{16.67}\right)°$$

$$\text{and} \quad 12 \text{ ms} = 21.596 \times 12$$

$$12 \text{ ms} = 259°$$

$$\text{but} \quad v = V_p \times \sin \theta$$

$$= 260 \times \sin 259°$$

$$= 260 \times -0.982$$

$$= -255 \text{ V}$$

PROBLEM 15-8

A 1-kHz audio signal has a peak amplitude of 24 V.

1. What is the instantaneous value of voltage 150 μs after zero angle?

444

2. What is the instantaneous voltage 45 μs before it reaches maximum positive voltage?

SOLUTION   (cover solution and solve)

f = 1000 Hz

V_p = 24 V

T = 150 μs

1. Period of 1000 Hz

$$\therefore T = \frac{1}{f} = \frac{1}{1000} \text{ s}$$

$$\text{or } T = \frac{1}{10^3} \times 10^6 = 1000 \ \mu\text{s for 1 cycle}$$

$$1000 \ \mu\text{s} = 1 \text{ cycle} = 360°$$

$$1000 \ \mu\text{s} = 360°$$

$$1 \ \mu\text{s} = \frac{360}{1000} \text{ degrees}$$

$$1 \ \mu\text{s} = (360 \times 10^{-3}) \text{ degrees}$$

Degrees after 150 μs

$$\text{degrees} = (360 \times 10^{-3}) \times 150$$

$$= 54°$$

$$v = V_p \times \sin 54°$$

$$= 24 \times 0.809 = 19.4 \text{ V}$$

2. Instantaneous voltage (v) 45 μs before V_p

period of 1000 Hz = 1000 μs

$$1000 \ \mu\text{s} = 1 \text{ cycle} = 360°$$

$$1 \ \mu\text{s} = \frac{360}{1000} = (360 \times 10^{-3}) \text{ degrees}$$

$$45 \ \mu\text{s} = (360 \times 10^{-3}) \times 45$$

$$= 16.2°$$

Voltage reaches maximum positive voltage at 90°.

$$16.2° \text{ before } V_p = 90 - 16.2 = 73.8°$$

$$\text{from} \quad v = V_p \times \sin \theta$$

$$= 24 \times \sin 73.8°$$

$$= 24 \times 0.96 = 23 \text{ V}$$

What is the value of instantaneous voltage 54 ms after a 400-Hz two-pole 24-V peak volt generator has gone through 0 V?

SOLUTION   *(cover solution and solve)*

$$f = 400 \text{ hz}$$

$$\therefore T = \frac{1}{f} = \frac{1}{400} \times 10^3 \text{ ms}$$

$$= 2.5 \text{ ms/cycles}$$

$$2.5 \text{ ms} = 1 \text{ cycle} = 360°$$

$$1 \text{ ms} = \frac{360}{2.5} = 144°$$

$$54 \text{ ms} = (144 \times 54)°$$

$$= 7776°$$

$$7776° = \frac{7776}{360} = 21.6 \text{ cycles}$$

The generator goes through 0 V every complete cycle (360°). Therefore, the angle of cut must be calculated as a decimal part of the cycle. Therefore, only the 0.6 of the 21.6 cycles is used.

$$\text{angle of cut} = 360 \times 0.6 = 216°$$

$$v = V_{\text{max}} \times \sin 216°$$

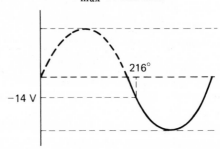

FIGURE 15-19

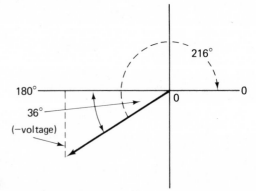

FIGURE 15-20

$$= 24 \times (-36°) \quad \text{(see Fig. 15-20)}$$
$$= 24 \times (-0.58778)$$
$$= -14 \text{ V}$$

15-8.1 *Alternate Method*

An alternate to the above method but possibly not as easily understood is:

$$v = V_{\max} \times \sin (360° \times f \times t)$$

where f is c/s
t is seconds

15-9 Frequency and Wavelength

15-9.1 *Wavelength*

Frequency is the number of complete cycles occurring every second.

Period is the time to make one cycle.

Wavelength is the actual length of one cycle. In the expression *wavelength*, wave is another word for cycle. In other words, what is the length of one cycle?

Wavelength is an important measurement in electronics. Notice the length of television antennas and CB antennas. Channel 4 television has a frequency of 66 Mc, which gives each cycle a length (wavelength) of 4 m. Television antennas are half a wavelength long, which makes a channel 4 antenna about 2 m long.

Color travels through space to the eye at frequencies of 430 to 750 $\times$ 10^{12} Hz. Blue has a wavelength of approximately 470 nm (nanometers). Red has a wavelength of approximately 650 nm.

the symbol for wavelength is the Greek letter lambda λ

$$\text{wavelength} = \frac{\text{velocity}}{\text{frequency (Hz)}}$$

$$\lambda = \frac{v}{f}$$

where v is velocity of the sine
wave
f is cycles per second

15-9.2 *Wavelength of Radio Waves*

Radio wave is a general term used to express any and all transmissons—from kHz to GHz in frequency—which travel through space. Radio waves travel at the speed of light, i.e., 300,000,000 m/s.

$$\lambda = \frac{v}{f} = \frac{300,000,000 \text{ m/s}}{f \text{ (Hz)}}$$

PROBLEM 15-10

What is the wavelength of an AM radio wave which has a frequency of 1200 kHz?

SOLUTION   (cover solution and solve)

$$\lambda = \frac{v}{f} = \frac{300{,}000{,}000 \text{ m/s}}{f \text{ (Hz)}}$$

$$= \frac{300{,}000{,}000}{1200 \times 10^3}$$

$$= 250 \text{ m}$$

Many people have trouble understanding this solution. The following logic may help.

Radio waves travel 300,000,000 m every second. If there are 1,200,000 cycles (occurring every second) what is the length of each one? Answer: 250 m long.

The problem is similar to cars in a city block. If a city block is 300 m long and 12 cars are parked there bumper to bumper, what is the length of each car? Answer: Each car is 25 m long.

PROBLEM 15-11

An FM radio station transmits at 110 MHz. What would be the length of a full wave antenna to receive this frequency?

SOLUTION   (cover solution and solve)

$$\lambda = \frac{v}{f} = \frac{300{,}000{,}000 \text{ m/s}}{110 \times 10^6}$$

$$= 2.73 \text{ m}$$

PROBLEM 15-12

Short-wave radio and amateur radio transmission is generally stated in meters. What is the transmission frequency of a transmitter which is broadcasting at 2 m?

SOLUTION   (cover solution and solve)

$$\text{from} \quad \lambda = \frac{v}{f} = \frac{300{,}000{,}000 \text{ m/s}}{f \text{ (Hz)}}$$

$$\text{then} \quad f = \frac{300{,}000{,}000 \text{ m/s}}{\lambda \text{ (m)}}$$

$$\therefore f = \frac{300{,}000{,}000}{2}$$

$$= 150 \text{ MHz}$$

Note that the high-frequency FM radio wave is much shorter in length than the low-frequency AM radio wave.

Conclusion: The higher the frequency, the shorter the wavelength.

15-9.3 Wavelength of Sound Waves

The velocity (speed) of sound waves is very low compared to radio waves. This explains why sound travels only a short distance. Sound waves travel through air at about 340 m/s, depending upon the temperature of the air.

$$\text{wavelength of sound} = \lambda_{snd} = \frac{v}{f} = \frac{340 \text{ m/s}}{f \text{ (Hz)}}$$

PROBLEM 15-13

What is the length of a 25-Hz sound wave (free air)?

SOLUTION   *(cover solution and solve)*

$$\lambda = \frac{340 \text{ m/s}}{25 \text{ Hz}}$$

$$= 13.6 \text{ m}$$

15-10 Angular Velocity (ω); ($2\pi f$)

15-10.1 Pi (π)

The Greek letter pi (π) is an important value in ac calculations. It is a symbol for the constant value 3.14159. The value 3.14 is the number of times which the length of the radius of a circle will divide into one-half the length of the circumference of a circle. In other words, pi (π) designates distance. It states that the arc composed of one-half the circumference of a circle is 3.14 times longer than the length of the radius of that circle.

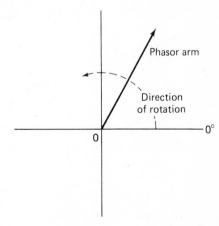

FIGURE 15-21 Phasor arm and direction of rotation (positive rotation).

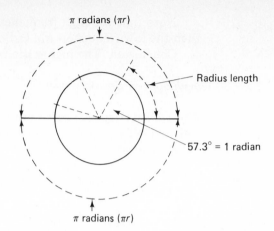

π radians (πr)

Radius length

57.3° = 1 radian

π radians (πr)

FIGURE 15-22   Circle divided into radians. 2π radians = 360°.

A complete cycle consists of 360°, not 180° (or one-half the circumference). Therefore, in its rotation, a loop will travel a distance of 2π or 6.28 times the length of the radius in a complete cycle. (See Figs. 15-21 and 15-22.)

15-10.2 Radian

A radian is the angle formed at the center of a circle by the two arms (phasors) coming out from the center and connecting to an arc (marked off on the circumference) which is equal in length to the radius of the circle.

$$\text{the angle formed } = 2\pi \text{ radians} = 360°$$

$$1 \text{ radian (rad)} = \frac{360°}{2\pi} = 57.3°$$

Vector diagrams are used to represent voltage values and angles. The positive rotation (counterclockwise) of the phasor arm is representative of the radius of the travel of the angles.

15-10.3 Deriving Angular Velocity (Speed) Formula

The velocity of rotation which determines the rate at which a sine wave angle changes in value is called angular velocity (ω).

the symbol for angular velocity is the Greek letter omega (ω)

We have seen that during the production of each sine wave ac voltage cycle, the imaginary radius of the rotating conductor travels

$$2\pi \text{ radians} = 6.28 \times 57.3°$$

$$= 360° = 1 \text{ cycle} \qquad (1)$$

An ac sine wave voltage with a frequency of f hertz will complete an f number of cycles per second; therefore, the angular velocity (ω) will be

$$\text{angular velocity } (\omega) = 2\pi \text{ radians} \times \text{number of cycles}$$

$$= 2\pi f \text{ rad/s} \tag{2}$$

Another way of explaining angular velocity is that velocity is the ratio of distance to time.

$$v = \frac{D}{T} \tag{3}$$

where D is distance in 2π
radians   (4)
T is the time (period) of
one cycle   (5)
v is speed of rotation
(angular velocity, ω)   (6)

Substitute 2π for D and ω for v in (3).

$$\text{then} \quad \omega = \frac{2\pi}{T}$$

$$\text{but} \quad T = \frac{1}{f}$$

$$\text{Therefore} \quad \omega = \frac{2\pi}{\left(\dfrac{1}{f}\right)} = 2\pi f$$

$$\text{angular velocity} = \omega = 2\pi f$$

EXERCISES

1. Who discovered the principle of producing ac voltage?

2. State two conditions which must exist in order to induce a voltage into a single conductor.

3. What two factors determine the magnitude of the voltage induced into a single conductor?

4. What determines the polarity of the voltage induced into a conductor?

5. State the left-hand generator rule.

6. Will voltage be induced into a conductor if it is moving parallel to the lines of force?

7. Name the three factors which determine the magnitude of ac voltage developed in an ac generator.

8. List the three units of instantaneous values.

9. State the equation for determining instantaneous values of voltage when peak voltage values are known.

10. A 60-Hz ac voltage has a peak value of 170 V. Calculate the instantaneous voltage values at the following angles of cut:

    a. 30°
    b. 45°
    c. 65°
    d. 90°
    e. 115°
    f. 165°
    g. 210°
    h. 255°
    i. 285°
    j. 340°

11. An 8.5-Ω load is connected across a 170-V source of voltage. Calculate the instantaneous current values at the following angles of cut of the conductor:

    a. 30°
    b. 45°
    c. 90°
    d. 120°
    e. 180°
    f. 195°
    g. 260°
    h. 270°
    i. 315°

12. Complete the following table for voltage relationships.

Table P15-1 Voltage Relationships

	Effective V	V_{rms}	V_p	V_{avge}
a.			140	
b.		120		
c.				260
d.	90			
e.			210	
f.				56
g.		208		
h.			310	
i.	180			

13. What is the value of peak voltage which would create the same amount of heat as a 170-V dc voltage?

14. Complete the following table for frequency and period relationships.

Table P15-2   Frequency and Period Relationships

	Frequency	Period			
		ms	μs	ns	ps
a.	50 kHz				
b.		0.08			
c.			80		
d.				60,000	
e.					50
f.	50 MHz				
g.	40 GHz				
h.			2		
i.				25,000	

15. What is the value of an instantaneous voltage 10 ms after a 60-Hz two-pole generator producing 220 V peak voltage has started?

16. A 2000-Hz signal has a peak amplitude of 60 V. Calculate:
    a. The instantaneous value of voltage 210 μs after zero angle
    b. The instantaneous value of voltage 40 μs before it reaches maximum positive voltage

17. Determine the value of instantaneous voltage 60 ms after a 400-Hz two-pole 48-V peak voltage generator has gone through 0 V.

18. Determine the wavelength of an AM radio signal which has a frequency of 1400 kHz.

19. A television station transmits at a center frequency of 90 MHz. What would be the length of a half-wavelength antenna constructed to receive this station?

20. Determine the angular velocity in radians per second of a conductor rotating at the following velocities:
    a. 60 Hz
    b. 100 Hz
    c. 400 Hz
    d. 1000 Hz
    e. 3000 Hz

chapter 16

AC CIRCUITS

16-1 Introduction

The subjects of inductance, capacitance, and resistance have been dealt with in dc circuits. They are now to be studied in ac circuits in which the voltage and current are continually changing. It must be pointed out that it is impossible to construct a pure resistive, inductive, or capacitive circuit; although with the exception of inductive circuits, which in many cases have a large inherent resistance, these components are close to being pure and are generally accepted as such. Hypothetical circuits only can be considered when discussing pure capacitive or pure inductive components.

The solution to problems involving resistance, inductance, and capacitance can be solved by using the same techniques as used in dc circuits with only minor changes.

16-2 Reactance

Inductance and capacitance, each acting alone, oppose ac currents in much the same manner as resistance opposes these currents. Their opposition is stated in ohms. The chief difference is that the current and voltage are in phase (in step) in resistance circuits, but are out of phase in capacitive and inductive circuits. Inductance causes the current which flows through it to lag the voltage across it by $90°$ ($90°$ phase angle). Capacitance causes the current which flows into it to lead the voltage developed across the capacitance by $90°$ ($90°$ phase angle).

The opposition which an inductance offers to ac voltage and current is called

$$\text{inductive reactance } (X_{\mathrm{L}}) = 2\pi f L$$

$$\text{or } = \omega L$$

where ω is angular velocity = $2\pi f$
X_{L} is inductive reactance in ohms
f is frequency in cycles per second
L is inductance in henries

Notice that the inductive reactance is directly proportional to the angular velocity (ω) and the inductance value.

The opposition that a capacitor offers to ac voltage and current is called

$$\text{capacitive reactance } (X_{\mathrm{C}}) = \frac{1}{2\pi f C}$$

$$\text{or } = \frac{1}{\omega C}$$

where ω is angular velocity = $2\pi f$
X_{C} is capacitive reactance in ohms
f is frequency in cycles per second
C is capacitance in farads

16-3 Vector Diagrams

Phase angles are shown by vector diagrams. A vector or phasor is a line representing magnitude and direction (Fig. 16-1).

The *horizontal line* to the right of origin (0) is always the *reference*

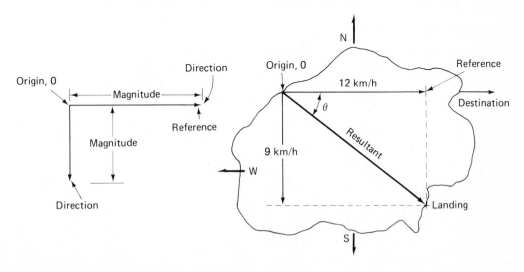

FIGURE 16-1 Pictorial description of a phasor diagram.

phasor arm onto which the *common value* is placed. The position of all other arms is placed with respect to it. The *length of each arm is drawn in proportion to its value.*

EXAMPLE

Suppose you were to paddle due east across a lake at 12 km/h but a north wind was blowing due south at 9 km/h. The point of landing would be at point L. Your speed of travel would have been 15 km/h (the resultant arm) and you would land 36.8° (phase angle: -36.8°) to the south of your intended spot. This is a vector diagram. The phasor arms show the magnitude of 12 km/h of the paddle, 9 km/h of the wind and the resultant speed of 15 km/h in proportion to their values. The intended direction as shown by the direction of the vector arm was due east (reference arm), the wind direction was due south (i.e., from the north). The wind was blowing at right angles to the boat's intended direction. The final destination had a phase angle of -36.8° with respect to the intended destination.

16-4 Phase Angles

When two or more quantities are in step (occur simultaneously) in a circuit, they are said to be in *phase*. Such is the case with resistance in ac circuits (Fig. 16-2). When two or more quantities occur at different times, they are out of step with one another and are said to be out of phase as shown in Fig. 16-3a and b. The current and voltage are out of phase with one another in all reactive circuits (circuits consisting of both resistance and capacitance or inductance). Sine waves are difficult to interpret mathematically; therefore, phase angles are shown on vector diagrams as in Figs. 16-2b and 16-3b. Each component value is represented by a separate phasor arm.

Phasors can be out of phase by any angle, and they rotate in a ccw direction (positive direction) (Fig. 16-3b).

It is as correct to say that V_b leads V_a by 90° in vector diagrams as it is to say that V_a lags V_b by 90° (Fig. 16-3).

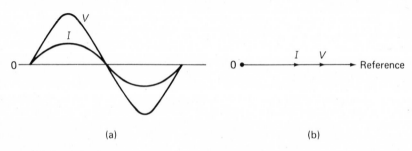

(a)   (b)

FIGURE 16-2

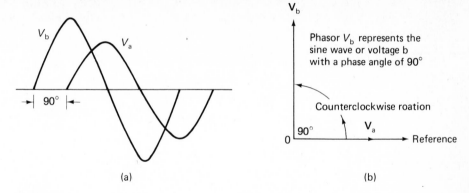

(a)   (b)

FIGURE 16-3 V_b leading V_a by $90°$ (a) Sine-wave representation. (b)
Phasor representation. Phasor V_a with a $0°$ phase angle.

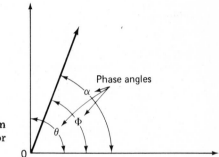

FIGURE 16-4   Phasor   diagram
showing different symbols used for
phase angles.

The reference axis is always the horizontal line to the right of "0"
value.

The angle by which phasors are out of phase is called the *phase angle*
(Fig. 16-4).

**The symbol for phase angle is the Greek letter phi (ϕ) or theta (θ)
or alpha (α)**

16-4.1   Leading Phase angle

the current leads the voltage in capacitive circuits

The above statement means that the current is present and at a peak
value before the voltage begins to develop across the capacitor (Fig. 16-5).
Figure 16-5b is a sine wave representation of this statement. Phasor arms
rotate counter clockwise in vector diagrams (Figs. 16-3b and 16-4); therefore, the applied current (I_a) is shown ahead of the capacitor voltage by
$90°$ in Fig. 16-5c.

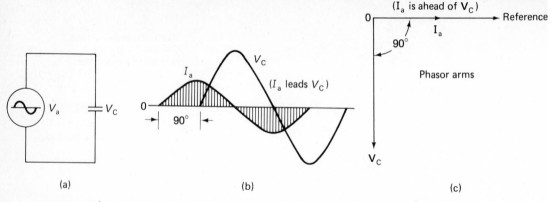

FIGURE 16-5 Phase relationships. I_a is the reference value. The source current (I_a) leads the voltage across the capacitor by 90°. (a) A capacitive circuit. (b) Sine-wave representation. (c) Phasor diagram (vector) representation.

16-4.2 Lagging Phase Angle

An inductance opposes any change (increase or decrease) in current flow through it because the induced voltage (v_L) produced within it creates an opposing current to the source current. This statement means that the induced voltage (v_L) appears across the inductance before circuit current flows. This relationship between induced voltage and the source current may be depicted by sine-waves (see Fig. 16-6a) or by a vector diagram (see Fig. 16-6b).

16-4.3 Summary: Phase Relationship Between the Voltage Across a Capacitor, Voltage Across an Inductance, and Voltage Across a Resistor

The circuit current in a series circuit is common to all the components; therefore, it is placed on the reference arm. The voltage across a resistor is always in phase with the current; therefore, it is always shown on the

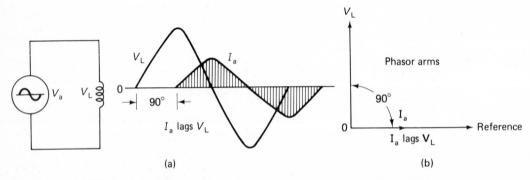

FIGURE 16-6 Phase relationships. I_a is the reference value. The source current (I_a) lags the induced voltage (V_L) by 90°. (a) Sine-wave representation. (b) Phasor diagram (vector) representation.

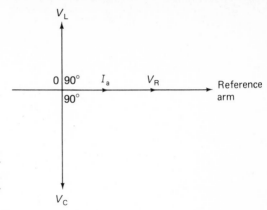

FIGURE 16-7 Phase angles (relationships) between various voltages. The current is always used as reference in a series circuit.

same phasor arm as the circuit current. The other voltages are placed in their respective leading or lagging relationships with the circuit current as shown in Fig. 16-7.

V_L has a leading phase angle with respect to the reference arm.

V_C has a lagging phase angle with respect to the reference arm.

V_R and I_a are placed on the reference arm because I_a is common to all the components and V_R is in phase with I_a.

16-4.4 Phasor Arm Lengths

The length of a phasor arm must always be drawn in proportion to the magnitude of each of the voltages represented.

PROBLEM 16-1

Three pure electronic components are connected in series with an ac voltage. Draw the vector diagram to scale for the following values and phase angles:

$$\text{Current } I_a = 2 \text{ A}; \quad V_R = 10 \text{ V}$$
$$V_L = 20 \text{ V}; \quad V_C = 30 \text{ V}$$

SOLUTION   *(cover solution and solve)*

Current is common in series circuits; therefore, the phase angle of all components is plotted against the current (Fig. 16-8).

PROBLEM 16-2

Three impure electronic components are connected in series with an ac voltage. (Impure means that the capacitance and the inductance each have some resistance.) Draw the vector diagram to scale and at the correct phase angles for the following values:

$$I_a = 3 \text{ A}; \quad V_R = 15 \text{ V}; \quad V_L = 15 \text{ V leading by } 70^\circ$$
$$V_C = 10 \text{ V lagging by } 50^\circ$$

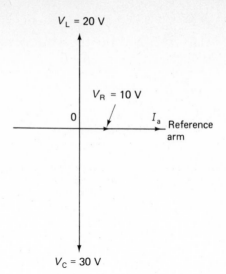

FIGURE 16-8

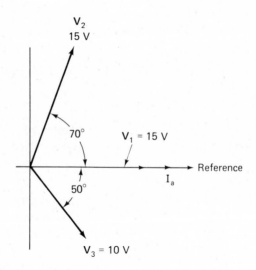

FIGURE 16-9

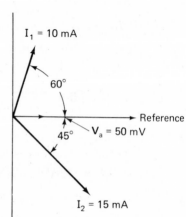

FIGURE 16-10 Phasor diagram for a parallel circuit. Voltage is common in a parallel circuit; therefore, the currents are plotted against the voltage.

SOLUTION   *(cover solution and solve)*

Current is common in series circuits; therefore, the phase angle of the voltage of all components is plotted against the current (Fig. 16-9).

16-5   Resistance in ac Circuits

Resistance produces the same effect on ac voltages and currents as it does on dc voltages and currents. The relationship between current, voltage, and resistance is the same. $V = I \times R$. The only difference is that there are instantaneous voltage and current values in resistive ac circuits.

Recall the voltage and current symbols for ac voltage as discussed in Chap. 15.

V_p is peak ac voltage

v is voltage at one particular instant in time during the ac cycle

V_rms is effective voltage

I_p is peak ac current

i is current at one particular instant in time during the ac cycle

I_rms is effective current

16-5.1   *Instantaneous Current and Voltage Values in Resistive Circuits*

The instantaneous value of current in a resistive circuit may be determined by Ohm's Law.

$$i_\mathrm{R} = \frac{v_\mathrm{R}}{R}$$

The instantaneous value of voltage in a resistive circuit may be determined by Ohm's Law.

$$v_\mathrm{R} = i_\mathrm{R} \times R$$

These formulas are an alternative to the sine wave formulas shown below to calculate the instantaneous value of currents or voltages in resistive ac circuits.

The sine wave formulas (instantaneous values) are

$$v_\mathrm{R} = V_\mathrm{p} \times \sin \theta$$

$$\text{or}\quad i_\mathrm{R} = I_\mathrm{p} \times \sin \theta$$

16-5.2   *Peak and rms Values*

Peak values of ac current in resistive ac circuits may be derived from the peak value of voltage and Ohm's Law.

$$I_\mathrm{p} = \frac{V_\mathrm{p}}{R}$$

The rms values of ac current in resistive ac circuits may be derived with Ohm's Law.

$$I_{\text{rms}} = \frac{V_{\text{rms}}}{R}$$

16-6 Inductive Reactance

16-6.1 Inductive Reactance

X_{L} **is the letter symbol for inductive reactance**

Inductive reactance is the opposition which an inductance offers to ac current and voltage.

The induced voltage (v_{L}) which prevents current from flowing in an inductance (coil) the first instant that voltage is connected to it is also called counter voltage (CEMF) (see Sec. 13-4).

In other words, inductance opposes *current change*. The magnitude of this CEMF (v_{L}) is proportional to the rate of change of the current; $v_{\text{L}} = L\,(di/dt)$. Time, therefore, affects the opposition which an inductance offers to current change. In ac circuits, the rate of change (di/dt) of the current depends upon the angular velocity (ω).

angular velocity (ω) = $2\pi f$ (see Sec. 15-10)

where f is frequency in Hz of the V_{a}
2π is number of rad/s of 1 cycle

Voltage is also dependent upon the value of inductance as well as the angular velocity. Therefore, the complete formula for the opposition inductive reactance which an inductance offers to ac voltage and current is

inductive reactance (X_{L}) = $2\pi fL$

$= \omega L$

where ω is angular velocity = $2\pi f$
X_{L} is inductive reactance in ohms
f is frequency in c/s
L is inductance in H

Notice that

X_{L} is directly proportional to frequency.

X_{L} is directly proportional to inductance.

X_{L} is not affected by the magnitude of the applied voltage but only by its frequency (rate of change).

The current in a pure inductance always lags the v_{L} (CEMF) by 90°, and the applied voltage is always 180° out of phase with the v_{L} (CEMF) voltage (Figs. 16-11 and 16-12).

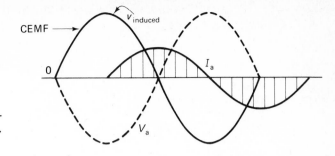

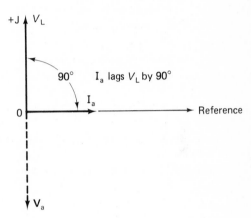

FIGURE 16-11 Sine-wave representation for an inductive circuit. Circuit current is the reference.

FIGURE 16-12 Phasor representation for the same pure inductive circuit as Fig. 16-11.

Note that although there is applied voltage in inductive circuits, the phase angle is considered to be between the current flowing in the inductance and the voltage developed across the inductance.

16-6.2 Inductive Circuits and Ohm's Law

Ohm's Law is the same in ac inductive circuits as it is in ac resistive circuits. However, inductive reactance may be used in place of resistance in the Ohm's Law formula in pure inductive circuits. The symbol used for voltage may be V for rms, peak, or peak to peak voltage, and v for instantaneous voltage.

$$V_{\mathrm{L}} = I_{\mathrm{L}} X_{\mathrm{L}}$$

$$\text{or} \quad I_{\mathrm{L}} = \frac{V_{\mathrm{L}}}{X_{\mathrm{L}}}$$

$$\text{or} \quad X_{\mathrm{L}} = \frac{V_{\mathrm{L}}}{I_{\mathrm{L}}}$$

16-6.3 Phasor Arm, Inductive Reactance, and Voltage Across a Coil

The equation $X_{\mathrm{L}} = V_{\mathrm{L}}/I_{\mathrm{L}}$, shows that inductive reactance and the voltage across a coil are directly proportional to one another. Inductive reactance

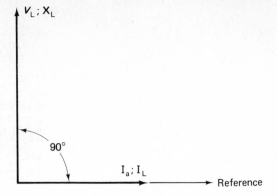

FIGURE 16-13 Phasor diagram. The inductive reactance and the voltage across the coil are represented by the same phasor arm. They lead the current by 90°. The current is placed on the reference arm because it is the common component.

and the voltage across a coil can be represented by the same phasor arm on the vector diagram (Fig. 16-13).

16-6.4 Inductive Susceptance

Inductive susceptance (B_L) is a unit in inductive circuits similar to conductance in resistive circuits. In other words, inductive susceptance represents the ease with which an inductive circuit will allow ac current to flow. Inductive susceptance (B_L) is the reciprocal of inductive reactance.

$$B_L = \frac{1}{X_L}$$

the unit of inductive susceptance is siemens (S)

PROBLEM 16-3

What is the inductive reactance of a 40-mH coil connected across a 10-kHz signal?

SOLUTION   *(cover solution and solve)*

$$X_L = 2\pi f L$$

$$= 6.28 \times (10 \times 10^3) \times (40 \times 10^{-3})$$

$$= 2512\ \Omega$$

PROBLEM 16-4

The current in an inductive circuit is 100 mA when 40 V and 400 Hz is connected across it. What is the inductance of the circuit?

SOLUTION   *(cover solution and solve)*

$$X_L = \frac{V_L}{I_L} = \frac{40}{(100 \times 10^{-3})}$$

$$= 400\ \Omega$$

$$\text{from} \quad X_L = 2\pi f L$$

$$\text{then} \quad L = \frac{X_L}{2\pi f} = \frac{400}{6.28 \times 400} = 0.159 \text{ H}$$

$$= 159 \text{ mH}$$

PROBLEM 16-5

What is the value of current which will flow through a 75-mH coil when a 3200-Hz, 25-V signal is applied to it?

SOLUTION   *(cover solution and solve)*

$$X_L = 2\pi f L$$

$$= 6.28 \times 3200 \times (75 \times 10^{-3})$$

$$= 1507.2 \ \Omega \quad {\scriptstyle 1507.9 \Omega}$$

$$\text{current} (I_L) = \frac{V_L}{X_L} = \frac{25}{1507.2} \times 10^3$$

$$= 16.587 \text{ mA}$$

16-7 Capacitance Reactance

16-7.1 Capacitive Reactance

X_C **is the letter symbol for capacitive reactance**

Capacitive reactance is the opposition which a capacitor offers to ac voltage and current.

The voltage developed across a capacitor (V_C) is the result of current flowing into one plate and an equal amount of current flowing out of the opposite plate, producing a potential difference across the capacitor plates. The current must, therefore, first flow into a capacitor before the voltage (V_C) can be developed across it. It is for this reason that it is said that the current leads the voltage developed across a capacitor by 90°. The amount of the current which flows into the capacitor is proportional to

1. The rate of change of the applied voltage (angular velocity)
2. The capacitance of the capacitor

The higher the frequency (rate of change) of the applied voltage and the resultant current flowing into a capacitor, the less time a capacitor has to develop the voltage across it and oppose the applied voltage, and the larger will be the capacitor current.

The larger the capacitor, the longer will be the time required for the current which is flowing into the capacitor to charge it and produce the capacitor voltage which opposes the applied voltage. In other words, the opposition (X_C) which a capacitor offers to an applied voltage is inversely proportional to capacitance and frequency.

$$\text{capacitive reactance } (X_C) = \frac{1}{2\pi f C}$$

$$= \frac{1}{\omega C}$$

where ω is angular velocity $= 2\pi f$

X_C is capacitive reactance in ohms

f is frequency in cycles per second

C is capacitance in farads

Note that the opposition (X_C) which a capacitor offers to ac voltages and currents is inversely proportional to the capacity of the capacitor and the frequency of the applied voltage. This fact can be demonstrated dramatically with an experiment in which a capacitor is connected in series with a light bulb and a source of voltage. The bulb will light brightly if a 200-μF capacitor is connected in series with the bulb, but it will not light if the 200-μF capacitor is replaced with, for example, a 0.2-μF capacitor. This is because the capacitive reactance of the 0.2-μF capacitor is so large at 60 Hz (1326 Ω) that the current flow in the bulb is too small to produce light, but the reactance of the 200-μF capacitor will be only 13 Ω and will allow a large current to flow.

If the frequency of the 110-V ac source were increased from 60 Hz to 60 kHz, the bulb would light brightly because the reactance of the capacitor would drop dramatically at the higher frequency.

16-7.2 Capacitive Reactance and Ohm's Law

Ohm's Law is the same in ac capacitive circuits as it is in ac resistive and inductive circuits. Capacitive reactance is used in place of resistance in the Ohm's Law formula in pure capacitive circuits. The symbol used for voltage may be V for rms, peak, or peak to peak voltage and v for instantaneous voltage.

$$V_C = I_C \times X_C$$

$$\text{or } I_C = \frac{V_C}{X_C}$$

$$\text{or } X_C = \frac{V_C}{I_C}$$

16-7.3 Phasor Arm, Capacitive Reactance, and Voltage Across a Capacitor

The equation $X_C = V_C/I_C$ shows that capacitive reactance and the voltage across a capacitor are directly proportional to one another. Capacitive

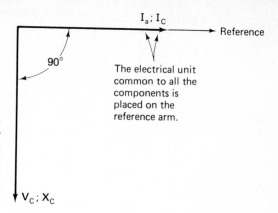

FIGURE 16-14 Phasor diagram. The capacitive reactance and the capacitor voltage are represented by the same phasor arm. They lag the current by 90°. The current is placed on the reference arm because it is the common component.

reactance and the voltage across the capacitor, therefore, can be represented by the same phasor arm on the vector diagram as in Fig. 16-14 (the current leads V_C by 90°).

X_C is inversely proportional to frequency.

X_C is inversely proportional to capacitance.

X_C is not affected by the magnitude of the applied voltage, but only by the frequency of the ac voltage.

16-7.4 Capacitive Susceptance

Capacitive susceptance (B_C) is a unit in capacitive circuits similar to conductance in resistive circuits. In other words, capacitive susceptance represents the ease with which a capacitive circuit will allow ac current to flow. Capacitive susceptance (B_C) is the reciprocal of capacitive reactance.

$$B_C = \frac{1}{X_C}$$

the unit of capacitive susceptance is siemens (S)

PROBLEM 16-6

What is the capacitive reactance of a 0.0001-μF capacitor to a 1-MHz signal?

SOLUTION   (cover solution and solve)

$$X_C = \frac{1}{2\pi f C}$$

$$X_C = \frac{1}{6.28 \times (1 \times 10^6) \times (0.0001 \times 10^{-6})}$$

$$= 1592 \ \Omega$$

PROBLEM 16-7

A current of 500 mA flows when a 250-V rms voltage is applied to a 0.02-μF capacitor.
Find:

1. The capacitive reactance
2. The frequency of the applied voltage

SOLUTION   *(cover solution and solve)*

1.

$$X_C = \frac{V_C}{I_C} = \frac{250}{500 \times 10^{-3}} = 500 \ \Omega$$

2.

from $X_C = \dfrac{1}{2\pi f C}$

$$f = \frac{1}{2\pi C X_C}$$

$$= \frac{1}{6.28 \times (0.02 \times 10^{-6}) \times 500}$$

$$= 15.923 \text{ kHz}$$

PROBLEM 16-8

A 240-V rms 20-kHz source of voltage is applied to a 0.05-μF capacitor. Calculate:

1. X_C of the capacitor
2. The effective current value

SOLUTION   *(cover solution and solve)*

1. $X_C = \dfrac{1}{2\pi f C}$

$$= \frac{1}{6.28 \times (20 \times 10^{3}) \times (0.05 \times 10^{-6})}$$

$$= 159.235 \ \Omega$$

2. $I_{\text{effective}} = \dfrac{240}{159.235} = 1.5 \text{ A}$

16-8 Power in ac Circuits

16-8.1 Types of Power

There are three types of power in ac circuits.

1. Reactive power, $Q*$
2. Active power, $P*$
3. Apparent power, $U*$

The current which flows in pure reactive components in ac circuits is stored in those components in the form of electrostatic or electromagnetic fields. This stored energy is not dissipated in another form of energy but is returned back to the circuit from which it came when the field collapses. The current energy stored in a reactive component is the result of voltage which caused it to flow. The product of this current and voltage would make it appear that power is being produced and work is being done, so it is called *reactive power*.

It is important to realize that reactive power is the power stored in a capacitor or inductance. True power, on the other hand, is converted energy which is dissipated in the form of heat, mechanical energy, or other forms of energy, and is lost forever. Reactive power can travel from the power source to the place of storage only by way of the wire which connects the two points. Wire has resistance which produces $I^2 \times R$ heat loss; therefore, true power is lost in the connecting wires during the process of storing reactive power.

16-8.2 Reactive Power

Reactive power is a general expression describing the power stored in any form of reactance. The word "power" is a misleading word when used with reactance, because the expression "power" denotes the conversion of one form of energy to another, and this does not take place in any pure reactive component.

Another indication that there is no power produced in a reactive component is the fact that the current and voltage in reactive components are 90° out of phase in ac circuits (see Fig. 16-15). This makes the current zero when the voltage is maximum and the voltage zero when the current is maximum; thus, the product of the voltage and current will always be zero.

$$P = VI$$

when V is maximum voltage
I is 0 V
$\therefore$ P is $V_{max} \times 0 = 0$ W.

*International System of Units (SI units), 1979.

when V is 0 V
I is maximum voltage
$\therefore$ P is $0 \times V_{max} = 0$ W.

the symbol for reactive power is Q

the unit of reactive power is vars (volt-amp reactive)

16-8.3 Active (True) Power

Electrical power or energy which is converted into, for example, heat or mechanical energy is lost forever and cannot be returned to the circuit. The conversion into this type of energy from electrical energy in ac circuits is called *true*, *real*, or *active power*. The resistor symbol is used to represent true power even though an electric motor (mechanical energy) is converting the electrical energy into mechanical energy. One practical example of energy conversion is an electric motor moving the weight of an elevator in a building; the active power is responsible for the mechanical power which lifts the elevator weight.

    Although any one of the three power formulas may be used to calculate *true power*, because the current and voltage are in phase (Fig. 16-2), it is best to use the first two equations incorporating R. This suggestion is made so that true power equations and calculations may be conclusively isolated from the reactive power or apparent power equations.

the unit of active power is the watt

the symbol for true power is P

$$P = I_a{}^2 R$$

$$\text{or} \quad = \frac{V_a{}^2}{R}$$

$$\text{or} \quad = I_R V_R \quad \text{(watts)}$$

16-8.4 Apparent Power

Circuits containing reactance and resistance are called impedance circuits, and the energy supplied to them by a source of voltage is stored as well as converted (dissipated) into another form of energy. It would appear to the casual observer that the product of the current and the voltage going to these circuits are being employed usefully, but the reactive power portion is actually being stored. It is because some of the power is being stored in electrostatic or electromagnetic fields that the product of voltage and current is called *apparent power*.

$$\text{apparent power } (U) = I_a V_a \quad \text{(volt amps)}$$

$$\text{or} \quad = I_a{}^2 Z \quad \text{(volt amps)}$$

where V_a is rms voltage

I_a is rms current

Apparent power is also referred to as wattless power because some of it is not being converted into working energy.

the unit of apparent power is VA*

the symbol for apparent power is U*

16-8.5 Power Factor

Power factor is the ratio of the active power to the apparent power in a circuit. It describes the percentage of apparent power that is being used as true power.

the symbol for power factor is cos θ

power factor is stated in decimal or percentage values

$$\text{power factor} = \frac{\text{true power}}{\text{apparent power}}$$

$$\cos \theta = \frac{P}{U}$$

where P is in watts

U is in volt-amps

Electrical utilities and electrical contractors often base their calculations on apparent power, since this unit will safely determine the required size of conductors, transformers, and other equipment needed for a particular installation.

FIGURE 16-15 A phasor diagram showing the relationship between true power, reactive power, and apparent power.

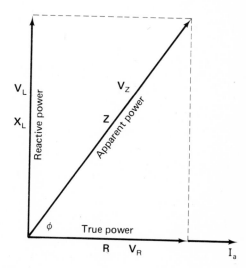

*International System of Units (SI unit), 1979.

The wattmeters used by electrical utilities supplying power to the home or the factory always measure active (true) power.

The value of power factor can never be greater than 1 because P can never be greater than the value of the product of the voltage and the current supplied to the circuit ($U = V_a I_a$).

Zero is the smallest possible value which power factor can have, but this value can occur only in nonexistent pure reactive circuits in which there would be zero resistance.

$$\text{power factor } \frac{P}{U} = \frac{I^2 R}{VI}$$

$$\text{if } R = 0$$

$$\text{then power factor} = 0$$

Since power factor is the ratio of the resistive component to the impedance component, power factor can be determined by the cosine of the angle ($\cos \theta$) in both parallel and series circuits. It is desirable to keep the power factor as low as possible in electronic components such as capacitors and coils. A low power factor means less loss from the unwanted resistance of these components.

In power applications in industry and in electrical utilities, it is desirable to keep the power factor as large as possible. This is because a large power factor means that most of the volt-amperes (apparent power) is being used as true power in the motors, and so on. Reactive power causes large heat loss in the transmission lines going to the energy converters. This loss is caused by the electrical energy travelling back and forth in the lines in the process of building up and collapsing (storage) in the energy converters.

SUMMARY:

Practical circuits consist of a portion of *reactive power* and *active (true* or *real) power*. The combination of the two is called *apparent power*.

the ratio of true power to apparent power is called power factor.

$$\text{power factor } (\cos \theta) = \frac{\text{true power (P)}}{\text{apparent power (U)}}$$

$$\cos \theta = \frac{P}{U} \quad \text{or} \quad = \frac{R}{Z}$$

Table 16-1   Power Symbols and Units

Power Type	Symbol	Unit
Active power	P	W
Reactive power	Q	VARS
Apparent power	U	VA
Power factor	$\cos \phi$	Decimal or (%)

Each of the above types of power has a different identifying symbol and unit which has been adopted by the electrical engineering profession and the electrical industry. See Table 16-1.

PROBLEM 16-9

A 240-V motor draws 40 A from the line. A wattmeter across the motor registers 9200 W. (A wattmeter registers true power.) What is the power factor of the motor?

SOLUTION *(cover solution and solve)*

$$P = 9200$$

$$U = VI$$

$$= 240 \times 40$$

$$= 9600 \text{ VARS}$$

$$\cos \theta = \frac{P}{U} = \frac{9200}{9600}$$

$$= 0.958$$

$$= 95.8\%$$

PROBLEM 16-10

A load has an impedance of 60 Ω. Part of the impedance consists of a 40-Ω resistance. The Z is connected to a 120-V 60-Hz source of voltage. Calculate the power factor.

SOLUTION *(cover solution and solve)*

Power factor

$$\cos \theta = \frac{R}{Z} = \frac{40}{60} = 0.667$$

$$= 66.7\%$$

EXERCISES

1. Define the following terms:
   a. Reactance
   b. Phase angle
   c. Lagging phase angle

2. What is the phase relationship between the voltage and current in an inductive circuit?

3. Why is there a phase difference between the voltage and current in capacitive and inductive circuits but not in pure resistive circuits?

4. Write the equation for
   a. Angular velocity
   b. Counter voltage (CEMF) (induced voltage)
   c. Inductive reactance
   d. Capacitive reactance
   e. True power
   f. Apparent power
   g. Reactive power
   h. Power factor
   i. Capacitive susceptance

5. Write the letter symbol for
   a. Apparent power
   b. Reactive power
   c. True power
   d. Power factor

6. a. What is the inductive reactance of an inductor if it permits 50 mA of current to flow in a series circuit to which an 80-V ac generator is connected?
   b. How much current would flow if the ac generator were replaced by a 120-V ac generator? (The resistance of the coil is 2 Ω.)

7. What is the value of inductance of a coil which, when connected in series with a 20-V 400-Hz generator limits the circuit current flow to 25 mA?

8. What value of current will flow in an ac circuit if a 200-mH inductor is connected in series with a 120-V 60-Hz generator?

9. What is the phase relationship between the current flow and the applied voltage in a pure capacitive circuit?

10. Calculate the capacitive reactance of the following capacitors when connected in series with a 60-Hz source of voltage:
    a. 0.002-μF
    b. 10-μF
    c. 270-pF
    d. 3000-nF

11. a. What value of current will flow in a pure capacitive circuit when a 0.01-μF capacitor is connected in series with a 120-V 60-Hz source of voltage?
    b. What value of current would flow if the 120-V 60-Hz generator in part a. were replaced by a 240-V dc generator?

12. What value of capacitance would be required in series with a light bulb to limit the current flow in the light bulb to 1.5 A when the bulb and capacitor are connected in series with a 120-V 60-Hz voltage source?

13. What happens to the current which flows in a pure reactive component?

14. Define:
    a. Reactive power
    b. Apparent power
    c. True power
    d. Active power
    e. Power factor

15. a. Does reactive power produce heat?
    b. Explain your answer to part a.

16. a. What is the unit of reactive power?
    b. What is the unit of active power?
    c. What is the unit of apparent power?

17. A 120-V single phase motor draws 40 A from the power line. A wattmeter to the motor registers 4600 W. What is the power factor of the motor?

18. A load consisting of a 40-Ω dc resistance and a 30-Ω inductive reactance is connected to a 208-V 60-Hz power line. Calculate the power factor.

19. Write the two equations for power factor for series circuits.

20. An impedance of 75 Ω $\underline{/50°}$ is connected to a 120-V single-phase voltage source (refer to Fig. 16-15). Calculate:
    a. The true power
    b. The power factor
    c. The reactive power

$Y_c = \frac{1}{2\pi f c}$

chapter 17

IMPEDANCE

17-1 Introduction

Every reactive circuit has some resistance. Whether added or inherent, the total opposition offered by resistance and reactance in an ac circuit is called impedance (Z) which is expressed in *ohms*.

the unit of impedance is ohms

the symbol for impedance is the letter Z

Impedance is the total opposition offered to voltage by the inductive reactance, the capacitive reactance, and the resistance in an ac circuit. The current in theoretically pure inductive or capacitive circuits is out of phase with the applied ac voltage by 90° (Figs. 16-5 and 16-6). The resistance which is present in all practical circuits causes the current to be out of phase with the voltage by less than 90°. This angle will decrease toward 0° as the resistance value relative to the other two reactive component values increases (Fig. 17-3).

If the circuit is a series type, the impedance may be determined by one or more of the following equations:

1. Ohm's Law

$$Z = \frac{V_a}{I_a}$$

2.

$$Z = \sqrt{R^2 + (X_L - X_C)^2}$$

Impedance may be expressed in either polar form or rectangular form. Impedance stated in polar form (PF) is expressed in ohmic value and degrees, as in the example below.

$$Z = 40\ \Omega\ \underline{/\ 30°}$$

The same rule applies to applied voltage.

Impedance expressed in rectangular form (RF) states the actual value of resistance and the actual value of the predominant reactance. If the reactance is capacitive, the value is preceded by $-J$. If the reactance is inductive, the value is preceded by $+J$.

EXAMPLES

1. If $R = 20\ \Omega$ and $X_C = 60\ \Omega$,
   then $Z = 20 - J60$
2. If $R = 50\ \Omega$ and $X_L = 80\ \Omega$,
   then $Z = 50 + J80$

The same rule applies to applied voltage and to individual related voltages.

17-2 Phasor Diagrams

17-2.1 Phasor Diagram Construction

Impedance is the vector sum of all the reactive and resistive component values in the circuit. A phasor diagram is a tool used to observe the phase relationships of these reactive component values and to show the relative reactive values of each component by the length of its phasor arm.

Phasor diagram construction was discussed in Chap. 16. Recall:

1. The first step in the construction of a phasor diagram is to determine the electrical unit—voltage or current—which is common to all the components in the circuit.
2. Current is common to all the components in series circuits.
3. Voltage is common to all components in parallel circuits.
4. The common electrical unit is always placed on the reference arm. All other values are placed on the phasor diagram in phase angle relationship with the common electrical unit.
5. The length of each phasor arm is drawn in proportion to the relative value of each reactance and resistance.

17-2.2 Determining the Length of the Resultant Arm

A reactive voltage and the voltage across a resistor will be 90° out of phase, and will have a total value equal to the vector sum of the two values. The

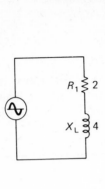

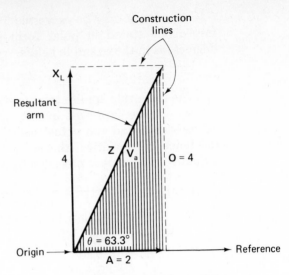

(a) A resistive inductive circuit.

(b) A parallelogram for a resistive inductive circuit.

FIGURE 17-1 (b) shows construction of a parallelogram and the resultant arm for the inductive-resistive circuit shown in (a).

vector sum, called the resultant arm (**R**), is the value of the applied voltage (V_a).

A reactance and a resistance value will also be 90° out of phase and will have a total value equal to the vector sum of the two values. The vector sum, called the impedance, is represented by the resultant arm of the phasor diagram.

The resultant arm (Figs. 17-1b and 17-2b) whether it represents

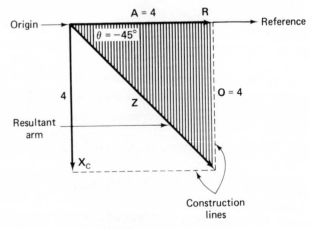

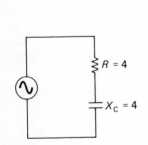

(a) A resistive-capacitive circuit.

(b) A parallelogram for a resistive-capacitive circuit.

FIGURE 17-2 (b) shows the construction of a parallelogram and the resultant arm for the capacitive-resistive circuit shown in (a).

Ch. 17 Impedance

voltage or impedance is obtained by drawing a diagonal line from the origin of the phasor arms to the opposite corner of the parallelogram formed by the construction lines and the two phasor arms.

Figure 17-1b shows the phasor diagram for the resistive inductive circuit in Fig. 17-1a. Since side O in Fig. 17-1b is equal in length to arm X_L, the three sides O, A, and Z enclose a right-angled triangle which includes the value of the resistance, the inductive reactance, and the impedance of the circuit. The angle between the impedance and the resistance is the phase angle between these two values or between the circuit current and the applied voltage. The value of the phase angle or of any of the three reactive component values may be obtained by an appropriate trigonometric function or by the use of the square of the sides (Pythagoras' Law).

17-2.3 The Effect of Resistance on the Phase Angle θ

The resistance in reactive circuits reduces the angle between the applied voltage and the applied current to less than 90°. Increasing the value of the resistance in comparison (relative) to the value of the reactance will reduce the angle between the reference arm and the resultant phasor arm in a series circuit. Compare Fig. 17-3b, which has a resistance of 2 Ω and a reactance of 8 Ω, with Fig. 17-3c, in which the reactance is 8 Ω and the resistance is 16 Ω. Observe the angle is much less in Fig. 17-3c than in Fig. 17-3b because the resistance is much larger in Fig. 17-3c.

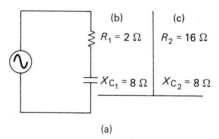

(a)

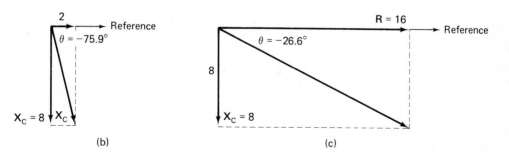

FIGURE 17-3   Phasor diagrams showing how an increase in the resistance in the circuit in (a) decreases the angle of lag.

17-2.4 Trig (Trigonometric) Functions

A trig function is a ratio between any two sides of a right-angled triangle (one angle is 90°).

Each ratio between any two specific sides is identified by giving it a name in order to identify it from other ratios. Each side must also be given a name so as to identify which specific side goes into each specific ratio and in what relationship. The triangle in Fig. 17-4 may be used as an example.

One of the angles other than the right-angled triangle is also named and always used as the reference to name the sides. In the following explanation, the reference angle will be called theta (θ).

The *adjacent side* is that side next to the reference angle (θ).

The *opposite side* is the side opposite to the reference angle (θ).

The *hypotenuse* is always the longest side of a right angled triangle.

$$\text{Ratio of } \frac{A}{H} = \frac{\text{adjacent side}}{\text{hypotenuse}} = \text{cosine } \theta$$

$$\text{Ratio of } \frac{O}{H} = \frac{\text{opposite side}}{\text{hypotenuse}} = \text{sine } \theta$$

$$\text{Ratio of } \frac{O}{A} = \frac{\text{opposite side}}{\text{adjacent side}} = \text{tangent (tan) } \theta$$

The reciprocal of these ratios produces three other functions:

$$\text{Ratio of } \frac{H}{A} = \frac{\text{hypotenuse}}{\text{adjacent side}} = \text{secant } \theta$$

$$\text{Ratio of } \frac{H}{O} = \frac{\text{hypotenuse}}{\text{opposite side}} = \text{cosecant } \theta$$

$$\text{Ratio of } \frac{A}{O} = \frac{\text{adjacent side}}{\text{opposite side}} = \text{cotangent (cotan) } \theta$$

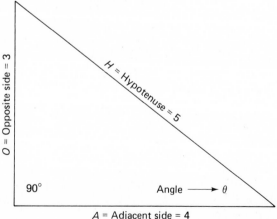

FIGURE 17-4   A right-angled triangle showing the name of each side with reference to the angle θ.

O = Opposite side = 3

H = Hypotenuse = 5

90°   Angle ⟶ θ

A = Adjacent side = 4

The name of each of these functions is usually abbreviated as follows:

$$\text{sine} = \sin \qquad \text{cosecant} = \csc$$

$$\text{cosine} = \cos \qquad \text{secant} = \sec$$

$$\text{tangent} = \tan \qquad \text{cotangent} = \text{cotan}$$

The ratio of any two sides may be obtained, if the reference angle is known, by using the appropriate trig function of the reference angle. For example (use trig tables or calculator),

$$\sin 36.9° = 0.6 \quad = \frac{3}{5} = \frac{\text{side } O}{\text{hypotenuse}}$$

$$\text{or} \quad \cos 36.9° = 0.8 \quad = \frac{4}{5} = \frac{\text{side } A}{\text{hypotenuse}}$$

$$\text{or} \quad \tan 36.9° = 0.75 = \frac{3}{4} = \frac{\text{side } O}{\text{side } A}$$

INVERSE TRIG FUNCTIONS

The reference angle may be obtained if the value of the sides is given, by taking what is known as the arc (identified by $^{-1}$) of the trig function (e.g., arcsin; arccos; arctan). This operation is just opposite to that of finding the trig function. For example find the angle theta (θ) (use trig tables or calculator) if the:

Ratio of $\dfrac{\text{opposite side}}{\text{hypotenuse}}$ is $\dfrac{3}{5}$

The trig function is $\sin \theta$

$$\text{operation is } \sin^{-1} \text{ of } 0.6 = 36.9°$$

$$\text{or} \quad \text{Ratio of } \frac{\text{adjacent side}}{\text{hypotenuse}} \text{ is } \frac{4}{5} = 0.8 = \theta$$

The trig function is $\cos \theta$

$$\text{operation is } \cos^{-1} \text{ of } 0.8 = 36.9° = \theta$$

$$\text{or} \quad \text{Ratio of } \frac{\text{opposite side}}{\text{adjacent side}} \text{ is } \frac{3}{4} = 0.75$$

The trig function is $\tan \theta$

$$\text{operation is } \tan^{-1} \text{ of } 0.75 = 36.9° = \theta$$

17-2.5 Application of Trigonometric Functions

The angle of lead or lag between a rotating phasor arm and the reference value may be calculated by the use of any trigonometric function that incorporates the given values.

The adjacent and opposite sides of the angle are given in Fig. 17-3; therefore, the most appropriate trigonometric function is tangent of the angle = O/A or cotangent of the angle = A/O. The tangent (tan) of the angle is generally preferred to the cotangent (cotan) of the angle.

If the value of the resultant arm and the value of O or A are known, the phase angle may be obtained by

$$\sin \theta = \frac{O}{H}$$

or

$$\cos \theta = \frac{A}{H}$$

PROBLEM 17-1

1. Refer to Fig. 17-5 and calculate the angle between the resistance and the impedance.
2. Draw the phasor diagram (Fig. 17-6).

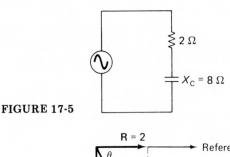

FIGURE 17-5

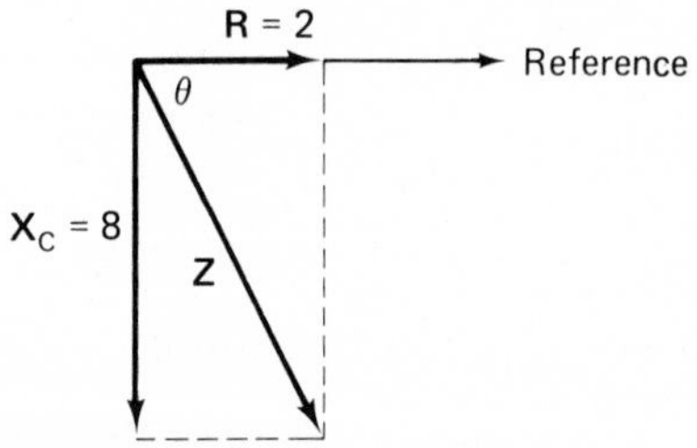

FIGURE 17-6

SOLUTION   (cover solution and solve)

$$\underline{/\theta} = \tan^{-1} \text{ of } \left(\frac{O}{A} = \frac{8}{2} = 4 \right)$$

$$= -75.96°$$

PROBLEM 17-2

Refer to Fig. 17-7 and calculate the phase angle between the resistance and the impedance.

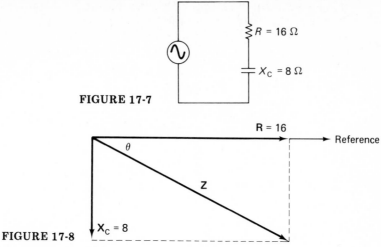

FIGURE 17-7

FIGURE 17-8

SOLUTION   (cover solution and solve)

$$\angle\theta = \tan^{-1} \text{ of} \left(\frac{O}{A} = \frac{8}{16} = 0.5\right)$$

$$= -26.56°$$

Figure 17-8 shows the phasor diagram for the circuit in Fig. 17-7 and the angle θ.

17-3 Series Circuits

17-3.1 Applied Voltage in Series L and R Circuits

Current is the common electrical unit in series circuits. Current lags the voltage across the inductance by 90°. However, the voltage across the resistance is in phase with the current. Since the applied voltage is the vector sum of the two component values it must have a phase angle of less than 90°.

PROBLEM 17-3

A coil and a resistor are connected in series with an ac voltage source, V_a. The circuit current is 0.5 A. The voltages across the inductor and resistor are each 50 V. Draw the circuit and:

1. Draw the phasor diagram.
2. Calculate the value of the applied voltage.
3. What is the value of phase angle between the V_a and I_a?
4. What is V_a in rectangular form?
5. Find the Z.

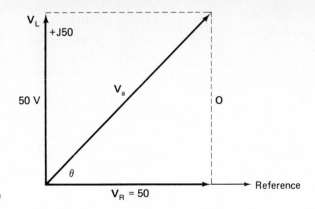

FIGURE 17-9

6. Find the X_L.
7. Find the R.

SOLUTION   *(cover solution and solve)*

1. See Fig. 17-9.

2. The solution may be given in one of two forms:
   a. Polar form (PF)
   b. Rectangular form (J-operator) (RF)
   The length of side **O** in Fig. 17-9 is equal in length to V_L. Therefore, the value of **O** must be equal to 50 V. The impedance triangle is now enclosed by **O**, V_R, and V_a. The value of V_a is

$$V_a = \sqrt[2]{V_R^2 + V_L^2}$$
$$= \sqrt{50^2 + 50^2}$$
$$= 70.7 \text{ V}$$

3. Now that the value of the applied voltage (V_a) is known, the phase angle θ may be obtained by any one of the three more common trigonometric functions:

$$\text{arc sin } \theta = \sin^{-1} \theta$$
$$\text{arc tan } \theta = \tan^{-1} \theta$$
$$\text{arc cos } \theta = \cosin^{-1} \theta$$

$$\theta = \sin^{-1} \text{ of } \frac{V_L}{V_a} = \frac{50}{70.7} = 45°$$

$$\theta = 45°$$

The V_a leads I_a by 45°. The traditional method of writing the complete answer in polar form is

$$V_a = 70.7 \text{ V} \underline{/45°}. \text{ (PF)}$$

4. Rectangular Form Method

The rectangular form does not give the phase angle nor the actual value of V_a. Rectangular form states the applied voltage (V_a) as the sum of its resistance and reactance values.

$$\text{Applied voltage } (V_a) = 50 + J50 \text{ (RF)}$$

Note: The resistance value is always stated first.

5. The impedance of the circuit may be calculated by

$$Z = \frac{V_a}{I_a} = \frac{70.7}{0.5} = 141.4 \ \Omega = 141.4 \ \Omega \ \underline{/45°} \text{ (PF)}$$

6.
$$X_L = \frac{V_L}{I_a} = \frac{50}{0.5} = 100 \ \Omega$$

which is written as

$$100 \ \Omega \ \underline{/90°} \text{ (PF)}$$

7.
$$R = \frac{V_R}{I_a} = \frac{50}{0.5} = 100 \ \Omega$$

which is written as

$$100 \ \Omega \ \underline{/0°} \text{ (PF)}$$

PROBLEM 17-4

A resistor and an inductor are connected in series with a 100-V ac generator. The voltage across the resistor measures 60 V.

1. Draw the phasor diagram.
2. What is the value of V_L?
3. What is the phase angle between V_a and I_a?
4. What is the value of V_a in rectangular form?
5. Find X_L if I_L = 10 A.
6. Find Z if I_L = 10 A.
7. Find R if I_L = 10 A.
8. Find Z (in RF).

SOLUTION   (cover solution and solve)

1. See Fig. 17-10.

2.
$$\text{From } V_a = \sqrt{V_R^2 + V_L^2}$$
$$\text{then } V_a^2 = V_R^2 + V_L^2$$
$$\text{and } V_L = \sqrt{V_a^2 - V_R^2}$$
$$\therefore V_L = \sqrt{100^2 - 60^2}$$
$$= 80 \text{ V}$$

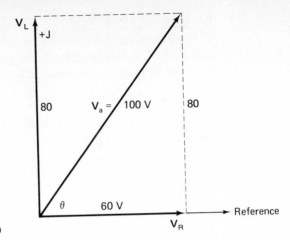

FIGURE 17-10

3. Phase angle

$$\theta = \text{arc cos of } \frac{V_R}{V_a}$$

$$= \cos^{-1} \frac{60}{100} = 53°$$

4. Rectangular form

$$V_a = V_R + JV_L = 60 + J80$$

5. $$X_L = \frac{V_L}{I_L} = \frac{80}{10\ A} = 8\ \Omega = 8\ \Omega\ \underline{/90°}\ (PF)$$

6. $$Z = \frac{V_a}{I_L} = \frac{100}{10\ A} = 10\ \Omega = 10\ \Omega\ \underline{/53°}\ (PF)$$

7. $$R = \frac{V_R}{I_L} = \frac{60}{10\ A} = 6\ \Omega = 6\ \Omega\ \underline{/0°}\ (PF)$$

8. $$Z = RF = 6 + J8$$

17-3.2 Impedance (Z) in Series L and R Circuits

Impedance of a series circuit is proportional to the applied voltage, V_a, and the values of X_L and R are proportional to the voltages across them. Therefore, they are placed on the same phasor arms as are their respective voltages (Fig. 17-11).

PROBLEM 17-5

A 900-Ω resistor R_1 is connected in series with a coil having an inductive reactance of 600 Ω. The circuit current is 0.5 A. Draw the circuit.

Ch. 17 Impedance

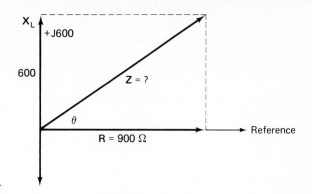

FIGURE 17-11

1. Draw the phasor diagram (see Fig. 17-11).
2. What is the total impedance of the circuit?
3. What is the phase angle between the impedance and the applied current (I_a)?
4. Calculate the Z in polar form (PF).
5. What is the Z in rectangular form (RF).
6. Find V_a (in PF).
7. Find V_R (in PF).
8. Find V_L (in PF).
9. Find V_a (in RF).

SOLUTION   (cover solution and solve)

1. See Fig. 17-11.
2. The solution may be given in either the PF or RF.

$$Z = \sqrt[2]{R^2 + X_L^2}$$
$$= \sqrt{900^2 + 600^2}$$
$$= 1082 \ \Omega$$

3. Phase angle
   Any of the three common trigonometric functions may be used to calculate the phase angle.

$$\theta = \text{arc tan of } \frac{X_L}{R}$$
$$= \tan^{-1} \text{ of } \frac{600}{900}$$
$$= 33.69°$$

   The Z leads the R by $33.69°$.

4. $$Z = 1082 \ \Omega \ \underline{/33.69°} \ \text{(PF)}$$

5. In the RF, the Z is given in its separate resistance and reactive values.

$$Z = R + JX_L$$

$$= 900 + J600 \ \Omega$$

The applied voltage may be determined if the current is known.

6. $\quad\quad V_a = I_a \times Z$

$$= 0.5 \times 1082 = 541 \text{ V} = 541 \text{ V} \underline{/33.69°} \text{ (PF)}$$

7. $\quad\quad V_R = I_a \times R$

$$= 0.5 \times 900 = 450 \text{ V} = 450 \text{ V} \underline{/0°} \text{ (PF)}$$

8. $\quad\quad V_L = I_a \times X_L$

$$= 0.5 \times 600 = 300 \text{ V} = 300 \text{ V} \underline{/90°} \text{ (PF)}$$

9. $\quad\quad V_a = 450 + J300 \text{ V}$

PROBLEM 17-6

A 20-kHz generator is connected in series with a 1-mH coil (L_1) and a 100-Ω resistor (R_1). The circuit current is 2 mA. Draw the circuit and:

1. Draw the phasor diagram.
2. Calculate the Z for the circuit.
3. What is the phase angle between the impedance and applied current (I_a)?
4. What is the value of the impedance in polar form?
5. State the value of the Z in rectangular form.
6. Find V_a (in PF).
7. Find V_R.
8. Find V_L.
9. State V_a in RF.

SOLUTION   *(cover solution and solve)*

1. See Fig. 17-12.

2. $\quad\quad Z = \sqrt{R^2 + X_L^2}$

$$\text{but} \quad X_L = 2\pi f L$$

$$= 6.28 \times (20 \times 10^3) \times (1 \times 10^{-3})$$

$$= 125.6 \ \Omega$$

$$\therefore Z = \sqrt{100^2 + 125.6^2}$$

$$= 160.55 \ \Omega$$

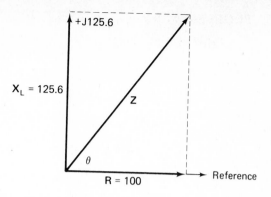

FIGURE 17-12

3. Phase angle θ

$$\theta = \text{arc tan} \frac{X_L}{R}$$

$$\theta = \tan^{-1} \frac{125.6}{100}$$

$$= 51.4°$$

4. Complete answer

$$Z = 160.55 \ \Omega \ \underline{/51.4°}$$

5. Rectangular form
$$Z = R + JX_L$$

$$= 100 + J125.6 \ \Omega$$

6. $\quad V_a = I_a Z$

$$= (2 \times 10^{-3}) \times 160.55 = 321 \ \text{mV} = 321 \ \text{mV} \ \underline{/51.4°}$$

7. $\quad V_R = I_a R$

$$= (2 \times 10^{-3}) \times 100 = 200 \ \text{mV} = 200 \ \text{mV} \ \underline{/0°}$$

8. $\quad V_L = I_a X_L$

$$= (2 \times 10^{-3}) \times 125.6 = 251 \ \text{mV} = 251 \ \text{mV} \ \underline{/90°}$$

9. $\quad V_a = 200 + J251 \ \text{mV} \ \text{(RF)}$

17-3.3 Applied Voltage in Series C and R Circuits

The applied voltage in capacitive-resistive circuits is also the vector sum of the voltages across the components. The current in a capacitive-resistive circuit leads the voltage across the capacitor by 90°. This is the opposite of what occurs in inductive-resistive circuits

Refer to Fig. 17-13.

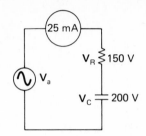

FIGURE 17-13

1. Draw the phasor diagram for Fig. 17-14.
2. What is the value of the applied voltage for the circuit in Fig. 17-13?
3. What is the phase angle of the V_a with respect to the applied current I_a of 25 mA?
4. What is the applied voltage in polar form?
5. State the V_a in RF.
6. State the impedance in polar form if I_C = 25 mA.
7. Find the X_C in PF if I_R = 25 mA.
8. Find the resistance in polar form if I_a = 25 mA.

SOLUTION   *(cover solution and solve)*

1. See Fig. 17-14.
2. $V_a = \sqrt{V_R^2 + V_{X_C}^2}$
   $= \sqrt{150^2 + 200^2}$
   $= 250$ V

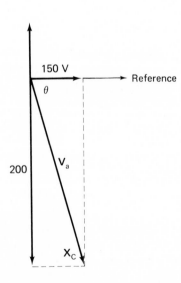

FIGURE 17-14

3. Phase angle
   Choose one of the three trigonometric functions.

$$\theta = \text{arc cosin } \frac{V_R}{V_a}$$

$$= \cos^{-1} \frac{150}{250} = -53°$$

The applied voltage is lagging the I_a by $-53°$.

4. $V_a = 250 \text{ V} \underline{/-53°}$

5. Rectangular form

   $V_a = 150 - J200 \text{ V}$ (from the phasor diagram)

6. $Z = \dfrac{V_a}{I_a} = \dfrac{250}{25 \times 10^{-3}} = 10 \text{ k}\Omega = 10 \text{ k}\Omega \underline{/-53°}$ (PF)

7. $X_C = \dfrac{V_C}{I_a} = \dfrac{200}{25 \times 10^{-3}} = 8 \text{ k}\Omega = 8 \text{ k}\Omega \underline{/-90°}$ (PF)

8. $R = \dfrac{V_R}{I_a} = \dfrac{150}{25 \times 10^{-3}} = 6 \text{ k}\Omega = 6 \text{ k}\Omega \underline{/0°}$ (PF)

PROBLEM 17-8

A capacitor is connected in series with a resistor and a 50-V ac generator (V_a). The circuit current is 2 A and the capacitor voltage is 40 V. Draw the circuit and:

1. Draw the phasor diagram.
2. Find the V_R in the circuit.
3. What is the phase angle between V_a and I_a?
4. State the V_a in polar form.
5. State the V_a in rectangular form.
6. Calculate Z in PF.
7. Calculate X_C in PF.
8. Calculate R in PF.

SOLUTION   (*cover solution and solve*)

1. See Fig. 17-15.
2. Find V_R.

   $$\text{from } V_a = \sqrt{V_R^2 + V_C^2}$$

   $$V_R = \sqrt[2]{V_a^2 - V_C^2}$$

   $$= \sqrt{50^2 - 40^2}$$

   $$= 30 \text{ V}$$

FIGURE 17-15

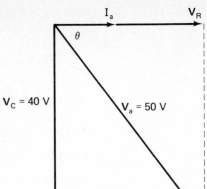

3. Phase angle

$$\theta = \text{arc tan of } \frac{V_C}{V_R}$$

$$= \tan^{-1} \frac{40}{30} = -53°$$

$$= -53°$$

The V_a lags I_a by $-53°$.

4. $V_a = 50 \text{ V } \underline{/-53°}$ (PF)

5. $V_a = 30 - \text{J}40 \text{ V}$

6. $Z = \dfrac{V_a}{I_C} = \dfrac{50}{2} = 25 \ \Omega = 25 \ \Omega \underline{/-53°}$ (PF)

7. $X_C = \dfrac{V_C}{I_C} = \dfrac{40}{2} = 20 \ \Omega = 20 \ \Omega \underline{/-90°}$ (PF)

8. $R = \dfrac{V_R}{I_C} = \dfrac{30}{2} = 15 \ \Omega = 15 \ \Omega \underline{/0°}$ (PF)

17-3.4   Impedance (Z) in Series C and R Circuits

The Z in series C and R circuits is calculated in the same way as in L and R circuits, except that the I_a leads V_C by $90°$.

PROBLEM 17-9

An ac generator (V_a) is connected in series with a 20-Ω resistor (R_1) and a capacitor which has a capacitive reactance of 60 Ω. The circuit current is 2 A. Draw the circuit and:

1. Draw the phasor diagram.
2. Calculate the value of the impedance.
3. What is the phase angle between the impedance and I_a in PF?
4. What is the value of Z in rectangular form?
5. Find V_a in PF.
6. Find V_R in PF.
7. Find V_C in PF.

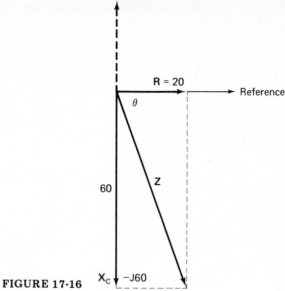

FIGURE 17-16

SOLUTION   *(cover solution and solve)*

1. See Fig. 17-16.
2. As in L and R circuits, the impedance of a series C-R circuit is proportional to the applied voltage V_a. The values of X_C and R are proportional to the voltages across them.

$$Z = \sqrt[2]{R^2 + X_C^2}$$
$$= \sqrt{20^2 + 60^2}$$
$$= 63.24 \ \Omega$$

3. Phase angle θ

$$\theta = \text{arc tan} \ \frac{X_C}{R}$$

$$= \tan^{-1} \frac{60}{20} = -71.56°$$

The Z lags R by $-71.56°$.
Final answer

$$Z = 63.24 \ \Omega \ \underline{/-71.56°}$$

4. <u>Rectangular Form</u>
    The RF states the Z in its separate resistance and reactive values.

$$Z = R - JX_C$$
$$= 20 - J60 \ \Omega$$

5. The applied voltage may be determined if the current is known.

$$V_a = I_a Z$$
$$= 2 \times 63.24$$
$$= 126.48 \text{ V} = 126.48 \text{ V} \underline{/-71.56^\circ} \text{ (PF)}$$

6.
$$V_R = I_a R$$
$$= 2 \times 20 = 40 \text{ V} = 40 \text{ V} \underline{/0^\circ} \text{ (PF)}$$

7.
$$V_C = I_a X_C$$
$$= 2 \times 60 = 120 \text{ V} = 120 \text{ V} \underline{/-90^\circ} \text{ (PF)}$$

PROBLEM 17-10

A 97.2-pF capacitor (C_1) is connected in series with a 1500-Ω resistor (R_1), a current meter, and a 1630-kHz ac generator (V_a). Draw the circuit and:

1. Draw the phasor diagram.
2. Calculate the capacitive reactance (X_C).
3. Calculate the phase angle between the impedance and I_a.
4. State the Z in PF.
5. State the impedance in rectangular form.
6. Find V_a in PF when I_C = 30 mA. (Ans. in PF.)
7. Find V_R in PF when I_C = 30 mA. (Ans. in PF.)
8. Find V_C in PF when I_C = 30 mA. (Ans. in PF.)

SOLUTION   *(cover solution and solve)*

1. See Fig. 17-17.

2. $$X_C = \frac{1}{2\pi f C}$$

$$= \frac{1}{6.28 \times (1630 \times 10^3) \times (97.2 \times 10^{12})}$$

$$= 1005 \ \Omega$$

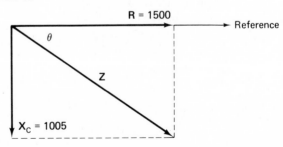

FIGURE 17-17

3. $$\theta = \tan^{-1} \frac{X_C}{R}$$

$$= \tan^{-1} \frac{1005}{1500}$$

$$= -33.82°$$

4. $$Z = \sqrt{R^2 = X_C^2}$$

$$Z = \sqrt{1500^2 + 1005^2}$$

$$Z = 1805\ \Omega\ \underline{/-33.82°}\ (PF)$$

5. $$Z = 1500 - J1005\ \Omega\ (RF)$$

6. Applied voltage

$$V_a = I_C Z$$

$$= (30 \times 10^{-3}) \times 1805 = 54.15 = 54.15\ V\ \underline{/-33.8°}$$

7. $$V_R = I_a R$$

$$= (30 \times 10^{-3}) \times 1500 = 45\ V = 45\ V\ \underline{/0°}$$

8. $$V_C = I_a X_C$$

$$= (30 \times 10^{-3}) \times 1005 = 30.15\ V = 30.15\ V\ \underline{/-90°}$$

17-3.5 Adding Reactances in Series Circuits

Capacitive reactances in series are added, as are resistors in series (Fig. 17-18).

$$X_{C_T} = X_{C_1} + X_{C_2} + X_{C_3} + \cdots + X_{C_n}$$

Inductive reactance in series is added the same way as are resistances in series (Fig. 17-19).

$$X_{L_T} = X_{L_1} + X_{L_2} + X_{L_3} + \cdots + X_{L_n}$$

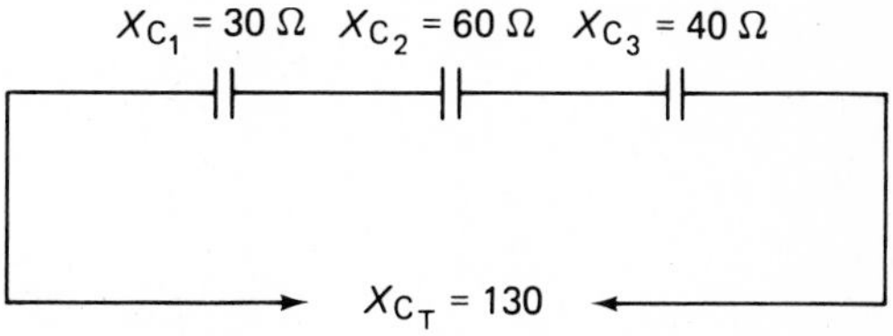

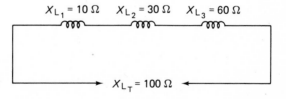

$X_{C_1} = 30\ \Omega$ $X_{C_2} = 60\ \Omega$ $X_{C_3} = 40\ \Omega$

FIGURE 17-18   Capacitive reactances are added when they are connected in series.

$X_{C_T} = 130$

$X_{L_1} = 10\ \Omega$ $X_{L_2} = 30\ \Omega$ $X_{L_3} = 60\ \Omega$

FIGURE 17-19   Inductive reactances are added when they are connected in series.

$X_{L_T} = 100\ \Omega$

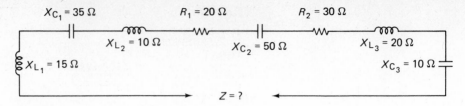

FIGURE 17-20 A series circuit consisting of resistance, capacitive reactance, and inductive reactance.

17-3.6 Reactances and Resistors in Series

In analyzing a series circuit, it is helpful to add all of the X_Cs together, all of the X_Ls together, and all of the resistance values together, and to place the totals in a phasor diagram and add them vectorially.

PROBLEM 17-11

See Fig. 17-20.

$$X_{C_T} = X_{C_1} + X_{C_2} + X_{C_3}$$
$$= 35 + 50 + 10 = 95 \ \Omega$$
$$X_{L_T} = X_{L_1} + X_{L_2} + X_{L_3}$$
$$= 15 + 10 + 20 = 45 \ \Omega$$
$$R_T = R_1 + R_2$$
$$= 20 + 30 = 50 \ \Omega$$
$$Z = \sqrt[2]{R^2 + (X_C - X_L)^2}$$
$$= \sqrt{50^2 + (95 - 45)^2}$$
$$= 70.71 \ \Omega$$

The phasor diagram is shown in Fig. 17-21.

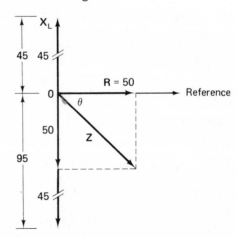

FIGURE 17-21

PROBLEM 17-12

Refer to the circuit in Fig. 17-22.

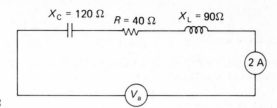

FIGURE 17-22

1. Draw the phasor diagram.
2. Calculate the **Z**.
3. What is the phase angle between $\mathbf{V_a}$ and $\mathbf{I_a}$?
4. Calculate the **Z** in PF.
5. Calculate the **Z** in RF.
6. Find $\mathbf{V_a}$. (Ans. in PF.)
7. Find $\mathbf{V_R}$. (Ans. in PF.)
8. Find $\mathbf{V_L}$. (Ans. in PF.)
9. Find $\mathbf{V_C}$. (Ans. in PF.)

SOLUTION   *(cover solution and solve)*

1. See Fig. 17-23.

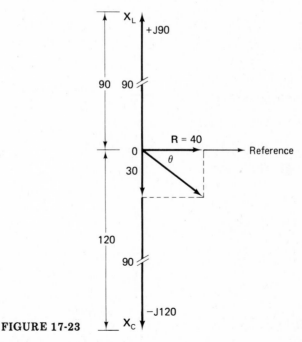

FIGURE 17-23

2. $Z = \sqrt{R^2 + (X_C - X_L)^2}$

   $= \sqrt{40^2 + (120 - 90)^2}$

   $= 50 \ \Omega$

3. Phase angle θ

   $\theta = \text{arc tan} \ \dfrac{X_C}{R}$

   $\theta = \tan^{-1} \ \dfrac{30}{40} = -36.87°$

4. $Z = 50 \ \Omega \ \underline{/-36.87°}$ (PF)

5. $Z = R - JX_C$

   $= 40 - J30 \ \Omega$

6. $V_a = I_a R = 100 \ V \ \underline{/-36.87°} \ V$

7. $V_R = 2 \times 40 = 80 \ V \quad = 80 \ V \ \underline{/0°}$ see Fig. 17-24

8. $V_L = 2 \times 90 = 180 \ V \quad = 180 \ V \ \underline{/90°}$ see Fig. 17-24

9. $V_C = 2 \times 120 = 240 \ V = 240 \ V \ \underline{/-90°}$ see Fig. 17-24

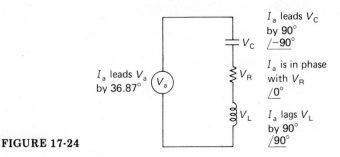

FIGURE 17-24

PROBLEM 17-13

Refer to the circuit in Fig. 17-25.

1. Calculate the value of the Z.
2. Calculate the phase angle.
3. State the Z in PF.
4. State the Z in RF.

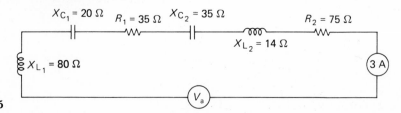

FIGURE 17-25

5. Calculate and state in PF the value of V_a and the voltage across each component.

SOLUTION   *(cover solution and solve)*

1. Add the values of all the common components.

$$R_T = R_1 + R_2$$
$$= 35 + 75 = 110 \ \Omega$$

$$X_{C_T} = X_{C_1} + X_{C_2}$$
$$= 20 + 35 = 55 \ \Omega$$

$$X_{L_T} = X_{L_1} + X_{L_2}$$
$$= 80 + 14 = 94 \ \Omega$$

$$X_T = (X_{L_T} - X_{C_T})$$
$$= (94 - 55) = 39 \ \Omega$$

Draw the phasor diagram (Fig. 17-26).
Calculate the **Z.**

$$\mathbf{Z} = \sqrt{R^2 + (X_L - X_C)^2}$$
$$= \sqrt{110^2 + (94 - 55)^2}$$
$$= 116.7 \ \Omega$$

2. Phase angle θ

$$\theta = \text{arc tan} \ \frac{X_T}{R}$$

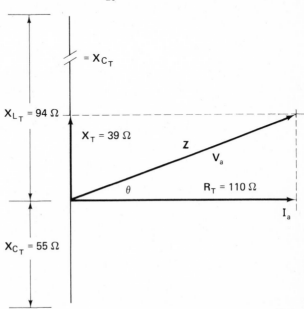

FIGURE 17-26 Phasor diagram for the circuit in Fig. 17-25.

$$\theta = \tan^{-1} \frac{39}{110}$$

$$= 19.5°$$

The R lags Z by 19.5°

3. Final answer

$$Z = 116.7 \ \Omega \ \underline{/19.5°}$$

4. Rectangular form

$$Z = R + JX_L$$

$$= 110 + J39 \ \Omega$$

5. The voltage of V_a

$$V_a = I_a \times Z$$

$$= 3 \times 116.7$$

$$= 350.13 \ V$$

The V_a and Z are represented by the same phasor arm. Therefore, I_a lags V_a by 19.5°.
V_a in polar form

$$V_a = 350.13 \ V \ \underline{/19.5°}$$

The voltage across each component is calculated with Ohm's Law in the same manner as is the V_a.

PROBLEM 17-14

Refer to the circuit in Fig. 17-27.

1. Calculate the value of the Z (in PF).
2. Determine the phase angle of the Z.
3. State the Z in PF.

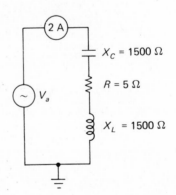

FIGURE 17-27

Ch. 17 Impedance

4. State the **Z** in RF.
5. Find the value of V_a (I_a = 2 A, an exaggerated value).
6. Find the value of V_R.
7. Find the value of V_C.
8. Find the value of V_L.

SOLUTION   *(cover solution and solve)*

1. Impedance (**Z**)

$$Z = \sqrt{R^2 + (X_C - X_L)^2}$$
$$= \sqrt{5^2 + (1500 - 1500)^2}$$
$$= 5\ \Omega$$

2. Phase angle = $0°$.
3. The inductive reactance is equal to, but $180°$ out of phase with the capacitive reactance (see Fig. 17-28). Therefore, the two values cancel, leaving only the resistance to oppose the applied voltage. Thus, the I_a and V_a are in phase.

$$Z = 5\ \Omega\ \underline{/0°}$$

4. Rectangular form

$$Z = R \pm JX$$
$$= 5\ \Omega$$

5. The value of V_a

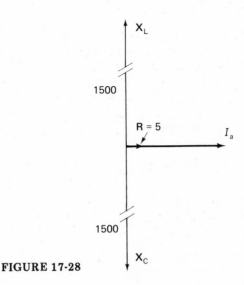

FIGURE 17-28

$$V_a = I_a Z$$

$$= 2 \times 5 = 10 \text{ V} = 10 \text{ V} \underline{/0°}$$

6. $$V_R = I_a R$$

$$= 2 \times 5 = 10 \text{ V} = 10 \text{ V} \underline{/0°}$$

7. $$V_C = I_a X_C$$

$$= 2 \times 1500 = 3000 \text{ V} = 3000 \text{ V} \underline{/-90°}$$

8. $$V_L = I_a X_L$$

$$= 2 \times 1500 = 3000 \text{ V} = 3000 \text{ V} \underline{/90°}$$

17-3.7 Series Resonance Circuits

Any resonant circuit is a circuit in which the $X_C = X_L$ and the dc resistance of the coil wire is the only opposition to the V_a voltage. The circuit in Fig. 17-27 is called a series resonant circuit because the components are in series with one another. The X_L is equal to the X_C and is 180° out of phase with it. Therefore, the two values cancel one another and the circuit is said to be resonant. Note the large voltage developed across each reactive component. Each voltage is 300 times larger than the supply voltage, but the component voltages cancel and therefore are not added into the vector sum.

Table 17-1   Summary of C-R and L-R Series Circuits

C-R Series	L-R Series
I_a is common	I_a is common
$Z = \sqrt[2]{R^2 + X_C^2}$	$Z = \sqrt[2]{R^2 + X_L^2}$
$Z = \dfrac{V_a}{I_a}$	$Z = \dfrac{V_a}{I_a}$
$Z = R - JX_C$	$Z = R + JX_L$
Phase angle $= -\theta$	Phase angle $= +\theta$
$\theta = \text{arc tan } \dfrac{X_C}{R}$	$\theta = \text{arc tan } \dfrac{X_L}{R}$
$V_a = \sqrt[2]{V_R^2 + V_C^2}$	$V_a = \sqrt[2]{V_R^2 + V_L^2}$
$V_a = V_R - JV_C$	$V_a = V_R + JV_L$
$X_C > R$ circuit is capacitive	$X_L > R$ circuit is inductive

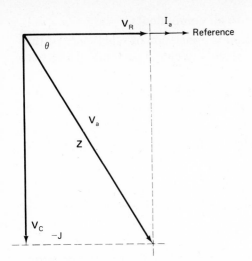

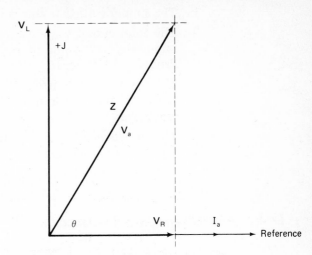

FIGURE 17-29 Phase relationship between current and voltages in a series resistive-capacitive circuit.

FIGURE 17-30 Phase relationship between current and voltages in a series resistive-inductive circuit.

17-3.8 Summary Equations for Series L, C, and R Circuits

1. $$Z = \sqrt[2]{R^2 + (X_L - X_C)^2} \quad (PF)$$

2. $$Z = \frac{V_a}{I_a}$$

3. $$Z = R \pm JX \quad (RF)$$

4. Phase angle $\quad \theta = \text{arc cos}\left(\dfrac{R}{Z}\right)$

$$= \text{arc sin}\left(\frac{X_L}{Z}\right)$$

$$= \text{arc tan}\left(\frac{X_L}{R}\right)$$

5. Resonance occurs when $X_C = X_L$.

17-4 Parallel Impedance Circuits

The most important fact to be realized about parallel impedance circuits is that impedance *cannot* be calculated by combining the reactance value with the resistance value, vectorially, as was done in series circuits.

The basic characteristic of parallel impedance circuits are:

1. The supply voltage is common to all components.

2. The total circuit current in circuits consisting of reactances of the same kind, is the sum of each of the individual currents. (Refer to Problems 17-15 and 17-16.)

3. The total circuit current in circuits with different types of reactances, is the vector sum of the currents through the individual reactances.

4. The impedance (Z) of a parallel reactive circuit is determined by the Ohm's Law equation.

$$Z = \frac{V_a}{I_a} \text{ (PF)}$$

17-4.1 Parallel Capacitive Reactance Circuits

PROBLEM 17-15

Two capacitors are connected in parallel with a 24-V ac voltage. Capacitor C_1 has a reactance of 6 Ω. Capacitor C_2 has a reactance of 4 Ω. Draw the circuit and the phasor diagram for the current values (see Fig. 17-31).

1. Find X_{C_T}.
2. Find I_a.

SOLUTION   *(cover solution and solve)*

1. Total X_C in parallel

$$X_{C_T} = \frac{X_{C_1} \times X_{C_2}}{X_{C_1} + X_{C_2}}$$

$$= \frac{6 \times 4}{4 + 6} = 2.4 \ \Omega$$

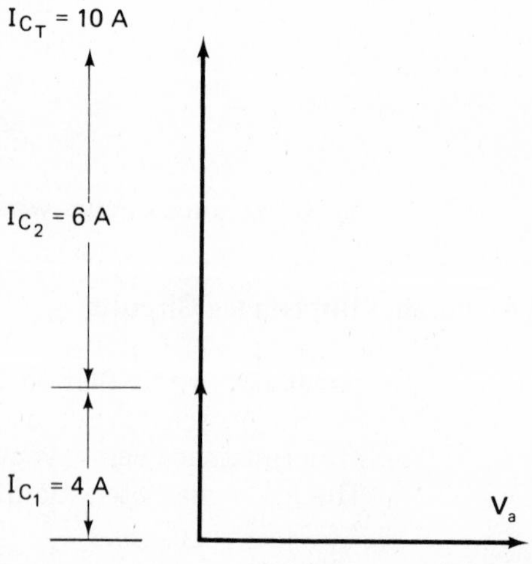

$I_{C_T} = 10$ A

$I_{C_2} = 6$ A

FIGURE 17-31   The phasor diagram for the current values.

$I_{C_1} = 4$ A

V_a

2.

$$I_a = \frac{V_a}{X_{C_T}} = \frac{24}{2.4} = 10 \text{ A}$$

or

$$I_a = I_{C_1} + I_{C_2}$$
$$= 4 + 6 = 10 \text{ A}$$

17-4.2 Parallel Inductive Reactance Circuits

PROBLEM 17-16

Refer to the circuit in Fig. 17-32.

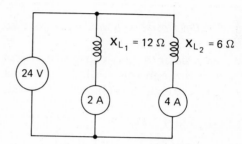

FIGURE 17-32

1. Find total X_L.
2. Find I_a.
3. Draw the phasor diagram for the currents and voltage.

SOLUTION   (cover solution and solve)

1.

$$\text{total } X_L = \frac{X_{L_1} \times X_{L_2}}{X_{L_1} + X_{L_2}}$$

$$= \frac{12 \times 6}{18} = 4 \text{ }\Omega$$

2.

$$I_a = \frac{V_a}{X_{L_T}} = \frac{24}{4} = 6 \text{ A}$$

or

$$I_T = I_{L_1} + I_{L_2}$$
$$= 2 \text{ A} + 4 \text{ A} = 6 \text{ A}$$

3. See Fig. 17-33.

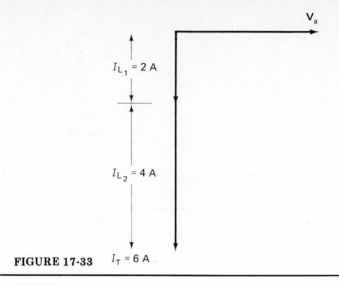

$I_{L_1} = 2\ A$

$I_{L_2} = 4\ A$

FIGURE 17-33 $\quad I_T = 6\ A$

17-4.3 Z of Parallel Resistive and Capacitive Circuits

When resistance and capacitive reactance are connected in parallel, the current through the resistance is in phase with the applied voltage, but the current through the capacitance leads the applied voltage by 90°.

PROBLEM 17-17

A 100-Ω resistor (R_1) and a capacitor are connected in parallel with a 200-V ac supply (V_a). The reactance of the capacitor is 50 Ω. Draw the circuit and:

1. Calculate I_a.
2. Calculate Z.
3. Calculate the phase angle.
4. State the Z in PF.
5. State the current value in PF.
6. State the current value in RF.
7. Is the current predominantly capacitive or resistive?

SOLUTION   *(cover solution and solve)*

1. The total current is the vector sum of I_R and I_C (Fig. 17-34).

$$I_a = \sqrt[2]{I_R^2 + I_C^2}$$
$$= \sqrt{2^2 + 4^2}$$
$$= 4.47\ A$$

Note that the applied current is the total current and is the result of $I_a = V_a/Z$. It is also called the impedance current (I_Z).

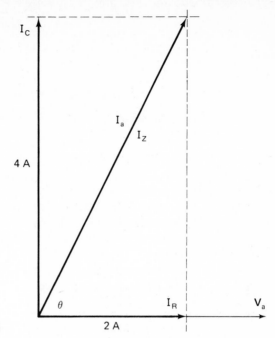

FIGURE 17-34

2.
$$Z = \frac{V_a}{I_a}$$

$$= \frac{200}{4.47} = 44.74 \ \Omega$$

3. Phase angle θ

$$\underline{/\theta} = \tan^{-1}\left(\frac{4}{2}\right)$$

$$= 63.4°$$

4.
$$Z = 44.74 \ \Omega \ \underline{/-63.4°}$$

5.
$$I_a = 4.47 \ A \ \underline{/63.4°}$$

6.
$$I_a = I_R + JI_C$$

$$= 2 + J4 \ A \quad \text{(capacitive circuit)}$$

7. The larger the capacitance in a parallel circuit, the smaller will be the X_C with respect to the resistance in the circuit. The result is that a larger amount of the total current will flow through the capacitor than through the resistor. The circuit is then said to be capacitive. Conversely, the smaller the capacitance, the larger will be the X_C with respect to the resistance, and a larger amount of the total current will flow through the resistor than through the capacitor. The circuit is then said to be resistive.

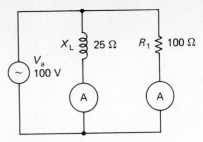

FIGURE 17-35 A parallel resistive-inductive circuit. Note that the voltage is common to all components.

17-4.4 Z of Parallel Resistance and Inductive Circuits

The applied voltage is common to all components in a parallel circuit (Fig. 17-35), and each branch current can be calculated from Ohm's Law. The current through the resistance is in phase with the applied voltage, but the current flow in the inductance is 90° out of phase with the applied voltage. (Fig. 17-36).

Note that the total current is the applied current and is equal to V_a/Z. It is also called the impedance current.

PROBLEM 17-18
_____

Refer to Fig. 17-35.

1. Draw the phasor diagram.
2. Find I_a.
3. Find the Z value.
4. Find the phase angle θ.
5. Find I_a in PF.
6. Find I_a in RF.
7. State the value of the Z in PF.

SOLUTION   (*cover solution and solve*)

1. See Fig. 17-36.
2. The total current is the vector sum of the I_R and I_L (Fig. 17-36).

$$I_a = \sqrt[2]{I_R^2 + I_L^2}$$
$$= \sqrt{1^2 + 4^2}$$
$$= 4.12 \text{ A}$$

3. $$Z = \frac{V_a}{I_a} = \frac{100}{4.12} = 24.27 \ \Omega$$

4. Phase angle θ

$$\theta = \tan^{-1} \frac{4}{1} = -75.96°$$

5. $$I_a = 4.12 \text{ A} \underline{/-75.96°}$$

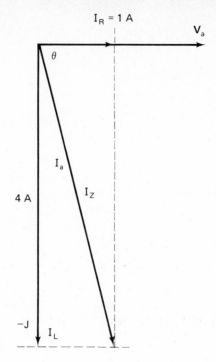

FIGURE 17-36

6. $$\mathbf{I_a} = \mathbf{I_R} - \mathbf{JI_L}$$

$$= 1 - \mathrm{J4~A} \quad \text{(inductive circuit)}$$

7. $$Z = 24.27~\Omega~\underline{/+75.96^\circ}$$

The smaller the inductance in a parallel circuit, the smaller the inductive reactance and the greater the current flow through the inductance. The circuit is thus said to be inductive. The larger the inductance in a parallel circuit, the larger the inductive reactance with respect to the resistance. The circuit is then said to be resistive.

Table 17-2 Features of *C-R* and *L-R* Parallel Circuits

C-R Parallel Circuits	L-R Parallel Circuits
$I_a = I_{C_1} + I_{C_2}$	$I_a = I_{L_1} + I_{L_2}$
$I_a = \sqrt[2]{I_R^2 + I_C^2}$	$I_a = \sqrt[2]{I_R^2 + I_L^2}$
Phase angle $= +\theta$	Phase angle $= -\theta$
$\mathbf{I_a} = \mathbf{I_R} + \mathbf{JI_C}$	$\mathbf{I_a} = \mathbf{I_R} - \mathbf{JI_L}$
$Z = \dfrac{V_a}{I_a}$	$Z = \dfrac{V_a}{I_a}$
$R > X_C$ circuit is capacitive	$R > X_L$ circuit is inductive

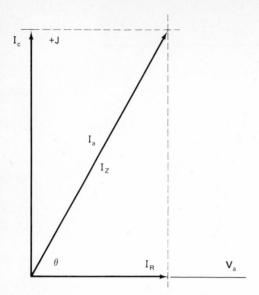

FIGURE 17-37 Phase relationship between the voltage and currents in a parallel resistive-capacitive circuit.

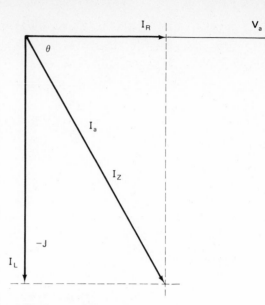

FIGURE 17-38 Phase relationship between the voltage and currents in a parallel resistive-inductive circuit.

17-4.5 Impedance of R, C, and L Parallel Circuits

The impedance of C, L, and R circuits is calculated in the same manner as in the C-R and L-R circuits. The impedance of C, L, and R circuits cannot be determined by directly combining the values. The procedure of solution is:

1. Determine the current flow through each component.
2. Add the currents vectorially.
3. Divide the total current into the supply voltage.

Capacitive current is $180°$ out of phase with inductive current. The value of the larger current will cancel the value of the smaller current. The larger current will lag or lead the V_a by $90°$; thus, the reactive current must be added vectorially to the resistive current.

PROBLEM 17-19

A 36-Ω resistor (R_1) is connected in parallel with a 36-V ac generator (V_a), an inductor, and a capacitor. The capacitor has a reactance of 18 Ω. The coil has a reactance of 9 Ω. Draw the circuit and:

1. Draw the phasor diagram.
2. Find the total current.
3. Find the impedance value.
4. Find the phase angle θ.
5. State I_a in PF.
6. State I_a in RF.
7. State Z in PF.

SOLUTION   *(cover solution and solve)*

1. See Fig. 17-39.
2. Total current, I_a or I_z

$$I_a = \sqrt{I_R^2 + (I_L - I_C)^2}$$
$$= \sqrt{1^2 + (4 - 2)^2}$$
$$= 2.236 \text{ A}$$

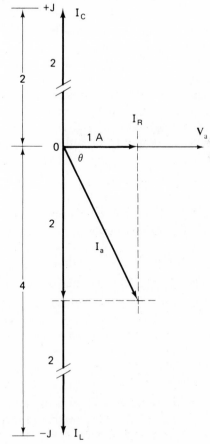

FIGURE 17-39

3.
$$Z = \frac{V_a}{I_a} = \frac{36}{2.236} = 16.1 \ \Omega$$

4. Phase angle θ

$$\theta = \tan^{-1} \frac{I_X}{I_R} = \tan^{-1} \frac{2}{1}$$

$$= -63.43°$$

5. I_a in PF

$$I_a = 2.236 \ \text{A} \ \underline{/-63.43°}$$

6. Rectangular form

$$I_a = 1 - J2$$

7. Z in polar form

$$Z = 16.1 \ \Omega \ \underline{/+63.43°}$$

17-4.6 Current Decides Type of Circuit

The circuit takes its name from the component which has the largest current flow through it.

If I_R is largest (small R compared to X), the circuit is said to be resistive.

If I_C is largest, the circuit is said to be capacitive. (A large C means that there is a small X_C compared to R or X_L.)

If I_L is largest (small L, so that X_L is small compared to R and X_C), the circuit is said to be inductive.

PROBLEM 17-20 ────────────────────

A 100-Ω resistor (R_1) is connected in parallel with a coil, a capacitor, and an ac voltage supply. The coil has an inductive reactance of 150 Ω. The capacitor has a reactance of 60 Ω. Draw the circuit and:

1. Draw the phasor diagram for the branch currents.
2. Find the impedance of the circuit in PF.
3. What type of circuit is it (inductive or capacitive)?

SOLUTION   (*cover solution and solve*)

1. See Fig. 17-40.
2. There is no applied voltage stated for this circuit, so the problem must be solved by using an assumed voltage. No matter what voltage value is assumed, the same value of impedance will always be produced. Assume, for example, that $V_a = 300$ V.

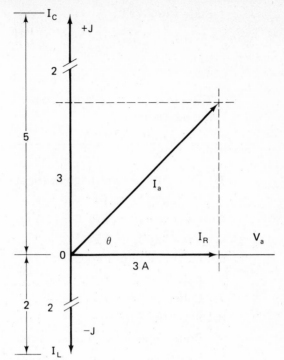

FIGURE 17-40

Calculate the currents.

$$I_R = \frac{V_a}{R} = \frac{300}{100} = 3 \text{ A}$$

$$I_L = \frac{V_a}{X_L} = \frac{300}{150} = 2 \text{ A}$$

$$I_C = \frac{V_C}{X_C} = \frac{300}{60} = 5 \text{ A}$$

$$\mathbf{I_a} = \sqrt[2]{\mathbf{I_R^2} + (\mathbf{I_C} - \mathbf{I_L})^2}$$
$$= \sqrt{3^2 + (5 - 2)^2}$$
$$= \sqrt{9 + 9}$$
$$= \sqrt{18} = 4.24 \text{ A}$$

Calculate the Z value.

$$Z = \frac{V_a}{I_a} = \frac{300}{4.24} = 70.75 \ \Omega$$

Determine the phase angle.

$$\text{phase angle } \theta = \tan^{-1} \frac{I_X}{I_R}$$

$$\theta = \tan^{-1} \frac{3}{3} = 45°$$

State the Z in PF.

$$Z = \frac{V_s}{I_a} = \frac{300}{4.24 \text{ A} \underline{/+45°}}$$

$$= 70.75 \ \Omega \ \underline{/-45°}$$

3. The circuit is said to be capacitive because the capacitor current is the largest current.

PROBLEM 17-21

1. Draw the phasor diagram for the circuit in Fig. 17-41.
2. Calculate I_R in PF.
3. Calculate I_C in PF.
4. Calculate I_L in PF.
5. Calculate I_a in PF.
6. Calculate Z.
7. Calculate the phase angle θ.
8. State the current in polar form.
9. State impedance in polar form.
10. State current in rectangular form.
11. State impedance in rectangular form.
12. What type of circuit is it?

SOLUTION   *(cover solution and solve)*

1. See Fig. 17-42.

2. $I_R = \dfrac{V_R}{I_R} = \dfrac{1500 \times 10^{-3}}{15 \times 10^3} = 100 \times 10^{-6}$

$$= 100 \ \mu\text{A} \ \underline{/0°}$$

3. $I_C = \dfrac{V_C}{X_C} = \dfrac{1500 \times 10^{-3}}{1500} = 1 \text{ mA} \ \underline{/+90}$

4. $I_L = \dfrac{V_L}{X_L} = \dfrac{1500 \times 10^{-3}}{1500} = 1 \text{ mA} \ \underline{/-90°}$

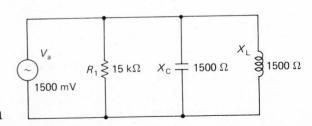

FIGURE 17-41

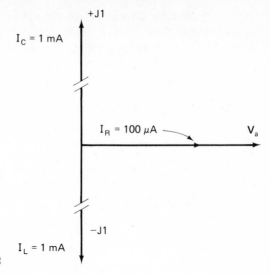

FIGURE 17-42

5. $\quad \mathbf{I_a} = \sqrt[2]{\mathbf{I_R^2} + (\mathbf{I_C} - \mathbf{I_L})^2}$

$\quad\quad = \sqrt[2]{(100 \times 10^{-6})^2 + [(1 \times 10^{-3}) - (1 \times 10^{-3})]^2}$

$\quad\quad = 100 \ \mu A$

6. $\quad Z = \dfrac{V_a}{I_R} = \dfrac{1500 \times 10^{-3}}{100 \times 10^{-6}} = 15 \ k\Omega$

7. $\mathbf{I_a}$ and $\mathbf{Z}$ are in phase with $\mathbf{V_a}$. Phase angle = $0°$.

8. $\quad \mathbf{I_a} = 100 \ \mu A \ \underline{/0°}$

9. $\quad \mathbf{Z} = 15 \ k\Omega \ \underline{/0°}$

10. $\quad \mathbf{I_a} = 100 + J0 \ \mu A$

11. $\mathbf{Z}$ rectangular form

$\quad \mathbf{Z = R \pm JX}$

$\quad\quad = 15 \ k\Omega + J0$

12. It is a resistive circuit. Note that $X_C = X_L$, so the circuit in Fig. 17-41 is a resonant circuit.

17-4.7 Parallel Resonant Circuits

The circuit in Fig. 17-41 is said to be a parallel resonant circuit because $X_C = X_L$. The reactive currents I_C and I_L will also be equal in value because the equal reactances have a common supply voltage. The I_C current will cancel the I_L current because they are $180°$ out of phase. The only current which will flow in the circuit, from the supply voltage, will be that permitted by the resistance (R) of the circuit. A parallel resonant circuit is said to be a resistive circuit because the resistive current is the only current flow, and I_a is in phase with the V_a. If R_1 were removed from the circuit (a hypothetical condition) the current would be zero, which would indicate that the parallel resonant circuit had infinite impedance.

EXERCISES

1. Define the term impedance.

2. Write the following equations for impedance:
    a. Ohm's Law
    b. The vector diagram sum
    c. Impedance expressed in complex numbers (rectangular form)

3. What value does the length of a phasor arm represent?

4. How does an increase in dc resistance affect the phase angle of an impedance circuit?

5. A 4-Ω resistance is connected in series with a 16-Ω capacitive reactance. Calculate:
    a. The phase angle which will exist between the two components
    b. The phase angle between the resultant impedance and the resistance
    c. The phase angle between the circuit current and the impedance value
    d. The phase angle between the circuit current and the applied voltage

6. An inductance is connected in series with a 1200-Ω resistor and a source of voltage. The voltage across the inductance is 40 V. The voltage across the resistor is 60 V.
    a. Draw the phasor diagram for the circuit.
    b. What is the value of the applied voltage in polar form?
    c. What is the phase angle between the applied voltage (V_a) and the circuit current (I_a)?
    d. What is the value of the impedance in polar form?
    e. How much is the inductive reactance?

7. State the total impedance in polar form for the series circuit in Fig. 17-43.

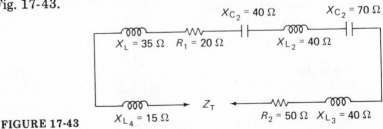

FIGURE 17-43

8. A capacitor with a reactance of 120 Ω, an 80-Ω resistor, and an inductor with a 180-Ω reactance are all connected in series to a source of voltage. The current through the 80-Ω resistor is 2 A. Draw the circuit and the phasor diagram, and calculate:

a. The total impedance of the circuit in polar form
b. The total impedance of the circuit in rectangular form
c. The phase angle between the impedance and the applied current
d. The value of applied voltage in rectangular form
e. The value of applied voltage in polar form
f. The value of V_L in polar form
g. The value of V_C in polar form

9. Define the term "resonance."

10. The capacitive reactance is larger than the resistance in a series circuit. Is the circuit a resistive or a capacitive circuit?

11. The inductive reactance in a series inductive-resistive circuit is smaller than the resistance. Is the circuit a resistive or an inductive circuit?

12. A 50-Ω resistor is connected in parallel with a capacitor possessing 25-Ω capacitive reactance and a 100-V ac generator. Draw the circuit and the phasor diagram, and calculate:

a. The total impedance to the generator in polar form
b. The current flow from the generator in polar form
c. The generator current in rectangular form
d. The current flow in polar form through each component
e. Is the circuit a capacitive or resistance circuit?

13. An inductor (L_1) having 36 Ω of inductive reactance is connected in parallel with a capacitor (C_1) possessing 144 Ω of capacitive reactance, a 72-Ω resistor (R_1), and a 144-V ac generator. Draw the circuit and the phasor diagram, and calculate:

a. The total impedance in polar form
b. The total current in polar form
c. The total current in rectangular form
d. The current in polar form through R_1
e. The current in polar form through L_1
f. The current in polar form through C_1
g. Is the circuit a capacitive, inductive, or resistive circuit?

chapter 18

RESONANT CIRCUITS

18-1 Introduction

Resonant circuits make it possible to select and remove one particular signal from the thousands of frequencies which exist in space. Resonant circuits make possible radio and television, two-way radio communication systems of all types, radar and navigation systems, and the latest applications of satellite transmission and reception.

In general, resonant circuits are used to select desired signals and to reject unwanted signals.

A resonant circuit is one in which $X_L = X_C$. In series resonant circuits, the only opposition to the signal current is the dc resistance of the circuit, i.e., the dc resistance of the wire of which the inductance is wound, because the reactances cancel out. The circuit current is in phase with the applied voltage and the power factor of the circuit is one.

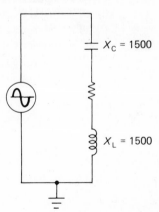

$X_C = 1500$

$X_L = 1500$

FIGURE 18-1   A series-resonant circuit.

The L-C-R circuits which comprise all resonant circuits are frequency-dependent. The X_C of the capacitance varies inversely with the frequency, while the X_L of the inductance varies directly with frequency. If the frequency of a signal to an L-C-R circuit is of such a value that $X_C = X_L$, the L-C-R circuit is then resonant. The resonant frequency for a particular capacitor and a particular inductor can be determined by the following formula:

$$f_r = \frac{1}{2\pi\sqrt{LC}}$$

where f_r is in hertz
L is in henries
C is in farads
Note: f_r is not
dependent upon
the resistance of
the circuit.

DERIVING THE RESONANCE FREQUENCY FORMULA ———————

Resonance frequency is that condition when

$$X_L = X_C$$

$$\text{or} \quad 2\pi fL = \frac{1}{2\pi fC}$$

multiply $(2\pi fL)$ by $(2\pi fC) = 1$.

$$4\pi^2 f^2 LC = 1$$

$$f^2 = \frac{1}{4\pi^2 LC}$$

Square root both sides and

$$f = \frac{1}{2\pi\sqrt{LC}}$$

Analysis of the resonance frequency formula shows that for one particular resonance frequency, the capacitance value will have to increase if the inductance is decreased, and vice versa.

There are only two types of circuits which can be made resonant:

1. L-C-R series circuits (series resonant)
2. L-C-R parallel circuits (parallel resonant)

18-2 Series Resonant Circuits

It was shown in Chap. 17 that an L-C-R circuit is resonant when

$$X_L = X_C$$

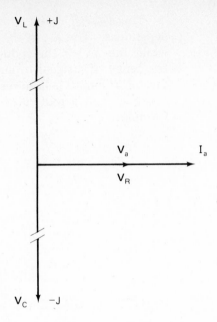

FIGURE 18-2 A phasor diagram for a series-resonant circuit $V_L = V_C$. The V_a are in phase. Power factor = 1.

Because $X_L = X_C$, the voltage across the inductance will be equal to the voltage across the capacitor. Because the two voltages are 180° out of phase and are equal, they will cancel. This will cause all the supply voltage to be developed across the coil resistance as shown in the following equation.

$$V_a = \sqrt{V_R^2 + (V_L^2 - V_C^2)}$$

$$\text{or} \quad V_a = V_R \pm J(V_L - V_C) \quad (\text{RF})$$

$$\text{but} \quad V_L = V_C$$

$$\therefore V_a = V_R$$

Because the resistance of a coil (e.g., an antenna coil) is only the resistance of the wire of which the coil is wound, the resistance is very small. The coil resistance is virtually the only opposition to signal voltage; thus, the signal current may be relatively large. The result is a comparatively large voltage developed across each of the large reactances. These voltages are 180° out of phase with one another; they cancel and become ineffective in the total circuit. See Fig. 18-2.

PROBLEM 18-1

A 200-mV radio wave signal is applied to a series L-C-R antenna circuit in which $L = 106$ μH, the coil resistance is 5 Ω, and $C = 106$ pF.

1. Draw the circuit.
2. Calculate the frequency at which the circuit will be resonant.
3. Determine the value of X_L and X_C at resonance.
4. Determine the circuit current at resonance.
5. Calculate the voltage developed across each of the three components at resonance.

SOLUTION   *(cover solution and solve)*

1. See Fig. 18-3.

2. $f_R = \dfrac{1}{2\pi\sqrt{LC}}$

$\qquad = \dfrac{1}{6.28 \times (106 \times 10^{-6}) \times (106 \times 10^{-12})}$

$\qquad = \dfrac{1}{6.28 \times 11236 \times 10^{-18}}$

$\qquad = \dfrac{1}{6.28 \times 106 \times 10^{-9}}$

$\qquad = 1502 \times 10^3$

$\qquad = 1502$ kHz (high end of AM radio band)

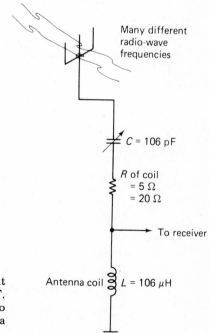

Many different radio-wave frequencies

$C = 106$ pF

R of coil = 5 Ω = 20 Ω

To receiver

Antenna coil $L = 106$ μH

FIGURE 18-3   A series-resonant antennae circuit. The selected R.F. wave signal induces 200 mV into the antennae (equivalent to a 200-mV generator).

3. $X_L = 2\pi f L$

$\qquad = 6.28 \times (1502 \times 10^3) \times (106 \times 10^{-6})$

$\qquad = 1000 \ \Omega$

$\quad X_C = X_L$

$\qquad = 1000 \ \Omega$

4. Circuit current

$$I_a = \frac{V_a}{R}$$

$$= \frac{200 \times 10^{-3}}{5}$$

$$= 40 \times 10^{-3} \ A$$

$$= 40 \ mA$$

5. Voltages

$$V_C = I_a \times X_C$$

$$= (40 \times 10^{-3}) \times 1000$$

$$= 40 \ V$$

$$V_L = V_C = 40 \ V$$

$$V_R = I_a \times R$$

$$= (40 \times 10^{-3}) \times 5 = 200 \ mV$$

18-2.1 Observations

Several observations about series resonant circuits can be made from Tables 18-1 and 18-2. These observations may be stated as characteristics of series resonant circuits.

CHARACTERISTICS OF SERIES CIRCUITS AT RESONANCE

1. Resonance occurs when $X_L = X_C$.
2. Circuits must have an inductance and a capacitor to produce resonance.
3. Z is minimum at resonance; $Z = R$.
4. Circuit current is maximum at resonance.
5. The reactances are large but equal in value at resonance; reactance is lower or higher in value on either side of resonance depending upon the frequency of the signal and depending upon whether the component is a capacitor or an inductor.

6. The voltage across the capacitor and the inductance is maximum at resonance. In Prob. 18-1, V_C and V_L are each 40 V, which is 200 times greater than the signal voltage value of 200 mV (voltage amplification). The V_C and V_L voltages are less than 40 V on either side of resonance. For example:

At 1450 kHz, V_C = 2.88 V and V_L = 3.08 V (XI_a).

At 1500 kHz, V_C = 40 V and V_L = 40 V.

At 1550 kHz, V_C = 3.08 V and V_L = 2.88 V.

7. The large voltage across the inductance and capacitance is stored energy. The inductance stores energy for one-half of the complete cycle, and the capacitor stores energy for the other half of the cycle. The quantities of energy stored by both reactances are equal and are transferred back and forth between them.

8. It is only at resonance that V_C cancels V_L.

9. The value of the (coil) resistance in the circuit does not affect the resonance frequency. Compare Table 18-1 at 1500 kHz and a coil resistance of 5 Ω with Table 18-2 at 1500 kHz and a coil resistance of 20 Ω.

10. a. The value of the coil resistance does affect the value of circuit current at resonance. For example:

When R_L = 5 Ω, circuit current = 40 mA at resonance (Table 18-1).

When R_L = 20 Ω, circuit current = 10 mA at resonance (Table 18-1).

   b. The greater the coil resistance (smaller diameter of coil wire), the less the circuit current.

   c. An increase in the coil resistance reduces the voltage across the capacitor and the coil at resonance. For example:

When R_L = 5 Ω, V_C and V_L = 40 V at resonance (XI_a).

When R_L = 20 Ω, V_C and V_L = 10 V at resonance (XI_a).

11. Resonance frequency

$$f_R = \frac{1}{2\pi\sqrt[2]{LC}}$$

12. The resonance frequency can be changed by changing either the capacitor value or the inductance value.

13. The solution to 2. in Prob. 18-1 shows that a 106-μH coil and a 106-pF capacitor are resonant at 1502 kHz, or approximately 1500 kHz.

Table 18-1 shows the value of reactance at different frequencies for a 106-μH coil and 106-pF capacitor when the coil resonance is 5 Ω.

Table 18-1

Frequency kHz	X_L Ω	X_C Ω	R Ω	Z Ω	I_a when signal = 200 mV
1450	969	1036	5	67.19	2.98
1460	975	1029	5	54.23	3.69
1470	980	1020	5	40.31	4.96
f_1 1480	987	1013	5	26.47	7.56
1490	993	1007	5	14.86	13.46
1495	996	1003	5	8.5	23.5
f_R 1500	1000	1000	5	5.0	40 mA
1505	1003	996	5	8.5	23.5
f_2 1510	1007	993	5	14.86	13.46
1520	1013	987	5	26.47	7.56
1530	1020	980	5	40.31	4.96
1540	1029	975	5	54.23	3.69
1550	1036	969	5	67.19	2.98

Table 18-2 shows the same values as in Table 18-1 except that the coil resistance (R_C) is 20 Ω.

Table 18-2

Frequency kHz	X_L Ω	X_C Ω	R Ω	Z Ω	I_a when signal = 200 mV
1460	975	1025	20	57.58	3.47
1470	980	1020	20	44.72	4.47
1480	987	1013	20	32.80	6.10
f_1 1490	993	1007	20	24.41	8.19
1495	996	1003	20	21.19	9.43
f_R 1500	1000	1000	20	20.0	10 mA
1505	1003	996	20	21.19	9.43
f_2 1510	1007	993	20	24.41	8.19
1520	1013	987	20	32.80	6.10
1530	1020	980	20	44.72	4.47
1540	1029	975	20	57.58	3.47

Table 18-1 shows the reactance of each component at the resonant frequency and in step frequencies of 10 kHz on either side of resonance.

A comparison of Tables 18-1 and 18-2 shows that at resonance, an increase in the coil resistance (R_C) does not change the resonant frequency, but does greatly reduce the circuit current. This reduced current also occurs at frequencies close to resonance. At resonance (Fig. 18-5) the circuit is resistive (Fig. 18-4b); below resonance the circuit becomes capacitive (Fig. 18-4a), and above resonance the circuit becomes inductive (Fig. 18-4c).

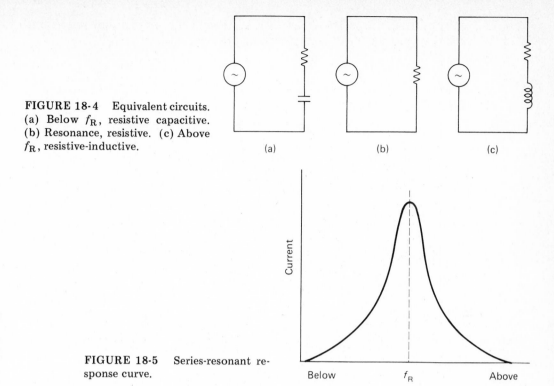

FIGURE 18-4 Equivalent circuits. (a) Below f_R, resistive capacitive. (b) Resonance, resistive. (c) Above f_R, resistive-inductive.

(a)   (b)   (c)

FIGURE 18-5   Series-resonant response curve.

Current

Below f_R Above

18-2.2 Series Resonant Response Curve

The circuit current is maximum at resonance and decreases on either side of resonance. The current values in Tables 18-1 and 18-2 can be plotted against their respective frequencies on linear graph paper. The result is the series resonant response curve as shown in Fig. 18-6.

Notice in Fig. 18-6 that the resistance does not change the resonance frequency, but increased resistance results in a smaller current flow at resonance and a lower response curve (see Curve B). Also note that the change in current flow is less pronounced with each change in frequency away from resonance in curve B.

Curve A is called a sharp response curve because the sides of the curve are very steep slightly above and below resonance.

Curve B is called a broad response curve because there is a gradual change in the slope of the curve on either side of resonance.

18-2.3 Selectivity

Curve A in Fig. 18-6 compared to curve B is a very selective curve because it responds sharply to the resonant frequency but responds very poorly to frequencies on either side of resonance.

Selectivity is the ability of a tuned circuit to accept, or pass currents of one frequency and to exclude all others. Curve A satisfies this defini-

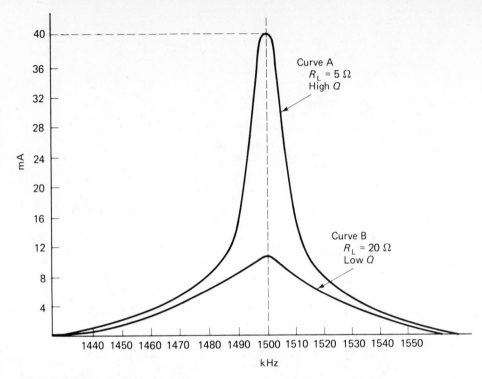

FIGURE 18-6   Frequency-response curve of a series-resonant circuit (current plotted against frequency).

tion, as it is responding to the frequency of 1500 kHz but not to the other frequencies.   Curve B responds equally well to a wide range of frequencies and thus it has very poor selectivity and would not reject unwanted signals on either side of resonance.   The selectivity of a tuned circuit depends upon the Q of the tuned circuit.

18-2.4   Factors Affecting Selectivity

Q OF THE CIRCUIT

The selectivity of a series resonant circuit is dependent upon the Q of the circuit which in turn is determined chiefly by the Q of the coil.

Q is called the figure of merit of a coil. Q is a measure of the circuit's ability to store energy.   The energy stored alternately in the reactance is reactive power $(I^2 X_L)$ and the energy dissipated in the form of heat is dissipated power $(I^2 R)$.

Q is the ratio of reactive power to dissipated power

$$Q = \frac{I^2 X_L}{I^2 R}$$

The I^2 cancel. Therefore

$$Q = \frac{X_L}{R} = \frac{\text{energy stored}}{\text{energy dissipated}} = \text{figure of merit}$$

$\qquad\qquad\qquad\qquad$ where $\ \ X_L$ is inductance reactance
$\qquad\qquad\qquad\qquad\qquad\qquad\qquad$ at resonance
$\qquad\qquad\qquad\qquad\quad\ \ R\ $ is resistance of the
$\qquad\qquad\qquad\qquad\qquad\qquad\qquad$ circuit

Another definition of Q is that it is a measure of quality or purity of an inductance.

The resistance of a coil is considered to be in series with its inductive reactance.   Curve A in Fig. 18-6 has a coil resistance of 5 Ω, whereas curve B has a coil resistance of 20 Ω. Curve A has high selectivity because of the low value of coil resistance. The coil in curve A is said to have a high Q. For example the Q of the coils in Fig. 18-6 are:

$$\text{Curve A:} \quad Q = \frac{X_L}{R_1} = \frac{1000}{5} = 200$$

$$\text{Curve B:} \quad Q = \frac{X_L}{R_2} = \frac{1000}{20} = 50$$

As seen in Fig. 18-6, the higher the Q—due to a lower coil resistance— the larger the circuit current at resonance and the higher the circuit selectivity. This is because the ratio of the amount of energy stored in the magnetic field in the coil to the amount of heat energy lost in the dc resistance is greater.

L/C RATIO

The Q of a circuit can also be increased by increasing the inductance of a coil, by winding the coil with a greater number of turns of wire or by using a coil core of a higher permeability. In any case, because the resistance of the coil should be kept constant, the coil wire used to rewind the coil should be of a larger diameter.

If the coil inductance is increased, the value of capacitance used to resonate at the same frequency would have to be reduced. In other words, the *L/C* ratio of the combination is increased.

In the example used in Table 18-1, a 106-μH inductance and a 106-pF capacitor produced resonance at 1500 kHz. From the resonant frequency formula,

$$f_R = \frac{1}{2\pi\sqrt[2]{LC}}$$

it can be seen that the same resonant frequency would be maintained if the value of the inductance were increased 4 times (106 μH $\times$ 4 = 424 μH) while the capacitor value were decreased to $\frac{1}{4}$ of its value ($\frac{1}{4}$ $\times$ 106 pF = 26.5 pF). (See Table 18-3.)

Table 18-3   Some *L* and *C* Combinations Which Resonate
at 1500 kHz

Equation	L in μH	C in pF	X_L	X_C
(1)	26.5	424	250	250
(2)	53	212	500	500
(3)	106	106	1000	1000
(4)	212	53	2000	2000
(5)	424	26.5	4000	4000

As can be seen from Table 18-3, there are many combinations of L and C values which will produce a resonance frequency of 1500 kHz, even though for each combination there is only one resonance frequency.

The combination of L in μH and C in pF in Eq. (3) produces an L/C ratio of

$$L/C = 106/106 = 1 \quad \text{(disregard } \mu H \text{ and pF values)}$$

The combination of L in μH and C in pF in Eq. (5) produces an L/C ratio of

$$L/C = 424/26.5 = 16/1$$

The L and C combination in Eq. (5) (Table 18-3) will produce the largest Q, if the dc resistance has not increased. This is because the X_L is larger in Eq. 5 than in Eq. 1. The X_L in Eq. 5 is 4000 Ω but is only 1000 Ω in Eq. 3. This means that the X_L is four times larger in Eq. 5 than the X_L in Eq. 3. The result is that the $Q(X_L/R)$ of the coil in Eq. 5 will be 4 times larger than that in Eq. 3 if the dc resistance of the 424-μH coil is kept at the same value as in the 106-μH coil. Theoretically the Q of a coil could be made to approach infinity by increasing the value of inductance with respect to the capacitance. In practical situations, however, a value of inductance is reached beyond which the Q can no longer be increased, because the dc resistance of the wire increases in proportion to the inductance increase. The optimum value of X_L for best selectivity in radio frequencies is 1000 Ω.

Figure 18-7 shows the frequency response curves for three different L/C ratios. Curve C has the highest L/C ratio and is the most selective.

18-2.5   Total Q of a Circuit

It has been shown that the Q of a tuned circuit is affected by

1. Ratio of X_L/R
2. The L/C ratio of the capacitor and coil of the tuned circuit

The formula for the total Q of a tuned circuit incorporates the L/C ratio and X_L/R ratio equations.

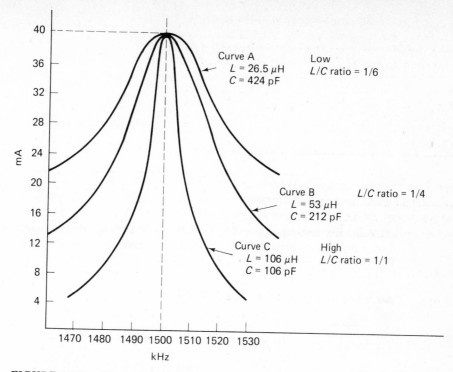

FIGURE 18-7 Series-resonant curves showing the effect of L/C ratios on selectivity.

DERIVING THE FORMULA FOR TOTAL Q OF A TUNED CIRCUIT

(Incorporating L/C ratio and X_L/R ratio)

$$Q = \frac{X_L}{R} = \frac{2\pi f_R\, L}{R}$$

$$\text{but}\quad f_R = \frac{1}{2\pi\sqrt{LC}}$$

$$\therefore Q = \frac{2\pi \left(\dfrac{1}{2\pi\sqrt{LC}}\right) L}{R}$$

$$= \frac{L}{\sqrt{LC}} \times \frac{1}{R}$$

Square both sides of the equation and

$$Q^2 = \frac{\overset{L}{\cancel{L^2}}}{\cancel{L}C} \times \frac{1}{R^2}$$

Square root both sides and

$$Q = \sqrt{\frac{L}{C} \times \frac{1}{R}}$$

$$Q = \frac{1}{R} \sqrt{\frac{L}{C}}$$

where R is in ohms
L is in henries
C is in farads

18-2.6 Q and Voltage Relationship

The Q of a circuit has been defined as the ratio of the reactive power to the dissipated power. A relationship between Q and voltage can be derived from this definition and stated in the form of an equation.

DERIVING THE FORMULA FOR Q AND VOLTAGE

Reactive power is $I_a V_L$. (6)

Power dissipation is $I_a V_R$. (7)

$$Q = \frac{\text{reactive power}}{\text{dissipated power}} = \frac{I_a V_L}{I_a V_R} \qquad (8)$$

The currents cancel and Q becomes

$$Q = \frac{V_L}{V_R} \qquad (9)$$

At resonance $V_R = V_a$. (10)

Substitute Eq. (10) in Eq. (9) and

$$Q = \frac{V_L}{V_a} \qquad (11)$$

At resonance, $V_L = V_C$. (12)

Substitute Eq. (12) in Eq. (11) and

$$Q = \frac{V_C}{V_a} \qquad (13)$$

If the value of V_a and Q are known, then

$$V_L = QV_a \quad \text{or} \quad Q = \frac{V_L}{V_a} \qquad (14)$$

$$\text{and} \quad V_C = QV_a \quad \text{or} \quad Q = \frac{V_C}{V_a} \qquad (15)$$

Ch. 18 Resonant Circuits

SUMMARY:

The purpose for which a circuit is used will determine the required *L/C* ratio of the two components. Series tuned (resonant) circuits are generally used in applications requiring high selectivity (steep slopes). Therefore, they usually have a high *L/C* ratio.

Typical *Q* values of series resonant circuits range from 100 to 300.

18-2.7 Bandwidth (Δf) (Sometimes Called Bandpass)

The frequency response curve, Fig. 18-8, of a series resonant circuit shows maximum current at resonance. Maximum power ($I^2 \times R$) is also developed at resonance. Figure 18-8 shows that the current and power diminish at frequencies below and above resonance. The frequencies above and below resonance at which the current is 0.707 of the I_p (I_{max}) current are those points at which *bandwidth* (BW) is measured.

Bandwidth is the difference in frequency between the two frequency points on an *L-C-R* resonance response curve at which the current is 70.7% of the resonance frequency current.

bandwidth (BW) (Δf) = $f_2 - f_1$

where   Bw is the frequency difference in hertz

f_1 is the frequency below resonance at which the current is $0.707\, I_{max}$.

f_2 is the frequency above resonance at which the current is $0.707\, I_{max}$.

18-2.8 Bandwidth Measurement of a Response Curve

To determine the BW of Curve A in Fig. 18-8:

1. Determine value of the maximum current (I_p).
2. Calculate the value of $0.707\, I_{max}$.
3. From the $0.707\, I_{max}$ current value, draw a horizontal line to cross both sides of the response curve.
4. From the two points on the response curve where the horizontal line crosses the curve, draw a vertical line to the frequency coordinate.
5. Read the frequency value where the vertical lines cross.
6. The bandwidth (BW or Δf) is

$$f_2 - f_1 = 1505 - 1495$$

$$= 10 \text{ kHz}$$

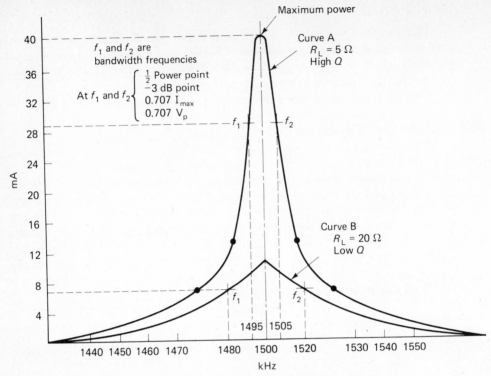

FIGURE 18-8 Shows bandwidth measurements on a series-resonant response curve. Note the name of the points where bandwidth is measured.

The Δf of curve B is determined in the same manner. Curve B bandwidth (BW) is

$$\Delta f = f_2 - f_1$$

$$= 1520 - 1480$$

$$= 40 \text{ kHz}$$

18-2.9 Bandwidth, $\frac{1}{2}$ Power and – 3 dB

The frequencies f_1 and f_2 at which the current is 0.707 $I_{\max}$ are also known as $\frac{1}{2}$ power points. (See Fig. 18-8.) The $\frac{1}{2}$ power point and – 3 dB values are related to bandwidth as shown in the following equations.

DERIVING $\frac{1}{2}$ POWER POINT AND – 3 dB FOR BANDWIDTH FREQUENCIES

$$P_{1(\max)} = I_{(\max)}^2 \times R \quad (R = 1 \text{ as reference})$$

$$\text{but at } f_1, \quad I = 0.707\, I_{\max}$$

Therefore at f_1

$$P_2 = (0.707\, I_{\max})^2 \times 1$$

$$= 0.5 \times 1 = 0.5$$

The $\frac{1}{2}$ power point can also be stated as -3 dB down, from the peak current (at resonance).

$$\therefore dB = 10 \log \frac{P_2}{P_1}$$

$$= 10 \log \frac{0.5}{1}$$

$$= 10 \times -0.3010$$

$$= -3$$

18-2.10 Bandwidth Frequencies and Impedance

The current at the frequencies at which bandwidth is measured is 0.707 times the current at resonance. Note that current is inversely proportional to impedance: $I_a = 1/Z$.

EXAMPLE

$$\text{from} \quad I_a = \frac{1}{Z} \tag{16}$$

$$\text{then} \quad Z = \frac{1}{I_a} \tag{17}$$

but at BW frequency $I_a = 0.707$. $\qquad (18)$

Substitute Eq. (18) in Eq. (17) and

$$Z = \frac{1}{0.707} \tag{19}$$

Therefore, at bandwidth frequencies f_1 and f_2,

$$Z = 1.414 \, Z \text{ at } f_R \tag{20}$$

$$\text{but} \quad Z = R \text{ at } f_r \tag{21}$$

Therefore

$$Z = 1.414 \, R \text{ at the bandwidth frequencies}$$

$$\text{or} \quad Z = 1.414 \, R \text{ at } f_1 \text{ and } f_2$$

18-2.11 Relationship Between X_T and R at Bandwidth Frequencies

At bandwidth frequencies, the impedance is 1.414 times the resistance of the circuit. This is because the reactances no longer cancel. The difference between X_L and X_C adds vectorially to the resistance value. The total reactance at bandwidth frequencies is equal to the circuit resistance ($X_T = R$). The proof of this statement is derived as follows:

PROOF

At either BW frequency

$$Z = 1.414\, R \tag{22}$$

$$\text{where } R \text{ is } 1 \tag{23}$$

$$\text{but } Z = \sqrt{R^2 + X_T^2} \tag{24}$$

From Eq. (24)

$$Z^2 = R^2 + X_T^2 \tag{25}$$

and

$$X_T^2 = Z^2 - R^2 \tag{26}$$

$$X_T = \sqrt{Z^2 - R^2} \tag{27}$$

Substitute Eq. (22) and Eq. (23) in Eq. (27) and

$$X_T = \sqrt{1.414^2 - 1^2} \tag{28}$$

$$= \sqrt{2 - 1} \tag{29}$$

$$X_T = 1 \tag{30}$$

$X_T = R$ at either bandwidth frequency.

$$\text{but } R = 1 \quad \text{(reference)}$$

$$\therefore X_T = R = 1 \quad \text{(reference)} \tag{31}$$

18-2.12 Phasor Diagram, X_T, R, and Z

The phasor diagrams, Figs. 18-9a and 18-9b, are another way of showing that the Z at BW frequencies is 1.414 times the impedance at resonance.

At f_1 or f_2

$$X_T = (X_{C_1} - X_{L_1}) = R \tag{32}$$

(See Sec. 18-2.11.)

Draw the phasor diagram as shown in Fig. 18-9a.

$$Z = \sqrt{R^2 + X_T^2} \tag{33}$$

but at either f_1 or f_2

$$X_T = R = 1 \quad \text{(reference)} \tag{34}$$

Substitute Eq. (34) in Eq. (33) and

$$Z = \sqrt{1^2 + 1^2} \tag{35}$$

$$= \sqrt{2} \tag{36}$$

$$= 1.414\, R \tag{37}$$

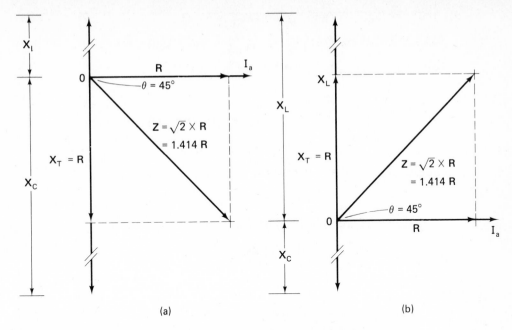

(a)   (b)

FIGURE 18-9   Phasor diagram for reactances and circuit resistance at (a) frequencies below resonance where bandwidth is measured (at 0.707 on the curve). (b) is for frequencies above resonance where bandwidth is measured.

At the frequencies where bandwidth is measured,

$$Z = 1.414\,R$$

18-2.13   Characteristics of a Series Resonance Circuit at BW Frequency

The characteristics of a series resonance circuit at the bandwidth frequencies are now known.

At f_1 or f_2 :

1. $X_T = (X_C - X_L) = R$
2. Current is $0.707\,I_{max}$
3. Power is at $\frac{1}{2}$ power ($\frac{1}{2}$ power point)
4. Amplitude of curve is -3 dB loss
5. The phase angle between the V_a and I_a is $45°$

None of these characteristics provide a convenient method of calculating the bandwidth when the Q of the circuit and the resonance frequency are known. The following equations show this relationship:

DERIVING THE RELATIONSHIP BETWEEN BW, f_R, AND Q

$$BW = (f_2 - f_1) \tag{38}$$

$$\text{and} \quad X_T = (X_{C_1} - X_{L_1}) \tag{39}$$

$$\text{At } f_1, \quad (X_{C_1} = X_{L_2}) \quad \text{(see Table 18-1)} \tag{40}$$

Substitute X_{L_2} in Eq. (40) for X_{C_1} in Eq. (39)

$$\text{but} \quad X_T = (X_{L_2} - X_{L_1}) \tag{41}$$

$$= (2\pi f_2 L) - (2\pi f_1 L) \tag{42}$$

$$\therefore X_T = 2\pi L (f_2 - f_1) \tag{43}$$

Substitute R for X_T, because $R = X_T$ at f_1 or f_2.

$$\text{then} \quad R = 2\pi L (f_2 - f_1) \tag{44}$$

Substitute BW for $(f_2 - f_1)$ in Eq. (44).

$$\text{and} \quad R = 2\pi L\, \text{BW} \tag{45}$$

$$\text{or} \quad \text{BW} = \frac{R}{2\pi L} \tag{46}$$

Divide both sides by f_R and

$$\frac{\text{BW}}{f_R} = \frac{\dfrac{R}{2\pi L}}{f_R} \tag{47}$$

$$\text{or} \quad \frac{\text{BW}}{f_R} = \frac{R}{2\pi L} \times \frac{1}{f_R} \tag{48}$$

$$\text{and} \quad \frac{\text{BW}}{f_R} = \frac{R}{X_L} = \frac{1}{Q} \tag{49}$$

$$\therefore \text{BW} = \frac{f_R}{Q} \tag{50}$$

$$\text{bandwidth} = \frac{\text{resonant frequency}}{Q}$$

$$\text{BW} = \frac{f_R}{Q}$$

where BW is bandwidth in hertz
f_R is the resonance frequency
Q is factor of merit

also from Eq. (46) $\text{BW} = \dfrac{R}{2\pi L}$

where R is resistance of the coil

PROBLEM 18-2

The resonant frequency of a series circuit is 1500 kHz, at which the X_L of the coil is 1000 Ω.

1. Calculate the Q of the coil and the bandwidth when the resistance of the coil is 5 Ω.
2. Calculate the Q of the coil and the bandwidth when the resistance of the coil is 20 Ω.

SOLUTION   (cover solution and solve)

1. When resistance of the coil is 5 Ω

$$Q = \frac{X_L}{R} = \frac{1000}{5} = 200$$

$$\text{but} \quad BW = \frac{f_R}{Q} = \frac{f_R}{\dfrac{X_L}{R}}$$

$$= \frac{1500}{200} = 7.5 \text{ kHz}$$

2. When coil resistance is 20 Ω

$$Q = \frac{X_L}{R} = \frac{1000}{20} = 50$$

$$\therefore BW = \frac{f_R}{Q} = \frac{1500}{50} = 30 \text{ kHz}$$

PROBLEM 18-3

A series resonant antenna, consisting of a 30-μH coil, a 5-Ω resistance, and a 360-pF capacitor has a 250-mV radio signal applied to it.

1. Draw the circuit.
2. Calculate the resonant frequency (f_R).
3. Calculate the total Q of the circuit.
4. Calculate the circuit current at resonance.
5. Determine the voltage across the coil at resonance.
6. Determine the bandwidth.
7. Will the circuit have good selectivity? Give your reason.
8. State two methods of increasing the selectivity.

SOLUTION   (cover solution and solve)

1. See Fig. 18-10.

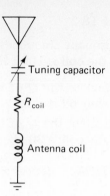

FIGURE 18-10

2. $$f_R = \frac{1}{2\pi\sqrt{LC}}$$

$$= \frac{1}{6.28 \times (30 \times 10^{-6}) \times (360 \times 10^{-12})}$$

$$= \frac{1}{6.28 \times 103.9 \times 10^{-9}}$$

$$= \frac{10^9}{652.49}$$

$$= 1532 \text{ kHz}$$

3. $$\text{total } Q = \sqrt{\frac{L}{C}} \times \frac{1}{R}$$

$$= \sqrt{\frac{30 \times 10^{-6}}{360 \times 10^{-12}}} \times \frac{1}{5}$$

$$= \sqrt{\frac{30 \times 10^6}{360}} \times \frac{1}{5}$$

$$= \frac{288}{5} = 57.7$$

4. Circuit current at f_R

$$I_a = \frac{V_a}{R} = \frac{250 \times 10^{-3}}{5} = 50 \text{ mA}$$

5. Voltage across coil

$$\text{from} \quad Q = \frac{V_L}{V_a}$$

$$V_L = QV_a$$

$$= 57.7 \times 250 \times 10^{-3}$$

$$= 14400 \times 10^{-3} \text{ V}$$

$$= 14.4 \text{ V}$$

6. $$\mathrm{BW} = \frac{f_\mathrm{R}}{Q} = \frac{1532 \times 10^3}{57} = 27 \text{ kHz}$$

7. The circuit has poor selectivity because the minimum Q for good selectivity is 100.

8. a. Use a larger inductance and smaller capacitor.
   b. Use a coil wound with larger diameter (thus lower resistance) wire.

PROBLEM 18-4

A series tuned trap consists of a variable capacitor, with a range of 40 pF to 375 pF, connected in series with a 240-μH coil which has a dc resistance of 10 Ω. The signals applied have a voltage of 8 mV.

1. Find the frequency range over which the trap is tunable.
2. Find the total Q of the circuit at each end of the frequency range.
3. Find the impedance of the circuit at resonance.
4. Calculate the voltage across the coil at the high end of the tuning range.
5. Find the bandwidth (BW) at the high end.
6. Does the largest or smallest value of capacitance tune in the highest frequency?
7. Will the variable capacitor be fully open or fully closed at the highest frequency?
8. What end of the tuning range will have the highest selectivity?

SOLUTION   *(cover solution and solve)*

1. f_R at high-frequency end

$$f_\mathrm{R} = \frac{1}{2\pi\sqrt{LC}}$$

$$= \frac{1}{6.28 \times \sqrt{240 \times 10^{-6} \times 40 \times 10^{-12}}}$$

$$= \frac{1}{6.28 \times 97.98 \times 10^{-9}}$$

$$= 1625 \text{ kHz}$$

f_R at the low-frequency end

$$f_\mathrm{R} = \frac{1}{2\pi\sqrt{LC}}$$

$$= \frac{1}{6.28 \times \sqrt{240 \times 10^{-6} \times 375 \times 10^{-12}}}$$

$$= \frac{1}{6.28 \times 300 \times 10^{-9}}$$

$$= 530 \text{ kHz}$$

2. Q at the high end

$$Q = \frac{1}{R} \sqrt{\frac{L}{C}}$$

$$= \frac{1}{10} \times \sqrt{\frac{240 \times 10^{-6}}{40 \times 10^{-12}}}$$

$$= \frac{1}{10} \times 2450 = 245$$

Q at the low end

$$Q = \frac{1}{R} \sqrt{\frac{L}{C}}$$

$$= \frac{1}{10} \times \sqrt{\frac{240 \times 10^{-6}}{375 \times 10^{-12}}}$$

$$= \frac{1}{10} \times 800 = 80$$

3. Z at f_R

$$Z = R = 10 \ \Omega$$

4. V_L at high end

$$V_L = QV_a$$

$$= 245 \times 8 \times 10^{-3}$$

$$= 1.96 \text{ V}$$

5. $\text{BW} = \dfrac{f_R}{Q} \dfrac{\text{(at high end)}}{\text{(at high end)}}$

$$= \frac{1625 \times 10^3}{245}$$

$$= 6.63 \text{ kHz}$$

6. The smallest capacitance tunes in the highest frequency.
7. The capacitor will be fully open at the high frequency end.
8. The high end has the highest selectivity.

Series resonant circuits are generally used as acceptor traps when it is necessary to short out one particular frequency to ground. The short will not be present at frequencies on either side of the resonant frequency, because the reactance values will change and will no longer cancel. In other applications, the very large voltage developed across either the coil or the capacitor may be removed and used to operate a transistor or vacuum tube.

18-3 Parallel Resonance

When a coil and a capacitor are connected in parallel, there will be a frequency at which the reactances are equal and cancel. At that frequency, the circuit is said to be resonant, but sometimes it is referred to as an anti-resonant circuit because the impedance across the circuit is maximum. The parallel resonant circuit is also referred to as a *tank circuit*.

The resonance frequency of a practical parallel resonant circuit, if coil resistance is small, is determined by

$$f_R = \frac{1}{2\pi\sqrt[2]{LC}}$$

18-3.1 Z of a Theoretical Parallel Resonant Circuit

In a hypothetically pure *L-C* parallel resonant circuit, there will be zero resistance in the circuit, and the only current flow will be I_C and I_L. The capacity current (I_C) will be $180°$ out of phase with I_L and the two currents will cancel. The total current therefore will be zero as shown in the following example.

EXAMPLE ————————————————————————————————

Currents in a theoretical parallel resonant circuit (Figs. 18-11 and 18-12).

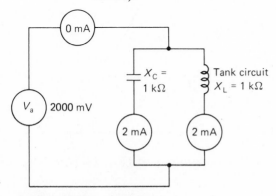

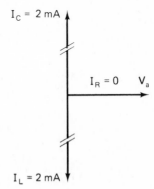

FIGURE 18-11 Currents in a theoretical parallel-resonant circuit. Note that there is no coil resistance.

FIGURE 18-12 Phasor diagram of a theoretical parallel-resonant circuit.

$$I_C = \frac{V_a}{X_C} = \frac{2000 \times 10^{-3}}{1 \times 10^3} = 2 \text{ mA}$$

$$I_L = \frac{V_a}{X_L} = \frac{2000 \times 10^{-3}}{1 \times 10^3} = 2 \text{ mA}$$

$$\therefore I_T = I_a = I_C - I_L$$

$$= 2 \text{ mA} - 2 \text{ mA} = 0 \text{ mA}$$

Since the line current in a theoretical parallel resonant circuit is zero, the impedance must be infinite.

$$Z = \frac{V_a}{I_a}$$

$$\text{but } I_a = 0$$

$$\therefore Z = \frac{V_a}{0} = \infty$$

The relationship between impedance and frequency in a theoretical parallel resonant circuit can be plotted on a graph as shown in Fig. 18-13.

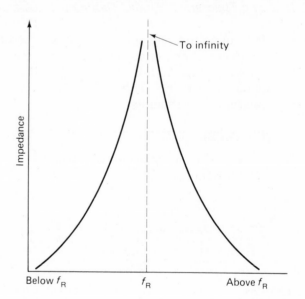

FIGURE 18-13 Impedance curve of a theoretical ideal parallel-resonant circuit.

PROBLEM 18-5

Calculate for a theoretical parallel resonant circuit (Fig. 18-14):

1. Resonant frequency
2. Inductive current
3. Capacitive current

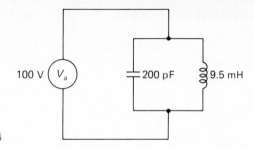

FIGURE 18-14

4. Tank current
5. Source current

SOLUTION   *(cover solution and solve)*

1. Resonant frequency f_R

$$f_R = \frac{1}{2\pi\sqrt[2]{LC}}$$

$$= \frac{1}{6.28 \times \sqrt[2]{9.5 \times 10^{-3} \times 200 \times 10^{-12}}}$$

$$= \frac{1}{6.28 \times \sqrt[2]{1.9 \times 10^{-12}}}$$

$$= \frac{1}{6.28 \times 1.378 \times 10^{-6}}$$

$$= 115.55 \text{ kHz}$$

2. $$I_L = \frac{V_a}{X_L}$$

but $X_L = 2\pi f L$

$$= 6.28 \times 115.5 \times 10^3 \times 9.5 \times 10^{-3}$$

$$= 6893 \ \Omega$$

$$\therefore I_L = \frac{V_a}{X_L} = \frac{100}{6893} = 14.5 \text{ mA} = -J14.5 \text{ mA}$$

3. $$I_C = I_L = 14.5 \text{ mA} = +J14.5 \text{ mA}$$

(+ sign because I_C is $180°$ from I_L)

4. Tank current (circulating current)

$$I_L = I_C = 14.5 \text{ mA}$$

5. Source current (line current)

$$I_a = I_L - I_C \quad (I_C \text{ is } 180° \text{ from } I_L)$$

$$= 14.5 - 14.5 = 0 \text{ mA}$$

18-3.2 Operation of a Practical Parallel Resonant Circuit

The hypothetical parallel resonant circuit does not exist, because coils are wound with wire and wire has dc resistance. The practical resonant circuit is depicted in Fig. 18-15.

The parallel resonant circuit is often called a tank circuit because the current in the charged capacitor discharges into the coil, where the energy is stored until the capacitor is discharged. The electromagnetic energy stored in the coil then collapses and charges the capacitor. This back and forth flow of current between the capacitor and inductor is called the fly-wheel effect or oscillation. It takes place once every complete cycle; thus, the L-C parallel resonant circuit acts as a tank to store electrical energy.

The above description may sound like perpetual motion, but in a practical situation coils have resistance, and current flow through the resistance produces an I^2R heat loss, resulting in energy loss and a cessation of the circulation. The purpose of the supply voltage (V_a) is first to charge the capacitor which starts the oscillation in motion, and then to replace the small energy loss taking place in the coil resistance. The higher the resistance of the coil, the greater the energy loss and the greater the amount of energy demanded from the source voltage. The source voltage need not be sinusoidal, because it replaces the lost energy only in spurts. The operation of the tank circuit, however, produces a sinusoidal voltage waveform across each component and thus across itself.

18-3.3 Current and Impedance in a Practical Parallel Resonant Circuit

The presence of the resistance in the coil in the practical parallel resonant circuit in Fig. 18-15 causes the current through the coil to lag the voltage by less than 90° at resonance, as shown in the phasor diagram in Fig. 18-16. The result is that the capacitor current can never be 180° out of phase with the inductive current so they cannot cancel one another. Since the two currents do not cancel, some small amount of current (line current) will flow fɪ   the generator or voltage source, indicating that the

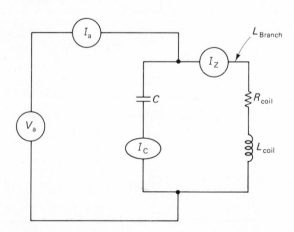

FIGURE 18-15   The practical par-allel-resonant circuit (tank circuit).

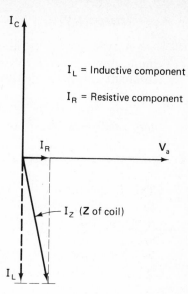

I_L = Inductive component

I_R = Resistive component

I_Z (**Z** of coil)

FIGURE 18-16 Phasor diagram for a practical parallel-resonant circuit. The resistance of the coil shifts the I_L phasor arm away from $180°$.

impedance of the parallel resonant circuit is extremely large. (See Tables 18-4 and 18-5.)

Table 18-4 The Effect of Frequency on Impedance and Current in a Parallel L-C-R Circuit when L = 106 μH, C = 106 pF, and Coil Resistance = 5 Ω

f kHz	X_L Ω	X_C Ω	R_C Ω	Z Ω	I_a when V_a = 200 mV mA
1470	980	1020	5	25000	8.0
1480	986	1012	5	38000	5.26
1490	992	1006	5	67000	2.98
(f_R) 1500	1000	1000	5	200000	1
1510	1006	992	5	67000	2.98
1520	1012	986	5	38000	5.26
1530	1020	980	5	25000	8.00

Table 18-5 When R_L = 10 Ω

f kHz	X_L Ω	X_C Ω	R_C Ω	Z Ω	I_a when V_a = 200 mV mA
1470	980	1020	10	24250	8.25
1480	986	1012	10	36000	5.55
1490	992	1006	10	58000	3.45
(f_R) 1500	1000	1000	10	100000	2
1510	1006	992	10	58000	3.45
1520	1012	986	10	36000	5.55
1530	1020	980	10	24250	8.25

SUMMARY:

At resonance, the line current to a parallel resonant circuit is minimum, but the tank current is maximum. The impedance of a parallel resonant circuit is maximum at resonance but it is not infinite (see Tables 18-4 and 18-5).

Note the drastic reduction in circuit impedance at resonance with a very small increase of coil resistance.

18-3.4 Currents and Impedance at Frequencies Above and Below Resonance

The frequency of the signal increases above resonance. Therefore, X_L ($2\pi fL$) will increase and the I_L will decrease. The X_C, however, will decrease and the I_C will become greater. The result is:

1. A decrease in tank current flow (not shown in Tables 18-4 or 18-5)
2. A decrease in the tank circuit impedance
3. An increase in the line current flow (I_a) (see Tables 18-4 and 18-5)

The parallel circuit is said to be capacitive above resonance because the line current (I_a) is predominantly capacitive current.

The frequency of the signal is lower below resonance. Therefore, the capacitive reactance becomes larger than the inductive reactance. The line current increases (see Tables 18-4 and 18-5) and the circuit is said to be inductive below resonance because the line current consists predominantly of inductive current.

A graph plotting the value of line current against the resonant frequency and the frequencies above and below resonance will produce a curve as shown in Fig. 18-17.

A plot of the resonant frequency and the frequencies above and below resonance against impedance will produce the curve shown in Fig. 18-18.

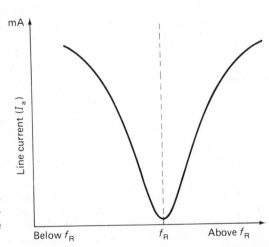

FIGURE 18-17 Line current response curve of a practical parallel-resonant circuit. Note that the line current does not go quite to zero.

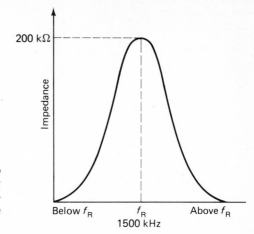

FIGURE 18-18 The relationship between impedance and frequency for a practical parallel-resonant circuit (from values given in Table 18-4; $R_{coil} = 5\ \Omega$).

18-3.5 Impedance Formula for a Parallel Resonant Circuit

The impedance at resonance for a parallel tank circuit may be calculated by the following formula:

$$\text{impedance} = \frac{\text{inductance}}{\text{capacity} \times \text{resistance}}$$

$$Z = \frac{L}{C \times R}$$

The equation is derived with the use of the phasor diagram shown in Fig. 18-20. The phasor diagram in Fig. 18-20 is developed by first drawing the phasor diagram for the inductive branch of the tank circuit (Fig. 18-19).

The current is common to the coil resistance and the inductance in the inductive branch; therefore, the current is placed on the reference arm and the V_L and V_R phasor arms are drawn with respect to the current I_Z (Fig. 18-19). The resultant vector diagram in Fig. 18-19 shows the supply voltage leading the inductive branch current (I_Z) by less than 90°. The

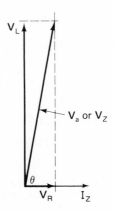

FIGURE 18-19 The phasor diagram for the inductive branch only of the parallel-resonant circuit shown in Fig. 18-15.

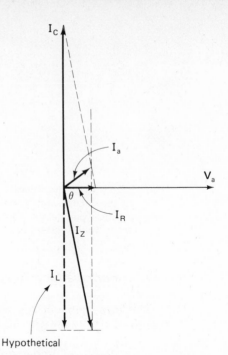

FIGURE 18-20   Accurate phasor diagram of parallel-resonant circuit. (This is an accurate but seldom considered diagram.)

Hypothetical

supply voltage is common to both the capacitive leg and the inductive leg of the tank circuit. The supply voltage, therefore, is placed on the reference arm when drawing the final vector diagram for the tank circuit (Fig. 18-20). The inductive branch current (I_Z) in turn will lag the supply voltage by less than 90°. The capacitive current, however, still leads the supply voltage by 90°. The resultant phase relationship is shown in Fig. 18-20. The phasor diagram in Fig. 18-20 may now be used to assist in deriving the equation for the impedance of a tank circuit.

PROBLEM 18-6

A parallel circuit consists of a 12-μH inductance, a coil resistance of 8 Ω and a 180-pF capacitor connected across a 400-mV signal.

1. Draw the circuit and label the values.
2. Calculate the resonant frequency (f_R).
3. Calculate the impedance (Z) of the circuit.
4. Calculate the line current at resonance (I_a).
5. Determine the capacitive current at resonance (I_C).
6. Determine the inductive current at resonance (I_Z).
7. Compare the line current and tank current values.
8. Compare impedance and reactance values.
9. What is the relationship between the quantities in 7 compared with those in 8?

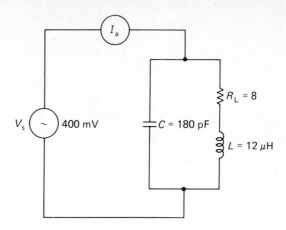

FIGURE 18-21

SOLUTION   (cover solution and solve)

1. See Fig. 18-21.

2. $$f_R = \frac{1}{2\pi\sqrt[2]{LC}}$$

$$= \frac{1}{6.28 \times \sqrt[2]{12 \times 10^{-6} \times 180 \times 10^{-12}}}$$

$$= \frac{1}{6.28 \times \sqrt[2]{2160 \times 10^{-18}}}$$

$$= \frac{1}{6.28 \times 46.48 \times 10^{-9}}$$

$$= 3426 \text{ kHz}$$

3. $$Z = \frac{L}{CR}$$

$$= \frac{12 \times 10^{-6}}{180 \times 10^{-12} \times 8} = \frac{12 \times 10^{6}}{1440}$$

$$= 8333 \ \Omega$$

4. Line current at resonance

$$I_a = \frac{V_a}{Z}$$

$$= \frac{400 \times 10^{-3}}{8333} = 0.048 \text{ mA} = 48 \ \mu A$$

5. Capacity current (I_C)

$$I_C = \frac{V_a}{X_C}$$

$$\text{but } X_C = \frac{1}{2\pi f C}$$

therefore

$$X_C = \frac{1}{6.28 \times 3426 \times 10^3 \times 180 \times 10^{-12}}$$

$$= 258 \ \Omega$$

and $\quad I_C = \dfrac{400 \times 10^{-3}}{258} = 1.55 \ \text{mA}$

6. Inductive current in coil (I_Z)

coil impedance $(Z) = \sqrt[2]{R^2 + X_L^2}$

$$X_L = 2\pi f L$$

$$= 6.28 \times 3426 \times 10^3 \times 12 \times 10^{-6}$$

$$= 258 \ \Omega$$

At resonance $X_L \cong X_C$ but the coil resistance affects the current value in the coil.

coil impedance $(Z) = \sqrt[2]{R^2 + X_L^2}$

$$= \sqrt[2]{8^2 + 258^2}$$

$$= \sqrt[2]{64 + 66{,}564}$$

$$= 258.124 \ \Omega$$

inductive branch current $I_Z = \dfrac{V_a}{Z_L} = \dfrac{400 \times 10^{-3}}{258.1}$

$$= 1.549 \ \text{mA} \quad \text{(slightly smaller than } I_C)$$

When the coil resistance is small, the I_Z current is nearly that of the capacitor.

7. The line current is 48 μA (should be $I_C - I_Z$). The capacity (circulating) current is 1.55 mA. The I_a/I_C current ratio is $(1.55 \times 10^{-3})/(0.048 \times 10^{-3}) = 32$. The tank current is 32 times larger than the line current.

8. The Z is 8333 Ω.

The $X_L \cong X_C = 258 \ \Omega$.

The Z/X_C ratio is $\dfrac{8333}{258} = 32$.

9. The impedance is 32 times larger than the reactance which accounts for the tank current being 32 times larger than the line current.

18-3.6 Characteristics of Parallel L-C Circuits at Resonance

The characteristics of L-C parallel circuits can now be listed from the information obtained from the preceding discussions.

1. $X_L = X_C$. Therefore, the tank circuit, internally, is resistive.
2. $I_L = I_C$ but they are 180° out of phase.
3. The I_C and I_L are called the tank current because tank current circulates between the capacitance and the inductive branches.
4. The I_a is smallest at resonance but increases in value as the resistance of the coil (R_C) increases (Tables 18-4 and 18-5).
5. The Z approaches infinity at resonance as the coil resistance (R_C) decreases in value (Tables 18-4 and 18-5).
6. The values of X_L and X_C are affected by the frequency in the same manner as in series circuits (Tables 18-4 and 18-5).
7.
$$f_R = \frac{1}{2\pi\sqrt[2]{LC}} \quad \text{(accepted equation)}$$

where L is in henries
        C is in farads
        f_R is in hertz

8. At resonance

$$Z = \frac{L}{CR}$$

where Z is impedance in ohms
        L is inductance in henries
        C is capacitance in farads
        R is resistance of coil in ohms

18-3.7 Selectivity

Selectivity is the ability of a circuit to respond to the resonant frequency and to exclude all others.

In parallel resonant circuits, selectivity may also be defined as the ability of the circuit to offer a large impedance to the resonant frequency but a small impedance to other frequencies.

The Q of the circuit in parallel L-C circuits determines the bandwidth and thus the selectivity of the circuit as it did in series circuits. The Q is also affected by the L/C ratio and the dc resistance of the circuit as it is in series circuits.

18-3.8 Factors Affecting the Selectivity of Parallel Circuits

Q OF THE COIL

The Q of the coil in a parallel circuit is exactly the same as in a series circuit.

$$Q = \frac{X_L}{R}$$

L/C RATIO

The *L/C* ratio requirement for parallel circuits is the same as that for series circuits. Parallel circuits require a high *L/C* ratio for good selectivity. This is because X_L increases as L increases if the frequency is kept constant. Increasing X_L with respect to the resistance of the circuit increases the impedance. Good selectivity in parallel circuits requires high impedance (high *L/C* ratio). (See Fig. 18-22.)

TOTAL Q OF A PARALLEL RESONANT CIRCUIT

The formula for Q, incorporating the *L/C* ratio and X_L/R is the same as for the series circuits.

$$Q = \frac{1}{R} \sqrt[2]{\frac{L}{C}}$$

where R is the series resistance in the inductive branch in ohms

L is inductance in henries

C is capacity in farads

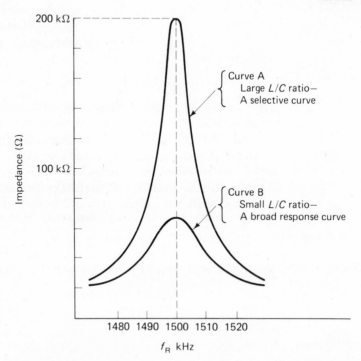

FIGURE 18-22 Frequency-response curves for different relationships between *L/C* ratio and selectivity.

The value of the Q of the circuit will be higher at the high frequencies than it will be at the low frequencies if L is kept constant, as is the case in radio receiver tuning circuits.

DAMPING (LOADING) RESISTOR

The equation X_L/R for Q includes only the resistance in series with the coil. Often it is necessary to place a resistor in parallel with the coil in order to decrease the Q of the circuit still further and thus increase its bandwidth. The equation for Q which includes the parallel resistor is much different from the equation for Q which includes the series resistor.

The resistor added in parallel with an L-C tank circuit decreases the Q and increases the bandwidth of the tank circuit for the following reasons. The line current to a parallel resonant circuit increases when a resistor is connected in parallel with the tank circuit. This increase in current occurs because the current has an added path in which to flow. If the resistance of the shunt resistor is less than the Z of the tank circuit (before the resistor was added) the total current value will be predominantly resistive. The resistance of a resistor does not change with frequency (resistors are not frequency conscious). Therefore, the impedance and the voltage developed across a tank circuit with a shunt resistor connected across it will remain at approximately the same value regardless of the frequency. The result is a response curve with a wide bandwidth. Since the band-

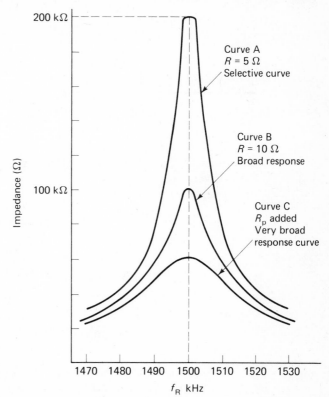

FIGURE 18-23 Three response curves showing the effect of coil resistance and shunt resistance on selectivity.

width increases as the value of the shunt resistor decreases, the Q of the circuit must be directly proportional to the value of R_p. (See Curve C in Fig. 18-23.)

high value shunt resistor—high Q

low value shunt resistor—low Q

DERIVING THE FORMULA FOR $Q = R_p/X_L$

The voltage is common in a parallel circuit. Therefore, the Q can be defined as the ratio of the inductive current (I_L) to the resistive current (I_R) through the shunt resistor.

$$Q = \frac{I_L}{I_R} = \frac{\left(\dfrac{V}{X_L}\right)}{\left(\dfrac{V}{R_p}\right)} = \frac{\cancel{V}}{X_L} \times \frac{R_p}{\cancel{V}}$$

The voltages cancel and

$$Q = \frac{R_p}{X_L}$$

$$Q = \frac{\text{shunt resistance}}{\text{inductive reactance}}$$

or $\quad Q = \dfrac{R_p}{X_L}$

18-3.9 Relationship between Q, X, Z, and Currents

The Q of parallel resonant circuits determines the Z of the circuit, and since selectivity of a parallel circuit is dependent upon the impedance, the Q is very important.

At resonance:

$$\text{impedance} = Q \times \text{inductive reactance}$$

$$Z = QX_L$$

or $\quad \text{impedance} = Q \times \text{capacitive reactance}$

$$Z = QX_C$$

and $\quad \text{inductive reactance} = \dfrac{\text{impedance}}{Q}$

$$X_L = \frac{Z}{Q}$$

18-3.10 Relationship Between I_{tank}, V_a, and Q

The current in a parallel resonant tank circuit is also dependent upon the Q of the circuit, and it increases in direct proportion to the Q of the circuit.

$$I_{circulating} \text{ (tank)} = Q \times \text{applied current}$$

$$I_{cir} = QI_a$$

18-3.11 Parallel Resonant Response Curve

Figure 18-24 shows the frequency response curve as may be seen on an oscilloscope of a parallel resonant circuit. Note that voltage is plotted against frequency. It may be difficult to understand how the voltage across a parallel circuit can change when the circuit is connected in parallel with a source of voltage. In practice, the parallel resonant circuit is not connected in parallel with the source of voltage as indicated in previous discussions, but is actually connected in series with the voltage source and its internal resistance (internal resistance of a transistor) as shown in Fig. 18-25. The internal resistance of the voltage source is relatively fixed in value, but the value of the tank circuit impedance varies from maximum at resonance to a small value above and below resonance.

Recall that the open circuit voltage of any source of voltage is called Thevenin's Voltage, and the internal resistance of that source of voltage is called Thevenin's Resistance. Recall also from Thevenin's Circuits that the voltage developed across the load connected to the terminals of most voltage sources will decrease as the load impedance (resistance) decreases.

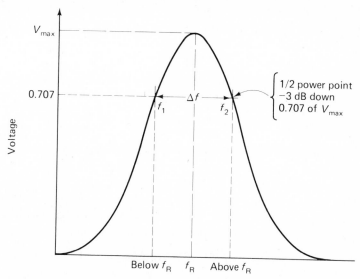

FIGURE 18-24 Parallel-resonant response curve. Note that voltage is plotted against frequency.

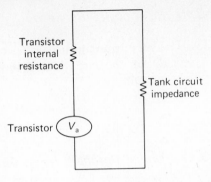

Transistor internal resistance

Tank circuit impedance

Transistor V_a

FIGURE 18-25 The impedance of the tank circuit is in series with the internal resistance of the source of the signal (transistor or tube).

The voltage developed across the load decreases because the increased line current drawn through the internal resistance of the voltage source drops the voltage supplied to the load. Such is the case with a tank circuit. The impedance of a tank circuit is very large at resonance and the line current drawn from the voltage source is small. The result is a large voltage developed across the parallel resonant circuit.

The line current drawn from the voltage source increases considerably at frequencies above and below resonance and the resultant voltage supplied to the tank circuit drops off proportionally with increases in line current.

A graph plotting the voltage developed across a tank circuit against the frequency of the source voltage will produce a voltage response curve for the parallel resonant circuit similar to that depicted in Fig. 18-24.

18-3.12 Bandwidth Measurements

The bandwidths of parallel resonant circuits are measured the same manner as in series resonant circuits (see Sec. 18-2.7). The same bandwidth equation is used, with the same $\frac{1}{2}$ power point, -3 dB point, and 0.707 point on the curve.

$$BW = \frac{f_R}{Q} \qquad BW = \frac{R}{2\pi L}$$

18-3.13 Impedance at Bandwidth Frequencies

The parallel response curve is produced by plotting voltage against frequency. The bandwidth is measured at the 0.707 point on the response curve (Fig. 18-26). Therefore, the voltage at bandwidth frequencies must be 0.707 of the peak voltage. Voltage and impedance are directly proportional to one another, so the impedance at bandwidth frequencies must also be 0.707 of the impedance at the resonant frequency.

impedance at BW frequencies $= 0.707\ Z_{f_R}$

$$Z_{BW} = 0.707\ Z_{f_R} = \frac{Z_{f_R}}{1.414}$$

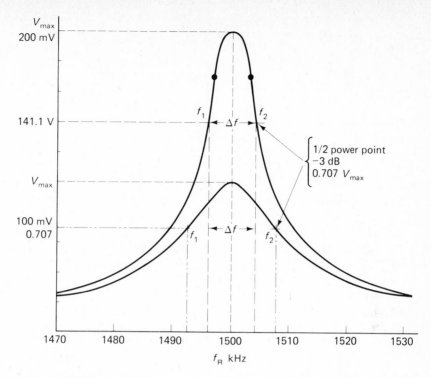

FIGURE 18-26 Parallel-resonant response curve showing the bandwidth of the selective and broad response curves.

PROBLEM 18-7

A parallel resonant circuit consists of a 400-pF capacitor and a 400-μH coil, the resistance of which is 4 Ω. A signal voltage of 250 mV is connected across the tank circuit.

1. Find the resonant frequency (f_R).
2. Find the Q of the circuit at resonance.
3. Find X_L at resonance.
4. Find the Z of the circuit at resonance.
5. Determine the supply current (I_a).
6. Determine the circulating current $I_{(circ)}$.

SOLUTION   (cover solution and solve)

1.
$$f_R = \frac{1}{2\pi\sqrt{LC}}$$

$$= \frac{1}{6.28 \times (400 \times 10^{-6}) \times (400 \times 10^{-12})}$$

$$= \frac{1}{6.28 \times 400 \times 10^{-9}}$$

$$= 398 \text{ kHz}$$

2.
$$Q = \frac{1}{R} \sqrt[2]{\frac{L}{C}}$$

$$= \frac{1}{4} \times \sqrt[2]{\frac{400 \times 10^{-6}}{400 \times 10^{-12}}}$$

$$= 250$$

3.
$$X_L = 2\pi f L$$

$$= 6.28 \times 398 \times 10^3 \times 400 \times 10^{-6}$$

$$= 1000 \ \Omega$$

4.
$$Z = \frac{L}{CR}$$

$$= \frac{400 \times 10^{-6}}{400 \times 10^{-12} \times 4}$$

$$= 250{,}000 \ \Omega$$

or $Z_{f_R} = QX_L$

$$= 250 \times 1000$$

$$= 250{,}000 \ \Omega$$

5.
$$I_a = \frac{V}{Z}$$

$$= \frac{250 \times 10^{-3}}{250 \times 10^3}$$

$$= 1 \ \mu A$$

6.
$$I_{circ} = \frac{V}{X_L}$$

$$= \frac{200 \times 10^{-3}}{1000}$$

$$= 0.2 \ mA$$

or $I_{circ} = QI_a$

$$= 250 \times (1 \times 10^{-6})$$

$$= 250 \ \mu A$$

18-3.14 Precise Formula for Resonance of Practical Parallel Resonant Circuits

The formula $f_R = 1/(2\pi\sqrt[2]{LC})$ is not an accurate formula because it does not include the coil resistance. The coil resistance shifts the phase angle of the coil current so that it is less than 90° out of phase with the

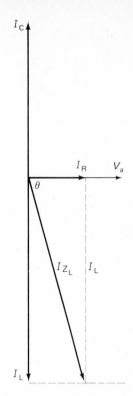

FIGURE 18-27 The phasor diagram for the currents for the precise formula for parallel resonance derived from Fig. 18-20.

applied voltage (Fig. 18-27). This brings the I_L to less than $180°$ out of phase with I_C; therefore, the two reactive currents do not cancel at resonance. The precise resonance formula incorporates resistance but it is seldom used in circuit design because:

1. The final answer differs little from the simple formula answer.
2. Most components have wide tolerances which make precise calculations unnecessary.
3. Trimmer and padder capacitors are always used to compensate for tolerances and other variations.

The precise resonant formula for parallel resonant circuits is

$$f_R = \frac{1}{2\pi} \sqrt[2]{\frac{1}{LC} - \frac{R^2}{L^2}}$$

Table 18-6 Comparison of Series and Parallel Resonant Circuits

Series	Parallel
1. L, C, and R_C are in series (R_C is coil resistance	L, C, and R_p are in parallel (R_p is shunt resistance)

continued

Table 18-6 (*Continued*)

Series	Parallel
2. $f_R = \dfrac{1}{2\pi\sqrt[2]{LC}}$	$f_R = \dfrac{1}{2\pi\sqrt[2]{LC}}$ (practical formula)
3. I_a supply current is maximum at resonance	I_a supply current is minimum at resonance
4. $I_a = \dfrac{V_a}{Z}$	$I_a = \dfrac{V_a}{Z}$
5. I_a leads V_a at frequency below f_R (capacitive circuit)	I_a lags V_a at frequency below f_R (inductive circuit)
6. Circuit is inductive above f_R	Circuit is capacitive above f_R
7.	$I_{tank} = QI_a$
8. V_L and V_C are large at f_R	V_L and V_C are large at f_R
9. Circuit Q $Q = \dfrac{1}{R}\sqrt[2]{\dfrac{L}{C}}$	Circuit Q $Q = \dfrac{1}{R}\sqrt[2]{\dfrac{L}{C}}$
10. At resonance $V_L = QV_a$ $V_C = QV_a$	At resonance $Z = QX_L$ (circuit Q) $Z = QX_C$ (circuit Q)
11. At resonance $Z = R$	At resonance $Z = \dfrac{L}{CR}$ $Z = QX_L = QX_C$
12. At BW frequencies $Z_{BW} = 1.414\,R$ or $= 1.414\,Z_{f_R}$	At BW frequencies $Z_{BW} = 0.707\,Z_{f_R}$ or $= \dfrac{Z_{f_R}}{1.414}$
13. Impedance is minimum at resonance	Impedance is maximum at resonance
14. High selectivity, if L/C ratio is large and R_C is small	High selectivity, if L/C ratio is large and R_C is small

Table 18-6 (*Continued*)

Series	*Parallel*

15. An acceptor trap circuit

$Z = 0\ \Omega$ at resonance, and the signal is shorted to ground.

A rejector trap circuit

High Z at resonance, and the signal is blocked.

---

16. Bandwidth (Δf)

$$BW = \frac{f_R}{Q}$$

or

$$= \frac{R}{2\pi L}$$

Bandwidth (Δf)

$$BW = \frac{f_R}{Q}$$

or

$$= \frac{R}{2\pi L}$$

---

17. Response curve is plotted as frequency vs. current

Response curve is plotted as frequency vs. voltage

---

18. Bandwidth is measured at

$0.707\ I_{max}$
-3 dB down
$\frac{1}{2}$ power point

Bandwidth is measured at

$0.707\ V_{max}$
-3 dB down
$\frac{1}{2}$ power point

EXERCISES

1. What is the most important use of resonant filters?

2. What is the one parameter of a reactive circuit which determines that it is resonant?

3. Write the equation for resonance.

4. Name the only two types of circuits which can be made resonant.

5. Draw the phasor diagram for a series resonant circuit.

6. What is the phase relationship between the applied signal and the circuit current in a series resonant circuit?

7. What is the power factor of a series resonant circuit at the resonant frequency?

8. Are resonant circuits used in radio frequency circuits or in audio circuits?

9. a. What is the frequency of a radio signal received by a radio when the capacity of the series resonant antenna circuit is 112 pF and the inductance of the antenna that has 12 Ω resistance is 103 μH?
   b. What will be the value of inductive reactance of the antenna coil at resonance?
   c. What will be the value of capacitive reactance at resonance?
   d. Determine the antenna circuit current at resonance if the signal is 24 mV.
   e. Determine the voltage developed across the capacitor at resonance.

10. Describe the difference between a sharp response curve and a broad response curve.

11. What type of resonant response curve is considered to be very selective?

12. Define the terms:
   a. Selectivity
   b. Q of a circuit
   c. Bandwidth

13. How does Q affect the response of resonant circuits?

14. How do Q and energy relate to one another?

15. Write the equation relating Q, inductive reactance, and resistance.

16. State two steps which could be taken to increase the value of Q of a coil.

17. Are there more than one or is there only one combination of inductance and capacitance which will resonate at, for example, 1200 kHz?

18. What is the optimum value of inductive reactance for best selectivity at a particular frequency?

19. Is a high or low L/C ratio necessary for greatest selectivity?

20. Write the equation for the total Q of a tuned circuit which incorporates L/C ratio and X_L/R ratio.

21. State three units to express the points where bandwidth is measured on response curves.

22. What is the value of the impedance at bandwidth frequencies compared to the impedance at resonance?

23. Write the equation relating bandwidth and frequency.

24. What does the impedance of a series resonant circuit equal at resonance?

25. Write the equation relating bandwidth, resonant frequency, and the Q of the inductance.

26. Calculate the Q of a coil possessing 10 Ω resistance when its inductive reactance is 800 Ω at a resonant frequency of 1600 kHz.

27. A series tuned antenna circuit consists of a variable capacitor which has a range of 50 pF to 380 pF and a 230 μH antenna coil which has a dc resistance of 15 Ω. The radio signals of interest have a signal strength of 12 mV. Draw the antenna circuit and determine:
   a. The frequency range of radio signals to which the radio is tunable
   b. The circuit Q at each end of the frequency range
   c. The impedance of the antenna circuit at resonance
   d. The voltage developed across the antenna coil at the high end of the tuning range at resonance
   e. Whether the capacitance is the largest or the smallest value at the high frequency range
   f. The bandwidth value at:
      (1) The high end
      (2) The low end
   g. Will the plates of the variable capacitor be fully opened or closed at the highest frequency?
   h. Which end of the tuning range will have the highest selectivity?

28. State the general equation for resonance of a parallel tank circuit.

29. Why is a parallel resonant circuit sometimes referred to as a tank circuit?

30. Is the impedance across a parallel resonant circuit a maximum or minimum value at resonance?

31. a. Is a parallel inductive-capacitive circuit considered to be inductive or capacitive below the resonant frequency?
    b. Explain your answer to part a.

32. a. Is the impedance of a practical parallel resonant circuit infinite in value at resonance?
    b. Explain your answer to part a.

33. A parallel antenna circuit consists of a variable capacitor with a range of 40 pF to 370 pF connected in parallel with a 240-μH antenna coil possessing 10 Ω resistance. The strength of the average radio signal is 18 mV. Draw the circuit and determine:
    a. The frequency range to which the radio is tunable
    b. Impedance of the antenna circuit at the high end of the tuning range at resonance
    c. Impedance of the antenna circuit at the low end of the tuning range at resonance

d. The Q of the circuit at the highest frequency

e. The Q of the circuit at the lowest frequency

34. Complete Table P18-1 for parallel and series resonant circuits with respect to the listed properties.

Table P18-1   Comparison of Series and Parallel Resonant Circuits

Particulars	Series Resonant	Parallel Resonant
a. Formula for resonance		
b. Line current at resonance (maximum or minimum)		
c. Formula to determine line current		
d. Is circuit inductive or capacitive above resonance?		
e. Formula for circuit Q		
f. Is Q measured above or below resonance?		
g. Impedance at resonance		
h. Impedance at BW frequencies		
i. What type of trap?		
j. Bandwidth formula		
k. Bandwidth is measured at—?		

chapter 19

FILTERS

19-1 Introduction

A filter is a circuit designed to separate a specific signal or band of signals from other signals. A filter can also be defined as a circuit designed to separate voltage or current of a particular frequency from other voltages or currents.

Filters are very common circuits. They are used to remove the pictures of strong adjacent television channels so that they don't interfere with the picture of the channel being viewed. All power supplies which convert ac voltage to dc voltage employ filters to smooth the rectified ac voltage so that it will appear as a dc voltage. The bass and treble controls in stereo and audio amplifiers are filter circuits. Stereo equalizer equipment consists of numerous filter circuits. A 10-channel stereo equalizer, for example, will have 10 filters on the right side, each operating at a different frequency, with 10 more filters on the left side which are duplicates of those on the right side.

19-2 Classifications

Filter circuits can be divided into five major classifications.

1. Resonant filters
   a. Series
   b. Parallel
2. Nonresonant filters
   a. High-pass
   b. Low-pass

3. Constant K filters
4. Hybrid (m-derived) filters
5. Active filters

19-3 Resonant Filters

Resonant filters are frequency-sensitive circuits, comprised of at least one capacitor and one inductance, either in a parallel or series arrangement. The operation of resonant circuits depends upon a change in frequency causing a change in the reactance of each component until the reactances are equal.

The acceptor and rejector traps mentioned in Table 18-6 are examples of series and parallel resonant filters. Resonant filters are used only in the radio frequency bands.

19-3.1 Series and Parallel Resonant Wave Traps

Resonant filters are merely applications of series or parallel resonant circuits as discussed in Chap. 18. Resonant filters depend upon a resonance frequency for their maximum response and the Q of the tuned circuit for the width of their response. The lower the Q, the wider the response. All the characteristics and parameters which apply to series or parallel resonant circuits also apply to resonant filters. Resonant filters are usually referred to as wave traps, or just as traps. The expression is derived from

Table 19-1   Series and Parallel Resonant Filters

Series Resonant Filter	Parallel Resonant Filter
Names	*Names*
1. Series trap 2. Acceptor trap 3. Bandpass or bandstop, depending upon the circuit arrangement	1. Parallel trap 2. Rejector trap 3. Bandpass or bandstop, depending upon the circuit arrangement
Formulas	*Formulas*
$f_R = \dfrac{1}{2\pi\sqrt{LC}}$	$f_R = \dfrac{1}{2\pi\sqrt{LC}}$
$BW = \dfrac{f_R}{Q}$	$BW = \dfrac{f_R}{Q}$
$BW = \dfrac{R}{2\pi L}$	$BW = \dfrac{R}{2\pi L}$
cutoff frequency $f_c = f_R \pm \dfrac{BW}{2}$	cutoff frequency $f_c = f_R \pm \dfrac{BW}{2}$

the fact that a filter effectively captures a specific frequency and its side bands and removes them from the circuit in which they are present.

The side band frequencies in bandpass filter circuits range from a minimum of 70.7% of the resonant frequency voltage to the maximum value of the frequency at resonance.

In bandstop filters, the side band voltages extend from a maximum of 70.7% of the resonant frequency voltage to a theoretical minimum of 0 V at the resonant frequency.

19-3.2 Series Resonant Filters

A series resonant circuit may be called a bandpass filter in one configuration (Fig. 19-5a), but a bandstop filter in a different configuration (Fig. 19-5b). The names bandpass and bandstop describe the function of a filter in a circuit.

BANDPASS FILTERS

Resonant filters are called bandpass filters when they pass only the resonant frequency plus frequencies within 3 dB above and below the resonant frequency (see Fig. 19-2).

The principle of a bandpass filter is shown in Figs. 19-1 and 19-2. A band of frequencies from 1.0 MHz to 3.0 MHz is injected into a bandpass filter. A narrow band of frequencies from 1.5 MHz to 1.6 MHz with a −3 dB loss are passed through the filter to successive circuits. The bandwidth is 0.1 MHz (100 kHz).

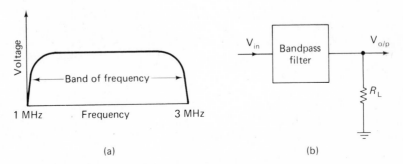

(a) (b)

FIGURE 19-1 (a) Band of input frequencies. (b) Bandpass circuit.

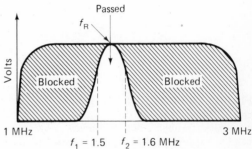

FIGURE 19-2 Output frequencies. A narrow band of frequencies are passed, but the remainder are prevented from passing to the output.

BANDSTOP FILTERS

A series resonant filter is a bandstop filter when it will stop a narrow band of frequencies from passing on to successive circuits, but will pass a wide range of other frequencies. The diagrams in Figs 19-3 and 19-4 show the principle of a bandstop filter. A range of frequencies from 1 MHz to 3 MHz is injected into a bandstop filter. All frequencies appear at the output except a narrow band of frequencies from 1.5 MHz to 1.6 MHz (within a - 3 dB loss).

If the series trap is placed in parallel with the load as shown in Fig. 19-6b, the trap will act as a short across the load at the resonant frequency and the bandwidth frequencies but will act as a high impedance at other frequencies on either side of it. The series circuit is now a bandstop filter. Note that R_L must be much larger than the coil resistance.

The center frequency of bandpass filters is the resonant frequency of the circuit (Figs. 19-2 and 19-4), and the frequency response or bandwidth depends upon the Q of the coil.

$$\text{BW} = (f_1 - f_2).$$

The resonant frequency of either a bandstop or bandpass circuit is

$$f_R = \frac{1}{2\pi\sqrt{LC}}$$

The bandwidth of a bandpass or bandstop response curve is deter-

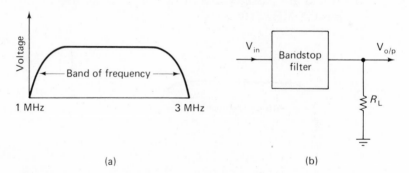

(a) (b)

FIGURE 19-3   (a) Input frequencies. (b) Bandstop filter circuit.

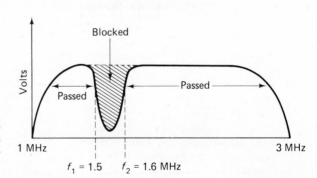

FIGURE 19-4   Output frequencies. The majority of the frequencies pass through, but a narrow band is blocked.

mined by

$$BW = \frac{f_R}{Q} \quad \text{or} \quad BW = \frac{R}{2\pi L}$$

where R is $R_C + R_L$

Q is $\frac{1}{R}\sqrt{\frac{L}{C}}$

PROBLEM 19-1

Calculate the following values for the circuit in Fig. 19-5 when $L = 180.5 \ \mu H$, $C = 200$ pF, $R_C = 2 \ \Omega$, and $R_L = 8 \ \Omega$.

1. The resonant frequency
2. Bandwidth
3. Bandwidth frequencies

SOLUTION   (cover solution and solve)

1. $f_R = \dfrac{1}{2\pi\sqrt{LC}}$

$= \dfrac{1}{6.28 \times \sqrt{180.5 \times 10^{-6} \times 200 \times 10^{-12}}}$

$= \dfrac{1}{6.28 \times \sqrt{361 \times 10^{-16}}}$

$= \dfrac{1}{6.28 \times 19 \times 10^{-8}}$

$= 839$ kHz

2. Bandwidth (BW)

$BW = \dfrac{R}{2\pi L}$

$= \dfrac{1}{6.28 \times (180.5 \times 10^{-6})}$

$= 8.82$ kHz

3. $f_1 = f_R - \dfrac{BW}{2}$

$= 839 - \dfrac{8.82}{2}$ kHz

$= 834.59$ kHz

$f_2 = f_R + \dfrac{BW}{2}$

$= 839 + 4.41$ kHz

$= 843.41$ kHz

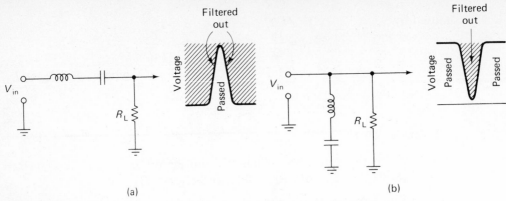

FIGURE 19-5   (a) A bandpass circuit (series resonant circuit). (b) A bandstop circuit (series resonant circuit).

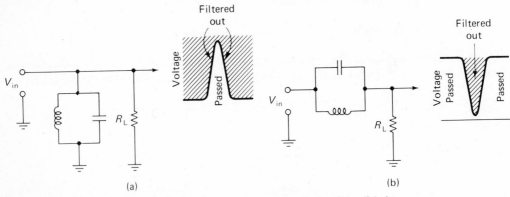

FIGURE 19-6   (a) A bandpass circuit (parallel resonant circuit). (b) A bandstop circuit (parallel resonant circuit).

19-3.3 Parallel Resonant Filters

Parallel resonant filters may also be classed as bandstop or bandpass filters. A parallel circuit is called a bandstop filter when it passes a wide range of frequencies but stops the resonant frequency and frequencies 3 dB above and 3 dB below resonance (Fig. 19-6b).

A parallel circuit is in a bandpass configuration (Fig. 19-6a) when it passes the resonance frequency plus the frequencies above and below resonance which have a voltage of at least 0.707 (–3 dB) of the resonance frequency.

The explanation of the way in which parallel resonant circuits operate as bandpass and bandstop filters is similar to that of series circuits, except that parallel circuits have a high impedance at resonance and a low impedance on either side of resonance.

BANDPASS AND BANDSTOP FILTERS

A parallel resonant circuit is called a bandpass filter or a bandstop filter depending upon the circuit arrangement in which it is used.

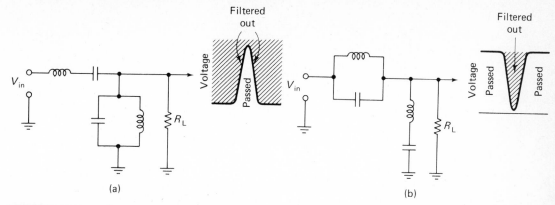

FIGURE 19-7 (a) A double bandpass circuit. (b) A double bandstop circuit.

19-4 Nonresonant Filters

Nonresonant filters are R-C or R-L filters, and consist of a minimum of one capacitor and one resistor, or one resistor and one inductance, respectively. Other types of nonresonant filters such as L-C types consist of a minimum of one inductance and one capacitor. These are generally of the constant K type.

19-4.1 Classifications of Nonresonant Filters

Nonresonant filters usually consist of a resistor and a capacitor, or a resistor and an inductance. Either combination will produce either a low-pass filter or a high-pass filter. The deciding factor is the arrangement of the two components in the circuit.

Nonresonant filters are classified as either:

1. Low-pass
2. High-pass

19-4.2 Low-Pass Filter

The frequencies at the output of a low-pass filter, from 0 Hz to the cutoff frequency (f_c), constitute the bandpass of a low-pass filter. The frequencies above the cutoff frequency (f_c) are effectively shorted to ground and are described as being in the attenuation band (see Fig. 19-9).

PRINCIPLE OF OPERATION OF LOW-PASS FILTERS

The 10-V, 100-Hz audio frequency shown in Fig. 19-8 produces a capacitive reactance of 1.6 MΩ across the capacitor C_1. The 1.6-MΩ capacitive reactance is one hundred times larger than the resistance of the 16-kΩ load resistor R_L, and will have no effect on the total resistance of the load

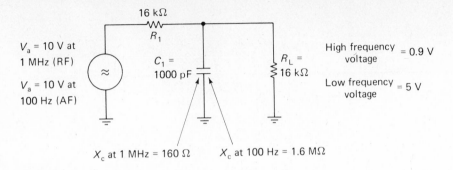

FIGURE 19-8   A math example of a low-pass circuit.

R_L. The 100-Hz audio signal therefore divides equally between the 16-kΩ resistance of R_1 and the 16-kΩ resistance of R_L, producing an output voltage of 5 V.

The 10-V 1-MHz radio frequency, on the other hand, produces a capacitive reactance of only 160 Ω across the capacitor C_1. The 160 Ω across the load resistor reduces its total resistance to effectively 160 Ω. The result is that the voltage divider R_1, R_L now consists of the 16-kΩ resistor R_1 in series with the total effective load resistance of 160 Ω. Nearly all the 10-V radio frequency signal will be developed across R_1, and a negligible voltage will be developed across R_L. It is thus said that the capacitor has shorted the radio frequency signal to ground. The circuit $R_1 C_1$ develops a voltage at low frequencies but shorts the high frequencies to ground. The filter is therefore called a low-pass filter and produces the response curve as depicted in Fig. 19-9b.

19-4.3   High-Pass Filter

The high-pass filter operation is opposite that of the low-pass filter. The frequencies above the cutoff frequency (f_c) are developed across the load in the output of the filter, but all frequencies below the cutoff frequency are effectively blocked and do not appear at the output of the filter (Fig. 19-10b). The frequencies above the cutoff frequency are called the band-pass frequencies, and those below the cutoff frequency are described as being in the attenuation band.

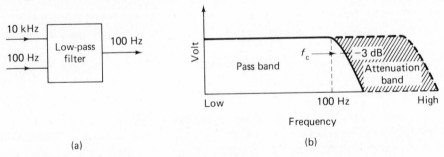

FIGURE 19-9   (a) Low-pass filter. (b) Low-pass filter response curve.

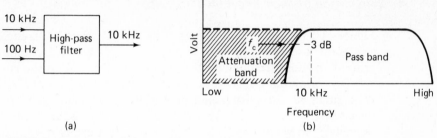

FIGURE 19-10   (a) High-pass filter. (b) High-pass filter response curve.

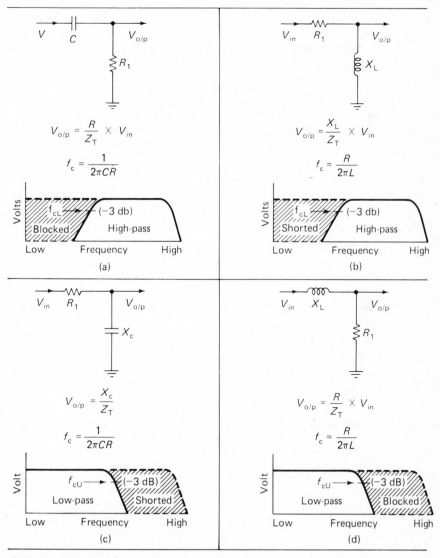

FIGURE 19-11   (a) A $C\text{-}R$ high-pass filter. (b) An L/R high-pass filter.
(c) A $C\text{-}R$ low-pass filter. (d) An L/R low-pass filter.

19-4.4 Comparison of High- and Low-Pass Filters

Figures 19-11a and b are high-pass filters. High-pass filters allow frequencies above a certain value to appear at their output, but block or short to ground all those frequencies below the -3 dB point on the curve.

The principle of operation of the filter in Fig. 19-11a is that X_c is very small at high frequencies; therefore, all the high frequencies are developed across R_1. The reactance of C_1 is very high at low frequencies, however, and all of the low voltage is developed across the capacitor, with a minimum voltage appearing across resistor R_1 from which the output is taken.

The circuit in Fig. 19-11b has a high X_L across the inductance at the high frequencies. The result is that the high frequencies produce a large output voltage across L. At low frequencies, X_L will be small, and the low frequencies will be effectively shorted to ground by the small value of X_L.

The circuits in Figs. 19-11c and d are low-pass filters. The principle of their operation is also based on the voltage divider concept. Note that the position of the components is the reverse of those in Figs. 19-11a and b. Therefore, the equivalent voltage divider components at high frequencies will be reversed, and will produce a low voltage at the output of the voltage divider at the high frequencies.

19-4.5 Upper and Lower Cutoff Frequencies (f_c)

Resonant circuits as described in Sec. 19-3 respond to the resonance frequency and to frequencies on either side of resonance (the bandwidth frequencies). The two bandwidth frequencies may be called cutoff frequencies. A nonresonant filter, on the other hand, has only one cutoff frequency on its response curve. The cutoff frequency on the high-pass response curve is called the low cutoff frequency, and the cutoff frequency on the low-pass response curve is called the upper cutoff frequency (Fig. 19-12).

19-4.6 Determining the Cutoff Frequency

Cutoff frequency in L-R low-pass and high-pass filters produces an inductive reactance equal to the resistance.

$$X_L = R$$

$$\text{or} \quad 2\pi f_c L = R$$

$$\therefore f_c = \frac{R}{2\pi L} \quad \text{(for } L\text{-}R \text{ filters)}$$

Cutoff frequency in C-R low-pass or high-pass filters produces a capacitive reactance equal to the resistance.

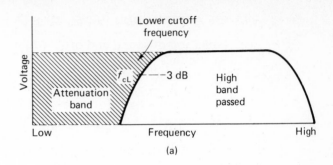

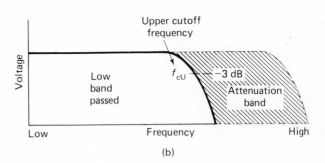

FIGURE 19-12 (a) Response curve for a high-pass circuit showing the lower cutoff frequency. (b) Response curve for a low-pass filter showing the upper cutoff frequency.

$$X_c = R$$

$$\text{or} \quad \frac{1}{2\pi f_c C} = R$$

$$\text{and} \quad 1 = 2\pi f_c CR$$

$$\therefore f_c = \frac{1}{2\pi CR} \quad \text{(for } C\text{-}R \text{ filters)}$$

19-4.7 Voltage Output Versus Voltage Input at Cutoff Frequency

Cutoff frequency is measured at the −3 dB point, which is also 0.707 of the $V_{\max}$ point on the response curve. The reference for the 0.707 value is the maximum voltage value, which is the input voltage value. All voltage values are compared with the input voltage, which is given a reference value of 1; therefore, the output voltage at the cutoff frequency is 0.707 of the input voltage.

$$V_o = 0.707 \, V_{in}$$

19-4.8 Attenuation

The frequencies which are not allowed to pass through a filter are said to be attenuated. High frequencies are attenuated in low-pass filters and low frequencies are attenuated in high-pass filters.

VOLTAGE ATTENUATION IN DECIBELS (dB)

Attenuation of sound and often of radio-frequency voltage is generally stated in decibels (dB). It has already been stated that the cutoff frequencies are present at -3 dB down on the frequency response curve. The following equations prove this statement:

$$\text{dB gain or loss} = 20 \log \frac{V_o}{V_{in}} \tag{1}$$

$$\text{but} \quad V_o = 0.707 \times V_{in} \quad \text{(Sec. 19-4.7)} \tag{2}$$

$$\text{If} \quad V_{in} = 1$$

$$\text{then} \quad dB = 20 \log \frac{0.707}{1} \tag{3}$$

$$= 20 \times -0.5058$$

$$= -3 \text{ dB} \tag{4}$$

Therefore, at cutoff frequency the voltage attenuation is -3 dB down on the response curve.

POWER ATTENUATION

Attenuation is sometimes quoted in power loss; therefore, it is necessary to know the relationships between voltage, decibels, and power.

$$P = \frac{V^2}{R} \tag{5}$$

But cutoff frequency is measured at $0.707 \, V_{max}$. Therefore, at f_c the voltage will be

$$V_{f_c} = 0.707 \, V_{max} \tag{6}$$

Substitute Eq. (6) in Eq. (5) and

$$P = \frac{V^2}{R} = \frac{(0.707 \, V_{max})^2}{R} \tag{7}$$

$$= \frac{0.5}{R} \tag{8}$$

Or, power at cutoff frequency $= \frac{1}{2}$ power.

The cutoff frequency is thus a frequency at which the power output of the circuit is $\frac{1}{2}$ of the power output at maximum output.

19-4.9   Value of R and X at Cutoff Frequencies

The output voltage from a filter is taken from across either the reactive component or the resistance in a C-R or L-R filter circuit (see Fig. 19-11). If the output voltage is $0.707 \, V_{max}$ at cutoff frequency, the reactance or resistance value of the component from which the output voltage is removed must be 1/1.414 (0.707) of the total impedance of the two components in the filter circuit (voltage ratio formula). (The total impedance is the opposition offered to the input voltage, or V_{max}.) An impedance value of 1.414 is obtained from a pure reactance and a resistance only when they are equal to one another, since they are 90° out of phase. The vectorial addition of two such equal values is:

$$Z = \sqrt[2]{R^2 + X^2}$$
$$= \sqrt[2]{1^2 + 1^2} = 1.414$$

Therefore, at cutoff frequency

$$X = R = \frac{1}{1.414} Z = .707\,Z$$

19-4.10   Use of R and X to Calculate Attenuation in dB at Cutoff Frequency f_c

The ratio of the value of the reactance of the capacitor from across which the voltage is removed to the input impedance of the low-pass C-R filter circuit (Table 19-2) may be used to determine the dB voltage output of the filter. This is a voltage ratio formula as discussed in Sec. 5-5 and is an alternative to using voltage output and voltage input values directly. At the cutoff frequency, the value of the input impedance of the C-R low-pass filter is 1.414 times the value of the reactance of the capacitor or the resistance in the filter circuit. (See Sec. 19-4.9.)   The proportion of the output voltage to the input voltage (voltage ratio formula) at the cutoff frequency may therefore be written as X_c/Z or 1/1.414. This voltage proportion is always stated in decibel values such as

$$dB = 20 \log \frac{X_c}{Z}$$
$$= 20 \log \frac{1}{1.414}$$
$$= -3 \text{ dB}$$

The voltage output or attenuation at the cutoff frequency, therefore, is −3 dB of the input voltage. The common expression is that the cutoff frequency is −3 dB down on the response curve.

The voltage output to voltage input ratio of a C-R high-pass filter at the cutoff frequency is

$$dB = 20 \log \frac{R}{Z}$$
$$= 20 \log \frac{1}{1.414} = -3 \text{ dB}$$

The methods of calculating dB voltage outputs for various filter circuits and frequencies are shown in Table 19-2.

Table 19-2   Low- and High-Pass Filters

General equation

$$dB = 20 \log \frac{X_c}{Z}$$

where $Z = \sqrt[2]{X_c^2 + R^2}$

General equation

$$dB = 20 \log \frac{R}{Z}$$

where $Z = \sqrt[2]{X_c^2 + R^2}$

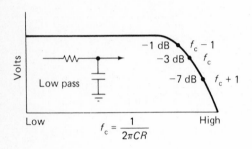

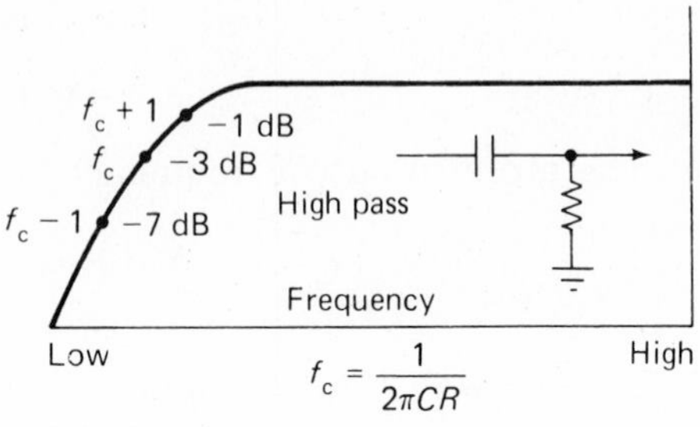

FIGURE 19-13a   A low-pass filter.

At f_c, $X_c = R$

$$dB = 20 \log \frac{1}{\sqrt[2]{1^2 + 1^2}}$$

At $f_c + 1$ octave

$$dB = 20 \log \frac{0.5}{\sqrt[2]{0.5^2 + 1^2}}$$

At $f_c - 1$ octave

$$dB = 20 \log \frac{2}{\sqrt[2]{2^2 + 1^2}}$$

FIGURE 19-13b   A high-pass filter.

At f_c, $X_c = R$

$$dB = 20 \log \frac{1}{\sqrt[2]{1^2 + 1^2}}$$

At $f_c + 1$ octave

$$dB = 20 \log \frac{1}{\sqrt[2]{0.5^2 + 1^2}}$$

At $f_c - 1$ octave

$$dB = 20 \log \frac{1}{\sqrt[2]{2^2 + 1^2}}$$

19-4.11   Octaves and dB Attenuation

An octave is the interval between any two frequencies when one frequency is twice the frequency of the other. If, for example, a cutoff frequency is 1000 Hz, the first octave above 1000 Hz would be 2000 Hz. In other words, the frequency at the first octave above the cutoff frequency is twice the cutoff frequency

Each octave above a reference frequency is equal to the frequency of the octave below it multiplied by two.

EXAMPLE

If $f_o = 1000$ Hz, octaves above f_o are

$$\text{Octave } 1 = f_o \times 2 = 1000 \times 2 = 2000 \text{ Hz}$$
$$\text{Octave } 2 = 2000 \times 2 = 4000 \text{ Hz}$$
$$\text{Octave } 3 = 4000 \times 2 = 8000 \text{ Hz}$$

Table 19-2 (Continued)

General equation

$$dB = 20 \log \frac{R}{Z}$$

where $Z = \sqrt[2]{X_L^2 + R^2}$

General equation

$$dB = 20 \log \frac{X_L}{Z}$$

where $Z = \sqrt[2]{X_L^2 + R^2}$

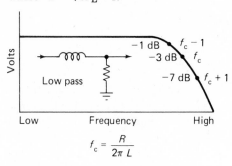

$$f_c = \frac{R}{2\pi L}$$

FIGURE 19-14a A low-pass filter.

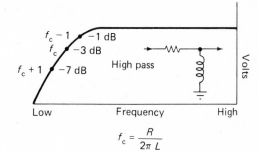

$$f_c = \frac{R}{2\pi L}$$

FIGURE 19-14b A high-pass filter.

At f_c, $X_L = R$

$$dB = 20 \log \frac{1}{\sqrt[2]{1^2 + 1^2}}$$

At $f_c + 1$

$$dB = 20 \log \frac{1}{\sqrt[2]{2^2 + 1^2}}$$

At $f_c - 1$

$$dB = 20 \log \frac{1}{\sqrt[2]{0.5^2 + 1^2}}$$

At f_c, $X_L = R$

$$dB = 20 \log \frac{1}{\sqrt[2]{1^2 + 1^2}}$$

At $f_c + 1$

$$dB = 20 \log \frac{1}{\sqrt[2]{2^2 + 1^2}}$$

At $f_c - 1$

$$dB = 20 \log \frac{0.5}{\sqrt[2]{0.5^2 + 1^2}}$$

Each octave below a reference frequency is equal to the frequency of the octave above it divided by two.

EXAMPLE

If f_o = 1000 Hz

Octave 1 below = 1000/2 = 500 Hz
Octave 2 below = 500/2 = 250 Hz

Voltage attenuation at frequencies on either side of the cutoff frequency may be measured in decibels (dB) per octave. The graph in Fig. 19-15 reveals that the frequency of the first octave above the cutoff frequency in an R-C low-pass circuit is attenuated -7 dB below the cutoff frequency voltage. The frequencies at the second octave above f_c are attenuated -13 dB below the maximum voltage value. See Fig. 19-15.

19-4.12 The dB Attenuation Value for the First Octave Above Cutoff Frequency for a C-R Low-Pass Filter (Fig. 19-15)

At cutoff frequency (f_c)

$$X_c = R \tag{9}$$

$$\text{and} \quad X_c = \frac{1}{2\pi f_c C} \tag{10}$$

$$\text{where} \quad f_c \text{ is cutoff}$$
$$\text{frequency}$$

At the first octave above the cutoff frequency (f_c + 1 octave), because the frequency is $2 f_c$, the reactance X_c is equal to $(1/2) X_c$ or $(1/2) R$ at f_c ($R = 1$ as reference), as shown in the following equation.

$$X_c = \frac{1}{2\pi \, 2f_c C} = 0.5 \, X_c \quad \text{or} \quad 0.5 \, R \text{ at } f_c \tag{11}$$

The general attenuation equation for a C-R low-pass filter in dB is equal to

$$dB = 20 \log \frac{X_c}{Z} \quad \text{where} \quad Z = \sqrt[2]{X_c^2 + R^2} \quad \text{(Table 19-2)} \tag{12}$$

but

$$X_c = 0.5 \, R \text{ at } f_c + 1 \text{ octave } (R = 1 \text{ as reference}) \tag{13}$$

Substitute Eq. (13) in Eq. (12) and

$$dB = 20 \log \frac{0.5}{\sqrt[2]{0.5^2 + 1^2}} \tag{14}$$

$$= -7 \text{ dB}$$

Therefore, attenuation at 1 octave above cutoff frequency is -7 dB for a low-pass C-R filter.

19-4.13 The dB Attenuation Value for the First Octave Below Cutoff Frequency for a C-R Low-Pass Filter (Fig. 19-15)

At cutoff frequency

$$X_c = R \tag{15}$$

$$\text{and} \quad X_c = \frac{1}{2\pi f_c C} \tag{16}$$

At the first octave below the cutoff frequency ($f_c - 1$ octave), because the frequency is $(1/2) f_c$, the reactance X_c is equal to $2 X_c$ or $2 R$ at f_c ($R = 1$ as reference), as shown in the following equation.

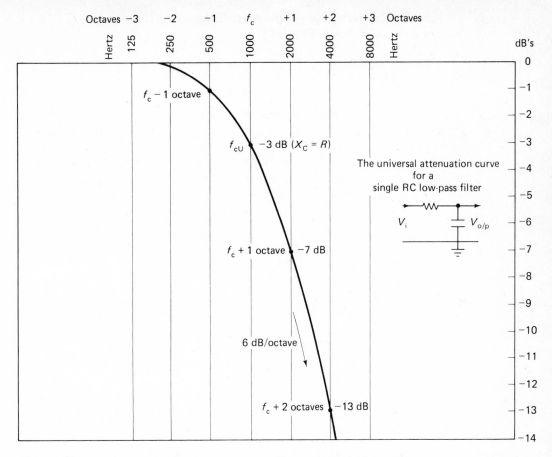

FIGURE 19-15 A universal attenuation curve for a single R-C low-pass filter. Note the attenuation per octave.

$$X'_c = \frac{1}{2\pi \left(\dfrac{f_c}{2}\right) C} = 2\,X_c \quad \text{or} \quad = 2\,R \text{ at } f_c \tag{17}$$

The general attenuation equation for a C-R low pass filter in dB is equal to

$$dB = 20 \log \frac{X_c}{Z} \quad \text{where} \quad Z = \sqrt[2]{X_c^2 + R^2} \tag{18}$$

but $\;X_c = 2\,R$ at $f_c + 1$ octave ($R = 1$ as reference) $\tag{19}$

Substitute Eq. (19) in Eq. (18) and

$$dB = 20 \log \frac{2}{\sqrt[2]{2^2 + 1^2}} \tag{20}$$

$$= -1 \text{ dB}$$

Therefore, attenuation at 1 octave below cutoff frequency is - 1 dB for a low-pass C-R filter (Fig. 19-15).

19-4.14 Attenuation per Octave, at Two or More Octaves Away from the Cutoff Frequency

The dB attenuation equation per octave, at two or more octaves away from the cutoff frequency for a low-pass C-R filter (Fig. 19-15), may be derived in the following manner. The reactance of the capacitor at one octave to the resistance of the circuit is compared with the reactance at the adjacent octave to the same resistance. This comparison can be placed in a voltage ratio formula.

$$\frac{V_o \text{ at adjacent octave}}{V_o \text{ at reference octave}} = \left(\frac{\dfrac{X'_c}{R}}{\dfrac{X_c}{R}}\right) = \frac{X'_c}{\cancel{R}} \times \frac{\cancel{R}}{X_c}$$

The R is common to both values. Therefore,

$$\frac{V_o \text{ at adjacent octave}}{V_o \text{ at reference octave}} = \frac{X'_c}{X_c}$$

The voltage attenuation from one octave to the adjacent octave will be

$$dB = 20 \log \frac{X'_c}{X_c}$$

The capacitive reactance at one octave higher than the reference octave (X'_c) is one-half as large as the reactance at the reference octave (X_c), because the frequency is twice as great. Therefore,

$$X'_c = 0.5 X_c \ (X_c = 1)$$

Attenuation in dB is

$$dB = 20 \log \frac{0.5}{1}$$

$$= 20 \times -0.30102$$

$$= -6 \text{ dB}$$

The attenuation per octave at two or more octaves away from the cutoff frequency is -6 dB for all low-pass and high-pass filters. (See Figs. 19-15 and 19-16.)

PROBLEM 19-2

A 40-Ω resistor and a 5-μF capacitor are connected as a low-pass filter.

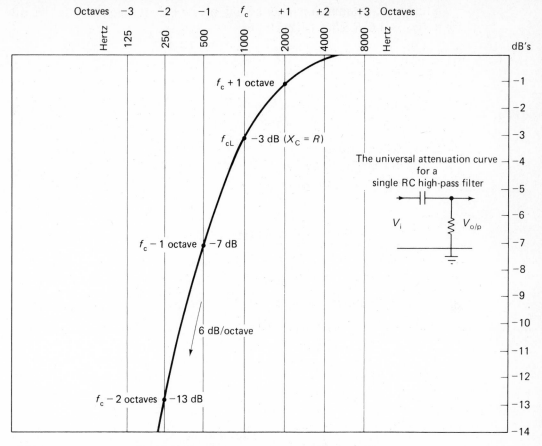

FIGURE 19-16   A universal attenuation curve for a single R-C high-pass
filter. Note the attenuation per octave.

1. Draw the filter circuit.
2. What is the upper cutoff frequency of the above filter?
3. What will be the frequency 1 octave below the cutoff frequency?
4. What would be the lower cutoff frequency if the resistor and capacitor were connected as a high-pass filter?
5. How many decibels will the frequency be attenuated 1 octave
   above the cutoff frequency in the high-pass filter?

SOLUTION   *(cover solution and solve)*

1. See Fig. 19-17.

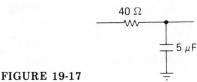

FIGURE 19-17

2. Upper cutoff frequency (f_c) will be

$$f_c = \frac{1}{2\pi RC} \quad \text{(Sec. 19-4.6)}$$

$$= \frac{1}{6.28 \times 40 \times 5 \times 10^{-6}}$$

$$= 796 \text{ Hz}$$

3. Cutoff frequency (f_c) = 796 Hz. $\therefore$ The frequency one octave below cutoff frequency (Fig. 19-15) will be

$$\frac{f_c}{2} = \frac{796}{2}$$

$$= 398 \text{ Hz}$$

4. The cutoff frequency will be the same; 796 Hz.

5. Attentuation at 1 octave above f_c (Fig. 19-15)

   At f_c $X_c = R$ (21)

$$\text{At } 2 f_c \quad X'_c = \frac{R}{2} \tag{22}$$

$$\therefore X'_c = 0.5 \, X_c \quad \text{(because } X_c = R \text{ at } f_c \text{)}$$

$$\text{dB} = 20 \log \frac{R}{\sqrt[2]{X_c^2 + R^2}} \tag{23}$$

Substitute Eq. (22) in Eq. (23) and

$$\text{dB} = 20 \log \frac{1}{\sqrt[2]{(0.5)^2 + 1^2}} \quad (R = 1 \text{ as reference})$$

$$= -1 \text{ dB} \tag{24}$$

PROBLEM 19-3

A low-pass filter in a radio circuit consists of a 100-Ω resistor and a 50-μH inductance having negligible resistance.

1. Draw the filter circuit.
2. Determine the cutoff frequency f_c of the low-pass filter.
3. How much would the fourth octave above the cutoff frequency be attenuated in the low-pass filter?
4. How much would the first octave above the cutoff frequency in the low-pass filter be attenuated?
5. How much would the first octave below the cutoff frequency be attenuated in the low-pass filter?

                                                                                       Ch. 19 Filters

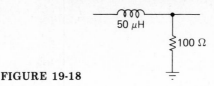

50 μH

100 Ω

FIGURE 19-18

SOLUTION   (cover solution and solve)

1. See Fig. 19-18.
2. Cutoff frequency (f_c) of the low-pass filter

$$f_c = \frac{R}{2\pi L} \quad \text{(Sec. 19-4.6)}$$

$$= \frac{100}{6.28 \times (50 \times 10^{-6})}$$

$$= 318.47 \text{ kHz}$$

3.   At f_c + 1 octave, attenuation = – 7 dB.   (Fig. 19-15)

At + 4 octaves above f_c, attenuation = 3 × –6 dB = – 18 dB.

Total attenuation = – 25 dB.

4. At f_c $X_L = R$

At 1 octave above the cutoff frequency (low-pass) (Fig. 19-15)

$$X'_L = 2 X_L \quad \text{(because frequency is twice)}$$

$$\therefore \ X'_L = 2 R \quad \text{(because } R = X_L \text{ at } f_c)$$

$$dB = 20 \log \frac{R}{\sqrt{X_L^2 + R^2}} \quad \text{(Table 19-2)}$$

$$dB = 20 \log \frac{1}{\sqrt{2^2 + 1^2}} \quad (R = 1 \text{ as reference})$$

$$= 20 \times -0.349$$

$$= -7 \text{ dB}$$

5. At f_c

$$X_L = R$$

At 1 octave below cutoff frequency (low-pass) (Fig. 19-15)

$$X'_L = \frac{X_L}{2} \quad \text{(because frequency is } \tfrac{1}{2})$$

$$X'_L = 0.5 R \quad \text{(because } R = X_L \text{ at } f_c)$$

$$dB = 20 \log \frac{R}{\sqrt[2]{X_L^2 + R^2}} \quad \text{(Table 19-2)}$$

$$= 20 \log \frac{1}{\sqrt[2]{0.5^2 + 1^2}} \quad (R = 1 \text{ as reference})$$

$$= 20 \log 0.8944$$

$$= -1 \text{ dB}$$

19-5 Constant-K Filters

19-5.1 Definition

Constant-K filters are another type of nonresonant filter. They consist of an inductance and a capacitance, but they are not resonant circuits. Constant-K filters are L-C circuits in which the product of the reactances at any frequency is constant. A crossover network between a woofer speaker and a tweeter speaker is an example of a constant-K filter.

The following examples may be used to show how the product of X_L and X_C is always constant regardless of frequency.

EXAMPLES

If the X_L is 1000 Ω and the X_C is 400 Ω, the product is 1000 $\times$ 400 = 400,000 Ω.

If the frequency is doubled, the X_L will be 2000 Ω and the X_C will be 200 Ω and the product will be 2000 $\times$ 200 = 400,000 Ω.

The product in the second combination is still 400,000 Ω. Therefore, the product of the two components is constant in value. Thus it is said that the K value of this type of filter remains constant and is independent of frequency.

19-5.2 Advantage of Constant-K Filters

Both the R-C and R-L nonresonant filter circuits have less sharp attenuation curves (Fig. 19-19) and produce more insertion loss than do reactive (constant-K) filters. Insertion loss is the difference between the power received at the load before the insertion (installation) of a filter and the power after insertion. Insertion loss is stated as the log of a ratio of power output to power input, and is the result of power loss due to resistance in the circuit. The resistance in the components of an L-C (constant-K) filter is negligible; therefore, there is no significant power loss. Constant-K (reactive filters) are, therefore, used when sharp cutoff (sharp attenuation) and minimal insertion loss is important.

Constant-K filters consisting of capacitors with negligible leakage and

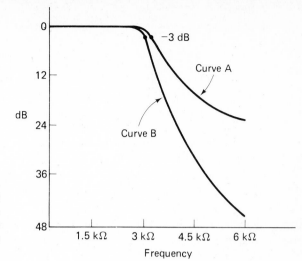

FIGURE 19-19   Curve A is the attenuation curve for a low-pass R-C or R-L filter. Curve B is the attenuation curve for a constant-K reactive filter. Notice that curve B has a sharper cutoff than curve A.

inductors with negligible resistance are used extensively in communication systems.

19-5.3 Types of Constant-K Filters

There are three main types of constant-K filters:

1. L-type
   a. Low-pass
   b. High-pass
2. T-type
   a. Low-pass
   b. High-pass
3. π-type
   a. Low-pass
   b. High-pass

19-5.4 L-Type Constant-K Filters

The filters shown in Figs. 19-20 and 19-21 are L-type filters. The filter in Fig. 19-21 is a low-pass filter and the filter in Fig. 19-20 is a high-pass filter.

One use of L-type constant-K filters is to attenuate a spectrum of frequencies, as shown in the response curves in Figs. 19-20 and 19-21. One common application of this use is the cross over network to the woofer and tweeter speakers in an audio amplifier, as shown in Fig. 19-22.

Another use of L-type constant-K filters is as an impedance step-up or step-down device, i.e., as an impedance-changing device. The arrangement of the reactive components in Figs. 19-24 and 19-26 makes the L-type

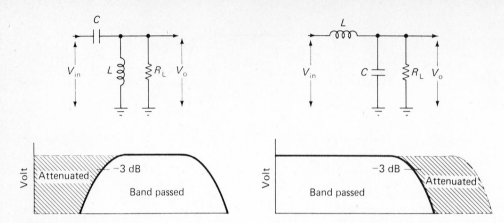

FIGURE 19-20 A simple *L*-type constant-K, high-pass filter.

FIGURE 19-21 A simple *L*-type constant-K, low-pass filter.

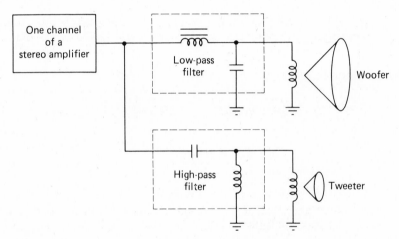

FIGURE 19-22 Two *L*-type constant-K filters used as a cross-over network in a stereo amplifier.

filter operate as a step-up impedance device. A reversal of the filter arrangement as shown in Figs. 19-23 and 19-25 makes the same filter act as an impedance step-down device.

A third application of L-type constant-K filters is as delay circuits.

The low-pass filter in Fig. 19-25 can also be used as a delay line. It takes time (a delay) for the current to increase in the series inductor (*L*) (because of counter voltage); therefore, when the capacitor charges and then discharge into the coil, there is a delay. A delay circuit is used in color television receivers to delay the black and white information, which travels faster than the color information by 0.8 ms. The delay permits the color information to arrive at the picture tube at the same time as does the black and white information.

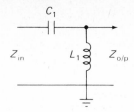

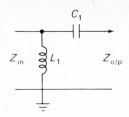

FIGURE 19-23 A *L*-type constant-K, high-pass filter working as an impedance step-down device.

FIGURE 19-24 A *L*-type constant-K, high-pass filter working as an impedance step-up device.

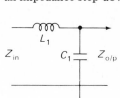

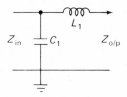

FIGURE 19-25 A *L*-type constant-K, low-pass filter acting as an impedance step-down device.

FIGURE 19-26 A *L*-type constant-K, low-pass filter acting as an impedance step-up device.

19-5.5 T-Type Filters

An L-type constant-K low-pass filter attenuates the higher frequencies. The sharpness of attenuation, however, is much better when the inductance or capacitance value is divided into two parts and placed on either side of the vertical component to form the symmetrical T-type filter as shown in Figs. 19-27 and 19-28.

The T-type filter produces a more symmetrical attenuation than does the L-type filter. A T-type filter with the same impedance value and cut-off frequency value as an L-type filter will have the same value of total inductance and capacitance as the L-type filter, but the value of the series component will be divided into two components and placed on either side of the shunt component (see Figs. 19-27 and 19-28). A T-type filter is generally used when the filter is to be connected to a low impedance source. A π-type filter (see Sec. 19-5.8) is used, however, when connecting to a high impedance source.

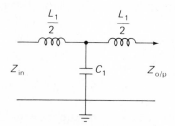

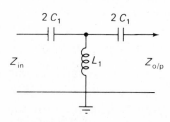

FIGURE 19-27 A *T*-type constant-K low-pass filter.

FIGURE 19-28 A *T*-type constant-K high-pass filter.

19-5.6 Cutoff Frequency and Constant-K Low-Pass Filters

The equation to determine the cutoff frequency for any one of the three constant-K low-pass filters is

$$f_c = \frac{1}{\pi\sqrt[2]{LC}} \text{ (low-pass)}$$

The equations to determine the value of inductance or capacitance to be used in a low-pass L-type constant-K filter to produce a specific cutoff frequency (f_c) for a constant load (R_L) are

$$L = \frac{R_L}{\pi f_c}$$

$$C = \frac{1}{\pi R_L f_c}$$

19-5.7 Cutoff Frequency and Constant-K High-Pass Filters

The cutoff frequency for any of the three constant-K high-pass filters is determined by

$$f_c = \frac{1}{4\pi\sqrt[2]{LC}} \text{ (high-pass)}$$

The equations to determine the value of inductance or capacitance to be used in a high-pass L-type constant-K filter to produce a specific cutoff frequency (f_c) for a constant load (R_L) are

$$L = \frac{R_L}{4\pi f_c}$$

$$C = \frac{1}{4\pi f_c R_L}$$

19-5.8 L and C in T-Type and π-Type Filters

The same equations are used to determine the total value of inductance and capacitance in a T-type and π-type filter as in an L-type filter, but with an additional calculation.

IN THE LOW-PASS CONSTANT-K FILTERS

1. The value of the series component in the T-type filter must be divided by 2 (Fig. 19-29).
2. The value of the shunt components in the π-type filter must be divided by 2 (Fig. 19-31).

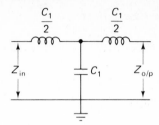

FIGURE 19-29 A low-pass T-type constant-K filter.

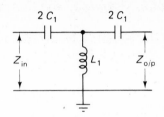

FIGURE 19-30 A high-pass T-type constant K- filter.

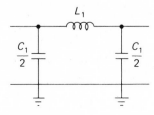

FIGURE 19-31 Low-pass π-type constant-K filter.

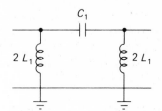

FIGURE 19-32 High-pass π-type constant-K filter.

IN THE HIGH-PASS CONSTANT-K FILTERS

1. The value of the series component in the T-type filter must be multiplied by 2 (Fig. 19-30).

2. The value of the shunt component in the π-type filter must be multiplied by 2 (Fig. 19-32).

PROBLEM 19-4

A 50-μH inductance and a 250-pF capacitor are connected in an L-type constant-K high-pass filter circuit.

1. Draw the circuit.

2. Calculate the low cutoff frequency.

3. What will be the value of each component if the same high-pass constant-K filter is connected in a T-type arrangement?

4. What will be the value of each component if the same constant-K filter is connected in a π-type arrangement?

5. What is the attenuation in decibels at cutoff frequency?
   a. L-type
   b. T-type
   c. π-type

6. Calculate the high cutoff frequency if the same components were connected as a T-type constant-K low-pass filter.

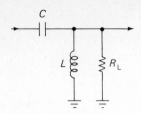

FIGURE 19-33

SOLUTION   *(cover solution and solve)*

1. See Fig. 19-33.

2.
$$f_c = \frac{1}{4\pi\sqrt{LC}}$$

$$= \frac{1}{4 \times 3.14 \times \sqrt{50 \times 10^{-6} \times 250 \times 10^{-12}}}$$

$$= \frac{1}{12.56 \times \sqrt{125 \times 10^{-16}}}$$

$$= \frac{1}{12.56 \times 11.18 \times 10^{-8}}$$

$$= 712.14 \text{ kHz}$$

3.
$$C_1 = 500 \text{ pF}$$
$$L = 50 \text{ } \mu\text{H}$$

4.
$$L_1 = 100 \text{ } \mu\text{H}$$
$$C_1 = 250 \text{ pF}$$

5. -3 dB for each type of filter

6. The same equation is used to calculate the cutoff frequency of a low-pass T-type filter as is used for a low-pass L-type filter. Cutoff frequency (f_c) of a low-pass filter

$$f_c = \frac{1}{\pi\sqrt[2]{LC}}$$

$$= \frac{1}{3.14 \times \sqrt[2]{50 \times 10^{-6} \times 250 \times 10^{-12}}}$$

$$= \frac{1}{3.14 \times \sqrt[2]{125 \times 10^{-16}}}$$

$$= \frac{1}{3.14 \times 11.18 \times 10^{-8}}$$

$$= 2.848 \text{ MHz}$$

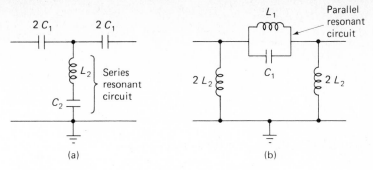

FIGURE 19-34   Some *m*-derived high-pass filters.

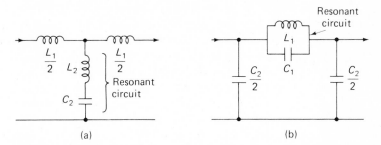

FIGURE 19-35   Low-pass *m*-derived filters.

19-6   m-Derived Filter

The m-derived filter is a hybrid filter, as it consists of a series or parallel wavetrap (resonant filter) added across the input or output of a single-section constant-K filter. The m-derived filter produces a sharper cutoff for the attenuation skirts of a response curve than is possible with a constant-K filter.

The m-derived filter can be divided in two main categories.

1. Low-pass
2. High-pass

There are many variations of each category, but only two variations are diagrammed for each of the low-pass and high-pass types (Figs. 19-34 and 19-35). The formulas for m-derived filters are rather complex and not within the scope of this chapter.

19-7   Active Filters

Resonant and nonresonant filters are called passive devices. Passive devices do not include amplification. Circuits with these same components plus an amplifier are called active filters. Active filters are not new devices, but they are being used in a new way; they have been made popular in recent years by the availability of low-cost operational amplifier I.C.

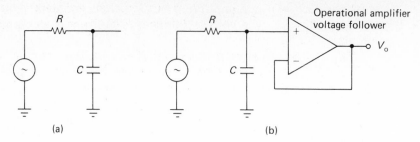

FIGURE 19-36   (a) A simple low-pass filter. (b) A low-pass active filter.

circuits. Active filters are inexpensive, easy to adjust, and can be cascaded so that a complex filter response can be reduced to simple factored blocks which do not interact. The filter circuits in a stereo equalizer are all active filters.

Active filters are rapidly replacing the ordinary R-C, the R-L, and the constant-K filters. Nearly all of the equalizing and tone control circuits being designed today are of the active-filter type because of their low cost and the versatility of the new operational amplifier I.C. circuits.

A basic active filter is merely an R-C, an R-L, or a constant-K filter coupled into an operational amplifier. An example of a low-pass active filter is shown in Fig. 19-36. An example of a bandpass active filter is shown in Fig. 19-37.

An operational amplifier consists of many direct-coupled transistors built into one I.C. circuit to produce a high gain dc-coupled voltage amplifier. Operational amplifiers have one output but two inputs. One input, called the inverting input and labeled (–), produces an amplified and inverted signal at the output. The second input, labeled (+), does not invert the amplified output. Operational amplifiers can be used to amplify dc voltages as well as ac signals, because the internal transistors are direct-coupled. The input impedance of operational amplifiers may have an input impedance of 1 MΩ or greater at each input terminal. They may provide gains as high as 100,000, or when they are connected as a unity noninverting voltage follower, the gain may be as low as 1.

Shorting the operational amplifier output to its inverting input con-

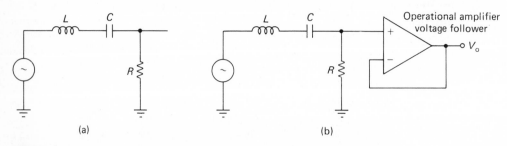

FIGURE 19-37   (a) A simple bandpass filter. (b) An active bandpass filter.

verts the I.C. to a nonamplifying but impedance-matching circuit called a voltage follower. The voltage follower effectively is a common collector circuit.

The design of active filters starts first with the design of the filter for the desired cutoff frequency, and then incorporates the operational amplifier circuits to which the filter is connected.

EXERCISES

1. Briefly describe the principle of operation of the following circuits:
   a. Resonant filter
   b. Constant-K filter
   c. m-derived filter

2. Draw the following filter circuits:
   a. Series resonant bandpass circuit
   b. An R-C low-pass filter
   c. A parallel bandstop filter
   d. A constant-K high-pass filter
   e. An m-derived filter
   f. An active bandpass filter
   g. Rejector trap

3. What does the Q of a resonant filter determine?

4. Write an equation for the following properties of resonant circuits:
   a. Bandwidth (incorporating resistance and inductance)
   b. Resonant frequency
   c. Cutoff frequency
   d. Resonance

5. a. Draw the response curve for an RF signal which is flat from 1 MHz to 3 MHz.
   b. Draw (and label) the response curve which will be produced after 1 MHz to 3 MHz range of frequencies has passed through:
      (1) A bandpass filter
      (2) A bandstop filter

6. A series resonant bandpass circuit consists of a 100-μH inductance coil with 3 Ω resistance, a 300-pF capacitor, and a load of 10 Ω. (R_L affects Q.) Calculate:
   a. Bandwidth frequencies
   b. Bandwidth

7. Define the term "attenuation."

8. Draw the response curve for a high-pass nonresonant circuit and label the cutoff frequency.

9. Draw:
   a. An *L-R* high-pass filter
   b. An *R-C* high-pass filter

10. What is the value of X_L at the cutoff frequency if the $R = 500 \ \Omega$ in:
    a. A high-pass *R-L* filter?
    b. A low-pass *R-L* filter?

11. What is the impedance of the circuit at cutoff frequency if $R = 500 \ \Omega$ in:
    a. A high-pass *R-L* filter?
    b. A low-pass *R-L* filter?

12. What is the value of the cutoff frequency of an *R-C* nonresonant filter if the capacity is 1000 nF and the resistance is 50 kΩ when the circuit is connected as a:
    a. High-pass *R-C* filter?
    b. Low-pass *R-C* filter?

13. The voltage input to an *R-C* filter is 20 V. What is the amount of voltage at the output at the cutoff frequency when it is connected as:
    a. A high-pass filter?
    b. A low-pass filter?

14. Refer to question 13. How much is the power attenuated and what is the decibel attenuation at cutoff frequency for:
    a. The high-pass *R-C* filter?
    b. The low-pass *R-C* filter?

15. Describe the term octave.

16. What is the total voltage attenuation in dB at the first octave above cutoff for:
    a. A low-pass filter?
    b. A high-pass filter?

17. A low-pass filter consists of a 50-Ω resistor and a 4-μF capacitor.
    a. Draw the filter circuit.
    b. Calculate the cutoff frequency.
    c. Is the cutoff frequency the upper or lower cutoff frequency?
    d. What is the frequency 1 octave below the cutoff frequency?
    e. What would be the value of the cutoff frequency if the circuit were connected as a high-pass filter?
    f. How many dB would the frequency be attenuated at 1 octave below the cutoff frequency, if the circuit were connected as a high-pass filter?

18. State the main advantage of constant-K filters over *R-C* or *L-R* circuits.

19. Define insertion loss.

20. a. Name three main types of constant-K filters.
    b. State the subtypes of constant-K filters.

21. Draw an L-type constant-K low-pass filter, acting as an impedance device.

22. State two uses of constant-K filters.

23. Draw a T-type constant-K high-pass filter.

24. A 40-μH and a 240-pF capacitor are connected as an L-type constant-K high-pass filter.
    a. Draw the circuit.
    b. Calculate the low cutoff frequency.
    c. What is the attenuation in dB at the cutoff frequency?

25. Draw an m-derived high-pass filter which incorporates a parallel resonant circuit.

26. Briefly describe an active filter.

chapter 20

COMPLEX NUMBERS

Reactance may be stated in +J or -J values on phasor diagrams. One reactance value plus a resistance value is known as a complex number and expresses impedance or voltage in ac circuits in what is known as rectangular form.

Values may be stated in rectangular form or their converted value, polar form. Polar form states the impedance in magnitude and phase angle. Both forms used alternately can make the calculation of otherwise very difficult or unsolvable complex series–parallel circuits relatively easy.

20-1 Complex Numbers and Impedance

Analysis of simple resistive reactive series circuits or simple parallel circuits can be done quite easily with the use of Pythagoras' triangle.

$$A = \sqrt[2]{B^2 + C^2}$$

Complex series–parallel impedance circuits and theorems are much more difficult to solve and require the J-operator for ease of solution. Complex numbers are especially useful when solving theorems in impedance circuits.

20-1.1 Conversions

Impedance and voltage are stated in one of two values:

1. Rectangular form.
2. Polar form.

An example of a polar form value is $Z = 50 \, \Omega \, \underline{/53.13°}$. An example of a rectangular form value is $Z = 30 + J40$. The methods of conversion from PF to RF and vice versa must be understood as well as knowing the mechanics of complex numbers. There are two types of conversion:

1. Rectangular form (RF) to polar form (PF)
2. Polar form (PF) to rectangular form (RF)

Impedance given in RF states the value of the separate components included in the impedance: resistance and reactance. Regardless of whether the original circuit is a series, parallel, or a complex circuit, its equivalent impedance, stated in rectangular form, is that of a series circuit consisting of one resistance and one reactance value.

CONVERSION FROM RECTANGULAR FORM TO POLAR FORM

The conversion from rectangular form (resistance and reactance are given) to polar form has been performed many times in Chap. 17 as an operation of Pythagoras' Triangle.

$$\text{impedance magnitude} = \sqrt[2]{R^2 + X^2}$$

The *phase angle* θ required as part of the polar form expression may be determined by taking the arc tan of the reactance divided by the resistance (Figs. 20-1 and 20-2).

$$\theta = \tan^{-1} \frac{X}{R}$$

PROBLEM 20-1

The impedance of a circuit is stated as, for example,

$$Z = 30 + J40 \text{ (rectangular form)}$$

What is the value of the same impedance stated in polar form?

SOLUTION   *(cover solution and solve)*

1. The value is obtained by determining the magnitude of the impedance from the equation

$$Z = \sqrt{R^2 + X^2}$$
$$= \sqrt{30^2 + 40^2}$$
$$= 50 \, \Omega$$

2. The phase angle between the impedance and the reference arm is equal to

$$\theta = \text{arc tan } (\tan^{-1}) \frac{X}{R}$$

$$\text{arc tan} (\tan^{-1}) = \frac{X}{R}$$

$$\tan^{-1} = \frac{40}{30}$$

$$\theta = 53.13°$$

The angle is negative only if the J-operator is negative.
The complete answer in polar form is

$$Z = 50 \ \Omega \ \underline{/53.13°}$$

Thus, impedance (Z) is expressed as $30 + J40 \ \Omega$ in rectangular form and as $50 \ \Omega \ \underline{/53.13°}$ in polar form. See Figs. 20-1 and 20-2.

CONVERSION FROM POLAR FORM TO RECTANGULAR FORM

Converting an impedance value expressed in polar form to rectangular form involves converting an ohmic value and phase angle into two component values:

1. Resistance
2. Reactance

The phasor diagram in Fig. 20-1 shows that the resistance value may be obtained by

$$R = \cos \theta \times Z$$

The reactance value may be obtained by

$$X = \sin \theta \times Z$$

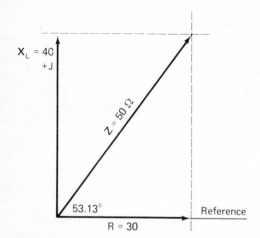

FIGURE 20-1   Phasor diagram and polar-form values.

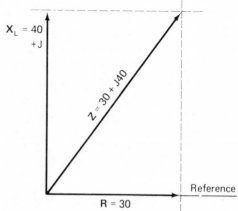

FIGURE 20-2   Phasor diagram and rectangular-form values.

Placing the two component values together produces an impedance in rectangular form as follows:

$$Z = R \pm J$$

The J-operator takes the negative sign if the angle is negative and takes the positive sign if the angle is positive.

The following problem shows the method of converting from polar form to rectangular form.

PROBLEM 20-2

Convert the impedance of 50 Ω $\underline{/53.13°}$ to rectangular form. Refer to Fig. 20-1.

SOLUTION   *(cover solution and solve)*

$$R = \cos 53.13° \times 50 = 30 \; \Omega$$

$$X = \sin 53.13° \times 50 = 40 \; \Omega$$

The angle is positive. Therefore, the J-operator will be a +J.

Final answer $Z = 30 + J40 \; \Omega$ (Fig. 20-2)

20-2   Complex Numbers and Series Circuits

20-2.1   J-Operator and Signs

Inductive reactance, whether in series or parallel circuits, is always represented by +J.

Capacitive reactance, whether in parallel or series circuits, is always represented by – J.

20-2.2   Rectangular Form Values in Series Circuits

The value of the total impedance in series circuits was calculated in Prob. 17-11 by taking the sum of the resistances and the sum of the same types of reactances and adding them vectorially.

The difference between the vectorial method used in Chap. 17 and complex numbers is that the latter uses the J-operator and its respective sign, preceding the reactance value. The answer obtained is in rectangular form, which can then be converted to the polar form.

PROBLEM 20-3

Calculate the total impedance of Fig. 20-3. (For purposes of comparison, the same values as in Fig. 17-20 are used.)

SOLUTION   *(cover solution and solve)*

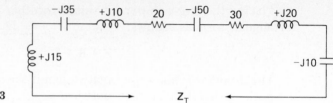

FIGURE 20-3

Add all like J values.

+J	−J	R
+J15	−J35	20
+J10	−J50	30
+J20	−J10	—
+J45	−J95	50

Final answer

$$Z = 50 - J50$$

$$= 70.7\ \Omega\ \underline{/-45°}$$

20-2.3 Polar Form Values in Series Circuits

Many impedances may be connected together in series circuits with their values shown in polar form, as in Fig. 20-4. Polar form values cannot be added because angles cannot be added. The polar form values must therefore first be converted to rectangular form, as shown in Prob. 20-4.

PROBLEM 20-4

Calculate the total impedance of Fig. 20-4.

SOLUTION   *(cover solution and solve)*

1. Convert all PF values to RF values.
2. Add all values of the same type. See Fig. 20-5.

	R	+J	−J
$Z_1 = 20\ \Omega\ \underline{/-90°}$ =	0	—	−J20
$Z_2 = 22.36\ \Omega\ \underline{/26.57°}$ =	20	+J10	—
$Z_3 = 58.31\ \Omega\ \underline{/-59°}$ =	30	—	−J50
$Z_4 = 20\ \Omega\ \underline{/+90°}$ =	—	+J20	—
Z_T =	50	+J30	−J70

$$Z_T = 50\quad +J30\quad -J70$$

$$= 50 - J40\ \text{(RF)}$$

$$Z = 64.03\ \Omega\ \underline{/-38.66°}\ \text{(PF)}$$

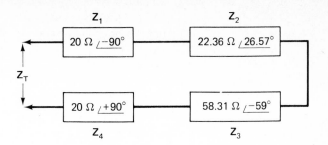

FIGURE 20-4

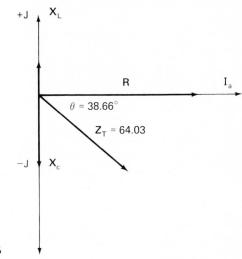

FIGURE 20-5

PROBLEM 20-5

A choke coil with an impedance (Z_L) of 22.5 Ω /63.4° is connected in series with a 25-Ω resistance (R_2). The circuit current (I_a) is 2 A.

1. Calculate the total circuit impedance in RF.
2. Draw the series equivalent circuit. (See Fig. 20-6.)
3. Calculate the total impedance of the circuit in polar form.
4. Draw the phasor diagram showing R_2, Z, X, R, and the phase angles. (See Fig. 20-7.)
5. Calculate the following voltages in polar form:

    a. V_a
    b. V_{R_L}
    c. V_{R_2}
    d. V_L (pure)

6. Enter the voltage values on the phasor diagram.

SOLUTION   *(cover solution and solve)*

1. Total circuit impedance in RF
   Z_L of coil in rectangular form

$$R = (\cos 63.4°) \times 22.5 = 10 \ \Omega$$

$$X_L = (\sin 63.4°) \times 22.5 = +J20$$

$$Z_L = 10 + J20 \ \Omega$$

   Total circuit impedance (Z_T) in RF

$$Z_T = Z_L + R_2$$

$$= (10 + J20) + 25$$

$$= 35 + J20 \ \Omega$$

2. Equivalent circuit (Fig. 20-6)
3. Total circuit impedance in polar form

$$\text{magnitude} = \sqrt{35^2 + 20^2} = 40.3 \ \Omega$$

$$\text{angle } (\theta) = \tan^{-1} \frac{20}{50} = 29.74°$$

$$Z_T = 40.3 \ \Omega \ \underline{/29.74°}$$

4. Phasor diagram (Fig. 20-7)
5. a. $V_a = ZI_a$

$$= 40.3 \ \Omega \ \underline{/29.74°} \times 2 = 80.6 \text{ V} \ \underline{/29.74°}$$

   b. $V_{R \text{ coil}} = R_L I_a$

$$= 10 \ \Omega \ \underline{/0°} \times 2 = 20 \text{ V} \ \underline{/0°}$$

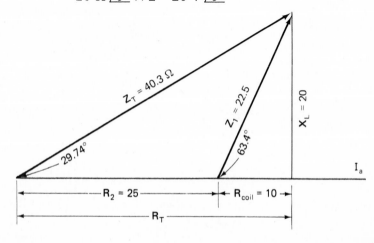

FIGURE 20-7

c. $$V_{R_2} = R_2 I_a$$
$$= 25\ \Omega\ \underline{/0^\circ} \times 2 = 50\ V\ \underline{/0^\circ}$$

d. $$V_{X_L} = X_L I_a$$
$$= 20\ \Omega\ \underline{/90^\circ} \times 2 = 40\ V\ \underline{/90^\circ}$$

PROBLEM 20-6

The applied voltage (V_a) to a circuit containing a coil with a large dc resistance in its windings in series with a resistor (R_2) is 80 V $\underline{/30^\circ}$. The voltage across the coil (includes the resistance of its wire) is 44.15 V $\underline{/65^\circ}$.

1. Draw the phasor diagram.
2. Calculate the voltage across each component part in the circuit.
3. Calculate the ohmic value of each component if the circuit current is 2 A.

SOLUTION   (cover solution and solve)

1. Draw a rough phasor diagram using the given values (Fig. 20-8).
    a. Draw the reference arm.
    b. Place V_a on the reference arm at 30°.
    c. Draw a vertical line from the end of V_a to the reference arm and label it V_L.
    d. Draw V_{Z_L} on the phasor diagram by drawing a line from the reference arm to the point of the phasor arm V_a so that it is at an angle of 65° to the reference arm.

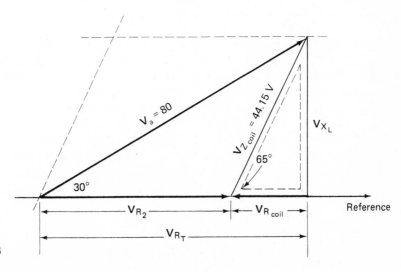

FIGURE 20-8

2. a. Voltage across R_T of circuit

$$V_{R\,total} = \cos \theta \times V_a$$
$$= \cos 30° \times 80 = 69.28 \text{ V}$$

b. Voltage across coil inductance

$$V_{X_L}(\text{pure}) = \sin \theta \times V_a$$
$$= \sin 30° \times 80 = 40 \text{ V}$$

c. Voltage across coil resistance

$$V_{R\,coil} = \cos \theta \times Z$$
$$= \cos 65° \times 44.15$$
$$= 0.4226 \times 44.15 = 18.66 \text{ V}$$

d. Voltage across series resistor (R_2)

$$V_{R_2} = V_{R\,total} - V_{R\,coil}$$
$$= 69.28 - 18.66 = 50.62 \text{ V}$$

3. a.
$$Z_T = \frac{V_a}{I_a} = \frac{80 \text{ V } \underline{/30°}}{2} = 40 \text{ }\Omega \text{ }\underline{/30°}$$

b.
$$R_2 = \frac{V_{R_2}}{I_a} = \frac{50.62 \text{ V } \underline{/0°}}{2} = 25.31 \text{ }\Omega \text{ }\underline{/0°}$$

c.
$$X_L = \frac{V_{X_L}}{I_a} = \frac{40 \text{ V } \underline{/90°}}{2} = 20 \text{ }\Omega \text{ }\underline{/90°}$$

d.
$$R_L = \frac{V_{R\,coil}}{I_a} = \frac{18.66 \text{ V } \underline{/0°}}{2} = 9.33 \text{ }\Omega \text{ }\underline{/0°}$$

20-3 Complex Number Operations

Expressing the total impedance of a series multiple-impedance circuit in rectangular form is merely a matter of adding together the same kinds of reactances (J signs) and connecting the real and imaginary values with the correct sign.

Determining the total impedance of parallel or series–parallel circuits involves all of the four arithmetic operations of addition, subtraction, multiplication, and division of complex numbers. All of these operations, therefore, must be understood.

20-3.1 The J-Operation

The letter "J," called the J-operator, represents all inductive-reactance values. The letter "–J" represents all capacitive-reactance values. The

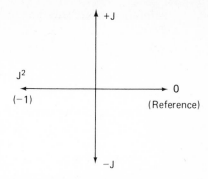

FIGURE 20-9   Positions of the J-operator and its value in each position.

actual meaning and importance of the J-operator, however, has not been discussed up until now. The J-operator is a mathematical sign used in ac theory to signify the direction of a quantity, and indicates the position of values, in terms of quadrants (Fig. 20-9).

20-3.2 Derivative of the J-Operator

The J-operator is also called the complex operator or imaginary number because it is the square root of -1 ($\sqrt[2]{-1}$). The square root of a given number is a number that, when multiplied by itself, equals that given number.

$$\sqrt{4} = \pm 2$$
$$\sqrt{9} = \pm 3$$
$$\sqrt{1} = \pm 1$$

The square root of -1 ($\sqrt{-1}$) is called an "imaginary" number because although -1 can exist, its square root must be imagined. The letter "+J" is a label used to represent the square root of -1 ($\sqrt{-1}$) and signifies a value positioned 90° counterclockwise from the reference arm. The letter "$-$J" is a label used to signify a value positioned 270° counterclockwise from the reference arm.

20-3.3 Signs of the J-Operator

If the arm rotates counterclockwise, the first quadrant (90°) is +J. The second quadrant is $J \times J = J^2$ and is 180° counterclockwise from the

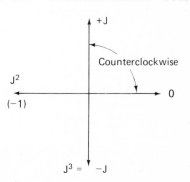

FIGURE 20-10   Direction of rotation of the J-operator.

reference arm. However, since $J = \sqrt{-1}$ and $J^2 = \sqrt{-1} \times \sqrt{-1}$, the value of J^2 is equal to -1 in numerical value (Fig. 20-10). Because it is constantly substituted in complex number multiplication problems, remember that $J^2 = -1$.

The $-J$ operator found in the third quadrant may be explained in two ways:

The first quadrant clockwise from the reference arm is a negative quantity; thus $-J$. Or, the third quadrant is

$$J^2 \times J = J^3$$

$$\text{but} \quad J^2 = -1$$

$$\therefore J^3 = -1 \times J$$

$$= -J$$

In summary, the values on the horizontal axis are referred to as the *real values* or *real axis* and the vertical axis is referred to as the *imaginary values*. A "J" appearing before a value (take note, *before* a value) indicates the number of degrees, in multiples of 90, at which a value is positioned with respect to the real value on the reference arm.

20-4 Mathematical Operations of Real and Imaginary Numbers

20-4.1 Adding Complex Numbers

EXAMPLE 1

Add $(3 + J5)$ and $(8 + J2)$.

$$\begin{array}{r} 3 + J5 \\ \underline{8 + J2} \\ 11 + J7 \end{array}$$

EXAMPLE 2

Add $(4 + J3)$ to $(3 - J4)$.

$$\begin{array}{r} 4 + J3 \\ \underline{3 - J4} \\ 7 - J1 \end{array}$$

EXAMPLE 3

Add $(5 - J9)$ and $(3 - J8)$.

$$\begin{array}{r} 5 - J9 \\ \underline{3 - J8} \\ 8 - J17 \end{array}$$

EXAMPLE 4

Add (4 − J3) and (3 + J4).

$$\begin{array}{r} 4 - J3 \\ \underline{3 + J4} \\ 7 + J1 \end{array}$$

20-4.2 Subtracting Complex Numbers

EXAMPLE 1

Subtract (3 − J2) from (3 + J4).

$$\begin{array}{r} 3 + J4 \\ \underline{-(3 - J2)} \\ 0 + J6 \end{array}$$

EXAMPLE 2

Subtract (3 − J2) from (6 − J4).

$$\begin{array}{r} 6 - J4 \\ \underline{-(3 - J2)} \\ 3 - J2 \end{array}$$

EXAMPLE 3

Subtract (3 − J5) from (5 − J3).

$$\begin{array}{r} 5 - J3 \\ \underline{-(3 - J5)} \\ 2 + J2 \end{array}$$

20-4.3 Multiplying Complex Numbers

EXAMPLE 1

(Foil method may also be used.) Multiply (9 + J5) by (3 − J2).

$$\begin{array}{r} 9 + J5 \\ \underline{3 - J2} \\ 27 + J15 \\ \underline{- J18 - J^2 10} \\ 27 - J3 - J^2 10 \end{array} \quad \text{but} \quad J^2 = -1$$

Therefore

$$27 - J3 - (-1)\,10$$
$$= 27 - J3 + 10$$
$$= 37 - J3$$

EXAMPLE 2

Multiply (3 – J2) by (6 + J3).

$$3 - J2$$
$$6 + J3$$
$$\overline{18 - J12}$$
$$+ J9 - J^2 6$$
$$\overline{18 - J3 - J^2 6} \quad \text{but} \quad J^2 = -1$$

Therefore

$$18 - J3 - (-1)\, 6$$
$$= 18 - J3 + 6$$
$$= 24 - J3$$

EXAMPLE 3

Multiply (5 – J2) by (4 – J3).

$$5 - J2$$
$$4 - J3$$
$$\overline{20 - J8}$$
$$- J15 + J^2 6$$
$$\overline{20 - J23 + (-1)\, 6 = 14 - J23}$$

20-4.4 Division of Complex Numbers

The division of complex numbers (rectangular form) may be accomplished by the rather cumbersome procedure of conjugating and rationalizing, as demonstrated in the following three examples. This method is usually avoided, however, in calculations by using the much easier method of converting the rectangular form to polar form and then dividing, as explained in Sec. 20-5.

The method of conjugating and rationalizing complex numbers is explained in the following three examples.

EXAMPLE 1

Divide (4 + J8) by (2 + J3).

$$\frac{\text{numerator}}{\text{denominator}} = \frac{4 + J8}{2 + J3}$$

The division involves dividing a real number by an imaginary (J) number, but it is impossible to divide real numbers by J-values (imaginary numbers); therefore, the J-value in the denominator must be eliminated.

The J-value may be eliminated in the denominator by multiplying (rationalizing) the numerator and denominator by the conjugate of the denominator.

The *conjugate* is formed by changing the sign of the J-operator in the denominator.

Example of conjugating and rationalizing

$$\frac{(4 + J8) \times (2 - J3)}{(2 + J3) \times (2 - J3)}$$

Numerator

$$\begin{array}{r} 4 + J8 \\ 2 - J3 \\ \hline 8 + J16 \\ - J12 - J^2 24 \\ \hline 8 + J4 - J^2 24 \end{array} \quad \text{but} \quad J^2 = -1$$

$$= 32 + J4$$

Denominator

$$\begin{array}{r} 2 + J3 \\ 2 - J3 \\ \hline 4 + J6 \\ - J6 - J^2 9 \\ \hline 4 \qquad +9 \end{array}$$

$$= 13$$

Note that the product of the denominator and its conjugate is equal to the sum of the squares of the numbers in the denominator.

Rationalized values

$$\frac{32 + J4}{13} = 2.46 + J0.308$$

EXAMPLE 2

Divide $(4 + J3)$ by $(2 - J2)$.

$$\frac{4 + J3}{2 - J2}$$

Conjugate and rationalize

$$\frac{(4 + J3) \times (2 + J2)}{(2 - J2) \times (2 + J2)}$$

Numerator

$$\begin{array}{r} 4 + J3 \\ 2 + J2 \\ \hline 8 + J6 \\ J8 + J^2 6 \\ \hline 8 + J14 + J^2 6 \end{array} \quad \text{but} \quad J^2 = -1$$

Therefore $8 + J14 + J^2 6 = 8 + J14 - 6$
$$= 2 + J14$$

Denominator $2 - J2$
$2 + J2$
$\overline{4 - J4}$
$\underline{\quad + J4 - J^2 4}$
$4 \qquad -J^2 4$ but $J^2 = -1$
$$= 8$$

Rationalized values

$$\frac{2 + J14}{8} = 0.25 + J1.75$$

EXAMPLE 3

Divide $(4 - J1)$ by $(1 + J2)$.

$$\frac{4 - J1}{1 + J2}$$

Conjugate and rationalize

$$\frac{(4 - J1) \times (1 - J2)}{(1 + J2) \times (1 - J2)}$$

Numerator $4 - J1$
$1 - J2$
$\overline{4 - J1}$
$\underline{\quad - J8 + J^2 2}$
$4 - J9 + J^2 2$ but $J^2 = -1$
$$= 2 - J9$$

Denominator: sum of the squares

$$1 + 4 = 5.0$$

Rationalized values

$$\frac{2 - J9}{5} = 0.4 - J1.8$$

20-5 Polar Form Mathematical Operations

20-5.1 Adding and Subtracting

It is not possible to add or subtract phasors expressed in polar form. Impedances expressed in polar form must first be converted into rectangular form, then added or subtracted, and then, if desired, finally converted into polar form.

20-5.2 Multiplication and Division

Multiplication and division of polar form values is such a simple process that it is usually simpler to convert RF values into PF values and divide them, than to leave values in RF and conjugate and rationalize.

MULTIPLICATION

EXAMPLE 1

Multiply $20\,\underline{/25°}$ by $10\,\underline{/14°}$.
Procedure:

1. Multiply the magnitudes.
2. Add the angles.

Answer $= 200\,\underline{/39°}$.

EXAMPLE 2

Multiply the magnitudes. Add the angles.

$$24\,\underline{/40°} \times 2\,\underline{/30°} = 48\,\underline{/70°}$$

$$24\,\underline{/40°} \times -2\,\underline{/30°} = -48\,\underline{/70°}$$

$$12\,\underline{/-20°} \times 3\,\underline{/+50°} = 36\,\underline{/+30°}$$

$$12\,\underline{/-20°} \times 4\,\underline{/-5°} = 48\,\underline{/-25°}$$

$$4\,\underline{/0°} \times 2\,\underline{/20°} = 8\,\underline{/20°}$$

$$4 \times 2\,\underline{/20°} = 8\,\underline{/20°}$$

(A real number has an angle of $0°$.)

DIVISION

EXAMPLE 1

Divide $20\,\underline{/35°}$ by $4\,\underline{/-15°}$.

$$\frac{20\,\underline{/35°}}{4\,\underline{/-15°}} = 5\,\underline{/50°}$$

In division of polar forms, the sign of the phase angle in the denominator is changed and then it is brought up to the numerator side where its value is added to the numerator value.

EXAMPLE 2

$$\frac{24\,\underline{/40°}}{2\,\underline{/30°}} = 12\,\underline{/10°}$$

$$\frac{12\,\underline{/-20^\circ}}{3\,\underline{/50^\circ}} = 4\,\underline{/-70^\circ}$$

$$\frac{12\,\underline{/20^\circ}}{3\,\underline{/50^\circ}} = 4\,\underline{/-30^\circ}$$

EXAMPLE 3

To divide polar forms by a real number, divide only the magnitudes by the real number.

$$\frac{12\,\underline{/30^\circ}}{4} = 3\,\underline{/30^\circ}$$

$$\frac{12\,\underline{/-30^\circ}}{3\,\underline{/0^\circ}} = 4\,\underline{/-30^\circ}$$

EXAMPLE 4

To divide a real number by a polar form:

1. Divide the magnitudes.
2. Change the sign of the angle in the denominator and place it in the numerator.

$$\frac{10}{5\,\underline{/30^\circ}} = 2\,\underline{/-30^\circ}$$

$$\frac{10}{5\,\underline{/-30^\circ}} = 2\,\underline{/30^\circ}$$

20-6 Complex Numbers and Parallel Circuits

The method used in Sec. 17-4 to determine impedance of parallel circuits was vectorial. Impedance was found by first determining the current in each leg, then vectorially adding these currents, and finally dividing the total current into the applied voltage which was either given or assumed. Note that the parallel resistance formula could not be used.

The advantage of the complex number (J-operator) method for finding impedance in parallel circuits is its simplicity. The complex number method uses the rectangular form for the values in each branch in the parallel resistance formula to calculate Z.

$$Z_T = \frac{Z_1 \times Z_2}{Z_1 + Z_2}$$

where All impedances are in either rectangular form (RF) or in polar form (PF).

614

20-6.1 Use of the Parallel Impedance Formula

$$Z_T = \frac{Z_1 \times Z_2}{Z_1 + Z_2} \tag{1}$$

Equation (1) can be used to determine the total impedance (Z) of a circuit whenever the value of each branch impedance is in either:

1. Polar form (e.g., 6 Ω $\underline{/20°}$)
2. Rectangular form (e.g., 3 + J8 Ω)

Whereas Eq. (1) cannot be used with straight vectorial values such as 10, 20, and so on, it can be used with polar form values or rectangular form values.

PROBLEM 20-7

1. Calculate the total impedance of Fig. 20-11 and state the impedance in RF.
2. Draw the equivalent series circuit. State the impedance in PF.

SOLUTION   *(cover solution and solve)*

1. Total impedance in RF
   The value of inductive reactance is always a +J, whether the circuit is parallel or series.
   a. Leg 1 has no resistance. Z_1 = 0 + J10 Ω.
      Leg 2 has no reactance. Z_2 = 10 + J0 Ω.
   b. Total impedance

$$Z_T = \frac{Z_1 \times Z_2}{Z_1 + Z_2} \tag{1}$$

$$= \frac{(0 + J10) \times (10 + J0)}{(0 + J10) + (10 + J0)} \tag{2}$$

$$= \frac{J100}{(10 + J10)} \tag{3}$$

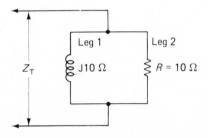

FIGURE 20-11

c. Division must now be done in Eq. (3) to obtain the impedance value (Z). The division may be done by one of two methods:

Method 1: Conjugate and rationalize. (A cumbersome method. Not recommended.)

Method 2: Convert to polar form and divide.

Method 1: (If preferred) conjugate and rationalize

$$Z = \frac{J100}{10 + J10}$$

$$Z = \frac{(J100) \times (10 - J10)}{(10 + J10) \times (10 - J10)}$$

Divide

$$Z = \frac{J1000 + 1000}{200}$$

$$Z = J5 + 5$$

The impedance of 5 + J5 may now be converted into polar form.

$$Z = 7.07 \ \Omega \ \underline{/45°}$$

Method 2: Convert the complex numbers to polar form and then divide to obtain the Z in polar form. Finally, convert to rectangular form if required.

(1) $$Z = \frac{J100}{10 + J10} = \frac{100 \ \underline{/+90°}}{14.14 \ \underline{/+45°}}$$

$$= 7.07 \ \Omega \ \underline{/+45°}$$

(2) Convert

$$R = \cos (+45°) \times 7.07 = 5 \ \Omega$$

$$X_L = \sin (+45°) \times 7.07 = 5 \ \Omega$$

The angle is positive, so

$$Z = 5 + J5$$

2. Equivalent series circuit (Fig. 20-12).

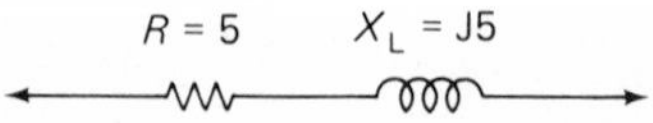

FIGURE 20-12

20-6.2 Complex Numbers and Parallel Resonant Circuits

Vectorial methods of computing the impedances of parallel resonant circuits are ponderous. The complex number method is simple and direct.

Compute the Z of Fig. 20-13

1. In RF at resonance
2. In PF at resonance

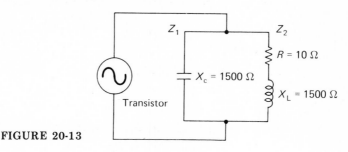

FIGURE 20-13

SOLUTION   (cover solution and solve)

1. Total impedance in rectangular form
    a. Find the impedance of each leg.

    $$Z_1 = 0 - J1500 \ \Omega$$

    $$Z_2 = 10 + J1500 \ \Omega$$

    b. Find total impedance in RF

    $$Z = \frac{Z_1 \times Z_2}{Z_1 + Z_2}$$

    $$= \frac{(-J1500) \times (10 + J1500)}{(-J1500) + (10 + J1500)}$$

    Numerator $\quad$ 10 + J1500

    $\underline{\quad\quad \times \quad (-J1500)}$

    $-J15{,}000 - J^2\,2{,}250{,}000$

    Denominator $\quad$ 10 + ~~J1500~~

    $\underline{+ \quad - ~~J1500~~}$

    10 + J0

    Divide $\quad \dfrac{(-J15{,}000) - J^2\,2{,}250{,}000}{10}$

    $$Z = -J1500 - J^2\,225{,}000$$

    but $\quad J^2 = -1$

    $\therefore$ Total impedance $= Z_T = -J1500 + 225{,}000 \ \Omega$

    $$= 225{,}000 - J1500 \ \Omega$$

The parallel resonant circuit in Fig. 20-13 is equivalent to a series circuit with a dc resistance of 225,000 Ω and 1500-Ω capacitive reactance. The circuit is predominantly resistive at resonance.

2. Total impedance in polar form

$$Z = 225,000 - J1500 \ \Omega \quad (RF)$$

$$Z = 225,005 \ \Omega \ \underline{/-0.39°} \quad (\text{polar form})$$

Practical value

$$Z = 225,005 \ \Omega \ \underline{/0°} \quad (PF)$$

20-7 Complex Numbers and Series-Parallel Circuits

PROBLEM 20-9

Find the impedance of the series–parallel circuit in Fig. 20-14.

SOLUTION   (cover solution and solve)

1. Find the series equivalent of the parallel circuit C_1 and L_1.

$$Z_{AB} = \frac{(-J40) \times (+J60)}{(-J40) + (+J60)} = \frac{-J^2 2400}{+J20}$$

$$= -J120$$

2. Add all the series impedances.

$$Z = R + Z_{AB} + X_L$$

$$= 10 + (-J120) + (+J20)$$

$$= 10 - J100 \quad (RF)$$

3. Convert to polar form.

$$Z = 100.5 \ \Omega \ \underline{/-84.3°}$$

The voltage is lagging the current by $-84.3°$

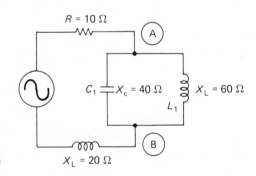

FIGURE 20-14

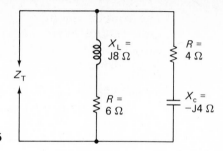

FIGURE 20-15

PROBLEM 20-10

Find the impedance of the series–parallel circuit in Fig. 20-15.

SOLUTION *(cover solution and solve)*

Figure 20-15 is an excellent example of the type of problem which can be solved easily only by complex numbers.

1. Find the impedance of each leg in RF and PF.

$$Z_1 = 6 + J8 = 10 \ \Omega \ \underline{/53°}$$

$$Z_2 = 4 - J4 = 5.66 \ \Omega \ \underline{/-45°}$$

2. Use the parallel formula.

$$Z = \frac{Z \times Z_2}{Z_1 + Z_2}$$

Method 1: (values in RF)
    Multiply the numerator and add the denominator in RF.

$$Z = \frac{(6 + J8) \times (4 - J4)}{(6 + J8) + (4 - J4)}$$

Numerator   6 + J8
             4 - J4
             ‾‾‾‾‾‾‾
            24 + J32
            $- J24 - J^2 32$ but $J^2 = -1$
            ‾‾‾‾‾‾‾‾‾‾‾‾‾
            24 + J8 + 32
            = 56 + J8 Ω (RF)

Denominator = (6 + J8) + (4 - J4)
             = 10 + J4

$$Z = \frac{56 + J8}{10 + J4}$$

It is easier to divide in polar form:

$$Z = \frac{56 + J8}{10 + J4} = \frac{56.6 \ \underline{/8°}}{10.8 \ \underline{/22°}} \quad \text{(divide in PF)}$$

$$= 5.24 \ \Omega \ \underline{/-14°} \quad \text{(PF)}$$

$$5.08 \qquad X_c = 1.27$$

FIGURE 20-16 ⟵——⟋⟍⟍——⊣⊢——⟶

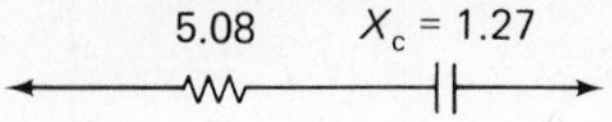

Method 2: (values in PF)

    a. Multiply the numerator in polar form and add the denominator in rectangular form.

$$Z = \frac{(6 + J8) \times (4 - J4)}{(6 + J8) + (4 - J4)}$$

$$Z = \frac{(10 \underline{/53°}) \times (5.66 \underline{/-45°})}{10 + J4}$$

    b. Divide in polar form.

$$Z = \frac{(10 \underline{/53°}) \times (5.66 \underline{/45°})}{10.8 \underline{/22°}}$$

$$= \frac{56.6 \underline{/8°}}{10.8 \underline{/22°}}$$

$$= 5.24 \ \Omega \underline{/-14°}$$

    c. Convert to rectangular form.

$$R = (\cos \underline{/14°}) \times 5.24 = 5.08 \ \Omega$$

$$X_c = (\sin \underline{/14°}) \times 5.24 = 1.27 \ \Omega$$

    d. Series equivalent circuit (Fig. 20-16)

20-8 Z of Multiple Parallel Circuits

Whenever one calculates the total impedance of a multiple parallel circuit, it is easier to use the *branch current* method than to calculate impedance in pairs. The procedure is as follows:

1. Find each branch Z in RF.
2. Convert each branch Z to polar form (for ease of division).
3. Assume an applied voltage if the applied voltage is unknown. (Voltage is common; therefore, voltage will have $0°$ phase angle.)
4. Obtain branch current values, e.g., $I_{Z_1} = V_a/Z_1$ (polar form).
5. Convert the branch current values from polar form to rectangular form so that they can be added for the I_T.
6. Add the branch currents.
7. To obtain Z_T, convert the rectangular form of I_T to the polar form of I_T for ease of division.
8. Find total impedance $(Z_T) = V_a/I_T$.
9. Draw the series equivalent circuit.

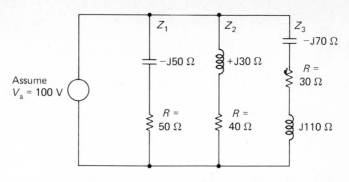

FIGURE 20-17

PROBLEM 20-11

Find the total impedance of the multiple parallel circuit of Fig. 20-17.

SOLUTION   (cover solution and solve)

$$Z_1 = 50 - J50 \qquad\qquad = 70.7\ \Omega\ \underline{/-45°}$$

$$Z_2 = 40 + J30 \qquad\qquad = 50\ \Omega\ \underline{/+37°}$$

$$Z_3 = 30 - J70 + J110 \qquad = 50\ \Omega\ \underline{/+53°}$$

$$I_1 = \frac{V_a}{Z_1} = \frac{100\ \underline{/0°}}{70.7\ \underline{/-45°}} = 1.414\ A\ \underline{/+45°} = 1 + J1\ A$$

$$I_2 = \frac{V_a}{Z_2} = \frac{100\ \underline{/0°}}{50\ \underline{/+37°}} = 2\ A\ \underline{/-37°} \qquad = 1.6 - J1.2\ A$$

$$I_3 = \frac{V_a}{Z_3} = \frac{100\ \underline{/0°}}{50\ \underline{/+53°}} = 2\ A\ \underline{/-53°} \qquad = 1.2 - J1.6\ A$$

$$I_T = I_1 + I_2 + I_3$$

$$I_T = (1 + J1) + (1.6 - J1.2) + (1.2 - J1.6)$$

$$I_T = 3.8 - J1.8 = 4.2\ A\ \underline{/-25.4°}$$

$$Z_T = \frac{V_a}{I_T} = \frac{100\ \underline{/0°}}{4.2\ \underline{/-25.4°}}$$

$$= 23.8\ \Omega\ \underline{/+25.4°}\quad (PF)$$

Rectangular form value

From $Z_T = 23.8\ \Omega\ \underline{/+25.4°}$

then $R = (\cos\ \underline{/25.4°})\ \times\ 23.8 = 21.49\ \Omega$

$X_L = (\sin\ \underline{/25.4°})\ \times\ 23.8 = J10.2\ \Omega$

$Z_T = 21.49 + J10.2\ \Omega \qquad$ (Fig. 20-18)

$$\overset{\displaystyle 21.49 \qquad +J10.2}{\text{FIGURE 20-18} \quad \longleftarrow \!\!-\!\!\bigwedge\!\!\bigwedge\!\!-\!\!\overset{\frown}{\bigcirc\!\bigcirc\!\bigcirc}\!\!-\!\!\longrightarrow}$$

20-9   Thevenin's Theorem and Impedance

20-9.1   Conditions of Application

Thevenin's equivalent circuits are particularly useful when it is necessary to calculate what value of current will flow into a variety of loads from a complex ac circuit containing one or two sources of constant voltage. Thevenin's theorem in ac networks is confined to linear network circuits in which all sources of voltage are constant and are operating at the same frequency.   The key word is constant.   The source of voltage may be considered constant when the value of the load resistance connected to the terminals of the voltage source is several times larger than the internal resistance (Thevenin's resistance) of the voltage source.   The constancy increases as the value of the load resistance relative to the internal resistance of the source increases.   Thevenin's voltage is the voltage which is measured across an open load.   A minimum ratio of R_L/R_{source} of 5/1 may be satisfactory in wide tolerance situations, yet a ratio of 10/1 may be unsatisfactory for critical circuits.   A voltage source may be considered constant if its output varies less than 1% with a variety of loads.

20-9.2   Instructions

Thevenin's Theorem as it applies to dc circuits was discussed in Chap. 10. The same rules and applications are used in ac circuits, with the exception that impedance values replace resistance values.   The open circuit (no-load) voltage between terminals A and B is still Thevenin's Voltage, and the internal value of impedance across the no-load terminals of the circuit is now Thevenin's Impedance instead of Thevenin's Resistance.

PROBLEM 20-12

Refer to Fig. 20-19.

1.   Find V_{TH}.
2.   Find Z_{TH}.
3.   Draw Thevenin's Equivalent Circuit.
4.   Find I_{Z_L}.
5.   Find V_{Z_L}.

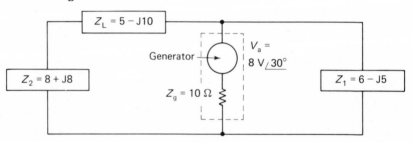

FIGURE 20-19

SOLUTION (*cover solution and solve*)

1. Thevenin's Voltage
    a. Remove Z_L and label the open terminals A and B.
    b. Redraw the equivalent circuit to a familiar format (Fig. 20-20).
    c. List the impedance values in both RF and PF.

$$Z_g = 10 = 10\ \Omega\ \underline{/0°}$$

$$Z_Z = 8 + J8 = 11.31\ \Omega\ \underline{/+45°}$$

$$Z_1 = 6 - J5 = 7.81\ \Omega\ \underline{/-39.8°}$$

$$Z_L = 5 - J10 = 11.18\ \Omega\ \underline{/-63.43°}$$

    d. Find Thevenin's voltage

$$V_{TH} = V_a \times \frac{Z_1}{Z_1 + Z_g} \quad \text{(Voltage ratio formula)}$$

$$V_{TH} = \frac{(8\ \underline{/30°}) \times (7.8\ \underline{/-39.8°})}{(6 - J5) + (10)} \left(\frac{\text{Easier to multiply in PF}}{\text{Must add in RF}} \right)$$

$$V_{TH} = \frac{62.48\ \underline{/-9.8°}}{16 - J5}$$

$$= \frac{62.48\ \underline{/-9.8°}}{16.76\ \underline{/-17.35°}} \quad \text{(Easier to divide in PF)}$$

$$V_{TH} = 3.73\ V\ \underline{/+7.55°}$$

2. Find Thevenin's Impedance (Z_{TH}). (Theveninize the circuit around Z_L.)
    a. Remove V_a and replace with a short.

$$Z_{TH} = \left(\frac{(Z_g) \times (Z_1)}{Z_g + Z_1} \right) + Z_2$$

$$Z_{TH} = \left[\frac{(10) \times (7.8\ \underline{/-39.8°})}{(10) + (6 - J5)} \right] + (8 + J8) \left(\frac{\substack{\text{Easier to mul-} \\ \text{tiply in PF}}}{\substack{\text{Must add in} \\ \text{RF}}} \right)$$

$$Z_{TH} = \left(\frac{78\ \underline{/-39.8°}}{16 - J5} \right) + (8 + J8)$$

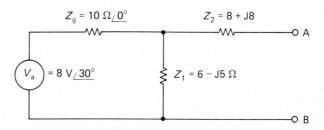

FIGURE 20-20 The standard format circuit for solving Thevenin's voltage and impedance values.

$$Z_{TH} = \left(\frac{78 \angle{-39.8°}}{16.76 \angle{-17.35°}}\right) + (8 + J8) \quad \text{(Easier to divide in PF)}$$

$$Z_{TH} = (4.65 \angle{-22.45°}) + (8 + J8)$$

b. Convert Z_{S_T} to RF before adding to Z_2.

$$R = (\cos{-22.45°}) \times 4.65 = 4.3 \ \Omega$$

$$X_L = (\sin{-22.45°}) \times 4.65 = -J1.78 \ \Omega$$

$$Z_{TH} = (4.3 - J1.78) + (8 + J8)$$

$$= 12.3 + J6.22$$

$$= 13.78 \ \Omega \ \angle{26.8°}$$

3. See Fig. 20-21.

4. Find current through the load.

a. $Z_{total} = Z_{TH} + Z_L$

$$= (12.3 + J6.22) + (5 - J10) \quad \text{(Must add in RF)}$$

$$= 17.3 - J3.78 \quad \text{(RF)}$$

$$= 17.7 \angle{-12.33°} \quad \text{(PF)}$$

b. $I_{Z_L} = \dfrac{V_{TH}}{Z_{total}}$

$$= \frac{3.73 \angle{+7.55°}}{17.7 \angle{-12.33°}} \quad \text{(Easier to divide in PF)}$$

$$I_{Z_L} = 0.21 \ A \angle{+19.88°}$$

5. Find voltage across the load.

$$V_{Z_L} = I_a Z_L$$

$$= (0.21 \angle{19.88°}) \times (11.18 \angle{-63.43°}) \quad \left(\begin{array}{l}\text{Easier to mul-}\\\text{tiply in PF}\end{array}\right)$$

$$= 2.34 \ V \angle{-43.55°}$$

Conversion: Thevenin's Voltage may be converted to Norton's Current with the same procedure as used in resistive circuits.

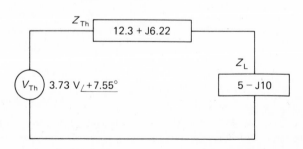

FIGURE 20-21   Thevenin's Equivalent Circuit.

$$I_N = \frac{V_{TH}}{Z_{TH}}$$

$$= \frac{3.73\ \underline{/7.55^\circ}}{13.78\ \underline{/26.8^\circ}}$$

$$= 0.27\ A\ \underline{/-19.25^\circ}$$

PROBLEM 20-13

Refer to Fig. 20-22.

1. Calculate Thevenin's Voltage.
2. Calculate Thevenin's Impedance. (Theveninize the circuit around Z_L.)
3. Draw Thevenin's Equivalent Circuit.
4. Voltage across Z_L.
5. Current through Z_L.

SOLUTION   *(cover solution and solve)*

1. Thevenin's Voltage.
   a. Redraw circuit to familiar format (Fig. 20-23) with Z_L removed and the open terminals labelled A and B.
   b. List all impedance values in RF and PF.

$$Z_1 = 2 - J1 = 2.236\ \Omega\ \underline{/-26.56^\circ}$$

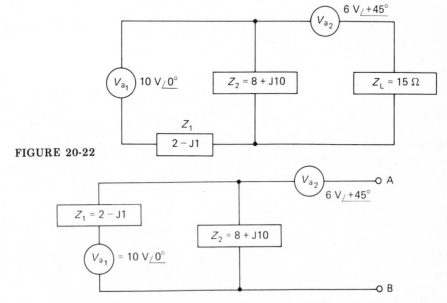

FIGURE 20-22

FIGURE 20-23 The standard format circuit for solving Thevenin's voltage and impedance values.

$$Z_2 = 8 + J10 = 12.8 \ \Omega \ \underline{/51.34°}$$

$$Z_L = 15 = 15 \ \Omega \ \underline{/0°}$$

The voltage across Z_2 plus the voltage of V_{a_2} is the voltage which will appear across terminals A and B, and is Thevenin's Voltage.

$$V_{TH} = V_{Z_2} + V_{a_2}$$

$$= \left(V_{a_1} \times \frac{Z_2}{Z_2 + Z_1}\right) + (V_{a_2})$$

$$= \left(\frac{(10 \ \underline{/0°}) \times (12.8 \ \underline{/51.34°})}{(8 + J10) + (2 - J1)}\right) + (6 \ \underline{/45°}) \qquad \left(\begin{array}{c}\text{Multiply in} \\ \text{PF} \\ \overline{} \\ \text{Must add} \\ \text{in RF}\end{array}\right)$$

$$V_{TH} = \left(\frac{128 \ \underline{/51.34°}}{10 + J9}\right) + (6 \ \underline{/45°})$$

$$= \left(\frac{128 \ \underline{/51.34°}}{13.45 \ \underline{/41.98°}}\right) + (6 \ \underline{/45°}) \quad \text{(Easier to divide in PF)}$$

$$V_{TH} = (9.516 \ \underline{/9.36°}) + (6 \ \underline{/45°}) \quad \text{(Cannot add in PF)}$$

$$= (9.389 + J1.5476) + (4.24 + J4.243)$$

$$V_{TH} = 13.629 + J5.79$$

$$V_{TH} = 14.8 \ V \ \underline{/23°}$$

2. Calculate Thevenin's Impedance. (Theveninize the circuit around Z_L.)

   a. Imagine the supply voltages V_{a_1} and V_{a_2} replaced with shorts. Thevenin's Impedance is the impedance across the open terminals A and B.

$$Z_{TH} = \frac{(Z_1) \times (Z_2)}{Z_1 + Z_2}$$

$$= \frac{(2.236 \ \underline{/26.56°}) \times (12.8 \ \underline{/51.34°})}{(2 - J1) + (8 + J10)} \qquad \left(\begin{array}{c}\text{Easier to mul-} \\ \text{tiply in PF} \\ \overline{} \\ \text{Must add} \\ \text{in RF}\end{array}\right)$$

$$Z_{TH} = \frac{28.635 \ \underline{/24.775°}}{10 + J9}$$

$$= \frac{28.64 \ \underline{/24.78°}}{13.45 \ \underline{/41.98°}} \quad \text{(Easier to divide in PF)}$$

$$Z_{TH} = 2.128 \ \Omega \ \underline{/-17.2°} \quad \text{(PF)}$$

$$= 2.033 - J.6298 \ \Omega \quad \text{(RF)}$$

3. Thevenin's Equivalent Circuit (Fig. 20-24)

4. Voltage across the load (V_{Z_L})

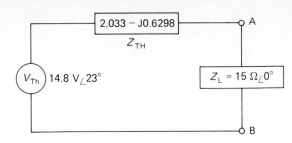

FIGURE 20-24   Thevenin's Equivalent Circuit.

$$\mathbf{V_{Z_L}} = \mathbf{V_{TH}} \times \left(\frac{\mathbf{Z_L}}{\mathbf{Z_L} + \mathbf{Z_{TH}}}\right) \quad \text{(Voltage ratio formula)}$$

$$= 14.8 \underline{/23°} \times \left(\frac{15 \underline{/0°}}{(15 + J0) + (2 - J0.63)}\right) \left(\begin{array}{c}\text{Easier to mul-}\\\text{tiply in PF}\\\hline\text{Must add in}\\\text{RF}\end{array}\right)$$

$$\mathbf{V_{Z_L}} = \frac{222 \underline{/23°}}{17 - J0.63}$$

$$= \frac{222 \underline{/23°}}{17 \underline{/-2.12°}} \quad \text{(Easier to divide in PF)}$$

$$\mathbf{V_{Z_L}} = 13.05 \text{ V } \underline{/25.12°}$$

5. Current through the load ($\mathbf{I_{Z_L}}$)

$$\mathbf{I_{Z_L}} = \frac{\mathbf{V_{Z_L}}}{\mathbf{Z_L}}$$

$$\mathbf{I_{Z_L}} = \frac{13.05 \underline{/25.12°}}{15 \underline{/0°}}$$

$$\mathbf{I_{Z_L}} = 0.87 \text{ A } \underline{/25.12°}$$

20-10 Norton's Theorem and Impedance

20-10.1   Conditions of Application

The conditions of application are similar to those for Thevenin's equivalent circuit. The circuit components must be linear; the voltage and current sources must be constant, and they must be operating at the same frequency. The device is said to be a constant source when the loads are approximately 1/10 to 1/100 the internal impedance of the current source depending upon the accuracy required.

20-10.2   Instructions

As in ac circuits, Norton's current is the current flowing through a current meter connected between terminals A and B. Norton's Impedance is the impedance across the open terminals A and B which is the same as Thevenin's Impedance.

Norton's Current may also be calculated by $I_N = V_{TH}/Z_{TH}$ if the open circuit voltage (V_{TH}) and the internal impedance (Z_{TH} or Z_N) of the voltage source are known.

PROBLEM 20-14

Refer to Fig. 20-25.

1. Calculate I_a.
2. Find Norton's Current (I_N).
3. Find Norton's Impedance.
4. Draw Norton's Equivalent Circuit.
5. Find the current through the load (I_{Z_L}).
6. Find the voltage across the load.

SOLUTION   (*cover solution and solve*)

Preliminary Steps
a. List all impedances in RF and PF.

$$Z_g = 10 + J0 = 10 \ \Omega \ \underline{/0°}$$

$$Z_1 = 6 - J5 = 7.81 \ \Omega \ \underline{/-39.8°}$$

$$Z_2 = 8 + J8 = 11.31 \ \Omega \ \underline{/+45°}$$

$$Z_L = 5 - J10 = 11.18 \ \Omega \ \underline{/-63.43°}$$

b. Remove Z_L and replace it with a current meter across terminals A and B. (A current meter is considered a short.)
c. Redraw Fig. 20-25 to a familiar format with Z_L replaced by a short (Fig. 20-26).

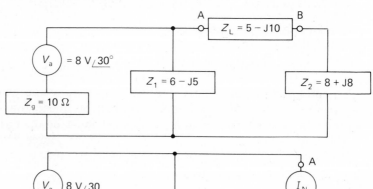

FIGURE 20-25

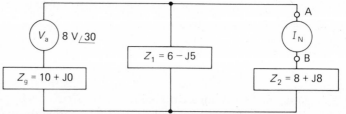

FIGURE 20-26   Standard format circuit for solving Norton's Current.

1. Calculate the circuit current (I_a).

$$I_a = \frac{V_a}{Z_T}$$

$$Z_T = \left(\frac{(Z_1) \times (Z_2)}{Z_1 + Z_2}\right) + Z_g$$

$$= \left(\frac{(7.81 \underline{/-39.8}) \times (11.31 \underline{/+45})}{(6 - J5) + (8 + J8)}\right) + 10 \quad \left(\begin{array}{c}\text{Easier to mul-}\\ \text{tiply in PF}\\ \hline \text{Must add}\\ \text{in RF}\end{array}\right)$$

$$= \left(\frac{88.36 \underline{/5.2^\circ}}{14 + J3}\right) + 10$$

$$= \left(\frac{88.37 \underline{/5.2^\circ}}{14.31 \underline{/12.09^\circ}}\right) + 10 \quad \text{(Easier to divide in PF)}$$

$$Z_T = (6.175 \underline{/-6.89^\circ}) + 10 \quad \text{(Must add in RF)}$$

$$= (6.149 - J0.741) + 10$$

$$= 16.149 - J0.741 \ \Omega \quad \text{(RF)}$$

$$= 16.16 \ \Omega \ \underline{/-2.62^\circ} \quad \text{(PF)}$$

$$I_a = \frac{V_a}{Z_T}$$

$$= \frac{8 \underline{/30^\circ}}{16.16 \underline{/-2.62^\circ}}$$

$$= 0.495 \text{ A} \underline{/32.62^\circ}$$

2. Calculate Norton's Current. (Use the inverse-current ratio formula).

$$I_N = I_a \times \left(\frac{Z_1}{Z_1 + Z_2}\right) \quad \text{(Inverse current ratio formula)}$$

$$= \frac{(0.495 \underline{/32.62^\circ}) \times (7.81 \underline{/-39.8^\circ})}{(6 - J5) + (8 + J8)} \quad \left(\begin{array}{c}\text{Easier to mul-}\\ \text{tiply in PF}\\ \hline \text{Must add}\\ \text{in RF}\end{array}\right)$$

$$I_N = \frac{3.865 \underline{/-7.18^\circ}}{14 + J3}$$

$$I_N = \frac{3.865 \underline{/-7.18^\circ}}{14.31 \underline{/12.09^\circ}} \quad \text{(Easier to divide in PF)}$$

$$I_N = 0.27 \underline{/-19.27^\circ}$$

Norton's Current (I_N) = 0.27 A $\underline{/-19.27^\circ}$

3. Determine Norton's Impedance.
   Norton's Impedance and Thevenin's Impedance are identical and are found with exactly the same procedure.

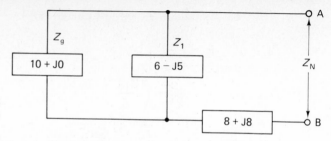

FIGURE 20-27 Standard format circuit for determining Norton's Impedance.

Redraw the circuit in Fig. 20-26 to Fig. 20-27 with the short (I_N) removed, the open terminals labelled A and B, and the battery replaced with a short.

$$Z_N = \left[\frac{Z_1 \times Z_g}{Z_1 + Z_g}\right] + Z_2$$

$$= \left[\frac{(6 - J5) \times (10 \,\underline{/0°}\,)}{16 - J5}\right] + Z_2$$

$$= \left[\frac{(7.81 \,\underline{/-39.81}\,) \times (10 \,\underline{/0°}\,)}{16.76 \,\underline{/-17.35}}\right] + Z_2$$

$$= \left[\frac{78.1 \,\underline{/-39.81}}{16.76 \,\underline{/-17.35}}\right] + Z_2$$

$$Z_N = (4.66 \,\underline{/-22.46}\,) + Z_2$$

$$Z_N = (4.66 \,\underline{/-22.46}\,) + (8 + J8)$$

$$Z_N = 12.29 + J6.22 \; \Omega \quad (RF)$$

$$Z_N = 13.78 \; \Omega \,\underline{/26.8°} \quad (PF)$$

4. Norton's Equivalent Circuit (Fig. 20-28)

5. Determine the current through the load. (Use the inverse current ratio formula.)

$$I_{Z_L} = I_N \times \left(\frac{Z_N}{(Z_N + Z_L)}\right) \quad \text{(Inverse current ratio formula)}$$

$$= \frac{(0.27 \,\underline{/-19.27°}\,) \times (13.78 \,\underline{/26.8°}\,)}{(12.3 + J6.22) + (5 - J10)}$$

$$= \frac{3.72 \,\underline{/+7.53°}}{17.3 - J3.78}$$

$$= \frac{3.72 \,\underline{/+7.53°}}{17.71 \,\underline{/-12.32°}}$$

$$= 0.21 \; A \,\underline{/+19.85°}$$

6. Determine the voltage developed across the load.

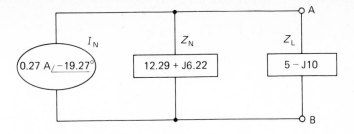

FIGURE 20-28 Norton's Equivalent Circuit.

$$V_{Z_L} = (0.21 \underline{/+19.85°}) \times (11.18 \underline{/-63.43°})$$
$$= 2.34 \text{ V} \underline{/-43.58°}$$

EXERCISES

1. Solve the following problems:
   a. $-J7 \times -J8 =$
   b. $-J6 \times J9 =$
   c. $J5 \times 2 =$
   d. $\dfrac{-J30}{-J3}$ (divide) $=$
   e. $\dfrac{-J60}{-J3} =$
   f. $\dfrac{J8}{4} =$

2. Add
   a. $4 + J7$
      $\underline{5 + J5}$
   b. $7 - J6$
      $\underline{2 + J4}$

   b. $3 + J4$
      $\underline{3 - J4}$
   d. $9 + J5$
      $\underline{3 - J8}$

   e. $30 \underline{/20°}$
      $\underline{40 \underline{/10°}}$

3. Subtract
   a. $8 + J9$
      $\underline{8 - J7}$
   b. $10 - J2$
      $\underline{-6 - J1}$

4. Multiply
   a. $9 + J7$
      $\underline{4 - J4}$
   b. $9 - J2$
      $\underline{7 + J4}$

5. Divide

    a. $\dfrac{4 - J5}{3 + J2}$ b. $\dfrac{2 + J6}{5 - J3}$

    c. $\dfrac{24\;\underline{/40°}}{2\;\underline{/30°}}$ d. $\dfrac{12\;\underline{/-20°}}{4\;\underline{/50°}}$

    e. $\dfrac{4 + J3}{2 - J2}$ f. $\dfrac{-J12}{-4}$

    g. $\dfrac{-J15}{-J3}$

6. Convert to rectangular form

    a. $40\ \Omega\ \underline{/-45°}$ b. $200\ \Omega\ \underline{/75°}$ c. $200\ \Omega\ \underline{/300°}$

7. Convert to polar form

    a. $300 - J175$ b. $85 + J90$

8. Use complex numbers to solve the following problem. A voltage of $150\ \text{V}\ \underline{/40°}$ is applied to an impedance of $177\ \Omega\ \underline{/-90°}\ \Omega$. Calculate the current value and degrees.

9. Calculate the voltage across each component for Fig. 20-29 and draw the phasor diagram showing the relationships among all values (current included).

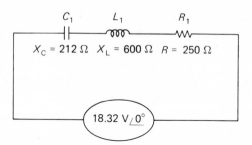

FIGURE 20-29

10. Determine the total impedance of the series circuit shown in Fig. 20-30.

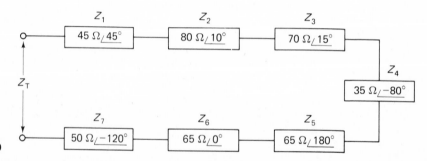

FIGURE 20-30

11. Determine the total impedance of the series circuit in Fig. 20-31. State the answer in:

    a. Rectangular form

    b. Polar form

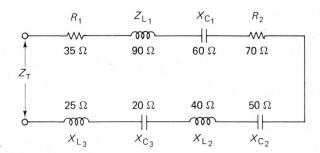

FIGURE 20-31

12. Calculate the total impedance of the circuit in Fig. 20-32 in RF.

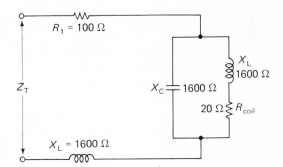

FIGURE 20-32

13. Determine the total impedance of the circuit shown in Fig. 20-33. State the answer in:

    a. Rectangular form

    b. Polar form

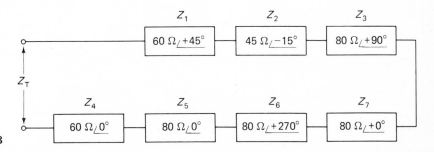

FIGURE 20-33

14. Determine the total impedance for the circuit in Fig. 20-34 in:

    a. Rectangular form

    b. Polar form

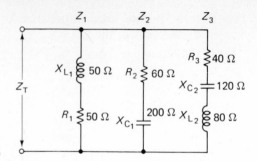

FIGURE 20-34

15. Determine for the circuit in Fig. 20-35 (state answers in PF):

   a. Thevenin's Voltage
   b. Thevenin's Impedance
   c. Thevenin's Equivalent Circuit for the circuit in Fig. 20-35
   d. Current through the load
   e. Voltage across the load

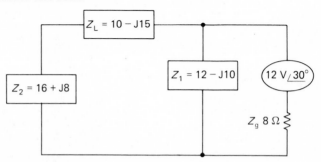

FIGURE 20-35

16. Determine for the circuit in Fig. 20-36 (state answers in PF):

   a. Generator current
   b. Norton's Current
   c. Norton's Impedance
   d. Norton's Equivalent Circuit for the circuit in Fig. 20-36
   e. Current through the load
   f. Voltage across the load

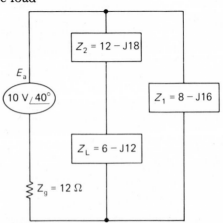

FIGURE 20-36

chapter 21

TEST EQUIPMENT

21-1 The Multimeter

Voltage, current, and resistance measurements are generally made with a measuring device called a VOM or multimeter. This meter is used in detecting defects in electronic equipment and in taking measurements in the lab. When the multimeter is used to measure current flow it is called an ammeter. When switched to the voltage mode it is called a voltmeter. The ohmmeter mode measures resistance. All three measurements depend upon current flow for their readings.

21-1.1 The Transducer (D'Arsonval Movement)

The majority of meters used to make electrical measurements employ the D'Arsonval, sometimes called a suspended moving coil meter movement (Fig. 21-1), or galvanometer, or just a dc meter.

21-1.2 Principle of D'Arsonval dc Meter Movement Operation

The essential parts of a D'Arsonval meter movement are:

1. Stationary element (permanent magnet)
2. Moving element (rotating coil)
3. Controlling element (spiral springs)
4. Mountings and bearings
5. Damping device
6. Pointer and scale

The dc meter operates on the electric motor action principle, which is based on the following four rules of magnetism.

1. A magnetic field is produced around a wire when current flows in that wire (rotating coil).
2. The polarity of the magnetic flux around the wire depends upon the direction of the current flow through it.
3. Magnetic lines of force leave the N-pole and enter the S-pole (the stationary element).
4. Like poles repel, unlike poles attract (motor action).

The pole pieces of the permanent magnet in a practical meter are curved in shape. Located between these pole pieces is a fixed cylindrical iron core. Although Fig. 21-1 indicates wide spacing between the coil and magnet, in an actual meter there is only enough space for the coil to rotate within the gap of the permanent magnet without striking the pole pieces. The iron core insures that the magnetic field is uniform in strength and constant in direction. Any torque sufficient to overcome the inertia and friction of the moving parts causes rotation until the magnetic fields are aligned. Uncontrolled rotation, however, would produce inaccurate current indications; therefore, the mechanism must include some method of controlling the coil rotation. The coil rotation is controlled by a special spring, one end of which is fixed and the other end of which is attached to the movable coil. The spring exerts a tension proportional to the amount of rotation and thus controls the rotation of the coil and insures accurate meter readings.

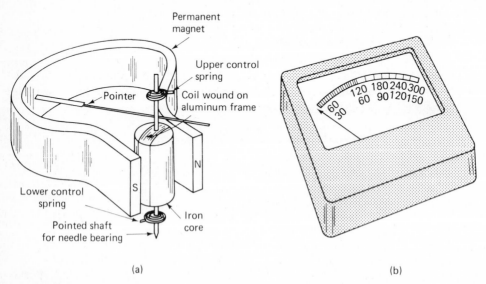

(a)   (b)

FIGURE 21-1   (a) The D'Arsonal meter movement.   (b) The meter movement made into a voltmeter.

Ch. 21 Test Equipment

The meter movement with the addition of an indicating pointer and a graduated scale becomes a complete current measuring instrument except for one thing—a damping device. Without a damping device, the pointer would swing back and forth several times at the indicated reading on the scale before coming to rest. Meters are generally damped by winding the movable coil on a lightweight aluminum frame. Aluminum is used since its low weight keeps inertia at a minimum. The frame itself is a closed circuit. The current which is being measured causes the coil to swing, and the frame is cut by the magnetic lines of force set up by the permanent magnet. Small currents and magnetic fields are induced into the frame which are opposite in direction to the field produced by the magnet itself. These fields in the frame oppose the permanent magnetic field and act as a brake to reduce the swing of the moving parts when the torque of the coil and the controlling force of the spring reach a position of balance.

Some other parameters of good meter movements include lightweight material used in the moving parts to keep inertia low. The coil is wound of many turns of small wire in order to keep down the weight and at the same time insure that there is sufficient torque. Any lightweight material is satisfactory for the pointer but usually very thin aluminum is used. Friction is reduced by making the pivot on which the coil shaft rotates out of a very hard metal and by setting the shaft in highly polished jewel bearings.

21-1.3 Characteristics of dc Meter

The dc meter movement has two important characteristics, the values of which must be known before the galvanometer can be designed as a voltmeter, an ammeter, or an ohmmeter. They are:

1. Meter sensitivity (I_M)
2. Meter internal resistance (R_M)

The values of these two characteristics often are not known and must be determined by the designer before the meter movement can be used.

METER SENSITIVITY (I_M)

The sensitivity of a meter movement is generally given as the amount of current flow required through the rotating coil of the meter to produce a full-scale deflection of the pointer.

Meter movements are available with sensitivities ranging from 30 mA to 10 μA.

The following procedure may be used to measure the sensitivity (I_M) of a meter movement.

1. Connect the series circuit (Fig. 21-2).
   a. M_M is the meter to be measured.

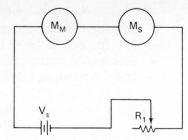

FIGURE 21-2 Test circuit for measuring the sensitivity of a meter movement.

b. M_S is a standard current meter set on the range which will measure the expected sensitivity of meter M_M.

c. R_1 is a 5000-Ω potentiometer set at maximum.

d. V_s is a 1-V to 2-V source of voltage.

2. The resistance of R_1 is reduced until the needle on meter M_M sets at maximum deflection.

3. The value of current as read on the standard meter M_S is the sensitivity (I_M) of the unknown meter (M_M).

The sensitivity of a meter movement is sometimes stated in ohms per volt instead of in milliamp or microamp values. This expression is often a difficult concept for a student to grasp because of the word "ohms." The expression means that if the value of the current required to make the pointer deflect to maximum is divided into the value of 1 V, a value of equivalent and only equivalent resistance is obtained. The value of equivalent resistance is that imaginary value of resistance which would be present to limit the current flow to the I_M value if 1 V were to be applied to the meter movement.

To convert meter sensitivity values from the "current" expression to the "ohms per volt" expression: Divide the amount of current to produce maximum deflection into 1 V and

$$\text{ohms per volt} = \frac{1\text{ V}}{\text{amount of current for maximum deflection}}$$

EXAMPLE

A meter pointer will deflect to maximum when 50 μA flows through it.

Meter sensitivity (I_M) in "current" term is 50 μA.

Meter sensitivity (I_M) in the "ohms per volt" term is

$$I_M = \frac{1\text{ V}}{50 \times 10^{-6}} = 20{,}000 \ \Omega/\text{V}$$

METER RESISTANCE (R_M)

Meter resistance is the resistance of the wire of which the rotating coil is wound. The more sensitive meters will have the greatest number of turns of wire and also the smallest diameter wire; in these a high resistance of several thousand ohms is not uncommon. Less sensitive meter movements have fewer turns of wire, larger diameter wire, and lower coil resistance.

A 30-mA meter movement may have a resistance of approximately 1 Ω, but a 50-μA movement may have a resistance of 1000 Ω. Meter movement coil resistance cannot be measured with an ohmmeter, however, because the battery current from the ohmmeter will cause the meter pointer to deflect violently past maximum travel. A damaged or destroyed meter movement results. Meter resistance must be measured by the following procedure:

1. Connect the circuit in Fig. 21-3 less the potentiometer R_2; this is actually the meter sensitivity circuit of Fig. 21-2.
2. Adjust potentiometer R_1 (Fig. 21-3) until the meter M_M reads maximum. The sensitivity reading of meter M_M is taken from the standard meter M_S and recorded.
3. Turn off the power and add potentiometer R_2 in parallel with M_M. (For meters with a sensitivity in the mA ranges, an R_2 of a few hundred ohms to 1500 ohms may be used. Higher sensitivity meters may require an R_2 of 5 kΩ or 10 kΩ.)
4. Turn on the power and adjust R_2 and R_1 until meter M_M is at maximum deflection when the standard meter M_S reads twice the sensitivity value of meter M_M (half the current is flowing through R_2).
5. Turn off the power and remove one lead from the potentiometer R_2.
6. Measure R_2 with an ohmmeter. This value is the resistance value of meter M_M.

Note: There are several ways to determine when the resistance of R_2 is equal to R_M, but because of the nonlinearity of especially cheap meter movements, it is best to adjust R_1 and R_2 until the M_M is set to its I_M and the meter M_S is reading twice the I_M of M_M.

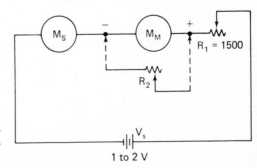

FIGURE 21-3 Test circuit for measuring the resistance of a meter movement.

21-1.4 Design Procedure for a Multimeter

It is not unusual for a person to have to build a voltmeter or current meter for use in industry, or for personal use. One very useful meter for personal use is a voltmeter to monitor the alternator and battery voltage in present day cars. The red light built into the dashboard, commonly referred to as the idiot light, usually warns of trouble only after it has already happened.

It is unlikely one will have to design an ohmmeter, but knowledge of the procedure will help one to understand its operation.

21-1.5 Design of a Voltmeter

There is no such thing as a voltmeter per se. A voltmeter is a current meter that has a resistance connected in series with it (see Fig. 21-4), which limits the current through the meter to its sensitivity value when a specific voltage is applied. That specific voltage is the maximum voltage, for a given voltage range, which will cause the meter needle to deflect to its maximum limit. The scale on the meter front is then labelled in volts instead of current values.

The design of a voltmeter range is a simple application of Ohm's Laws.

$$R_T = \frac{V}{I_M}$$

where I_M is the sensitivity value of the meter movement

V is the maximum value of voltage to be measured (range)

R_T is the total resistance of the circuit (R_M + R_{mult}) to limit the current to I_M when voltage (V) is applied

The current-limiting resistance in series with the meter movement is called the multiplier resistance (R_{mult}). The value of the (R_{mult}) is obtained by subtracting the resistance of the meter (R_M) from the total resistance (R_T) of the circuit.

$$R_{mult} = R_T - R_M$$

Some voltmeters have multiple ranges, as shown in Fig. 21-5. Depending upon the design, multiplier resistors may be connected in series and switched in and out, or each range may have its own separate multiplier

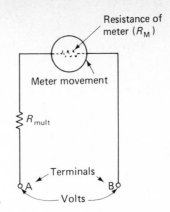

FIGURE 21-4   A voltmeter circuit.

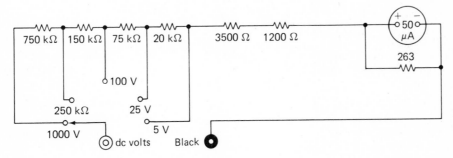

FIGURE 21-5   A multiple range dc voltmeter.

resistor. In any case, the resistors must be of the value calculated, or at least within 0.1% of that value. These close tolerance values are manufactured for meter companies, but for single design purposes, resistors may be selected by measuring several with an ohmmeter until the required value is found.

21-1.6 Current Meter Design

Current meter design entails placing a bypass resistance (R_S) in parallel (shunt) with a meter movement and then placing the combination in series with the circuit in which the current is to be measured. The shunting resistor enables the meter movement to measure currents much larger than those the meter movement could handle alone without being damaged.

The design of an ammeter (Fig. 21-6) is an application of Ohm's Law. Since R_S is in parallel with the meter movement, the voltage across the R_S will be the same as in the meter movement.

The voltage (V_M) which appears across the meter movement is always the same value at maximum deflection because

$$V_M = I_M \times R_M$$

where I_M is meter sensitivity (maximum current)

R_M is resistance of meter coil wire

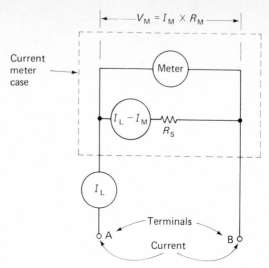

FIGURE 21-6 A current meter circuit to measure currents larger than the I_M of the meter movement.

The resistance value of R_S can now be obtained by

$$R_S = \frac{V_M}{I_L - I_M}$$

where I_L is the current to be measured in the line

R_S is the value of the bypass resistor

I_M is the meter sensitivity

$I_L - I_M$ is the current which will flow through R_S

21-1.7 Ohmmeter

SHUNT OHMMETER—TO READ X OHMS AT $\frac{1}{2}$ SCALE

There are several types of ohmmeter design. The one which will be discussed here is a design for testing low resistance values (less than 200 Ω).

A resistor has no voltage or current within itself. A meter movement is dependent upon current for its operation; therefore, an ohmmeter design must incorporate a battery.

SELECTION OF R_S

The principle of operation in Fig. 21-7 is that when terminals A and B are shorted together, 1.5 V is put in series with R_S. A voltmeter placed

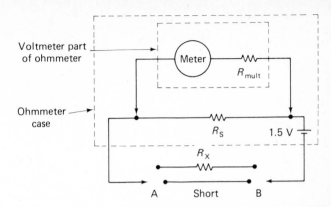

Voltmeter part of ohmmeter

Meter

R_{mult}

Ohmmeter case

R_S

1.5 V

R_X

A   Short   B

FIGURE 21-7 The basic circuit for a parallel-type ohmmeter.

across R_S will read 1.5 V at maximum deflection if it has been designed with a 1.5-V range.

If the short is removed and replaced by a resistor R_X, equal in value to R_S, the voltage across R_S will be one-half of 1.5 V, or 0.75 V, and the voltage across the meter connected in parallel with R_S will also be reduced to $\frac{1}{2}$ of the 1.5 V. The meter will now read $\frac{1}{2}$ scale.

To design an ohmmeter to read, for example, 20 Ω on $\frac{1}{2}$ scale, the value of R_S will have to be 20 Ω.

To design an ohmmeter to read, for example, 15 Ω, the value of R_S will have to be 15 Ω, and so on.

SELECTION OF R_M

The meter M and resistor R_M are actually a voltmeter circuit to which 1.5 V is applied when terminals A and B are shorted (Fig. 21-7). If the sensitivity of the meter movement is 50 μA and the battery voltage is 1.5 V, the total value of the limiting resistance will be

$$R_T = \frac{V_1}{I_M} = \frac{1.5}{50 \times 10^{-6}} = 30 \text{ k}\Omega$$

If the meter resistance is 1 kΩ the multiplier resistor will be

$$R_{mult} = R_T - R_{meter}$$

$$= 30 \text{ k}\Omega - 1 \text{ k}\Omega = 29 \text{ k}\Omega$$

Batteries weaken with age and use. If the battery voltage should reduce to 1 V, the multiplier resistor would have to be reduced (to compensate for the 0.5 V decrease). The reduction in resistance would be

$$\frac{V}{R} = \frac{0.5}{50 \times 10^{-6}} = 10 \text{ k}\Omega$$

The value of the R_{mult} therefore must be two resistors (Fig. 21-8).

$$R_1 = 19 \text{ k}\Omega \text{ fixed}$$

$$R_Z = 10 \text{ k}\Omega \text{ variable}$$

(R_Z is called the zero adjust on the ohmmeter.)

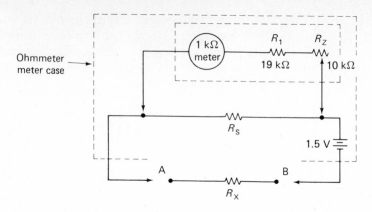

FIGURE 21-8   A parallel-type ohmmeter.

SCALE

A scale is marked on the face of the ohmmeter, with $0\ \Omega$ marked at the end of the scale to which the meter deflects to maximum when A and B are shorted. If R_S is $10\ \Omega$, the value $10\ \Omega$ will be marked on the center of the scale. The intermediate values may also be marked on the scale.

SERIES TYPE OHMMETER—WITH LOW SENSITIVITY METERS

The simplest type of ohmmeter, but one which will read only larger resistance values at center scale, is the series type ohmmeter (Fig. 21-9).

The principle of operation of the series type ohmmeter is that the unknown resistance values placed between terminals A and B will allow different values of current to flow through the meter movement and produce proportionate values of current readings on the meter. This current rating is converted into ohms by a scale placed behind the pointer and calibrated in ohms.

When terminals A and B are shorted together ($0\ \Omega$), maximum current

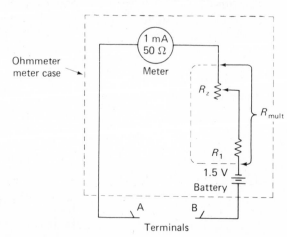

FIGURE 21-9   Series-type ohm-meter incorporating a 1-mA, 50-Ω meter movement.

flows through the meter, causing maximum pointer deflection which is read as 0 Ω on the ohmmeter scale behind the pointer. The largest resistance which can be measured is that which will allow some pointer deflection with the battery supplied in the ohmmeter. Larger resistances may be measured by using a larger battery.

LIMITING RESISTOR (R_M)

The standard 1.5-V dry cell is the smallest voltage supply available, but it will force more current through meter movements when terminals A and B are shorted than what they can handle without severe damage. A limiting resistor (R_{mult}), therefore, must be placed in series with the battery and meter movement. In other words, the meter and the limiting resistor R_M is actually a voltmeter circuit. The value of the total resistance of the circuit is determined by the sensitivity of the meter and the size of the battery when terminals A and B are shorted.

$$R_T = \frac{V_B}{I_M} = \frac{1.5}{1 \times 10^{-3}} = 1500 \ \Omega \tag{1}$$

The 1500 Ω is the total series resistance, but if the meter itself has a resistance of, for example, 50 Ω, the current limiting resistor will have a value of

$$R_{mult} = R_T - R_M$$
$$= 1500 - 50 = 1450 \ \Omega$$

ZERO ADJUST RESISTANCE

The above calculations are based on the value of a new dry cell (1.5 V). But batteries become weak with age and use, and this value may go down to 1 V or less. The meter pointer cannot deflect to maximum if the battery voltage (V_B) drops from 1.5 V to 1.0 V when a current limiting resistance of 1450 is used; therefore, a smaller R_M must be used. The simplest solution is to make part of R_M variable so that maximum deflection can be obtained whether the battery value is 1 V or 1.5 V. The variable resistor is called the "zero adjust" potentiometer (R_Z). The value of R_Z must be such that it will allow for a 0.5 difference in battery voltage.

$$R_Z = \frac{0.5 \ V}{I_M} = \frac{0.5}{1 \times 10^{-3}} = 500 \ \Omega$$

VALUE OF R_1

The value of the minimum circuit resistance R_1 is the difference between the value of R_M and R_Z.

$$R_1 = R_M - R_Z \quad \text{(R_M is the multiplier resistance)}$$
$$= 1450 - 500 = 950 \ \Omega$$

SCALE READINGS

1. The pointer goes to maximum deflection when terminals A and B are shorted together (zero ohms); therefore, the ohmmeter scale is calibrated 0 Ω at the right hand side of the meter.

2. The total resistance (R_T) of the meter circuit is 1500 Ω. Connecting the meter terminals across another 1500 Ω would allow one-half current flow and the meter needle would read one-half scale. The ohmmeter scale would be calibrated 1500 Ω at one-half scale.

3. Minimum deflection of this ohmmeter circuit may take place when a resistance as low as 3000 Ω is connected across the terminals.

21-1.8 Miscellaneous

HIGH SENSITIVITY METERS

High sensitivity meters such as a 50-μA movement will accommodate the measurement of much larger resistance values at center scale and at the minimum deflection end of the meter.

RESISTANCE RANGES

The range of resistance which a meter can measure can be changed by increasing or decreasing the size of the battery. In practice, a switch adds to or subtracts from the battery values.

There are many other types of ohmmeter circuits which are used, but they will not be discussed here.

21-2 Voltmeter Loading

Voltmeter loading is the bypassing of current around a component across which voltage is being measured, by using a meter whose input resistance is smaller than the resistance of the component. The consequence of meter loading is inaccurate voltage measurements. Most meter loading occurs with nonelectronic voltmeters in high impedance (resistance) and low voltage circuits.

Many people assume that a voltmeter measures voltage but draws no current and that the accuracy of a voltmeter is dependent mainly upon the cost of the instrument. The truth is, voltmeters do draw current, and the accuracy of a voltmeter is dependent primarily upon its input resistance which in nonelectronic voltmeters is dependent partly upon the sensitivity of the meter movement used in the voltmeter.

21-2.1 Meter Sensitivity and Input Resistance

Meter sensitivity is that value of current which is required to cause maximum meter pointer deflection. The less the amount of current required to produce maximum pointer deflection, the higher the sensitivity of the meter (see Sec. 21-1.3). The comparison of the following two meter movements illustrates how meter input resistance is determined by meter sensitivity.

METER 1: SENSITIVITY OF 1 mA

The total resistance of the meter circuit, including the multiplier, required to convert the meter to a 10-V range is

$$R_M = \frac{10}{1 \times 10^{-3}} = 10,000 \ \Omega$$

METER 2: SENSITIVITY OF 50 μA

The total resistance of the meter circuit, including the multiplier, required to convert the meter to a 10-V range is

$$R_M = \frac{10}{50 \times 10^{-6}} = 200 \ k\Omega$$

SUMMARY:

Meter 1 with the low sensitivity of 1 mA has an input resistance of 10 kΩ on the 10-V range.

Meter 2 with the higher 50-μA sensitivity has 200 kΩ of input resistance on the 10-V range.

Further comparison will reveal that the input resistance of each meter will be ten times greater on the 100-V range. For example, the 1-mA meter would be 100,000 Ω and the 50-μA meter would be 2 MΩ.

Electronic voltmeters have input resistances which vary from 1 MΩ to 10 or 15 MΩ for the FET types; these values remain relatively constant regardless of the range.

METER LOADING

The error in voltage measurement due to the low input resistance of a voltmeter in a circuit in which voltage is to be measured is due to meter loading. Now that meter input resistance is understood, the effect of meter loading can be illustrated in the following example.

1. Suppose each of the above 10-V range meters were to be used to measure the voltage in the circuit shown in Fig. 21-10a. The circuit has two 1-MΩ resistors connected in series with a 10-V source of voltage. Five volts are developed across each resistor as 5 μA of current flows through each resistor.

The circuit in Fig. 21-10b shows the 10-kΩ input resistance of Meter 1 connected in parallel with the 1-MΩ resistor R_2.

The use of Meter 1, with an input resistance of 10,000 Ω, to measure the voltage across R_2 would give a voltmeter reading across R_1 of

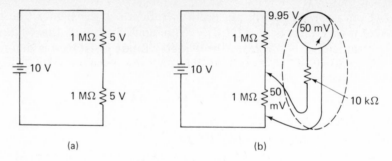

FIGURE 21-10 (a) Voltage developed across two 1-MΩ resistors. (b) The effect of meter loading on a high-value resistance.

approximately

$$10 \text{ k}\Omega \ (R_M) = \frac{1}{100} \times 1 \text{ M}\Omega \ (R_1)$$

The current through R_1 would then be

$$\frac{1}{100} \times 5 \ \mu\text{A} = \frac{5}{100} \ \mu\text{A}$$

Voltage across R_1 would be

$$I \times R = (5 \times 10^{-8}) \times (10^6) = 50 \text{ mV}$$

The meter would therefore read 50 mV instead of 5 V, which would lead the person taking the measurement to believe that a defect existed in the circuit.

The use of Meter 2, with an input resistance of 200 kΩ, to measure the voltage across resistor R_2 would give a voltmeter reading across R_2 of approximately

$$200 \text{ k}\Omega \ (R_M) = \frac{1}{5} \times 1 \text{ M}\Omega \ (R_2)$$

The current through R_2 would then be

$$\frac{1}{5} \times 5 \ \mu\text{A} = 1 \ \mu\text{A}$$

Voltage across R_2 would be

$$I \times R = (1 \times 10^{-6}) \times (10^6) = 1 \text{ V}$$

Meter 2 will read 1 V across the 1 MΩ resistor R_2, whereas the actual voltage across R_2 before the current was shunted was 5 V.

The correct voltage would be measured, however, with an electronic voltmeter, because its high input resistance of 10 or more MΩ would not bypass current around R_2, and the full 5 V would be produced across R_2.

SUMMARY:

Only electronic voltmeters register accurate voltage measurements in high impedance, low voltage circuits, because their high input resistance does not load the circuit.

EXERCISES

1. a. Draw a circuit which could be used to measure the sensitivity of a meter movement of unknown characteristics.
   b. Briefly explain how to measure the sensitivity of the unknown meter, using the circuit drawn for a.

2. A meter has a sensitivity characteristic of 20,000 Ω/V. Explain what is meant by this expression of sensitivity rating.

3. State the sensitivity value in Ω/V of
   a. 1-mA meter movement
   b. 500-μA meter movement

4. The sensitivity of a meter is rated at 20,000 Ω/V. What is its sensitivity in terms of current values?

5. A meter is rated at 100,000 Ω/V. What is sensitivity in current terms?

6. a. Draw a circuit which could be used to measure the internal resistance of a meter movement.
   b. Explain how to use the circuit drawn in a, to measure the resistance of the meter movement.

7. Why can an ohmmeter not be used to measure the resistance of a meter movement?

8. Design a current meter to measure 10 mA at full deflection with a 1-mA meter movement having a resistance of 200 Ω.
   a. Draw the circuit.
   b. Calculate the value of the shunt resistance.

9. Calculate the value of multiplier required to convert a meter movement with a sensitivity of 20,000 Ω/V and a 1000-Ω resistance into the following dc voltage ranges:

Range	Multiplier R
a. 1 V	_____
b. 3 V	_____
c. 50 V	_____
d. 150 V	_____
e. 300 V	_____
f. 50 kV	_____

10. Calculate the value of shunt resistance required to convert a 20,000-Ω/V sensitivity meter with a 1000 Ω resistance into the following current ranges:

Range	Shunt R
a. 50 μA	_____
b. 100 μA	_____
c. 1 mA	_____
d. 50 mA	_____
e. 1 A	_____

11. Must a current meter be connected in series or in parallel with the circuit being measured?

12. Must a voltmeter be connected in series or in parallel with the circuit being measured?

APPENDIX

Table A-1 Schematic Symbols

Symbol	Name
	Antenna (aerial)
	Ground
	Antenna (loop)
	Wiring method 1 Connection
	No connection
	Wiring method 2 Connection
	No connection
	Terminal
	One cell or "A" battery
	Multicell battery
	Resistor
	Potentiometer (volume control)
	Tapped resistor or voltage divider
	Rheostat
	Delay line
	Iron core choke coil
	Transformer (air core)
	Transformer (iron core)
	Power transformer P – 115 Volt primary; S_1 – Secondary; S_2 – Secondary; S_3 – Center-tapped high-voltage secondary
	Fixed capacitor
	Fixed capacitor (electrolytic)
	Adjustable or variable capacitor
	Adjustable or variable capacitors (ganged)
	I.F. Transformer (double-tuned)
	Power switch S.P.S.T.
	Switch SPDT
	Switch D.P.S.T.
	Switch D.P.D.T.
	Switch (rotary or selector)
	Spark gap
	Fuse
	Pilot lamp
	Headphones
	Loudspeaker
	Phono pick-up
	Non-polarized electrolytic (AC)
	Electrolytic
	Slide switch
	Multicontact switch
	Microphone (carbon)
	Microphone (crystal)
	Microphone (moving coil)
	Relay
	Phone jack
	Phone jack
	Phone jack
	Phono jack
	Shielded pair shield
	Shielded wire shield
	Shielded assembly
	Common ground
	Wires shielded between two points
	Lamps
	Neon lamps
	Meter
	Meter
	Powdered iron core tuned transformer
	Powdered iron core variable inductor
	Air core inductor
	Iron core inductor
	Magnetic P/B or recording head
	Relays
	Circuit breaker (Reset button)
	Phone plug
	RCA jack
	Interconnecting plug Male
	Interconnecting plug Female
	Line plug
	Piezoelectric crystal
	Piezoelectric crystal monaural phono cartridge
	Piezoelectric crystal stereo phono cartridge

652

International System of Units

The International System of Units (SI units) and quantity symbols are those established by the international scientific community in order that there be uniformity throughout the world.

Table A-2 Countries Already Metric or Changing

Afghanistan	Ecuador	Lebanon	Rwanda
Albania	El Salvador	Lesotho	San Marino
Algeria	Equatorial Guinea	Libya	Saudi Arabia
Andorra	Ethiopia	Liechtenstein	Senegal
Arab Republic of	Fiji	Luxembourg	Singapore
Egypt	Finland	Malagasy Republic	Somalia
Argentina	Formosa	Malawa	South Africa
Australia	France	Malaysia	Spain
Austria	Gabon	Maldives (Republic)	Sri Lanka (*formerly*
Bahrain	Germany (Federal	Mali	Ceylon)
Belgium	Republic)	Maita GC	Sudan
Bermuda	Ghana	Mauritania	Swaziland
Bolivia	Gibraltar	Mauritius	Sweden
Botawana	Greece	Mexico	Switzerland
Brazil	Guatemala	Monaco	Syria
Bulgaria	Guinea	Mongolian People's	Tanzania
Burundi	Guyana	Republic	Thailand
Cambodia	Haiti	Morocco	Togo
Cameroon	Honduras	Napal	Trinidad & Tobago
Canada	Hong Kong	Nauru	Tunisia
Central African	Hungary	Netherlands	Turkey
Republic	Iceland	New Zealand	Uganda
Chad	India	Nicaragua	U.S.A.
Chile	Indonesia	Niger	USSR
China (People's	Iran	Nigeris	United Arab
Republic)	Iraq	Norway	Emirates
Colombia	Ireland (Republic)	Oman	United Kingdom
Congo (Republic)	Israel	Pakistan	Upper Volta
Costa Rica	Italy	Panama	Uruguay
Cuba	Ivory Coast	Paraguay	Venezuela
Cyprus	Japan	Peru	Vietnam
Czechoslovakia	Jordan	Philippines	Western Samoa
Dahoney	Kenya	Poland	Yugoslavia
Denmark	Korea	Portugal	Zaire
Dominion Republic	Kuwait	Qatar	Zambia
East Germany	Laos	Romania	

Table A-3 SI Units Named After Scientists

Unit	Scientist	Country of Birth	Dates
ampere	Ampère, André-Marie	France	1775–1835
coulomb	Coulomb, Charles Augustin de	France	1736–1806
degree Celsius	Celsius, Anders	Sweden	1701–1744
farad	Faraday, Michael	England	1791–1867
henry	Henry, Joseph	United States	1797–1878
hertz	Hertz, Heinrich Rudolph	Germany	1857–1894
joule	Joule, James Prescott	England	1818–1889
kelvin	William Thomson, Lord Kelvin	England	1824–1907
newton	Newton, Sir Isaac	England	1642–1727
ohm	Ohm, Georg Simon	Germany	1787–1854
pascal	Pascal, Blaise	France	1623–1662
siemens	Siemens, Karl Wilhelm later Sir William	Germany (England)	1823–1883
tesla	Tesla, Nikola	Croatia (United States)	1856–1943
volt	Volta, Count Alessandro	Italy	1745–1827
watt	Watt, James	Scotland	1736–1819
weber	Weber, Wilhelm Eduard	Germany	1804–1891

Table A-4 SI Prefixes

Prefix	Symbol	Factor by Which the Unit is Multiplied	
exa	E	1 000 000 000 000 000 000	$= 10^{18}$
peta	P	1 000 000 000 000 000	$= 10^{15}$
tera	T	1 000 000 000 000	$= 10^{12}$
giga	G	1 000 000 000	$= 10^{9}$
mega	M	1 000 000	$= 10^{6}$
kilo	k	1 000	$= 10^{3}$
hecto	h	100	$= 10^{2}$
deca	da	10	$= 10^{1}$
		1	$= 10^{0}$
deci	d	0.1	$= 10^{-1}$
centi	c	0.01	$= 10^{-2}$
milli	m	0.001	$= 10^{-3}$
micro	μ	0.000 001	$= 10^{-6}$
nano	n	0.000 000 001	$= 10^{-9}$
pico	p	0.000 000 000 001	$= 10^{-12}$
femto	f	0.000 000 000 000 001	$= 10^{-15}$
atto	a	0.000 000 000 000 000 001	$= 10^{-18}$

Quantities and Their Symbols

Many SI units use sloping symbols or letters from the Greek alphabet as their symbols. Tables A-5 and A-6 show some of these letters

Table A-5 Quantities and their Sloping Symbols

Quantity	*Sloping Symbol*
Acceleration	a, g
Area	A
Breadth	b
Capacitance (Electric)	C
Charge	Q
Conductance (Electric)	G
Current (Electric)	I
Density	D
Diameter	d
Distance	d
Energy, Work	E, W
Flow (Volume Rate)	q
Force	F
Force of Gravity	G
Frequency	f
Height	h
Length	l
Mass	m
Moment of Force	M
Momentum	p
Period (Time of One Cycle)	T
Potential (Electric)	V
Power	P
Pressure	p
Quantity of Heat	Q

continued

Table A-5 *(Continued)*

Quantity	*Sloping Symbol*
Radius	r
Relative Density	d
Resistance (Electric)	R
Rotational Frequency	n
Specific Heat Capacity	c
Specific Latent Heat	l
Speed, Velocity	v, u, c
Temperature (Common)	t
Temperature (Thermodynamic)	T
Thickness	d
Time	t
Torque	T
Velocity	v, u, c
Volume	V
Width	w

Table A-6   The Greek Alphabet

Letter		Name	Designates
Small	*Capital*		
α	A	Alpha	Phase Angle, Coefficients, Attenuation Constant, Absorption Factor, Area
β	B	Beta	Phase Angle, Coefficients, Phase Constant, Magnetic Flux Density
γ	Γ	Gamma	Specific Quantity, Angles, Electrical Conductivity
δ	Δ	Delta	Density, Phase Angle, Increment or Decrement (Cap or Small)
ϵ	E	Epsilon	Permittivity, Base of Natural (Napierian) Logarithms, Illuminance, Energy
ζ	Z	Zeta	Coordinate, Coefficients
η	H	Eta	Intrinsic Impedance, Efficiency, Surface Charge Density, Magnetic Field Strength
θ	Θ	Theta	Angular Phase Displacement, Phase Angle
ι	I	Iota	Unit Vector
κ	K	Kappa	Coupling Coefficient
λ	Λ	Lambda	Wavelength, Attenuation Constant

continued

Small	Capital	Name	Designates
		Letter	
μ	M	Mu	Prefix *micro-*, Permeability, Amplification Factor
ν	N	Nu	—
ξ	Ξ	Xi	Coordinates
o	O	Omicron	—
π	Π	Pi	3.1416 (Circumference Divided by Diameter)
ρ	P	Rho	Resistivity, Volume Charge Density Momentum Coordinates, Active Power, Density
σ	Σ	Sigma	Surface Charge Density, Leakage Coefficient, Sign of Summation, Thickness
τ	T	Tau	Time Constant, Volume Resistivity, Time-Phase Displacement, Transmission Factor
υ	Υ	Upsilon	—
φ	Φ	Phi	Magnetic Flux, Phase Angle, Lumen
χ	X	Chi	Electric Susceptibility
ψ	Ψ	Psi	Dielectric Flux, Phase Difference, Coordinates
ω	Ω	Omega	Angular Velocity ($2\pi f$), Resistance in Ohms (Cap), Solid Angle

The following tables present those quantities most likely to be used in electronics, acoustics, instrumentation, and electricity. The quantities and their symbols, and the associated International System of Units (SI) and their multiples and submultiples, are updated to 1979.

NOTE: Some of the symbols and units used throughout this text may still be of the traditional types, as the text was written and in production before some of the following most recent SI units were published.

Table A-7   Electricity and Magnetism

Quantity Name and its Symbol	SI Unit and its Symbol	Some Common Multiples and Submultiples	Permitted Non-SI Units, Remarks, and Other Units
Active Power P	watt (W)	TW GW MW kW mW μW nW pW	—
Admittance Y	siemens (S)	MS kS mS μS nS	—

continued

Quantity Name and its Symbol	*SI Unit and its Symbol*	*Some Common Multiples and Submultiples*	*Permitted Non-SI Units, Remarks, and Other Units*
Apparent Power U	volt ampere (VA)	TVA GVA MVA kVA μVA nVA pVA	—
Capacitance C	farad (F)	μF nF pF	—
Conductance G	siemens (S)	MS kS mS μS	—
Electric Current I	ampere (A)	kA mA μA nA pA	—
Electric Charge, Quantity of Electricity Q	coulomb (C)	kC mC μC nC pC	—
Electric Potential V, φ, ϕ Potential Difference Tension U, V Electromotive Force E	volt (V)	MV kV mV μV nV	—
Electric Field Strength E, K	Volt per meter (V/m)	MV/m kV/m mV/m μV/m	—
Electric Flux ψ	coulomb (C)	MC kC mC	—
Electrical Energy	joule (J)	TJ GJ MJ kJ mJ	—

continued

Quantity Name and its Symbol	SI Unit and its Symbol	Some Common Multiples and Submultiples	Permitted Non-SI Units, Remarks, and Other Units
Electric Flux Density, Displacement D	coulombs per square meter (C/m^2)	kC/m^2 mC/m^2 $\mu C/m^2$	—
Electromagnetic Moment, Magnetic Moment m	amperes square meter (A/m^2)	—	—
Impedance Z	ohm (Ω)	$M\Omega$ $k\Omega$ $m\Omega$	—
Magnetization H_1, M	amperes per meter (A/m)	kA/m	—
Magnetic Flux ϕ	weber (Wb)	mWb	—
Magnetic Flux Density, Magnetic Induction B	tesla (T)	mT μT nT	—
Magnetic Field Strength H	ampere per meter (A/m)	kA/m	—
Magnetic Potential Difference U, U_m Magnetomotive Force F_m	ampere (A)	kA mA	—
Permeance A, P	henry (H)	—	—
Permeability μ	henries per meter (H/m)	mH/m $\mu H/m$ nH/m	—
Permittivity ϵ	farads per meter (F/m)	$\mu F/m$ nF/m pF/m	—
Power Level Difference D	—	—	When $Z_1 = Z_0$ $D = 20 \log_{10} V_0/V_1$ decibels (dB) or $D = 20 \log_{10} I_0/I_1$ decibels (dB)

continued

Quantity Name and its Symbol	SI Unit and its Symbol	Some Common Multiples and Submultiples	Permitted Non-SI Units, Remarks, and Other Units
Reactive Power Q	volt ampere reactive (VAR)	TVAR GVAR MVAR kVAR mVAR μVAR nVAR pVAR	—
Reactance X	ohm (Ω)	MΩ kΩ mΩ	—
Reluctance R, R_m	reciprocal henry (H^{-1})	—	—
Resistance R	ohm (Ω)	TΩ GΩ MΩ kΩ mΩ $\mu\Omega$	—
Self-Inductance L Mutual Inductance M, L_{mn}, L_{12}	henry (H)	mH μH nH pH	—
Surface Density of Charge σ	coulombs per square meter (C/m^2)	MC/m^2 kC/m^2 mC/m^2 μC/m^2	—
Susceptance B	siemens (S)	MS kS mS μS	—
Volume Density of Charge, Charge Density ρ, η	coulombs per cubed meter (C/m^3)	MC/m^3 kC/m^3 mC/m^3 μC/m^3	—

Quantity Name and its Symbol	SI Unit and its Symbol	Some Common Multiples and Submultiples	Permitted Non-SI Units, Remarks, and Other Units
Acceleration a Acceleration Due to Gravity g	meter per second squared (m/s^2)	cm/s^2 mm/s^2	Gal (gal)
Angular Acceleration α	radian per second squared (rad/s^2)	$mrad/s^2$	—
Angular Velocity ω, Ω	radian per second (rad/s)	—	—
Velocity u, v, c	meter per second (m/s)	cm/s mm/s	m/h km/h M/h kn (knot)
Frequency f, v	hertz (Hz)	THz GHz MHz kHz	—
Wavelength λ	meter (m)	mm μm nm pm	—
Time t (Time of One Cycle, Periodic time T)	second (s)	ks ms μs ns ps	day (d) hour (h) minute (min)
Length: Length l Breadth b Width w Distance d Path Length s Height h Thickness d, δ Radius r Diameter d	meter (m)	km cm mm μm nm	The angstrom is to be replaced by 0.1 nm: e.g., wavelength of 6000 Å = 600 nm.
Volume V, v	cubic meter (m^3)	dm^3 cm^3 mm^3	L (liter) GL ML kL

continued

Quantity Name and its Symbol	SI Unit and its Symbol	Some Common Multiples and Submultiples	Permitted Non-SI Units, Remarks, and Other Units
			hL mL μL
Area A, S	square meter (m^2)	km^2 dm^2 cm^2 mm^2	ha (hectare)
Solid Angle Ω, ω	Steradian (sr)	—	—
Plane Angle $\alpha, \beta, \gamma, \theta, \phi$	Radian (rad)	mrad μrad	$^\circ$ (degree) $'$ (minute) $''$ (second) r (revolution)

Table A-9　Acoustics

Quantity Name and its Symbol	SI Unit and its Symbol	Some Common Multiples and Submultiples	Permitted Non-SI Units, Remarks, and Other Units
Frequency f, v	hertz (Hz)	GHz MHz kHz	—
Wavelength λ	meter (m)	mm	—
Period, Periodic Time T	second (s)	ms μs	—
Velocity of Sound c	meter per second (m/s)	—	344 m/s at 20°C
Sound Intensity I, J, J_a	watt per square meter (W/m^2)	mW/m^2 nW/m^2 pW/m^2	Ref. = 1 pW/m^2 $I = 10 \log_{10} \dfrac{\text{actual intensity } (W/m^2)}{1 \times 10^{-12} \ (W/m^2)}$ = decibels (dB) (re 1 pW/m^2)
Sound Pressure Level L_p	—	—	Ref. = 20 μPa (in air) $L_p = 20 \log_{10} \dfrac{\text{actual pressure (Pa)}}{20 \times 10^{-6} \ (Pa)}$ = decibels (dB) (re 20 μPa)

continued

Quantity Name and its Symbol	SI Unit and its Symbol	Some Common Multiples and Submultiples	Permitted Non-SI Units, Remarks, and Other Units
Sound Power Level L_P	—	—	Ref. = 1 pW $$L_P = 10 \log_{10} \frac{\text{actual power (W)}}{1 \times 10^{-12} \text{ (W)}}$$ = decibels (dB) (re 1 pW)
Sound Energy Flux, Sound Power P, P_a	watt (W)	kW mW nW pW	—
(Instantaneous) Sound Pressure p	pascal (Pa)	mPa μPa	—
Static Pressure p_s	pascal (Pa)	kPa	—
(Instantaneous) Sound Particle Velocity u, v	meter per second (m/s)	mm/s	—
(Instantaneous) Volume Velocity q, U	cubed meter per second (m³/s)	—	—
Acoustic Impedance Z_a	pascal second per cubed meter (Pa–s/m³)	—	—
Mechanical Impedance Z, Z_m, w	newton second per meter (N–s/m)	—	—
Specific Acoustic Impedance Z_s	pascal second per meter (Pa–s/m)	—	—
Equivalent Absorption (Area of a Surface or Object) A	square meter (m²)	—	—
Reverberation Time T	second (s)	—	—

Quantity Name and its Symbol	SI Unit and its Symbol	Some Common Multiples and Submultiples	Permitted Non-SI Units, Remarks, and Other Units
Energy, Work E, W Potential Energy E_p Kinetic Energy E_k	joule (J)	TJ GJ MJ kJ mJ	—
Mass m	kilogram (kg)	Mg g mg μg	metric ton tonne (t) $1\ t = 10^3\ kg$
Density p (Mass Density)	kilogram per cubic meter (kg/m^3)	t/m^3 g/L mg/L	—
Momentum p	kilogram meter per second (kg–m/s)	—	—
Moment of Momentum L	kilogram square meter per second $(kg–m^2/s)$	—	—
Moment of Inertia of Mass I, J	kilogram square meter $(kg–m^2)$	—	—
Moment of Force M Torque T	newton meter (N–m)	MN·m kN·m mN·m μN·m	—
Force F, G, P	newton (N)	MN kN mN μN	$1\ N = 1\ kg–m/s^2$
Pressure p	pascal (Pa)	GPa MPa kPa mPa μPa	mbar (millibar) $1\ Pa = 1\ N/m^2$
Power P	watt (W)	TW GW MW kW mW μW	—

continued

Quantity Name and its Symbol	SI Unit and its Symbol	Some Common Multiples and Submultiples	Permitted Non-SI Units, Remarks, and Other Units
Stress: Normal σ Shear τ	pascal (Pa)	GPa MPa kPa	—
Modulus of Elasticity E Shear Modulus, Modulus of Rigidity G Bulk Modulus, Modulus of Compression K	pascal (Pa)	GPa MPa kPa	—
Compressibility κ	reciprocal pascal (Pa^{-1})	—	—
Viscosity (Dynamic) η, μ	pascal second (Pa–s)	mPa·s μPa·s	—
Viscosity Kinematic ν	square meter per second (m^2/s)	mm^2/s	—
Surface Tension σ, γ	newton per meter (N–m)	mN/m	—
Flow Rate: Mass q_m	kilogram per second (kg/s)	g/s	—
Volume q_v	cubed meter per second (m^3/s)	dm^3/s	L/s mL/s

Table A-11   Heat

Quantity Name and its Symbol	SI Unit and its Symbol	Some Common Multiples and Submultiples	Permitted Non-SI Units, Remarks, and Other Units
Common Temperature Celsius Temperature t, θ	degree Celsius ($^{\circ}$C)	—	Freezing point of water = $0\,^{\circ}$C Boiling point of water = $100\,^{\circ}$C
Thermodynamic Temperature T, Θ	kelvin (K)	mK	In general, degree Celsius [$^{\circ}$C] may be used instead of kelvin [K]
		—	Freezing point of water = 273.15 K Boiling point of water = 373.15 K
Linear Expansion Coefficient α	reciprocal kelvin (K^{-1})	MK^{-1}	—
	(reciprocal degree Celsius) $^{\circ}C^{-1}$	—	The degree Celsius [$^{\circ}$C] may be used instead of kelvin
Heat, Quantity of Heat Q	joule (J)	TJ GJ MJ kJ mJ	—
Thermal Resistance (Thermal Insulance) (Mechanical Usage) M	square meter kelvin per watt (m^2-K/W)	—	—
Thermal Resistance (Electronic Usage) R	kelvin per watt (K/W) ($^{\circ}$C/W)	— $^{\circ}$C/mW	— —
Thermal Conductivity λ, k	watt per meter Kelvin [W/(m-K)]	—	—
	watt per meter degree celsius [W/(m-$^{\circ}$C)]	—	—

Table A-12   Light and Related Electromagnetic Radiations

Quantity Name and its Symbol	SI Unit and its Symbol	Some Common Multiples and Submultiples	Permitted Non-SI Units, Remarks, and Other Units
Wavelength λ	meter (m)	μm nm pm	—
Radiant Energy Q, W, Q_e, U	joule (J)	—	—
Radiant Flux, Radiant Power P, ϕ, ϕ_e	watt (W)	mW μW	—
Luminous Intensity l, l_v	candela (cd)	mcd	—
Luminous Flux ϕ, ϕ_v	lumen (lm)	mlm μlm	—
Quantity of Light Q, Q_v	lumen second (lm–s)	—	—
Luminance L, L_v	candela per square meter (cd/m^2)	—	—
Illuminance E, E_v	lux (lx)	—	—
Light Exposure H, H_v	lux second (lx–s)	klx	lx–h
Luminous Efficacy K	lumen per watt (1 m/W)	—	—

ANSWERS TO SELECTED PROBLEMS

CHAPTER 2

2. Flows from negative terminal
   Returns to positive terminal
4. 4 C
5. 150 A
6. 4.5 A
7. 80 μs
8. 6 mC
9. 300,000,000 m/s
10. Series
12. a. 0.2 s
    b. 50 mC
    c. 500 C
    d. 4 s
    e. 384 A
    f. 10 A
    g. 1.53 s
    h. 0.2255
13. Oppositely charged objects
15. V
16. 3 mA
17. a. 80 mV
    b. 10 μA
    c. 100 mV
    d. 280 V
    e. 100 μA
    f. 1 kV

g. 3.9 Ω
h. 3 kΩ
18. a. Current
    b. Resistance
20. a. (1) point B (+)
       (2) point D (−)
    b. (+)
    c. (−)
22. Small resistance
24. a. 220 kΩ—10%
    b. 470 Ω—20%
    c. 38 Ω—5%
    d. 10 Ω—5%
    e. 56 MΩ—10%
    f. 9 MΩ—10%
    g. 730 kΩ—10%
25. a. Y, V, Br
    b. O, W, Br
    c. R, V, R
    d. O, O, O
    e. G, B, G
27. 4.01 C—neg
28. 8.01 C—pos
29. 4 A
30. 8 C
31. 25 S
    2 S
    1 mS

32. 1000 Ω
     500 Ω
     10 KΩ
33. 200.3 mA

CHAPTER 3

2. a. kV
    b. A
    c. M
    d. V
    e. S
    f. kW
    g. V
    h. M
3. b. 3 mA
    c. 3 mA
    d. 1 mA
4. a. 4 A
    b. 2
    c. 48 V
    d. 270 kΩ
5. 16 A
6. 4 A
7. a. 12×10^{-3}
    b. 5×10^{3}
    c. 500 kΩ
    d. 0.1 MΩ
    e. 500 mA
    f. 9×10^{-3} S
    g. 1 mA
    h. 5000 Ω
8. 20 kΩ
9. a. 2 Ω
    b. 10,000 MΩ
10. a. 0.66 mA
    b. 200 μA
11. a. 3.196 V
    b. 50.6 kV
    c. 2.84 V
12. with respect to ground:
    a. C +70 V
    b. A +140 V
    c. P +140 V
    with respect to point D:
    d. P +110 V
    e. B +90 V
    f. N -30 V

g. A +110 V
with respect to point N:
h. A +140 V
i. B + 120 V
j. C +70 V
k. Ground 0 V
with respect to point B:
l. D -90 V
m. E -120 V
n. Ground -120 V
o. P +20 V
p. N -120 V

CHAPTER 4

4. 32,400 J
5. 2.33×10^{5} C
6. 1.34×10^{6} J
7. a. 6.25 A
    b. 3.33 A
    c. 4.17 A
8. 9.6 Ω
11. 35.8¢
12. $2.16

CHAPTER 5

1. c.
2. One.
3. a. 100 Ω
    b. 1 A
    c. R_1 = 40 V
        R_2 = 60 V
    d. 40 V
    e. 0 V
6. a. (1) 100 V
        (2)   0 V
        (3)   0 V
    b. 0 V
7. a. 2 A
    b. Increase
    c. (1) Increased
        (2) Decreased
8. Will operate cooler
9. a. 8 mA
    b. R_1 = 8 V
        R_2 = 24 V
        R_3 = 48 V

c. $P_{R_1} = 64$ mW
$P_{R_2} = 192$ mW
$P_{R_3} = 384$ mW

d. $V_s = 80$ V

f. 800 mW

g. Point A = 0 V
B = +8 V
C = +8 V
D = +32 V
E = +32 V
F = +80 V

h. Largest

i. 48 V

10. a. $V_T = 75$ V
$V_{R_1} = 45$ V
$I_{R_1} = 3$ A
$V_{R_2} = 30$ V
$I_{R_2} = 3$ A
$R_2 = 10$ Ω

b. $R_1 = 20$ Ω
$V_{R_2} = 75$ V
$I_{R_2} = 3$ A
$V_T = 135$ V
$I_T = 3$ A
$R_T = 45$ Ω

c. $V_T = 64$ V
$R_T = 16$ Ω
$V_{R_1} = 24$ V
$I_{R_1} = 4$ A
$V_{R_2} = 40$ V
$I_{R_2} = 4$ A

d. $I_T = 1$ A
$R_T = 50$ Ω
$I_{R_1} = 1$ A
$R_1 = 30$ Ω
$V_{R_2} = 20$ V
$I_{R_2} = 1$ A

11. 16 Ω

12. a. 6 Ω
b. 60 V
c. 25%
d. 75%
e. Much larger
f. (1) Decreased voltage
(2) Overheating

13. a. 0.045 Ω
b. 0.018 Ω

c. 0.018 Ω

d. (1) 1800 W
(2) 2.4 HP

e. (1) 2205 W
(2) 2.95 HP

f. (1) 882 W
(2) 1.2 HP

14. 6.3 V

18. 19.3 Ω

CHAPTER 6

4. a. 1.2 kΩ
b. 2.4 kΩ
c. 0

5. 6 kΩ

6. a. 800 V
b. 800 V
c. $P_{R_1} = 1600$ W
$P_{R_2} = 3200$ W

7. a. 75 Ω
b. 1500 W
c. 300 Ω
d. Increase
e. Decrease to 8 A
f. 0
g. 150 V
h. 150 V

9. 10-Ω resistor

10. 1 mA

11. 12 kΩ

12. 40 V

13. $I_1 = 4$ mA
$I_2 = 2$ mA

14. 3 kΩ

15. 3 kΩ

16. Remain the same

17. a. 4 kΩ
b. 48 V

18. 21 V

19. a. 50 Ω
b. 50 V

20. $R_T = 2$ kΩ $I_3 = 6$ mA
$I_T = 30$ mA $I_4 = 2$ mA
$P_T = 1.8$ W $I_5 = 2$ mA
$I_1 = 15$ mA $P_1 = 0.9$ W
$I_2 = 5$ mA $P_2 = 0.3$ W

$P_3 = 0.36$ W $\qquad P_5 = 0.12$ W
$P_4 = 0.12$ W

CHAPTER 7

1. $R_T = 55.33\ \Omega$
2. $R_T = 10\ \text{k}\Omega \qquad V_{R_2} = 12$ V
   $I_T = 3$ mA $\qquad V_{R_3} = 12$ V
   $P_T = 90$ W $\qquad P_{R_1} = 54$ mW
   $I_2 = 2$ mA $\qquad P_{R_2} = 24$ mW
   $I_3 = 1$ mA $\qquad P_{R_3} = 12$ mW
   $V_{R_1} = 18$ V
3. $R_T = 12\ \Omega$
4. a. $R_4 = 600\ \Omega$
   b. Wattage of
      $R_1 = 4$ W
      $R_2 = 2$ W
      $R_3 = 8$ W
      $R_4 = 6$ W
5. a. $R_1 = 5\ \Omega$
   b. $I_{R_4} = 1.25$ A
6. $R_1 = 12\ \text{k}\Omega$
7. a. $I_1 = 1$ mA
   b. $I_T = 2.5$ mA
   c. $V_a = 62.5$ V
   d. $R_T = 25\ \text{k}\Omega$
   e. R_4 dissipates greatest
      amount of heat
8. $R_1 = 32\ \Omega$
   $I_1 = 1.25$ A
9. $R_T = 10\ \Omega$
   $I_T = 10$ A
   $P_T = 1000$ W
10. $V_a = 80$ V
11. a. $R_T = 1\ \text{k}\Omega$
    b. $I_T = 10$ mA
    c. $V_{R_1} = 6.75$ V
       $V_{R_4} = 1.458$ V
       $V_{R_6} = 1.789$ V
       $P_{R_1} = 67.5$ mW
       $P_{R_5} = 3.34$ mW
       $P_{R_2} = 4.25$ mW
12. $V_{R_7} = 92.436$ V
13. a. $R_T = 95\ \Omega$
    b. $I_T = 4$ A
    c. $I_{R_2} = 1$ A
14. $I_T = 6$ mA

15. R_2 least wattage
16. $R_T = 10\ \text{k}\Omega$
17. $R_T = 20\ \text{k}\Omega$
18. $R_T = 34\ \Omega$
19. $R_T = 560\ \Omega$

CHAPTER 8

1. $R_B = 500\ \Omega$
   $R_2 = 125\ \Omega$
   $R_1 = 111\ \Omega$
   $P_{R_B} = 50$ mW
   $P_{R_2} = 50$ mW
   $P_{R_1} = 225$ mW
2. $R_B = 2000\ \Omega$
   $R_2 = 1000\ \Omega$
   $R_1 = 375\ \Omega$
   $P_{R_B} = 50$ mW
   $P_{R_2} = 225$ mW
   $P_{R_1} = 600$ mW
3. $R_B = 1\ \text{k}\Omega$
   $R_2 = 363.6\ \text{k}\Omega$
   $R_1 = 300\ \text{k}\Omega$
   $P_{R_B} = 100$ mW
   $P_{R_2} = 176$ mW
   $P_{R_1} = 480$ mW
4. $R_4 = 8.1\ \Omega$
   $R_3 = 268\ \Omega$
   $R_2 = 73.5\ \Omega$
   $R_1 = 14.8\ \Omega$
   $P_{R_2} = 770$ mW
   $P_{R_3} = 210$ mW
   $P_{R_2} = 340$ mW
   $P_{R_1} = 420$ mW
5. $900\ \Omega$
6. $377\ \Omega$
7. $1600\ \Omega$

CHAPTER 9

8. a. 6 cells
12. 60 A-H
13. 1.2 V
14. $0.1\ \Omega$
15. a. 23.4 V
    b. 36 mW
    c. 2.4 A
16. No. Because the meter move-
    ment will be damaged.

17.　a.　(1)　2 V
　　　　(2)　240 A-H
　　　b.　(1)　12 V
　　　　(2)　40 A-H

CHAPTER 10

1. 　　　I
V_{TH} = 36 V
R_{TH} = 2 Ω
I_{R_L} = 3.6 A
V_{R_L} = 28.8 V

　　　II
V_{TH} = 50 V
R_{TH} = 2.67 Ω
I_{R_L} = 3.4 A
V_{R_L} = 40.9 V

2. 　　　I
V_{TH} = 72 V
R_{TH} = 4 Ω
I_{R_L} = 3.6 A
V_{R_L} = 57.6 V

　　　II
V_{TH} = 25 V
R_{TH} = 1.33 Ω
I_{R_L} = 3.4 A
V_{R_L} = 20.46 V

3. 　　　I
V_{TH} = 2 V
R_{TH} = 6.13 Ω
I_{R_L} = 47.47 mA
V_{R_L} = 1.709 V

　　　II
V_{TH} = 4 V
R_{TH} = 7.33 Ω
I_{R_L} = 127.67 mA
V_{R_L} = 3.064 V

4. 　　　I
V_{TH} = 4 V
R_{TH} = 12.267 Ω
I_{R_L} = 47.47 mA
V_{R_L} = 3.41 V

　　　II
V_{TH} = 4 V
R_{TH} = 7.33 Ω
I_{R_L} = 127.67 mA
V_{R_L} = 3.064 V

5. 　　　I
V_{TH} = 2 V
R_{TH} = 84 Ω
I_{R_L} = 1.557 A
V_{R_L} = 1.869 V

　　　II
V_{TH} = 2 V
R_{TH} = 1000 Ω
I_{R_L} = 0.1538 mA
V_{R_L} = 1.846 V

6. 　　　I
V_{TH} = 4 V
R_{TH} = 168 Ω
I_{R_L} = 1.56 mA
V_{R_L} = 3.74 V

　　　II
V_{TH} = 1 V
R_{TH} = 500 Ω
I_{R_L} = 0.1538 mA
V_{R_L} = 0.923 V

7. 　　　I
I_N = 17.5 A
R_N = 1.6 Ω
I_{R_L} = 7.78 A
V_{R_L} = 15.56 V

　　　II
I_N = 9 A
R_N = 2.67 Ω
I_{R_L} = 6.55 A
V_{R_L} = 6.55 V

8. 　　　I
I_N = 17.5 A
R_N = 0.8 Ω
I_{R_L} = 7.78 A
V_{R_L} = 7.78 V

　　　II
I_N = 9 A
R_N = 2.67 Ω
I_{R_L} = 6.55 A
V_{R_L} = 6.55 V

CHAPTER 11

19. 5.294 T
20. 0.3 μWb
23. Increased 60 times
27. 6×10^{-3} T
28. 113.34×10^{-6} m^2
29. 15 μWb

CHAPTER 12

10. 640 A
12. 2500 A/m
27. 1.47 Wb/m^2
28. 5.84×10^{-4}
29. 1.475×10^{-3} H/m
31. 75.6×10^{-6} H/m
34. 111.47×10^6 H^{-1}
36. 1.8×10^{-3} Wb

CHAPTER 13

18. 10 H
19. 4800 V
20. 62.5 mH
22. 2.52 μH
23. 12.6 mH
26. 450 mH
27. 24 mH
33. 1.386 mH
34. 0.556
35. 8 H
36. 445.96 mH
37. 4.8 mH
40. 5040 V
41. 0 V
42. 5 A
43. Turns ratio 22.9 to 1
48. 4.167 ms
49. 4 μs
51. 40 A
53. 1.2 TC
54. a. 95%
    b. 11.4 V
    c. 5.7 mA
55. 0.47 TC
56. 30 ms
58. a. 82 mA
    b. 82 mA

c. 4.1 V
d. 4.1 V
59. a. 1.5 TC
    b. 1.554 A
    c. 775 V
    d. 1.554 A
60. Long TC

CHAPTER 14

12. 0.36 μF
14.

pF	nF	μF
2500	2.5	✕
3700	✕	0.0037
✕	4.7	0.0047
4700	✕	0.0047
39000	39	✕
✕	400	0.4
1000000	1000	1
4000000	✕	4

16. 393.7 V
24. Largest capacitor
25. 546.66 V
26. 6×10^{-3} C
27. a. 8 μF
    b. V_{C_1} = 100 V
       V_{C_2} = 400 V
28. a. 50 μF
    b. 250 V
32. 1 TC
34. 5 TC
36. 1875 μs
37. 0 V
38. a. 1.389 TC
    b. 0.463 TC
39. a. 103.56 μs
    b. 370.67 μs
    c. 433.43 μs
    d. 84 V

e. 0.847 TC
    305 μs
f. 1.6 TC
    579.4 μs
42. a. (1) 138.35 V
        (2) 50 ms
    b. (1) 35.94 ms
        (2) C is negative
        (3) 38 V
        (4) X is positive
        (5) 152 μA
        (6) 3.8 μC

CHAPTER 15

10. a. 85 V
    b. 120 V

c. 154 V
d. 170 V
e. 154 V
f. 44 V
g. −85 V
h. −164 V
i. −164 V
j. −58 V
11. a. 10 A
    b. 14.14 A
    c. 20 A
    d. 17.3 A
    e. 0 A
    f. −5.18 A
    g. −19.7 A
    h. −20 A
    i. −14.14 A
12. (see table below)

Table P15-1   Voltage Relationships

	Effective V	V_{rms}	V_p	V_{avge}
a.	98.98	98.98	140	89
b.	120	120	169.7	108
c.	288.5	288.5	408	260
d.	90	90	127.3	81
e.	148.5	148.5	210	133.6
f.	62.16	62.16	87.9	56
g.	208	208	294.1	187.2
h.	219	219	310	197
i.	180	180	254.4	161.9

14.

Table P15-2 Frequency and Period Relationships

	Frequency	ms	μs	ns	ps
a.	50 kHz	0.02	20	20,000	
b.	12.5 kHz	0.08	80	80,000	
c.	12.5 kHz	0.08	80	80,000	
d.	16.67 kHz	0.06	60	60,000	
e.	20 GHz	50×10^{-9}		0.05	50
f.	50 MHz		0.02	20	20×10^3
g.	40 GHz		25×10^{-6}	25×10^{-3}	25
h.	500 kHz		2	2000	2×10^6
i.	40 kHz	0.025	25	25,000	

15. −129 V
16. a. 28.9 V
    b. 52.578 V
17. 0 V
18. 214.29 m
19. 1.67 m
20. a. 376.8 rad/s
    b. 628 rad/s
    c. 2512 rad/s
    d. 6280 rad/s
    e. 18,840 rad/s

CHAPTER 16

6. a. 1600 Ω
   b. 75 mA
7. 318.47 mH
8. 1.59 A
10. a. 1.327 MΩ
    b. 265.4 Ω
    c. 9.8 MΩ
    d. 884 Ω
11. a. 452 μA
    b. 0 V
12. 33 μF
17. 0.96
18. 0.8

20. a. 123.392 W
    b. 0.643
    c. 147 VAR

CHAPTER 17

5. a. 90°
   b. −75.96°
   c. −75.96°
   d. 75.96°
6. b. 72 V $\underline{/33.69°}$
   c. 33.69°
   d. 1440 Ω
   e. 800 Ω
7. 72.8 Ω $\underline{/15.95°}$
8. a. 100 Ω $\underline{/36.87°}$
   b. 80 + J60
   c. 36.87°
   d. 160 + J120
   e. 200 V $\underline{/36.87°}$
   f. 360 V $\underline{/90°}$
   g. 240 V $\underline{/-90°}$
10. Capacitive
11. Resistive
12. a. 22.37 Ω
    b. 4.47 A $\underline{/63.4°}$
    c. 2 + J4 A

d. I_R = 2 A
   I_C = 4 A
e. Capacitive

13. a. 39.94 Ω $\underline{/56.3°}$
    b. 3.6 A $\underline{/-56.3°}$
    c. 2 - J3
    d. 2 A $\underline{/0°}$
    e. 4 A $\underline{/-90°}$
    f. 1 A $\underline{/90°}$
    g. Inductive

CHAPTER 18

9. a. 1.48 MHz
   b. 957 Ω
   c. 957 Ω
   d. 2 mA
   e. 1.92 V

26. Q = 80

27. a. 1.48 MHz to 0.538 MHz
    b. Q = 143 at 1.48 MHz
       Q = 51.8 at 0.538 MHz
    c. 15 Ω
    d. 1.7 V
    e. Smallest
    f. (1) 10.38 kHz
       (2) 10.38 kHz
    g. Opened

33. a. 1.625 MHz to 0.534 MHz
    b. 600 kΩ
    c. 64.86 kΩ
    d. 244.94
    e. 80.5

CHAPTER 19

6. a. 929.697 kHz
      908.997 kHz
   b. 20.7 kHz

10. a. 500 Ω
    b. 500 Ω

11. 707 Ω

12. a. 3.184 Hz
    b. 3.184 Hz

13. a. 14.14 V
    b. 14.14 V

14. a. $\frac{1}{2}$ power or -3 dB
    b. $\frac{1}{2}$ power or -3 dB

16. a. -7 dB
    b. -1 dB

17. b. 796.1 Hz
    c. Upper
    d. 398 Hz
    e. 796.1 Hz
    f. 7 dB

24. b. 812.59 kHz
    c. -3 dB

CHAPTER 20

1. a. -56
   b. 54
   c. J10
   d. -10
   e. 20
   f. J2

2. a. 9 + J12
   b. 9 - J2
   c. 6
   d. 12 - J3
   e. Must convert to RF first.
      67.58 + J17.21

3. a. J16
   b. 16

4. a. 64 - J8
   b. 71 + J22

5. a. 0.169 - J1.77
   b. -0.235 + J1.06
   c. 12 $\underline{/10°}$
   d. 3 $\underline{/-70°}$
   e. 0.25 + J1.75
   f. -J3
   g. +5

6. a. 28.3 - J28.3 Ω
   b. 51.8 + J193.2 Ω
   c. 100 - J173.2 Ω

7. a. 347 $\underline{/-30.256°}$
   b. 124 $\underline{/46.6°}$

8. 847.45 mA $\underline{/130°}$

9. V_{C_1} = 8.414 V $\underline{/-147.2°}$
   V_{L_1} = 23.8 V $\underline{/32.8°}$
   V_R = 9.92 V $\underline{/-57.2°}$

10. Z_T = 159.27 - J13.94 Ω
       = 159.88 Ω $\underline{/-5°}$

11. a. Z = 105 + J25 Ω
    b. Z = 107.94 Ω $\underline{/13.39°}$

12. $Z = 128{,}099.9 - J8.57 \ \Omega$
13. a. $Z = 305.9 + J30.78 \ \Omega$
    b. $Z = 307.44 \ \Omega \ \underline{/5.75^\circ}$
14. a. $Z_T = 38.49 - J11.42 \ \Omega$
    b. $Z_T = 40.21 \ \Omega \ \underline{/-16.35^\circ}$
15. a. $V_{TH} = 8.38 \ V \ \underline{/16.76^\circ}$
    b. $Z_{TH} = 21.31 + J6.75 \ \Omega$
       $= 22.35 \ \Omega \ \underline{/17.58^\circ}$
    d. $I_{Z_L} = 0.258 \ A \ \underline{/31.52^\circ}$
    e. $V_{Z_L} = 4.67 \ V \ \underline{/-24.79^\circ}$
16. a. $I_a = 0.53 \ A \ \underline{/66.78^\circ}$
    b. $I_N = 0.24 \ A \ \underline{/62.88^\circ}$
    c. $Z_N = 19.61 - J21.51 \ \Omega$
       $= 29.11 \ \Omega \ \underline{/-47.65}$
    d. $I_{Z_L} = 0.17 \ A \ \underline{/67.84^\circ}$
    e. $V_{Z_L} = 2.28 \ V \ \underline{/4.41^\circ}$

CHAPTER 21

3. a. $1000 \ \Omega/V$
   b. $2000 \ \Omega/V$
4. $50 \ \mu A$
5. $10 \ \mu A$
8. b. $22.22 \ \Omega$
9. a. $19 \ k\Omega$
   b. $59 \ k\Omega$
   c. $999 \ k\Omega$
   d. $2.99 \ M\Omega$
   e. $5.99 \ M\Omega$
   f. $999.99 \ M\Omega$
10. a. No shunt
    b. $1 \ k\Omega$
    c. $52.63 \ \Omega$
    d. $1.001 \ \Omega$
    e. $0.05 \ \Omega$

INDEX

Ac resistance, 461
Ac voltage:
  average values, 439
  effective values of a sine wave, 438
  generation, 426
  rms values, 438
Air-core, coils, 238
Air gap, 208, 267
  effect on reluctance, 264
Alternating current, 423
  average value, 439
  effective value, 438
  instantaneous value, 434, 435, 436
  sine-wave, 434
  symbol, 423
Alternating-current circuits:
  impedance networks, 622, 627
  parallel, 614
  power in, 435, 436
Alternating voltage:
  average value, 439
  effective value, 438
  generation of, 426
  instantaneous values, 436
  magnitude of, 425
  peak value, 437, 461
  rms values, 438, 461
  sine-wave values, 437
Ammeter shunts, 641
Ampere:
  definition, 11
  in magnetism, 226
  quantity, 11
  symbol, 659
Ampere law, 256

Ampere per meter (magnetizing force), 227
Amplitude of a sine wave, 425, 427, 429
Angle:
  of cut and voltage, 427
  lagging, 458
  leading, 457
  phase, 456
  power factor, 471
  in radians, 450
  symbols, 457, 657, 663
Angular velocity, 449, 450, 451
  deriving, 450
  symbol, 662
Anion, 6
Antilog, natural, 328, 329
Antiresonance, 541
Apparent power, 470, 471
  symbol, 659
Area:
  rectangular, 257
  rod, 258
  toroid, 259
Atom:
  Bohr, 3
  and electrons, 3
  neutral, 3
  nucleus of atom, 3
  shell, 3
  structure of, 3
Attenuation, 575, 576, 578
  and octaves, 578, 580, 581, 582
Autotransformer, 312, 313
Average value:
  current, 440
  voltage, 439, 440

B - symbol for magnetic flux density, 210, 660
B - symbol for susceptance, 661
  capacitive, 467
  inductive, 464
Bandpass filters, 570
  design formula, 568
  response curves, 567
Bandstop filters, 570
  design formula, 568
  response curves, 568
Bandwidth, 531
  and impedance, 533, 556
  measurements, 556
  and power, 531
  and Q, 535
  and reactance, 533
  resonant circuit, 531, 535
  and voltage, 531
Bar magnet, magnetic field pattern, 201, 204, 205, 206, 207
Battery:
  alkaline, 143
  applications, 153, 154
  capacity, 153
  carbon zinc, 141
  cell operation, 134
  characteristics, 142, 154, 159
  charge, 136, 145, 152
  classifications, 135
  comparison, 142, 153, 154
  discharge, 137, 139, 152
  dry-cell, 135
  gelled, 149
  hydrometer test, 140
  high-performance, 148

Battery (*cont.*)
  iron-clad, 148
  lead-acid, 146
  lead-antimony, 147
  lead-calcium, 148
  lithium, 145
  manganese-alkaline, 143
  memory, 151
  mercury, 144
  nickel-cadmium, 149
  parallel, 138
  primary, 140
  principle operation, 134
  recharging, 153
  replacement, 153
  secondary, 146
  series, 137
  series–parallel, 138
  silver cadmium, 152
  silver oxide, 144
  silver zinc, 152
  sintered-plate, 150, 151
  solar, 145
  sulphation, 147
  tables, 142
  testing, 139
  types, 142, 154
  wet-cell, 146
  zinc chloride, 143
BH:
  characteristics, 228, 229, 230
  curves, 228, 229, 230
BH magnetizing curve, 228, 229
  interpretation, 234
Bleeder current, 124
  resistor, 123
Bohr atom, 3
Branched magnetic flux, 269
Breakdown voltage in capacitors, 371
Bridge, general equation, 130
Bridge, Wheatstone, 128

Capacitance:
  in ac circuits, 454
  in dc circuits, 361, 368
  conversion of units, 368
  C-R charge, 391, 403
  C-R discharge, 402, 404, 405
  definition, 360
  and developed voltage, 375
  and dielectric strength, 369, 370
  and dielectric constant, 365, 370
  energy stored by a capacitor, 361, 362, 371
  equation for, 367
  factors affecting, 365, 366, 367
  farad, 367, 368
  and leakage resistance, 378
  in parallel, 374
  and permittivity, 365
  resistive time constant, 384
  in series, 373
  susceptance, 467, 661

Capacitance (*cont.*)
  symbol, 659
  unit of, 368, 661
  voltage and charge, 371, 372
  working voltage, 371
Capacitive reactance, 465, 504
  definition, 454, 455, 465
  factors governing, 465
  and Ohm's law, 466
  resistive time constant, 384
  susceptance, 467
  symbol, 465
Capacitors:
  charge and time constants, 391
  charging, 384
  clarification, 363
  and conductance, 361
  construction, 360, 366, 378
  definition, 360
  dielectric constant, 365
  dielectric strength, 369
  dielectrics, 361
  discharging, 402
  electric flux density, 363
  electrolytic, 380
  electrostatic field, 363
  electrostatic field strength, 363
  energy stored, 361, 371
  leakage resistance, 378
  parallel, 374, 376
  and percentage of charge, 392
  permittivity, 364
  plastic dielectric types, 381, 382
  plate area, 366
  practical considerations, 384
  principle of operation, 361
  purpose of, 361
  series, 373, 375
  tantulum, 381
  types, 379
  variable, 381
  voltage breakdown, 371
  working voltage, 371
Capacity, battery, 153
Cation, 6
Cell:
  charge, 136
  discharge, 137
  operation of, 134
  parallel, 138
  primary, 135
  secondary, 136
  series-aiding, 137
  series–parallel, 138
  voltaic, 135
Charge:
  battery, 145
  in capacitors, 362, 371
  coulomb, 9
  definition, 9
  electron, 3
  law, 14
  like, 15
  positive and negative, 15

Charge (*cont.*)
  static, 14
  unlike, 15
Charging current, 362, 371
Chemical energy, 135
Circuit:
  equivalent, 114
  magnetic, power loss, 245
  open, 58
  parallel, 78
  resonant, 518
  series, 55
  series–parallel, 99
  shorted, 59
  simplifying, 112
  tank, 541
  tracing, 111
Circuit diagram, 27, 28
Coefficient:
  of coupling, 292
  temperature, 25
Coercive force, 248
Coils:
  air-core, 238
  choke, 275
Color code, 20
Complex numbers, 598
  division, 610
  and impedance, 598
  of multiple parallel circuits, 620
  parallel circuits, 614
  parallel resonant circuits, 616
  and series circuits, 601
  and series–parallel circuits, 618
Compound, 2
Conductance, 23, 82
  symbol, 25, 659
  unit, 25, 659
Conductors:
  electric, 6
  and flux, 221
Constant-current source, 73
Constant-*K* filters, 586
Constant-voltage source, 72
Conventional current-direction, 14
Conversion factors:
  multiples, 27, 655
  submultiples, 27, 665
Conversion of units:
  kilo, 27
  mega, 27
  micro, 27
  milli, 27
  polar form to rectangular form, 600
  rectangular form to polar form, 599
Converting voltages, 438
Coordinates:
  conversions, 599, 600
  polar, 477, 598, 599, 600
  rectangular, 477, 598, 599, 600
  vectors, 455, 477

Copper losses, 312
Coulomb, 9
  definition, 9
  law, 9
  quantity, 9
  symbol, 659
Counter (emf), voltage, 278, 280
  magnitude, 285
Coupled circuits, 290, 291, 292
  parallel, 299, 300
  series, 295
Coupling:
  close, 292
  coefficient of, 292
  loose, 292
  mutual, 295
  tight, 292
C-R time constant, 384
C-R waveshaping, 411 to 416
Current, 9
  alternating, 423
  ampere, defined, 11
  branch, 79, 80
  capacitive, 361, 362, 368, 371,
    384, 385, 389, 392, 393, 404
  carriers, 7, 14
  changing, symbol, 320
  charging, 384, 385, 389, 392, 393,
    402
  collapse, 340, 341, 349, 402, 404
  constant source, 73
  conventional, 14
  decay, capacitive circuit, 373,
    374, 402
  decay, inductive circuits, 340,
    341, 349
  definition, 11
  direction, 14
  discharge, 402, 404
  divider principle, 78, 79, 80
  eddy, 245, 252
  effective, 440
  flow, 9
  induced, 280, 281, 283, 284, 285,
    286
  inductive, 286
  instantaneous, 325, 331, 432,
    434, 435, 461
  and left-hand rule, 220
  magnetizing, 219
  measuring, 13
  peak, 440
  peak to peak, 440
  pulsating, 439
  rate of change, 286, 317, 318,
    319, 321
  ratio, inverse, 187
  reactive, 504, 506, 508, 509, 510
  requirements, 17
  resonant, 522, 541, 546
  rise, 313, 314, 316, 317, 318, 319,
    321, 322, 349
  rms, 440
  short-circuit, 59, 93, 110, 111

Current (cont.)
  steady-state, 315, 324
  tank, 543
  in transformers, 305, 306, 307
  units of, 11, 659
  velocity, 13
Cutoff frequency, 574, 576, 577,
    590
Cycle, 429, 441, 442

Damping:
  meter movements, 635, 637
  resistors, 553
  resonant circuits, 553
D'Arsonval meter movement, 635
Decay, current, 340, 341, 349, 373,
    374, 402
Decibel (dB), 532, 567, 576, 578,
    579, 580, 582
Dc voltmeter, 640
Decay current, current curve, 342,
    350
  of charge in capacitive circuit,
    386, 387
  of current in inductive circuit,
    342, 350
  of voltage in capacitive circuit,
    387, 403
  of voltage in inductive circuit,
    320, 342, 350
Degree–radian conversion, 450
Demagnetization, 208
Depolarizer, batteries, 136, 143
Design formula:
  for bandpass filters, 568
  for high-pass filters, 574, 575
  for low-pass filters, 574, 575
Diagram:
  phasor, 455, 459, 477, 598
  power factor, 471
  rectangular form, 598, 601
  schematic, 27, 28
  vector, 455
Diamagnetic materials, 213, 214
Dielectric:
  breakdown voltage, 369, 370,
    371
  constant, 365, 370, 371
  relative permittivity, 365
  strength, 369, 370, 371
Difference in potential, 16, 135
Differential permeability, 234
Discharge:
  battery, 135, 136, 137, 139, 140,
    152
  of capacitive circuit, 373, 374,
    402
  capacitor, 373, 374, 402
  of inductive circuit, 340, 341,
    349
Dissipation, power, 49, 61
Divider:
  current, 78, 79, 80
  voltage, 122

Domains, magnetic, 202, 203
Dry cell, 135

E - symbol for electric field
    intensity, 363, 659
E - symbol for electric field
    strength, 363, 659
Eddy current, loss, 245, 252, 312
Effective value:
  of current, 440
  of voltage, 438
Efficiency, power, 71
Electric:
  charge, 9, 659
  field, 363
  field intensity, 363
  field strength, 363, 659
  flux, 362, 659
  flux density, 363, 660
  lines of force, 362
  potential, 15, 659
  pressure, 15
  quantity, 9, 659
Electrical degrees, 450
Electricity:
  and energy, 1
  nature of 1, 151
  static, 14
Electrolyte, 135
Electromagnetic:
  circuit calculations, 225
  induction, 275, 276
  voltage generation, 426
Electromagnetism, 219
Electromotive force, counter, 278,
    659
Electron, 7
  characteristics, 8
  charge, 3
  energy levels, 3
  flow, 87
  free, 6, 7
  mass, 3
  maximum, 4
  quantity, 9
  shells, 3, 4
  size, 2
  theory, 3
  valence, 7
  velocity, 13
Electrostatic field, 362, 363
Element, 2
Energy:
  definition, 45, 46, 47
  electric, 47, 659
  and electricity, 1
  forms of, 45, 46
  heat, 48
  joule, 47, 659
  kinetic, 3
  law, 1
  levels in an atom, 3
  measurement, 46
  mechanical, 45

Energy (cont.)
  SI unit, 45
  storage by capacitors, 361
  storage by inductors, 470
  symbol, 45, 46
  transformation, 45
  and work, 45
Engineering prefixes, 28
Epsilon (ϵ), 326, 343
Equations:
  general bridge, 130
  general transformer, 307, 310
Equivalent circuits:
  circuit simplification, 111, 112
  Norton's, 185
  Thevenin's, 173
Exponential curves, 316, 319
Exponents, 28
Expressions:
  less negative, 41
  more negative, 41

f - symbol for frequency, 442, 656
F_m - symbol for magnetomotive
  force, 226, 660
Factor, power, 471
Farad, definition, 368
Faraday, 276
Faraday's law, 286, 425
Ferrite iron, 214
Ferromagnetic materials, 213, 214
  power loss, 245
Field:
  electric, 363
  intensity, 227, 264
  magnetic, 202, 205
Filters:
  active, 593
  attenuation band, 575, 576, 577,
    578, 580, 582
  bandpass characteristics, 567
  bandstop, 568
  classifications, 565
  constant K, 586, 587, 590, 591
  hybrid (m-derived), high-pass,
    572, 574
  hybrid (m-derived), low-pass,
    571, 574
  nonresonant, 565, 571
  nonresonant classifications, 571
  nonresonant comparisons, 574
  nonresonant high-pass, 571,
    572, 591
  nonresonant low-pass, 571
  pi (π)-type, 590
  resonant, parallel, 570
  resonant, series, 565, 566, 567
  R-C, 571
  R-L, 571
  T-type, 589
  wave traps, 566
Flux, density:
  electric, 363, 660
  and magnetizing force, 228

Flux, density (cont.)
  symbol, 210, 215, 660
Flux, magnetic, 210, 225, 243
  and coils, 222
  and conductors, 221
  definition, 210
  density, 210, 226, 660
  fringing, 264, 265
  leakage, 254
  lines of force, 205, 207
  linkage, 285
  mutual, 291, 661
  SI units, 210, 215, 660
  symbol, 210, 215, 660
  webers, 210, 660
Free electrons, 7
Frequency:
  attenuation, 575, 577
  bandwidth, 531, 532, 533
  cutoff, 574, 576, 590
  definition, 441, 442
  and period, 352, 412, 441, 442
  resonance, parallel, 518, 519
  resonance, series, 518, 541
  symbol, 662
  and wavelength, 447
Fringing, flux, 264, 265
Full wave rectified sine wave, 440
Fuse, 26
  buss numbers, 27
  failure, 26
  slow blow, 26
  symbol, 26

G - symbol for conductance, 23,
  659
General bridge equation, 130
General transformer equation,
  307, 310
Generator, ac, 426
Geometric construction, vector
  diagrams, 455
Graphic symbols, 652
Ground, 40
Growth of charge in capacitive
  circuits, 385, 387, 391, 392
Growth of current in inductive
  circuits, 313, 314, 316, 317,
  318, 319, 321
Growth of voltage in capacitive
  circuits, 385, 386, 387, 391,
  392

H - magnetizing force (field
  intensity), 227
H - symbol for magnetic field
  intensity, 227, 660
Half-cycle average, 439
Half-power points, 532
Half-wave rectification, 439
Hand-rule, magnetic field
  direction, 220, 223
Henry, 232, 286, 661
  definition, 286

Henry (cont.)
  unit of permeance, 232, 660
Hertz, frequency, 442, 656, 662
High-pass filter, 571, 572
  R-C, 573, 574
  R-L, 573, 574
Horsepower, definition, 50
Hysteresis:
  loop, 245, 250
  loss, 245, 250, 312

Imaginary component of a vector,
  607
Imaginary numbers, 607, 608
  and real numbers, 608
Impedance:
  and applied voltage, 489
  and bandwidth, 556
  and complex numbers, 598
  definition, 476
  matching, 71
  networks, 662, 627
  and Norton's theorem, 627
  and Ohm's law, 476
  parallel, 503, 506, 508, 510
  and parallel resonant circuits,
    541, 547
  and phase angle, 479
  and phasor diagrams, 477
  and Q, 554
  rectangular form, 485
  reflected, 309
  series circuits, 483, 486, 492
  series–parallel, 618
  symbol, 476, 660
  and Thevenin's circuit, 616
  transfer, 71, 310
  winding ratio, 309
Incremental permeability, 234
  symbol, 234
Induced voltage, 278, 320
  definition, 278
  and flux, 425
  formula, 286
  and magnitude, 425, 427
  and polarity, 426, 427
Inductance, 275
  in ac circuits, 234
  in dc circuits, 290
  dc resistance of, 290
  definition, 278
  equation for, 286
  factors governing, 288
  mutual, 291, 295, 661
  in parallel, 299
  physical determinants of, 288
  radio frequency resistance of,
    290
  self-, 285
  in series, 295
  symbol for, 287, 661
  unit of, 287, 661
Induction:
  electromagnetic, 276

Induction (*cont.*)
  magnetic, 201, 203
  self-, 285
Inductive reactance, 455
  definition, 462
  equation, 462
  factors governing, 462
  and Ohm's law, 463
  and phasor arm, 463
Inductive resistive time constant, 313
Inductors:
  energy stored in, 469, 470
  parallel, 291, 299
  practical considerations, 290
  series, 291, 295
  types of coupling, 290
Insertion loss, 586
Instantaneous:
  current, 434, 435
  current in resistive circuits, 461
  power, 435
  symbols, 433, 434, 435
  voltage, 433
  voltage in resistive circuits, 461
Intensity:
  electric field, 363, 364
  magnetic field, 227
Internal resistance:
  of current sources, 72, 173
  of voltage sources, 73, 184
International system of units, 653
Ion:
  definition, 5
  negative, 6
  positive, 6
Iron, ferrite, 214
Iron, magnetic properties, 213
Iron-core:
  coils, 253
  powdered, 254
  saturation, 246
  transformers, 301
Iron losses, 245

J, operator, 477, 601
Joule:
  definition, 47
  equation, 47
  symbol, 659

Kilowatt-hour, definition, 51
Kirchhoff's:
  current law, 81, 159
  in dc circuits, 158, 159
  voltage law, 57, 158

L - symbol for inductance, 286, 661
Lagging phase angle, 458
Laminations, 253
Laws:
  of charges, 14
  Coulomb's, 9
  energy, 1

Laws (*cont.*)
  Faraday's, 286
  inverse current ratio, 187
  Kirchhoff's, 157, 158, 159, 164
  left-hand, 220, 223, 276, 277, 426, 427
  Lenz's, 285
  Ohm's, 35
L-C filter:
  bandpass, 567, 570
  high-pass, 572, 573, 574
  low-pass, 571, 573, 574
L-C ratio, 527
Leading phase angle, 457
Leakage current, 378
Leakage flux, 254
Leakage resistance, 378
Lenz's law, 285
Lines of force:
  electric, 362, 363
  magnetic, 205
Linkage, flux, 285
Load resistance, 19
Loaded voltage divider, 123
Loading effect of meters, 647
Loading of parallel resonant circuits, 553
Logarithms, natural, 326, 327
Loop:
  current, 158, 167
  tracing, 158
Loss:
  eddy current, 245, 252
  hysteresis, 245, 250
  iron, 245
Low-pass filters:
  R-C, 573
  R-L, 573
L-R time constant, 313
L-R waveshaping, 349, 350

M - symbol for mutual inductance, 292
Magnetic:
  air gap, 208, 264, 265
  attraction, 205, 207
  circuits, 254, 255, 256, 268
  domain, 202
  field, 205
  field intensity, 227
  field strength, 227, 660
  flux, 210
  flux density, 210, 226
  induction, 201, 203
  lines of force, 205
  path, area, 257, 258, 259
  path, average length, 257
  poles, 199, 200, 205, 206, 207
  repulsion, 205, 207
  retentivity, 216
  saturation, 229
  shielding, 214
  table, 215, 255
  terms, 216

Magnetic (*cont.*)
  units, 215, 255
Magnetic materials:
  diamagnetic, 213, 214
  ferrite, 214
  ferromagnetic, 213, 214
  nonmagnetic, 214
  paramagnetic, 213, 214
"Magnetic Ohm's law," 243
Magnetism:
  artificial, 200
  atomic theory, 201
  by induction, 203
  left-hand rule, 220, 223
  natural, 199
  residual, 216, 258
  retentivity, 216
  rules of, 423
  and voltage, 276
Magnetization:
  curves, 229, 239
  definition, 227
  and flux density, 228
  force, 227
  SI units, 227, 660
  symbols, 227, 660
Magnetomotive force:
  definition, 226
  SI units, 226, 660
  symbols, 226, 660
Magnets:
  artificial, 200, 203
  bar, 204, 205, 208
  care of, 209
  electro-, 203, 204
  permanent, 203, 204
  ring, 208
  temporary, 203, 204
  toroids, 208
Magnitude of induced voltage, 285
Matching, impedance, 71
Matter, structure of, 2
Maximum power transfer, 48, 309
Memory, battery, 151
Meter movement:
  accuracy, 636, 646
  characteristics, 607
  D'Arsonval, 635, 636
  resistance, 635
  sensitivity, 635, 636
Meters:
  current, 641
  multimeter, 635
  multiplier resistor, 640
  ohmmeter, 642, 644
  shunt resistor, 641
  voltmeter, 16, 640
Microamp, 11, 28
  conversions, 28
Milliamp, 11, 28
Molecule, 2
  monatomic, 2
  polyatomic, 2
  triatomic, 2

Multimeter, 635
Multiples, submultiples, 655
Multiplier, voltmeter, 641
Multirange meters, 635, 640
Mutual coupling, 291, 295
Mutual inductance, 291, 295, 661

Natural log, 327, 330, 393
    antilog, 328
Negative charge, 14
    less, 41
    more, 41
Networks:
    dc, 157
    impedance, 622, 627
    Kirchhoff's, 158
    Norton's, 184, 627
    superposition, 165
    theorems, 157
    Thevenin's, 172, 622
Neutral, 4
Neutron, 3
Nonmagnetic materials, 213, 214
Norton's theorem:
    ac circuit, 627
    dc circuit, 184
    equivalent circuit, 185
    and impedance, 627
Notation, scientific, 28
Nucleus of the atom, 3

Octaves and attenuation, 578, 580,
    581, 582
Oersted, 219, 276
Ohm, definition, 35
Ohmmeter, 642
Ohm's law, 35
    and capacitive reactance, 466
    graphical, 37
    and impedance, 476
    magnetic circuits, 243
    wattage, 49, 50
Open-circuit voltage, 173
Operator, j, 477, 601
Opposing flux, 221, 295, 299
Orbits, electron, 3

P - symbol for power, 48, 469, 659
P_m - symbol for permeance, 232,
    660
Parallel capacitors, 374
Parallel circuits:
    applications, 78
    arrangements, 78
    characteristics, 79
    classical formula, 83
    equal valued, 84
    fault-finding, 91
Parallel impedance, 503
    inductors, 291, 299
    resistors, 78
    resonant circuit, 515, 550
Parallel magnetic circuits:
    principle of operation, 78

Parallel magnetic circuits (cont.)
    versus heat, 89
    wattage (power), 88
Paramagnetic materials, 213, 214
Peak value of a sine wave, 437
Percent:
    capacitor charge, 392
    capacitor discharge, 405
    instantaneous current and
        voltage (L/R), 325, 331
    instantaneous current and
        voltage (C-R), 393
Period, 352, 412, 441, 442, 663
Periodic table, 5
Permanent magnet, 203, 204
Permeability:
    absolute, 212, 231, 233
    air, 212, 233
    definition, 231, 233
    differential, 234
    and flux density, 231
    free space, 212, 233
    incremental, 234
    relative, 211, 212, 233, 239
    relationships, 212, 239
    reluctance, 241
    SI unit, 231, 660
    symbol, 231, 660
    tables, 233
Permeance, 232, 660
    and reluctance, 241
Permittivity:
    definition, 232, 364, 365
    SI unit, 232, 660
    symbol, 232, 660
Phase angle, 456
    and effect of resistance, 477
    lagging, 458
    leading, 457
    symbols, 457, 657, 663
Phase current:
    capacitive, 457
    inductive, 458
Phase voltage, 456
    capacitive, 458
    inductive, 457
Phasor, 455
    arm lengths, 459, 477
    construction, 477
    diagram, 477
    reference, 455
Plante, Gaston, 147
Polar coordinates, 598, 599, 600
Polar form of vector, 598, 599, 600
    adding and subtracting, 612
    multiplication and division, 613
    and series circuits, 602
Polarity of voltage, 16
    drops, 38
Poles, magnetic, 199, 200, 205,
    206, 207
Positive charge, 6
Potential, voltage, 16
    difference, 15, 659

Potential, voltage (cont.)
    drop, 58
    energy, 46
Potentiometer, 22
    tapers, 24
Power:
    in ac circuits, 469
    ac types, 469
    apparent, 469, 659
    in dc circuits, 48
    definition, 48
    dissipation, 49, 61
    efficiency, 71
    equation, 48
    factor, 471, 472
    instantaneous, 435, 436
    loss in ferromagnetic circuits,
        245
    maximum transfer, 68, 309
    reactive, 469, 661
    transfer, maximum, 68, 309
    true, 469, 470, 658
    var, 470, 661
    voltampere, 470
    watt, 48, 470, 658
Power factor, 471
    angle, 471
    definition, 471
Power triangle, 471
Powers of ten, 655, 28
Precise formula for parallel
    resonance, 558
Prefixes:
    engineering, 28, 655
    table of, 28, 655
Pressure, electric, 15
Primary, transformer, 301
Primary cell, 135, 140, 142
Proton, 3
Pulsating current, 439, 440
Pythagoras triangle, 479

Q - circuit quality, 527
Q - symbol for electric charge or
    quantity, 9, 656, 659
Q - symbol for reactive power, 470,
    661
Q factor, resonant circuits, 526
    bandwidth frequency and Q,
        535
    formula, 554
    parallel circuits, 551
    series circuits, 526
Q total, 528, 529, 530, 552
Q and voltage, 530
Quantity of electric charge, 9, 659

Radian:
    conversion, 450
    defined, 450
Radio frequency resistance, 290
Rate of change, curve, 317
    defined, 318
Ratio, voltage, 60

R-C filters:
  high-pass, 571, 573
  low-pass, 571, 573
Reactance, 454
  adding, 495
  capacitive, 465, 504
  inductive, 455, 505
  and Ohm's law, 461
  and *Q*, 554
  and resistors in series, 496
Reactive power, 469, 472, 661
Real component of a vector, 608
Reciprocal henry (H^{-1}), 241, 661
Rectangular form of impedance,
  598, 599, 600
Reference:
  phasor, 455
  point, 38
  voltage, 17
Reflected impedance, 309, 310
Regulation, voltage, 126
Relative permeability of:
  diamagnetic materials, 213
  ferromagnetic materials, 213
  paramagnetic materials, 213
Relay, 270
Reluctance, 240
  and permeability, 241
  and permeance, 241
  SI unit, 242, 661
  symbol, 242, 661
Residual magnetism, 216, 258
Resistance, 18
  in ac circuits, 461
  component of impedance, 476
  definition, 18
  internal, 71, 72, 73, 172, 184
  load, 19
  measurement of, 18
  nature of, 18
  networks, 157
  Ohm, 18, 661
  symbol, 18, 661
  unit, 18, 661
Resistors:
  adjustable, 22
  bleeder, 123
  carbon, 19
  carbon film, 19
  color code, 20
  construction of, 19
  damping, 553
  definition, 18
  graphic symbol, 18
  how to read color code, 20
  limiting, 640, 645
  metal film, 19
  multiplier, 640
  parallel, 78
  potentiometer, 22
  power rating, 61
  series-dropping, 58
  shunt, meter, 641
  standard values, 22

Resistors (*cont.*)
  and temperature, 25
  tolerance, 21
  types, 19
  variable, 22
  wattage, 20
  zero adjust, 645
Resonance:
  above and below currents, 546
  and bandwidth, 536
  definition, 502, 515, 518
  frequency formula, 519
  parallel, 515, 541
  parallel and impedance, 541,
    547
  parallel and precise formula,
    558, 559
  series, 502
  series, characteristics, 522, 535
Resonant, circuit, 502, 515, 518
  characteristic, 550
  filter circuits, 566
  frequency, 519
  parallel, 541
  parallel operation, 544
  parallel response curve, 555
  series response curve, 525
Retentivity, 216, 247, 251
Rheostat, 22
Rise, current, 317, 318
  pure inductance, 316
  L-R circuit, 315
Rise, potential, 385, 387
R-L filters:
  high-pass, 571, 573
  low-pass, 571, 573
rms value of a sine wave, 437, 438
Root mean square (rms), 438
Rotating vector, 437, 456

Saturation, magnetic, 229
Schematic drawings, 27, 28
Scientific notation, 28, 665
Secondary, transformer, 301
Secondary cell, 135, 136, 146, 154
Selectivity, resonant circuit, 525,
  551
  factors affecting, 526, 551
Self-inductance:
  definition, 285
  factors governing, 286
Sensitivity, meter, 637, 646, 647
Series capacitors, 373
Series circuit:
  applications of, 55
  characteristics, 55, 56
  and complex numbers, 601
  fault-finding, 58
  impedance, 483, 601
  magnetic, 256
  and Ohm's law, 55
  and polar form, 602
  resistors, 55
  summary, 65

Series circuit (*cont.*)
  voltage drop, 58
Series–parallel circuits, 99
  calculation of V and I, 103
  fault-finding, 108
  total resistance, 100, 101
Series resistors, 55
Series resonance, 502
Shells, electron, 3, 4
Shock, 51
  and current, 52
  rescue, 52
Short-circuit, 59, 93, 110, 111
Shunt, ammeter, 641
SI units, 655, 658
Siemens:
  definition, 23
  SI unit, 25, 658, 659
  symbol, 25, 658, 659
Sine of the angle, 429, 430, 431, 432
Sine wave:
  average value, 437, 439, 440
  definition, 429
  effective value, 437, 438, 440
  half-wave rectified, 439
  instantaneous value, 432
  peak to peak, 437, 440
  peak value, 437
  period, 441
  rms value, 437, 438, 440
Skin effect, 290
Solar battery, 145
Solenoid, 219
Source:
  constant-current, 73
  constant-voltage, 71
  energy, 45
Specific gravity, 140
Static electricity, 14
Steady-state values, 315, 324
Structure of matter, 2
Superposition theorem, 157
  for dc circuits, 165
Susceptance, 464, 661
  capacitive, 467
  inductive, 464
Symbols, table, 652

t - symbol for time, 662
T - symbol for time, 442, 662
Tables:
  battery applications, 154
  color code, 20
  comparison, magnetic, SI and
    CGS, 215
  conversion of units, 28
  C-R and *L-R* series circuits, 502
  dielectric constants, 366
  dielectric strengths, 370
  electric and magnetic circuit
    comparison, 255
  frequency versus time, 442
  graphic symbols, 652
  how to read color code, 20, 21

Tables (*cont.*)
  magnetic units, 271
  multiples and submultiples, 28, 655
  parallel resonant circuits, 545
  periodic, 5
  permeability, 213, 233
  prefixes, 28, 665
  primary cells, 142
  progressive percentage, 323, 392
  relative permeability, 213
  resonant frequency, 524
  secondary cells, 154
  series and parallel resonant characteristics, 524, 560
  severity of shock, 52
  standard resistors, 22, 23
  trigonometric functions, 480
Tank circuit, 541
  current, 543
Temperature:
  coefficient, 25
  and resistors, 25
Temporary magnet, 203, 204
Terminal voltage, 140, 172, 173
Tesla, definition, 210, 226
Theorems:
  Norton's, 157, 184
  superposition, 157, 165
  Thevenin's, 157, 172, 173, 181, 182, 183
Thevenin's theorem:
  in ac circuits, 622
  applications, 172, 181, 182, 183
  in dc circuits, 157, 172, 173
  and impedance, 622
Time, angle of cut, and induced voltage, 427
Time constant (*C-R*), 384
  and capacitor charge, 391
  and current progression percentage, 392
  classifications, 411
  decay, 402
  definition, 386
  and frequency, 411
  graphical method, 392
  long, 416
  medium, 415, 416
  and natural log, 394
  number of, 396
  percent of charge, 391
  percent of discharge, 405
  short, 413
  symbol, 388
  and time, 390
  universal curve, 405
Time constant (*L/R*), 313
  and current progression percentage, 322, 323
  decay, 340, 341, 342, 344, 345

Time constant (*L/R*) (*cont.*)
  definition, 314, 321
  and frequency, 349
  graphical method, 323
  long, 351
  medium, 354
  and natural log, 330
  number of, 333, 335
  practical considerations, 349
  short, 353
Toroid, 208
Transducer, energy, 1
Transfer, power maximum, 68
Transformer, 301
  auto, 312
  construction, 301
  current ratio, 306
  and impedance, 310
  iron-core, 301
  isolation, 302
  loading, 304
  power, 303, 304, 308
  power losses, 312
  and power transfer, 309
  practical considerations, 312
  principle of operation, 302
  ratio, 302, 306, 310
  step-down, 303
  step-up, 302
  voltage ratio, 302
Transformer equations:
  current, 306
  impedance, 309, 310
  voltage, 302
Traps, wave, 566
Trigonometric functions, 480
  application of, 481
  inverse functions, 481
True power, 469, 470
Tuned circuits, 518

Units:
  CGS of inductance, 287
  magnetic, 215
  multiples, 27
  submultiples, 27
  and symbols, 658
Universal graphs:
  C-R circuits, 387
  L-R circuits, 342
Unloaded voltage divider, 123

V - symbol for voltage, 15
Valence ring, 7
Vector:
  definition, 456
  diagrams, 455
  j-operator, 477, 601
  rectangular coordinates, 600
  rotating, 437, 456

Velocity:
  angular, 449, 450, 451
  current, 13
Volt, definition, 14, 16
Voltage, 14
  and angle of cut, 427
  average, 437, 439
  changing, 320
  constant source, 72
  converting ac voltages, 438
  counter, 278, 280
  definition, 16
  divider, 122
  drop, 58
  effective, 437, 438
  electric pressure of, 15
  induced, 423
  instantaneous, 433
  Kirchhoff's, 57
  measuring voltage, 3 rules, 17
  open-circuit, 140, 172, 173
  polarity, 16, 428
  and Q, 530
  rate of change, 385, 387
  ratio, 60
  reference point, 17, 38, 39
  regulation, 126
  rms, 437, 438
  rules, 17
Volta, 135
Voltaic cell, 135
Voltmeter, 16, 640
Voltmeter loading effect, 646
Voltmeter sensitivity, 646

W - symbol for energy and work, 46
Wattage:
  calculation, 61
  definition, 48, 49
  equation, 61
  Ohm's law, 49, 50
  practical, 62
Wave:
  length and frequency, 447, 449
  sine, 429
  traps, 566
Waveshaping circuit, 349, 411
Wavetraps, 566
Weber, definition, 210
Wet-cell battery, 146
Wheatstone bridge, 128
Work:
  measurement, 47
  SI unit, 45, 48, 665
  symbol, 45, 48, 665

X - symbol for reactance, 455, 661

Z - symbol for impedance, 476, 660